Disaster Risk Reduction

Methods, Approaches and Practices

More information about this series at http://www.springer.com/series/11575

About the Series

Scope of the Series

Disaster risk reduction is a process, which leads to the safety of community and nations. After the 2005 World Conference on Disaster Reduction, held in Kobe, Japan, the Hyogo Framework for Action (HFA) was adopted as a framework of risk reduction. The academic research and higher education in disaster risk reduction has made/is making gradual shift from pure basic research to applied, implementation oriented research. More emphasis is given on the multi-stakeholder collaboration and multi-disciplinary research. Emerging university networks in Asia, Europe, Africa and Americas have urged for the process-oriented research in disaster risk reduction field. Keeping this in mind, this new series will promote the outputs of action research on disaster risk reduction, which will be useful for a wider range of stakeholders including academicians, professionals, practitioners, and students and researchers in the related field. The series will focus on some of emerging needs in the risk reduction field, starting from climate change adaptation, urban ecosystem, coastal risk reduction, education for sustainable development, community based practices, risk communication, human security etc. Through academic review, this series will encourage young researchers and practitioners to analyze field practices, and link it to theory and policies with logic, data and evidences. Thus, the series emphasizes evidence based risk reduction methods, approaches and practices.

Series Editor

Editorial Advisory Group

Rajib Shaw
Editor

Recovery from the Indian Ocean Tsunami

A Ten-Year Journey

Springer

Editor
Rajib Shaw
Graduate School of Global Environmental Studies
Kyoto University
Kyoto, Japan

ISSN 2196-4106 ISSN 2196-4114 (electronic)
ISBN 978-4-431-55116-4 ISBN 978-4-431-55117-1 (eBook)
DOI 10.1007/978-4-431-55117-1
Springer Tokyo Heidelberg New York Dordrecht London

Library of Congress Control Number: 2014948896

Printed on acid-free paper

Springer is part of Springer Science+Business Media (www.springer.com)

Preface

Ten years have passed since the 2004 Indian Ocean Tsunami. For a recovery program, it is difficult to say with justification whether 10 years is a long or short duration. For the affected people, sometimes time passes very quickly. For some communities or families whose members lost their lives, time remains stagnant during the disaster period. From the recovery perspective, 10 years is a good time within which to complete physical recovery. It is also a good time during which to achieve socio-economic recovery. For psychosocial recovery, however, more time may possibly be needed.

These past 10 years have taught us many important lessons. In a post-disaster scenario, the Hyogo Framework for Action (HFA) was adopted in 2005. We have seen implementation of the HFA over these 10 years and how it influenced the recovery program in different ways: institutionalization or legal framework of risk reduction, different levels of risk assessment, several education-related programs, looking at risk reduction as a part of development of initiative, enhanced response through early warning systems, and so on. The past decade has also seen a demand for greater education in risk reduction. Specific targets and measurements of progress have been incorporated.

Recovery lessons are never completed. This is an on-going process, and 10 years is a good time in which to review the past achievements and progress and to design future agenda. Thus, while the HFA 2 process is under way and we are preparing for the next world conference in Sendai, Japan, it is an important juncture from which to look back and see different recovery lessons and to contribute to future actions. This book is a modest attempt at that process. I am very grateful and indebted to all the contributors, who have spent their valuable time making critical analyses of their own experiences related to the recovery process.

This book is written for students and young researchers aspiring to a career in disaster risk reduction and environmental studies including sustainable development. I hope that they will find the book useful and relevant to their work.

Kyoto, Japan — Rajib Shaw

Contents

Contributors

Asharose Graduate School of Global Environmental Studies, Kyoto University, Kyoto, Japan

Miwa Abe Center for Policy Studies, Kumamoto University, Kumamoto, Japan

Wignyo Adiyoso National Development Planning Agency (BAPPENAS), Jakarta, Indonesia

Iftekhar Ahmed Humanitarian Architecture Research Bureau (HARB), RMIT University, Melbourne, Australia

Dilanthi Amaratunga University of Salford, Manchester, UK

Monsinee Attavanich Graduate School of Global Environmental Studies, Kyoto University, Kyoto, Japan

Benigno Balgos Center for Disaster Preparedness Foundation, Inc., University of the Philippines—Diliman, Quezon City, Philippines

Patrizia Bitter GIZ Education for Social Cohesion Programme, Battaramulla, Sri Lanka

Esther Charlesworth Humanitarian Architecture Research Bureau (HARB), RMIT University, Melbourne, Australia

Parnasri Ray Choudhury Safe Citizen.Org, Kathmandu, Nepal

Jesus Dominic Dizon Earthquakes and Megacities Initiatives, Quezon City, Philippines

Peter Elstner Independent Researchers, Chiang Mai, Thailand

S.H.M. Fakhruddin Regional Integrated Multi-Hazard Early Warning System (RIMES), Pathumthani, Thailand

Politecnico Di Milano, Piacenza, Italy

Glenn Fernandez Graduate School of Global Environmental Studies, Kyoto University, Kyoto, Japan

Rajarshi DasGupta Graduate School of Global Environmental Studies, Kyoto University, Kyoto, Japan

Farhan Helmy National Council on Climate Change (DNPI), Jakarta, Indonesia

Adhitya Iskandar Tsunami and Disaster Mitigation Research Center (TDMRC), Civil Engineering Department of Syiah Kuala University, Banda Aceh, Indonesia

K. Kamala Benchmark Electronic Systems, Chennai, India

Hidehiko Kanegae Ritsumeikan University, Kyoto, Japan

Gayani Karunasena University of Moratuwa, Moratuwa, Sri Lanka

Hirohide Kobayashi Graduate School of Global Environmental Studies, Kyoto University, Kyoto, Japan

R.R. Krishnamurthy University of Madras Arts and Science College, Kanchipuram, India

Sudhir Kumar United Nations Development Programme, Tacloban, Philippines

Dyah Kusumastuti Institute of Technology Bandung, Bandung, Indonesia

Kuntoro Mangkusubroto President's Delivery Unit for Development Monitoring and Oversight, Jakarta, Indonesia

Rehabilitation and Reconstruction Agency (BRR) for Aceh and Nias, Banda Aceh, Indonesia

N. Vinod Chandra Menon National Disaster Management Authority (NDMA), Government of India, New Delhi, India

Satoru Mimura Fukushima University, Fukushima, Japan

Osamu Murao International Research Institute of Disaster Science, Tohoku University, Sendai, Japan

Andreas Neef Centre for Development Studies, School of Social Sciences, University of Auckland, Auckland, New Zealand

Hari Krishna Nibanupudi Safe Citizen.Org, Kathmandu, Nepal

Chiho Ochiai Graduate School of Global Environmental Studies, Kyoto University, Kyoto, Japan

Kenji Okazaki Graduate School of Global Environmental Studies, Kyoto University, Kyoto, Japan

Malka Older Centre de Sociologie des Organisations, Institut d'Études Politiques (Sciences Po), Paris, France

Arusa Panyakotkaew Independent Researchers, Chiang Mai, Thailand

Krishna S. Pribadi Institute of Technology Bandung, Bandung, Indonesia

Teuku Muhammad Rasyif Tsunami and Disaster Mitigation Research Center (TDMRC), Civil Engineering Department of Syiah Kuala University, Banda Aceh, Indonesia

Irene Stephen Ravindran New Delhi, India

Izuru Saizen Graduate School of Global Environmental Studies, Kyoto University, Kyoto, Japan

Rajib Shaw Graduate School of Global Environmental Studies, Kyoto University, Kyoto, Japan

Hari Darshan Shrestha Institute of Engineering, Kathmandu, Nepal

Ajay Srivastava GISP, Noida, UP, India

Nitin Srivastava Graduate School of Global Environmental Studies, Kyoto University, Kyoto, Japan

Hari Srinivas School of Policy Studies, Kwansei Gakuin University, Nishinomiya, Japan

Syamsidik Tsunami and Disaster Mitigation Research Center (TDMRC), Civil Engineering Department of Syiah Kuala University, Banda Aceh, Indonesia

Terdsak Tachakitkachorn Chulalongkorn University, Bangkok, Thailand

Poorna Yahampath Integrated Environment Disaster Risk Management, Colombo, Sri Lanka

About the Editor

Rajib Shaw is a Professor in the Graduate School of Global Environmental Studies of Kyoto University, Japan. He worked closely with the local communities, NGOs, governments and international organization, including United Nations, especially in the Asian countries. He is currently the President of Asian University Network of Environment and Disaster Management (AUEDM). His research interests are: community based disaster risk management, climate change adaptation, urban risk management, and disaster and environmental education. He has published several books in the field of disaster and environmental management. He is also the Chief Editor of Asian Journal of Environment and Disaster Management.

Part I
Overview

Chapter 1
Ten Years of Recovery Lessons from Indian Ocean Tsunami

Rajib Shaw

Abstract Ten years from the devastating tsunami of December 26, 2004, have accumulated a wealth of experiences in different fields. The disaster risk reduction subject has seen a big shift in institutionalizing the efforts through formulation of policy and legislation in different countries. The beginning of the implementation of Hyogo Framework for Action (HFA) from the following year (2005–2015) has provided a framework to implement risk reduction in a holistic way. Some of the key lessons and progress at the regional level include: (1) Legal provision and institutionalization of risk reduction, (2) Enhance community based risk reduction measures, (3) Enhance role of civil society organizations (CSO), (4) Linking land use changes and risk reduction, (5) Focusing on urban risk reduction, (6) Securing vital infrastructures like schools, hospitals, (7) Linking disaster risk reduction (DRR) and climate change adaptation (CCA), (8) Linking technology to people, (9) Promoting higher education in disaster risk reduction, and (10) Enhancing corporate sector role in disaster risk reduction.

Keywords 10 years recovery • Community-based response • Early warning system • Indian Ocean Tsunami • Legal framework of risk reduction

1.1 Introduction

On 26 December 2004, a major earthquake of magnitude 9.0 with an epicenter located 160 km off the western coast of north Sumatra, with a depth of 30 km triggered a gigantic tsunami, which affected the neighboring countries. It is reported that 1,200 km of the fault line affected by earthquake was in nearly north–south orientation, which is part of the Indo-Australian plate boundary. Thus, the wave movement was more prominent in the east west direction. The tsunami had different

R. Shaw (✉)
Graduate School of Global Environmental Studies, Kyoto University, Kyoto, Japan
e-mail: shaw.rajib.5u@kyoto-u.ac.jp

R. Shaw (ed.), *Recovery from the Indian Ocean Tsunami: A Ten-Year Journey*, Disaster Risk Reduction, DOI 10.1007/978-4-431-55117-1_1

lead time in different countries, the closest one was around 15 min in Aceh to close to 3.5 h in Maldives, and more than 5 h in East African countries.

The Indian Ocean Tsunami is a massive disaster, which affected people so many nationalities. The event killed more than 200,000 people, damaged homes of over 1 million people, and affected (directly or indirectly) more than several millions of people. Having its epicenter near Aceh, Indonesia, in the India Ocean, the waves affected Indonesia, Thailand, India, Sri Lanka, Maldives the most with some damages in other Bay of Bengal countries, as well in the East African countries. Indonesia reported a death toll of around 168,000 people. Followed by 35,000 in Sri Lanka, 24,500 in India, 8,000 in Thailand and close to 100 in Maldives. Over 9,000 foreign tourists from more than 50 nationalities were killed in the disaster, leaving it as one single disaster, which affected so many nationalities.

It is difficult and challenging to justify how 10 years can be put in perspective after a disaster. For some people, time passes very quickly, for several affected people, time is still stuck in 2004. As always seen after the disaster, 10 years is a pretty good time for physical recovery. Physical recovery has different stages, and it is possibly the most visible part of the recovery process. However, the challenges remain for the psychosocial recovery, which has possibly no time limit, and varies widely based on the nature of community and households.

This chapter provides an overview of some of the current observations, and issues discussed over past 10 years. By saying this, the author understands and admits the vast diversity of the subject. A quick search in Science Direct with the key words "Recovery from Indian Ocean Tsunami" usually creates around 700–800 academic articles and book chapters. Starting from morphological or physical changes of the coastal area, to mechanism of tsunami effects; from psychological care to relief coordination, from housing recovery to livelihood recovery, from governance to NGO coordination: there are wide ranges of topics on which innovative research have been done. This chapter provides only a quick glimpse on some of the issues, and then reviews the key findings of the chapters presented in the book.

1.2 Some Observations and Current Research

1.2.1 Concept of Total and Psychosocial Recovery

In Sri Lanka, a unique concept of "5 R program" of Sarvodaya (one of the largest developmental NGO in Sri Lanka) in post 2004 Indian Ocean Tsunami was the base of the recovery process initiative by the organization. First R is relief, followed by rehabilitation, the third R is reconciliation, fourth R is reconstruction, and the fifth R is reawakening (self-development). Relief started with water, food, medical supplies, temporary shelter, sanitation facilities etc. Rehabilitation had different components: psychological, socio-economic and cultural components etc. Reconciliation includes psychological healing, along with other aspects of daily lives, including livelihoods. Reconstruction phase provides lots of emphasis on governance and decision making, and different stakeholders have important roles of play. Land allocation, resettlement etc. are incorporated in this component. Reawakening is important to bring people back to their pre-disaster conditions (Ariyaratne 2011).

At the aftermath of the disaster, there were several teams from abroad who flew in the affected areas of Sri Lanka, and conducted PTSD (post traumatic stress disorder) counseling. Lahad et al. (2011) listed a few key lessons on a unique three country psychosocial interventions in Sri Lanka. One of the key point was the effectiveness of Buddhist Monk's psychosocial care at the very beginning after the disaster. There needs to be a strong cultural sensitivity for the overseas PTSD programs, which need to incorporate the local context. Local rituals often become important, and it is important that the overseas team makes a strong collaboration with their local team to understand the importance of these local rituals and their values. A mutual learning system needs to be developed with the local and overseas team to make the "imported" knowledge to be effective in local context. Sustainability of the efforts lie in the systematic development of training and capacity building initiatives.

Children and orphans are the critical vulnerable groups after the disaster, which need different types of care, including psycho-social care. In many cases, the psychological care become more important that providing physical facilities. It needs step-wise efforts with time. There is also a different opinion that children becomes stronger and resilient, when they experience a disaster. At the beginning, clean water, education environment, sanitation become important. The teachers can play important roles here. Children also need proper medical care including vaccinations to avoid secondary disasters. The timely vaccination becomes very important, as evident in Cidahu in Indonesia, where more than 4,000 children were vaccinated in a short span (Tan 2011). Providing homes and family care of orphans become equally important. Religious groups can play important roles, as evident by Buddhist monasteries in Sri Lanka. A number of people in Jakarta have expressed their desire to become foster parents to Acehnese children. Some private sectors also played important role here. Going back to school ("Back to School Campaign" in several cases) seems to be effective. However, this needs resources and care of the communities and other stakeholders. Sometimes, sports, drama, music and art play important role, which inspires the children.

1.2.2 Scope and Challenges of Disaster Diplomacy

Response, recovery and reconstruction strategies were challenged in several countries because: (1) some countries experienced this magnitude tsunami event for the first time, and (2) internal conflicts in some countries like Indonesia (Aceh) and Sri Lanka. In Sri Lanka, the decade long conflict between ethnic groups of dominant Singhalese and minority Tamil continued for years. Prior of tsunami, there was at least one disaster diplomacy incident in 2002 and 2003, when flooding hit the southern part of the country, killing 250 people. The LTTE (Liberation Tigers of Tamil Eelam) mobilized resources to help the Singhalese dominated southern part.

Kelman (2012) pointed out that tsunami diplomacy had failed entirely in post-tsunami Sri Lanka. It was difficult to pinpoint whether or not the beginning of the end of the LTTE was the tsunami, was the disinterest of the parties in conflict to use the tsunami disaster as an opportunity to seek peace, or was inevitable irrespective of other factors such as disasters.

In contrast, Aceh has more scope for disaster diplomacy. Aceh was a successful independent sultanate until the Dutch forcibly took control in early twentieth century. Aceh separatist movement continued even after the independence of Indonesia in 1945. GAM (Gerakan Aceh Merdeka or Free Aceh Movement) was established in 1976 and soon became the main leader for Aceh's independence movement. On 23rd January 2005, the Indonesian government and GAM announced that they had agreed to resume peace talks in Helsinki. On 15th of August, both parties signed a MOU that was heralded as a peace deal for Aceh. Although there were several complexities in post peace deal process, tsunami created the international interest and domestic political space that could make successful use of those conditions. That is, a basis for peace in Aceh existed before the tsunami, but the tsunami ensured that this basis did lead to peace rather than succumbing to failure, as had happened before.

1.2.3 Using Social Capital and Role of Civil Society in Disaster Recovery

Aldrich (2012) revealed that social capital played an important role in post tsunami recovery process. With detailed field survey in Tamil Nadu, India, he concluded that fisherman villages in the state have high levels of bonding and linking social capital, demonstrated better recoveries than those with only bonding connections or none at all. The local organizations "uur panchayets", which is part of the local governance system contributed to the systematic approach in the community. Strong social capital brought several benefits to the communities, including more robust mental health and greater access to logistic and financial resources for survivors. The strong local institutions of caste and parish councils served as focal points and mediators with the aid community during the relief efforts, ensuring that in-group members received aid. Social capital reduced the need for counseling and external intervention after the disaster and social support systems such as extended and joint families contributed to the community's resilience. From his analysis, Aldrich (2012) concluded that by relying heavily on uur panchayats, both the government of India and NGOs were dependent on these tribal councils to list needy recipient and deliver aid. Dalits, women, Muslims, windows, and other often marginalized groups slipped through the cracks and were often helped only after weeks or even months of waiting.

An analysis by Minamoto (2010) in the role of social capital in livelihood recovery suggests that formal network in the community, leadership and trust of community-based organizations are the key contributing factors. Negative factors are bonds with relatives and neighbors, participatory decision making process and community based NGOs. Bonds among close individuals, for example, bonds with relatives and neighbors were not always positive. Almost 80 % of the community based NGOs were created after the tsunami and with close to 90 % of their members responding that they were semi-forced into participating. She concluded that disaster aid needs to consider seriously about the social factors and power structure to avoid dark side of social capital, and participatory approach may not bring expected outcome, if the approach is utilized only as means to provide the aid.

As study by Srinivasan and Nagaraj (2011) shows that a vibrant interface and partnership between the State and civil society is critical to ensure that people's concerns are addressed by both policy and practice, as well as to minimize corruptions. This synergy needs to be fostered by both the State and civil society proactively. Also important is the need for effective communication channels, coordination, dialogue and information among and within civil society organizations to avoid duplication, wastage of resources and competition, and to optimize individual efforts. They also argue that a strong case for civil society organizations can effectively foster transparency in their interventions. It is evident from their analysis that responsive and proactive government, that devolves power and responsibilities to the administration, is a necessary condition for effective disaster response. Further institutional autonomy, ad decision making free of political interference, are critical in fostering people-sensitive and effective relief and rehabilitation. Exclusion of certain racial groups become a challenging issue. There is a pressing need for detailed and inclusive geo-demographic and socio-economic data collection, and vulnerability mapping as a resource base for disaster preparedness.

An analysis by Kumaran and Torris (2011) pointed out that NGOs have been very active in relief and recovery process in Tamil Nadu, southern India. They emphasized that for several NGOs, the recovery process was also linked to preexisting issues in the area, like health related issues, land rights, women's rights etc. A strong cultural context is observed through NGO activities. A few issues which they observed were: unsuitable assistance with good intentions (due to lack of unfamiliarity to local context), relocation of affected people (which is related to fishing communities specially), uniting the fisher folk village, need for networking and a coordinating entity, language problems faced by INGO representatives, changing role of women in fishing communities, working with or around the leaders of a community etc.

1.2.4 Economic and Social Impacts

A large amount of money and effort have been invested in the tsunami research and development, earthquake and tsunami monitoring and warning systems and a large number of educational outreach to young generations. The proper tsunami and earthquake mitigation of the central government collaborated with local governments integrated with the people's understanding of the nature of the earthquake and tsunami help increase the readiness of the Thai community for such disaster more than ever. It seems challenging to maintain the level of the awareness of the people about the great danger of the earthquake and tsunami and it s also a big challenge to increase the ability of the Thai society to have better critical scientific thinking about the natural disaster (Pananont and Srisomboon 2014).

1.2.5 Focusing on Housing Issues

Gujarat reconstruction after 2001 earthquake in India is considered as one of the successful recovery program, especially in the context of housing recovery. However, these lessons were not reflected properly in case of Tamil Nadu. Duyne Barenstein (2011) concluded that prejudices against fishing communities, the availability of unprecedented private donations, and an insufficient number of committed local NGOs led to culturally and environmentally insensitive contractor-driven reconstruction, the systematic demolition of undamaged vernacular houses, and cutting down of thousands of trees to make space for new houses. She also concluded that the contractor driven housing reduced people's resilience and well–being. The positive experience of owner-driven reconstruction in Gujarat and the negative experience with contractor-driven reconstruction in Tamil Nadu are clear evidence that reconstruct after disasters should be controlled by the people themselves, supported with financial and technical assistance rather than with ready-made contractor-built houses.

In an analysis of level of participation in community based recovery program, Ophiyandri (2011) stated that community-government link should be at the level of collaboration or empowerment. Post disaster housing reconstruction should not only propose to provide houses for the beneficiaries, but it needs to involve them in the process of the recovery, which can be a good vehicle to understand the need of the communities and enhance its social capital. This method is proven to be fast (compared to the overall duration of the recovery process), cost effective, creates fewer problems and provides high quality housing, with higher level of satisfaction. This method also helps in regaining community confidence and ease the trauma they have suffered.

1.2.6 Capacity Development

Rowlands (2011) argues that training is a vital component of disaster recovery interventions. Training programs can be designed for very varied target groups who will be providing personal, psychological or community based services in affected communities. While qualified professional mental health workers have a matured knowledge, values and skills base from which to draw for their practice, other recovery workers and volunteers mat require foundational input on human responses to disaster and basic intervention techniques. Tertiary institution social work courses are also appropriate places to mount training in crisis and disaster recovery work. Evaluation of training programs delivered is desirable to continue to improve the quality of training offered and to ensure that content is current and evidence-based.

Availability of authorities, institutions and human resources, policies and legislation, framework and guidelines, projects and programs, knowledge, skills, and experience of people and their proper implementation often become crucial factors for post disaster recovery of built environment (Ginige and Amaratunga 2011). In this context, synergy of resources of national and local governments, international community, people and communities, civil society, private sector, academic and professional association become important. Capacity gap assessment becomes very important at an early stage of the recovery process. Ginige and Amaratunga (2011), based on their study in Sri Lanka have identified following factors as key capacity gaps: lack of appropriate policies, deficiencies in policy implementation, deficiencies in state of the art technology for rapid and sustainable reconstruction, deficiencies in trained and skilled human resources for post disaster reconstruction, and lack of related knowledge, experiences and coordination among stakeholders. The framework identified by the authors identified four stages of capacity development: analysis, creation, utilization and retention.

1.2.7 Stakeholder Consultation, Community Participation and Governance

The tsunami has forced governments, NGOs, and international organizations to take stock and reflect on their own disaster preparedness readiness and know-how (Schreurs 2011). It has produced significant changes in local, national, regional and international disaster response programs and institutions. Governments around the region have introduced and strengthened national disaster management institutions and laws. They have also joined the International Indian Ocean Tsunami Early warning system and emergency and recovery plans. Many steps have been taken to enhance disaster management coordination and communication capacities between national and local governments and NGOs in the region.

Kenny (2010) made an analysis of different factors affecting participatory recovery as follow:

- Size and complexity of the reconstruction effort: Rebuilding lives, physically, socially and emotionally was always going to be very challenging. The larger the size is, more complex the management system becomes, and more challenges happen for participatory process.
- Tensions in contemporary delivery if aid: Several times, conflict happens between development imperatives (accountability to recipients) and institutional imperatives (accountability to donors). While accountability to donors compels aid agencies to push for fast recovery, the community recovery always takes time in decision making and implementation.

- Power factors affecting participation: Aceh, being a traditional society, with Islamic regulation strongly implemented, there are some irregularities in the recovery process and conflict with the traditional laws. Thus, configuration of power relations also affects participatory approach in Aceh.
- Cultural factors: A fourth factor affecting participation concerns the role of culture. The first issue is culturalism, the second point is deeply religious culture of Aceh, and the third point is authoritarian and fatalistic elements of Aceh culture (Kenny 2010).
- Engaging the state: In spite of increasing NGO involvement, State still remains as the key role player, and its participation remains as a key element.

Kenny et al. (2010) made an interesting attempt to link the physical and functions capacities of infrastructures. They argued that the dramatic physical, social, and economic changes, including the tremendous loss of life brought by the tsunami, meant that Acehenese life could never be returned to a state that existed prior to the tsunami. Thus, the reconstruction of Aceh has required the construction of a new concept of "normal", the path of which is full of tensions between the efforts to strengthen real or imagined past customs and beliefs, such as strengthening commitment to sharia law on one hand, and reclaiming women's rights through celebration of Acehnese traditional heroines and supporting the more radical new women's group on the other. A new concept of "normal" also underpins the acceptance of peace in Aceh and the integration of ex-combatants into the Acehnese economy. A balance needs to be made between the infrastructure construction and providing appropriate facilities and functions to this (like libraries without proper books, schools without appropriate number of students etc.). Less visible, but equally important is the reconstruction of the governance infrastructure, which includes both restoration of the governance systems in villages and the rebuilding of an administrative and regulatory framework, with political and legal systems in place, and people with the capacity to operate these systems.

Stakeholder consultation becomes an important element of disaster recovery process. Government, civil society, private sector, professional groups, academia, community groups and media become some of the important stakeholders. It is important to identify and classify stakeholders based on the local context, which can be both internal and external. A proper mapping is required, and the stakeholders can be identified as: dormant, discretionary, demanding, dominant, dangerous, dependent and definite (Siriwardena and Haigh 2011). There is always an expectation gap in the prost disaster housing recovery, especially when a community based recovery process is followed. Differences include location, design, space, safety, accessibility, changeability of the houses, the level of community empowerment, the level of consultation, livelihood reconstruction, cultural and social facets, village planning, and the presence of supporting infrastructures and other facilities. Stakeholder engagement strategy is important, which should provide guidance for identifying and engaging with stakeholders, ensure that stakeholder engagement activities are integrated and undertaken in a coordinated manner, and improve the effectiveness of the engagement efforts undertaken.

Based on this, and stakeholder map, an engagement plan needs to be formed, which needs to be seen an evolving and dynamic document, and needs to be updated based on the situation.

1.3 Key Issues and Recovery Lessons

This book is developed with 30 valuable chapters, apart from this first overview chapter. The book has five sections: 10 chapters on overview (including this one), 8 chapters on Indonesia, 6 chapters on India, 5 chapters on Sri Lanka and 2 chapters on Thailand. To understand the key lessons in a structured way, the GET framework of Shaw (2014a) is incorporated in this chapter. GET framework talks about governance, education and technology issue, and argues that a better recovery program needs to have a balance of three of these issues. Shaw (2014b) also argued that recovery can be a used or misused development opportunity. Community based application of technology, community knowledge, low cost technology, multi-stakeholder approaches, judicial approaches are some of the key lessons argued by Shaw (2014b). Table 1.1 shows the key messages from the chapters. The chapters are divided into three categories governance, education, and technology. Governance has broader perspective from regulations, legislation, to community governance and community decision making. Education has perspectives from formal school education to non-formal/informal community education. Technology has different dimensions, scientific technology, community knowledge, indigenous knowledge etc.

Indian Ocean tsunami was a landmark event just before the second World Conference on Disaster Reduction (WCDR: organized in Kobe, Hyogo in Japan in January 2015), where the Hyogo Framework for Action (HFA) got adopted. The recent scar of Indian Ocean tsunami brought several global leaders interested in the field of disaster risk reduction. The affected countries have taken special interest in the second WCDR in drafting, promoting the key issues of disaster response, recovery and early warning system. Through the recovery process, the countries have undertaken some crucial decisions, both at national and local level. A few major changes observed over last 10 years are as follow:

1. **Legal provision and institutionalization of risk reduction:** Several countries in Asia Pacific region has made the Disaster Management Law, and made specific legal provisions to incorporate risk reduction in the all national and local institutions. This is considered a significant trend to institutionalize the efforts of risk reduction into national and local development (often termed as "mainstreaming").
2. **Enhance community based risk reduction measures:** Community based risk reduction has become a common approach over last 10 years. While in the past, community based approaches have been dominated among the NGOs and local institutions, state ownership of community based approaches have been observed in recent years, with specific funding and institutional mechanism.

Table 1.1 Key recovery lessons

Category	Chapter no.	Key message
Governance	2	Pre-disaster urban planning is key to the post disaster recovery
	3	Mangrove ecosystem plays an important role in coastal resilience and needs a community based approach
	6	Quantitative recovery model needs qualitative evidences
	7	Culture, women's lives, community activities and infrastructure development influence positively post disaster resettlement
	8	Recovery model depends on scale, governance, capacity, stakeholder, socio-economic condition, timeline and transition to development phase
	9	A proper governance system and capacity is required to balance between excessive amount of aid from conventional and non-conventional sources
	11	Specific need based and effective governance mechanism is the key to the success of the recovery program in Aceh
	12	Link between government, communities in one hand, and government and donor agency on the other hand is required for aid efficiency
	14	Combination of policy, communication strategy, heavy equipment, temporary collection sites are some of the challenging issues in waste management in Aceh
	19	People led, people owned, and people managed social transformation initiative is the key to the recovery process
	31	Post disaster economy recovery can deepen income disparities and may disrupt the sense of community if local elites can take advantage of lack of sound governance system
Education	4	School resumption, reconstruction and relocation of schools, school community partnership, curriculum review, reaching to the disadvantageous are some of the key challenges in education sector
	10	Social protection is not only a means to aid survivor in recovery, but it also reduces poverty, improves lives and facilitate community adaptation and mitigation
	18	Religious teaching is one of the effective means to convey the message to the people, and therefore can be an important risk communication strategy in post as well as pre-disaster
	22	Local knowledge, capacity and priorities of communities, and its sustainability become key factors, which can be only captured through listening to the communities
	23	Involving communities in physical and social recovery through participatory process is important
	24	A unique post disaster higher education program was observed in Tamil Nadu, India through establishing strong bonding Local Resident's Alliance
	25	Collaboration with partners, coordination and involvement of CBOs, specific strategies and mechanisms in local and national level are some of the key lessons in Sri Lanka
	28	To make sustainable resettlement, it requires specific issue based program, which can be environment related consciousness through development of an eco-village
	29	With an innovative effort, national school safety guideline was prepared for Sri Lanka with participation from government, non-government and international agencies

(continued)

Table 1.1 (continued)

Category	Chapter no.	Key message
Technology	5	An effective early warning system (EWS) need to link risk knowledge, response capabilities, vulnerabilities of communities including temporary and long-term mitigation measures
	13	Morphological changes in the coastal areas need to be addressed through proper technology, policy and awareness of people
	15	Community based housing need a good balance of technology and process of community participation and adaptation
	16	Post disaster retrofitting needs training of masons and demonstration of simple, affordable technology
	17	Proper seismic design, building materials, quality of workmanship, good project management and supervision are some of the key safety issues of recovery of housing
	20	GIS based system and solutions allow the integration of data from various resources and provide a single window for viewing, querying and analyzing information
	21	Ecosystem based adaptation approach needs to be linked to management, adaptation and proper use of ecosystem services
	26	Establishment of regulatory body, enforceable rules and regulations with necessary level of capacities are important for waste management in Sri Lanka
	27	To develop multi-hazard housing is important, which needs considerations from different aspects, including a proper evaluation tool
	30	Indigenous knowledge is important for some coastal communities, non-incorporation of which makes the communities more vulnerable

3. **Enhance role of civil society organizations (CSO):** CSO or NGO played important role in some countries, however, their regional influence have been less so far. Over past 10 years, CSO/NGO, through their innovative approaches have influenced the regional and national policy significantly. CSO/NGO have become important actors in the global/regional and national policy forum, and therefore, it is considered as a good link to the local level actions.
4. **Linking land use changes and risk reduction:** Land use change, regulations and risk reduction measures came closer after the Indian Ocean Tsunami. The coastal regulatory zones were imposed, the need of ecosystem-based approaches of risk reduction has been preached widely, and people based co-management system of coastal and/or mountain resources were advocated and stressed in different countries and contexts.
5. **Focusing on urban risk reduction:** Urban population in the world has crossed the rural population in 2007. A few major recent disasters hit the urban areas badly, and therefore, the need of urban risk reduction has got significant attention. The need of neighborhood approach has become popular in case of urban risk reduction.
6. **Securing vital infrastructures like schools, hospitals:** Over past several disasters, vital infra like schools and hospitals have seen major damages.

Campaign and advocacy by UN ISDR (International Strategy for Disaster Reduction) and other partners have made significant changes in government approaches and policies to protect vital infrastructures.

7. **Linking disaster risk reduction (DRR) and climate change adaptation (CCA):** Recent impacts of climate change issue, the recognition of climate change work of IPCC (Inter-governmental Panel on Climate change), and increasing need of adaptation than mitigation has brought the two fields of DRR and CCA together in terms of policy formulation as well as implementation.
8. **Linking technology to people:** Technology develop has been a key aspects over last 10 years, not restricted to the risk reduction field only. However, in case of DRR, gradual recognition has been done for the need of people based technology, or the knowledge/technology, which is linked to the people and communities.
9. **Promoting higher education in disaster risk reduction:** There has been a demand of professionals in the disaster risk reduction. To grow high quality professionals, several universities have started professional higher degree programs in the disaster risk reduction. This is a significant development, which develops disaster risk reduction as a professional subject with education, research and practice.
10. **Enhancing corporate sector role in disaster risk reduction:** Increasingly, corporate sectors have been found to play important roles in disaster risk reduction. This is not just corporate social responsibility, but how to make disaster risk reduction as a business, in close cooperation with the government and communities. Thus,

The above ten points are possibly the only brief synopsis of the vast dynamics of risk reduction issues in the subject. Disaster risk reduction is a growing and dynamic field, and it evolves over time. The Indian Ocean Tsunami was a landmark event in many senses, starting from awareness to policy formulation and implementation, technology innovations and implementing several innovative programs in the community level. At this juncture of HFA 2, when the post HFA issues are discussed regionally and globally for 2015 and beyond, this chapter and book provides some the in-depth analysis of the lessons from the Indian Ocean Tsunami.

Acknowledgements This book chapter and the whole book is based on the cooperation from a wide range of practitioners, policy makers and academicians. This is highly acknowledged.

References

Aldrich DP (2012) Building resilience: social capital in post-disaster recovery. The University of Chicago Press, Chicago, 232 p

Ariyaratne AT (2011) Words of wisdom following the tsunami: lessons from Sri Lanka. In: Kaufman R, Edwards R, Mirsky J, Avgar A (eds) Crisis as an opportunity. University Press of America, Maryland, pp 106–111

Duyne Barenstein J (2011) Housing reconstruction in Tamil Nadu: the disaster after the Tsunami in India. In: Community disaster resiliency: exploring global opportunities and challenges. CRC, Boca Raton, pp 343–362

Ginige K, Amaratunga D (2011) Capacity development for post disaster reconstruction on the built environment. In: Amaratung D, Haigh R (eds) Post disaster reconstruction of the built environment. Wiley Blackwell, West Sussex, pp 13–29

Kelman I (2012) Disaster diplomacy: how disasters affect peace and conflict. Routledge, Oxon, 174 p

Kenny S (2010) Reconstruction through participatory practice? In: Clarke M, Fanany I, Keny S (eds) Post-disaster reconstruction. Earthscan, Washington, pp 79–104

Kenny S, Clarke M, Fanany I, Kingsbury D (2010) Deconstructing Aceh's reconstruction. In: Clarke M, Fanany I, Kenny S (eds) Post-disaster reconstruction. Earthscan, Washington, pp 3–28

Kumaran M, Torris T (2011) The role of NGOs in tsunami relief and reconstruction in Cuddalore district, South India. In: Karan P, Subbaiah S (eds) The Indian Ocean Tsunami. The Univeristy Press of Kentucky, Kentucky, pp 183–211

Lahad M, Baruch Y, Shacham Y, Niv S, Rougel R, Nacasch N, Rachamim L, Leykin D (2011) Cultural sensitivity in psychological interventions following a disaster. In: Kaufman R, Edwards R, Mirsky J, Avgar A (eds) Crisis as an opportunity. University Press of America, Maryland, pp 129–154

Minamoto Y (2010) Social capital and livelihood recovery: post-tsunami Sri Lanka as a case. Disast Prev Manag 19(5):548–564

Ophiyandri T (2011) Community-based post disaster housing reconstruction: examples from Indonesia. In: Amaratung D, Haigh R (eds) Post disaster reconstruction of the built environment. Wiley Blackwell, West Sussex, pp 91–116

Pananont P, Srisomboon R (2014) 2004 tsunami in southern Thailand, lessons learned for the Thai communities. In: Kontar YA, Santiago-Fandino V, Takahashi T (eds) Tsunami events and lessons learned: environmental and societal significance, vol 35, Advances in natural and technological hazard research. Springer, Netherlands, pp 399–416

Rowlands A (2011) Training for disaster recovery: a review of training programs for social workers after the tsunami. In: Tan NT, Rowlands A, Yuen FKO (eds) Asian Tsunami and social work practice. Routledge, New York, pp 109–126

Schreurs M (2011) Improving governance structures for natural disaster response. In: Karan P, Subbaiah S (eds) The Indian Ocean Tsunami. The University Press of Kentucky, Kentucky, pp 261–282

Shaw R (2014a) Post disaster recovery: issues and challenges. In: Shaw R (ed) Disaster recovery: used or misused development opportunity. Springer, Tokyo, pp 1–13

Shaw R (2014b) Disaster recovery as a development vehicle. In: Shaw R (ed) Disaster recovery: used or misused development opportunity. Springer, Tokyo, pp 411–429

Siriwardena N, Haigh R (2011) Stakeholder consultation in the reconstruction process. In: Amaratung D, Haigh R (eds) Post disaster reconstruction of the built environment. Wiley Blackwell, West Sussex, pp 117–132

Srinivasan K, Nagaraj VK (2011) The state and civil society in disaster response: post tsunami experiences in Tamil Nadu. In: Tan NT, Rowlands A, Yuen FKO (eds) Asian Tsunami and social work practice. Routledge, New York, pp 57–80

Tan NT (2011) Impact of the Indian Ocean Tsunami on the well-being of children. In: Tan NT, Rowlands A, Yuen FKO (eds) Asian Tsunami and social work practice. Routledge, New York, pp 43–56

Chapter 2
Institutional and Legal Arrangements and Its Impacts on Urban Issues in Post Indian Ocean Tsunami

Nitin Srivastava and Rajib Shaw

Abstract Disasters are seen as opportunities for enhanced institutional and legal arrangements. Steps have been taken in all the affected countries casting an influence on the urban issues such as governance, housing, land-use and transportation. In India, the government gave significant authority to the local governments while in Indonesia urban reconstruction plan, called Blueprint, was adopted and influenced housing, transportation and economic activities in the city. Maldives government had formulated a population consolidation plan and Sri Lanka focused on development approval policies for enhanced community safety. This chapter tries to summarize urban issues impacted by the new institutional and legal arrangements in the decade after the Indian Ocean Tsunami.

Keywords India • Indian Ocean Tsunami • Indonesia • Institutional and legal • Maldives • Sri Lanka • Urban issues

2.1 Introduction

Disasters are seen as opportunities for change in institutional and legal arrangements. The extent of change in societies, institutions, and organizations, is often associated with the scale of disasters. While smaller disasters might not bring a change in the existing systems, a low-frequency extreme event might change the way a government, a community or an individual thinks and performs. A major disaster challenges the existing framework of policies and legal arrangements, which might have proved insufficient to limit the damages to lives and infrastructure. Gorg (2003) (in Birkmann et al. 2008), taking a leaf out of political economy theory, regards disasters as threshold events which causes review and revision of dominant ways of thinking and acting, thereby causing change in institutional and organization setup and policies.

N. Srivastava (✉) • R. Shaw
Graduate School of Global Environmental Studies, Kyoto University, Kyoto, Japan
e-mail: imnitin@gmail.com; shaw.rajib.5u@kyoto-u.ac.jp

R. Shaw (ed.), *Recovery from the Indian Ocean Tsunami: A Ten-Year Journey*, Disaster Risk Reduction, DOI 10.1007/978-4-431-55117-1_2

The Indian Ocean Tsunami (IOT) of December 2004 highlighted the vulnerability of coastal communities living in urban areas to low frequency, extreme events. Approximately 230,000 deaths were reported and an unprecedented economic loss of US$ 10 billion relegated this disaster to one of the most destructive events in recent times. This colossal loss of lives and infrastructure demanded a self-scrutiny by the respective governments in all the affected countries, which saw the annihilating waves of Tsunami hit their shores. Immediately after the Indian Ocean Tsunami (IOT), then situations demanded new policies and laws to facilitate quick recovery and better preparedness against the disasters.

There can be numerous learning from the past disasters, especially for developing countries, where often the institutional and legal arrangements are inadequate and underprepared to face a disaster. To match up the rapid global environmental change, governments in general and communities in particular, need to learn from past disasters particularly with regards to their management strategies (Birkmann et al. 2008). These disasters can also be an opportunity to build up a stronger community and a resilient nation, as accepted by international community and laid in Hyogo Framework for Action (UN 2005).

This chapter is an attempt to identify and understand how the situations post disaster forced or enabled the respective governments to adopt new institutional and legal arrangements in some of the affected countries. Also, in what ways the new arrangements impacted the urban issues such as governance, housing, land use and transportation. The next background section throws light on the importance of addressing the urban issues, after a disaster. The third section, presents the cases of Indonesia, Sri Lanka, India, and Maldives, where various urban issues are directly or indirectly dealt with, by the introduction of a new policy or adoption of a new approach of recovery and preparedness. The ensuing fourth section summarizes the findings and tries to decipher the implications to other mega disasters, followed by the conclusion in the last section.

2.2 Background

At World Conference on Disaster Reduction (WCDR) in Kobe city, Hyogo prefecture in January 2005, 168 states adopted the "Hyogo Framework for Action" (HFA), and acknowledged the focus shift towards local implementation of institutional issues (UN 2005). HFA needs to be implemented at the local level and would be more practical in achieving targets for the concerned city as each city has its own concerns and needs. This substantiates the importance of focusing on the impacts of the institutional and legal arrangements on urban issues, and in contrast, the impacts of prevalent issues in shaping up of these policy changes.

The difference in density, diversity and dynamics of a city contributes to the character of an urban area (ALNAP 2012). Urban areas have to deal with multi-sectoral issues with focus on both urban development and reduction in impacts of disasters. There are various issues faced by urban areas in normal scenarios such as overcrowding, slum formation, inadequate housing, vulnerable buildings, and pollution. Even where the rural population suffers more loss, the linkage between urban centers and rural areas in developing countries, puts the onus of rural poor on urban areas.

The coastal urban vulnerability in Indonesia, Sri Lanka, India and Maldives is majorly determined by the following factors:

i. Location of settlements in low lying areas along the coast
ii. Infirm Building bye-laws, i.e. lack of Tsunami resistant buildings
iii. Lack of Implementation of existing bye-laws and other regulatory systems
iv. Lack of Coordination within different stakeholders exemplified by no proper warning systems and evacuation plan

Worldwide, over 1.2 billion people live within 100 km of the coast and this population is on the rise (Adger and Hughes 2005). Also, approximately ten million people globally feel the effects of coastal problems yearly and most are due to natural hazards. Most of the coastal cities are prone to multi-hazards such as cyclones, tsunami, sea level rise, storm surges, erosion or urban floods. With the aggregated effect of such disasters combined with the urbanization issues, it needs comprehensive policy formulation and planning. A disaster like IOT provides an opportunity to adopt a path of development which reduces the combined effects of such hazards. For example, Chennai is a coastal metropolitan city in India, which suffered a loss of 160 people on the beach and 150 fishermen in the sea from IOT 2004. It has a history of Tsunami events even prior to that. It lies in low to moderate seismic zone. It faces the typical problems of urbanization such as inadequate physical and social infrastructure, large slum population (28.5 % households) (Census 2011). Urban flooding has also been a constant phenomenon. With this kind of multi-hazard background it becomes difficult for a city to cope with a low frequency extreme event. Even though the damage to the urban area due to Tsunami was not as severe as in the rural areas of the country, Chennai came up with a climate action plan in 2011, a comprehensive way to limit the impact of climate related hazards. The climate action plan is discussed in Sect. 2.3.3.

2.3 Case Studies

The authors take up case studies of Indonesia, Sri Lanka, India, and Maldives, to find a linkage between the institutional and legal policies and the issues prevalent in the urban areas of those cities.

2.3.1 Indonesia

Indonesia started its reconstruction with the outline urban reconstruction plan, called Blueprint, which was established in January 2005. Later, the country passed a President Act No. 30/2005 on Master Plan of Rehabilitation and Reconstruction in Aceh and Nias, which proposed a buffer zone in tsunami affected areas through a master plan for the reconstruction of Aceh. It defined buffer zones of two types: a 300 m coastal strip to be kept free of buildings (Fitzpatrick 2005) and a 2-km wide secondary buffer zone in which re erection of a building was restricted (Matsumaru et al. 2012). Both Blueprint and the ensuing master plan followed the concept of minimizing housing and economic activities in the damaged areas. This exercise in disaster prevention through policing and laying restrictions did not work well, as Village Plans were proposed in cooperation of United States Agency for International Development (USAID), with the consent of the villagers who wanted to or were forced by the circumstances to stay in the damaged villages, especially the fishermen communities (Birkmann et al. 2008). Therefore the focus shifted from restrictive policies to disaster prevention measures. Several community buildings in Banda Aceh were designated as evacuation facilities for the population living at the waterfront. The Aceh Tsunami Museum was established to educate people about Tsunami and increase their awareness to preparedness measures (Sugiyasu et al. 2010).

Another fact from Banda Aceh which re-emphasizes on the importance of spatial plans is that they enabled quick reconstruction, and can also be a basis for better response to disasters in the future, by inclusion of specific measures in such urban areas. In the reconstruction master plan, the road network was utilized as escape routes at two levels: road network for the whole coastal area, and community level road network. It can also be used as a chance for land adjustment for critical physical infrastructure, such as escape routes, evacuation centers and housing for the vulnerable sections.

The plans not only targeted on recovery but also addressed the complex issues of urban poverty, creating assets and gender sensitive policies. In entirety, four issues of longer term reconstruction were addressed: managing water resources and environment, reaching out to the poorest, sustaining better services, and rebuilding economy for livelihood recovery (Dercon and Kusumawijaya 2007).

2.3.2 Sri Lanka

Indian Ocean Tsunami was the worst natural disaster ever experienced in Sri Lanka which killed 35,322 people, caused a loss of 150,000 livelihoods, damaging unprecedented number of houses, infrastructure, schools, tourist hotels, and other commercial buildings, and one-third of the island's coastline and outlying 13 districts (Jayawardane 2006). Social networks also were severely disrupted, compounding the previously existing vulnerabilities.

The steps taken by the governments were mostly reactive emphasizing relief and recovery rather than proactive with damage prevention or minimization strategies, because of lack of knowledge and lack of expectation of a low frequency extreme event. The major risk reduction measure included the introduction of coastal buffer zones like in Indonesia, which were areas of coastal land, where construction was prohibited to provide protection from hazards such as tsunami in the future (Fig. 2.1). The government declared a buffer zone of 100 m from the high water line in the

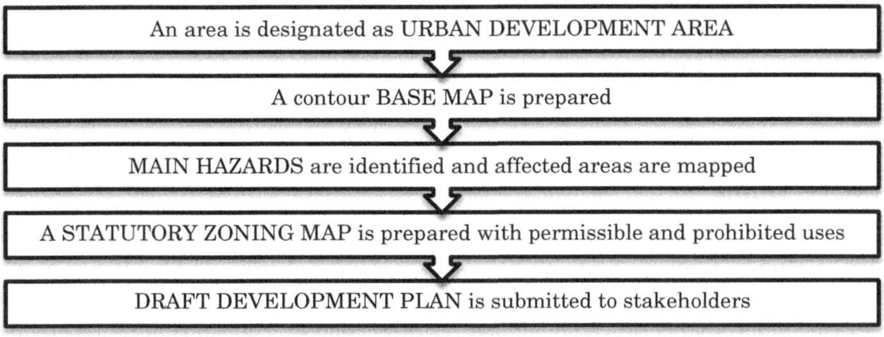

Fig. 2.1 Steps of Zoning system implemented in Sri Lanka after Tsunami. *Source*: Mannakkara and Wilkinson 2012

Table 2.1 Lessons from build back better for other disasters

Precautionary steps	Risk profiling through *risk-based land zones*
	Structural designs for buildings for each land zone with specific risk level
Recovery and reconstruction	Relocation for risk reduction to be implemented only through *comprehensive resettlement strategy*
	Participatory and consultative approaches used for reconstruction and socio-economic recovery (consideration for community's socio economic conditions, culture and traditional livelihoods)
	Simplified administrative *structures and permit* for reconstruction
	Local authorities to play a link between government and NGOs
	Monitoring and evaluation for the recovery process

Source: Adapted from Mannakkara and Wilkinson 2012

south and southwest 200 m in the north and east, where reconstruction of permanent houses was restricted (Department of Census and Statistics, Sri Lanka 2004). The buffer zone has been a critical issue in the recovery process. It is closely linked to the subject of resettlement (Birkmann et al. 2008). To provide shelter to the displaced families 98,000 permanent houses were required (Mannakkara and Wilkinson 2012). The scarcity of land to rehabilitate caused the relocation of vulnerable population from coastal vulnerable lands to flood prone zones. This was due to the absence of hazard assessment maps.

A new zoning system was introduced as a result of the lessons learnt after the Tsunami experience, where the steps, as illustrated in Fig. 2.1, were followed sequentially.

Programmes such as cash-for-work and micro-credit schemes implemented for livelihood and business recovery were criticized as people were dissatisfied as they had to deviate from their traditional types of employment such as fishing. A novel concept to post-disaster reconstruction and recovery was born out of the Indian Ocean Tsunami, i.e. Build Back Better (Table 2.1). It aims to use a holistic approach to improve a community's physical, social and economic situation during the post-disaster reconstruction and recovery process. It illustrates both precautionary steps and steps for recovery and reconstruction to be adhered in the case of a disaster.

2.3.3 India

Indian Ocean Tsunami affected the states of Tamil Nadu, Andhra Pradesh, and Kerala and Union Territory of Pondicherry in the mainland coast, while it also devastated the shores of Andaman and Nicobar islands. The tsunami inundated the coastal lands between 300 m and 3 km inland (Sridhar 2006). In all of these states the cities of Kochi, Tuticorin, Chennai, and Vishakhapatnam, did not suffer major loss (Sheth et al. 2006), and hence there were hardly any policy which addressed the urban issues prevalent in these major cities. However, Indian government did not fail to identify that the major cause of loss within its territories was due to lack of communication, and coordination at local level (Sridhar 2006). It enabled improved communication and coordination at local level with enhanced authority to local governments.

The coordination was exemplified by the set-up of District Resource Centers (DRCs) in tsunami affected districts of Tamil Nadu. DRCs helped in knowledge exchange, information collection and dissemination, and capacity building (UN 2008). This decentralized platform helped local governments to identify the potential of a coordinated effort and its value in times of future disasters. The coordination between relief and reconstruction agencies continued later in the recovery period for livelihood restoration, provision of temporary and permanent shelters, health monitoring, and psycho-social support (TGLLP 2009). This allowed equity based recovery in the urban areas amongst the vulnerable population, thereby maintaining the balance of the urban population.

As discussed earlier, Chennai came up with a Climate Action Plan in 2011. It proposed a series of actions by various government authorities, public and private agencies and individuals to boost the resilience of the city towards climate-related disasters. Most of the steps involve community as input source and work at neighborhood level. This allows the city to address the issues at the micro-level with lesser deployment of resources. The steps include development of a pre-disaster map denoting roads that are vulnerable to waterlogging during the monsoons, conducting disaster drills at ward level periodically, creation of an evacuation plan at the community level, and through coordinated efforts of local government, disciplined enforcement of building permits for all types of hazards for new buildings being constructed, and provision of insurance schemes for all houses from disasters. The mid-term goals included the implementation of already identified systems such as waste management, planning of urban land use in coherence with the environment and the threats, safeguarding the green space, enforcement of development control guidelines, and provisioning of budget for disaster reduction activities (CCA 2010; CNN-IBN 2011).

2.3.4 Maldives

The case of Maldives presents an interesting aspect of institutional arrangements to address the socio-economic challenges of rehabilitation, arising from a mega disaster. Maldives is a country with huge pressure on land resources. The stress can be understood by the fact that only 5 % of the total reef area in Maldives is land. Therefore conventional definitions of continental coastal area do not hold value in the case of Maldives and the whole of the reef are considered as a coastal area for planning and policy purposes. Approximately 200 of the 1,200 islands are inhabited with their size ranging from 0.1 to 5 km^2 by 290,000 people (UNEP 2013). The population is thinly distributed over a large number of small islands, and hence socio-economic development as well as disaster management faces challenges.

The Indian Ocean Tsunami (December 2004) displaced 29,577 residents from their original place of stay, 12,000 remained homeless; living in temporary shelter or with friends and relatives on their own or other islands (UNEP 2013), in addition to extensive damage to the island coastlines and infrastructure. Three of the inhabited islands were completely evacuated, not be inhabited again. This demanded from the government to rehabilitate the displaced to newer locations, primarily on islands where communities existed earlier. Therefore the government formulated a population consolidation plan, dividing the country into five regions for development. This particularly consolidates population for better social growth in larger and safer (from Tsunami) islands. After 2004, the government included land use instruments such as inclusionary zoning and quota for social housing (OHCHR 2013). This recognized the tenure rights of the urban poor.

Pre-Tsunami era recognized the importance of environmental degradation due to uncontrolled coastal mining for building purposes. However, with the protection provided by the vegetation during the catastrophe, the Government encouraged initiatives to raise environmental awareness through programmes, which have contributed to introduction of principles of conservation and sustainability at the urban and island level. These programmes included 'President of Maldives Green Leaf Award', the 'President of Maldives Green Resort Award', the 'Two Million Trees Programme', 'Independent Maldives, Clean Maldives', the establishment of ECO Clubs at schools, World Environment Day, and campaigns such as the Save the Turtles Programme (UNEP 2013).

2.4 Commonalities in Initiatives and Implications for Future Disasters

The regulations, after the disaster, deals with the coastal development and restrictions of constructions on the coast, conversion of spatial plans to legal framework, reconstruction as an opportunity for participatory processes, preservation of natural resources and environment, and utilizing the physical infrastructure as escape routes.

Table 2.2 Summary of initiatives in different countries

	Indonesia	Sri Lanka	India	Maldives
Initiatives	Urban Reconstruction Plan focused on (1) buffer zones, (2) minimizing housing and economic activities in disaster areas, and (3) addition of community buildings as designated evacuation shelters	Introduction of coastal buffer zones limiting construction. A statutory zoning system based on hazard mapping, was prepared and included in development plan. Also, focused on reviving livelihood through cash for work and micro credit schemes	Improved coordination and communication at local level was ensured. Chennai came up with Climate Action Plan which stressed on resilience towards climate-related disasters, in urban areas through development controls, urban land use planning, and urban system management	(1) Socio-economic development and rehabilitation through improved institutional arrangements at urban level and Population Consolidation Plan. (2) Environmental Awareness Programmes
Critical issues	Disaster prevention through policing and laying restrictions did not prove successful	The initiatives did not go beyond coastal buffer zones. Also, the traditional skills suffered as new livelihood methods were employed	Since urban areas did not suffer major loss, impetus was not laid on urban disaster management. No pro-active measure was taken except in Chennai	Land scarcity and thinly disaster population challenges the disaster management initiatives
Potential/ learning	Opportunity for (1) land readjustment for urban initiatives, (2) simultaneously address complex issues like urban poverty, gender issues, infrastructure development and livelihood recovery	Build Back better program is a learning experience for other disasters and has the potential to amalgamate the urban development and preparedness efforts	Example of Chennai Climate Action Plan highlights that the urban development should be seen in conjunction with disaster management, as issues can be addressed at micro level	Environmental Awareness Programmes can bring the communities together to work for conservation of vegetation for protection from future Tsunamis

Another common challenge was to channelize resettlement and rehabilitation of the affected households. This factor was instrumental in configuring the institutional and legal arrangements adopted by the countries.

Indonesia utilized this opportunity to mould the development of its coastal cities with proper urban planning (Table 2.2). Nevertheless, the housing reconstruction phase highlights certain lessons for professionals, particularly urban planners involved in disaster management. Disasters serve opportunity to enfranchise people in wider governance issues, and not only in micro-issues. Indonesia's Blueprint failed to utilize people's opinion in governance issues. If it would have been done it would have put the onus of recovery on the community, since they would be partners in majority of the decisions. This comes in handy as Twigg (1999) illustrates that the models of disaster management works only till an external support exists. It would utilize the participatory principles and consultative processes. Also, 'behavioral drivers of communities' (Dercon and Kusumawijaya 2007) need to be understood in a better way for taking decisions in a post-disaster context.

Maldives exhibit some of the typical issues often associated with dense urban areas, such as pressure on land resources and diminishing per capita land, and landless urban poor with no tenure rights. However, it also faces challenge of thinly distributed population over large expanse, as competent disaster management often requires close knit communities (Table 2.2). Indian case emphasizes on the coordination of different stakeholders in the recovery process. However, coordination alone would not be sufficient to achieve desired goals. The Chennai action plan exhibits how to deal with disasters and effects of climate change, through a single comprehensive strategy. Sri Lankan experience demonstrates the need to utilize the already existing regulatory systems in risk reduction measures for safe development. A simplified permit procedure and a simplified administrative structure are necessary to achieve a balance of quality control and fast process of construction.

Sri Lankan experience demonstrates the need to utilize the already existing regulatory systems in risk reduction measures for safe development. A simplified permit procedure and a simplified administrative structure are necessary to achieve a balance of quality control and fast process of construction. Mannakkara and Wilkinson (2012) presents the key lessons from 'Build Back Better' for urban planners which include: conducting thorough hazard and risk analyses of the land; based on the analyses divide areas into zones based on level of risk; determine safe land-uses for each zone; and impose controls on developments in each zone to adopt the required design regulations for optimum safety. Finally, the implementation requires a framework developed from past experiences to lay down the actions to be taken by different stakeholders involved; regulatory and legislative facilitations to be enacted to support reconstruction as the framework needs to be legalized; and context-specific arrangements to be made to support the economic and social recovery of the community, based on traditional livelihood.

The implication of such arrangements is crucial. The learning from the affected countries should be applied to other countries. This would not only prepare the countries for a mega disaster but also help them recover following the successful model. Even though the policies and legal arrangements are often made at the central government level, the local government is responsible for local delivery of services. Addressing the issues at the local level needs more impetus. Top to down approach may be beneficial for overall policy formulation at the country level, but cities and their government needs to step up and come up with management systems and strategies which can be readily used at a time of next Tsunami or any other mega disasters.

2.5 Conclusion

Indian Ocean Tsunami's affected countries with their diverse economic, social and geographical characteristics have provided an opportunity to understand the relevance of urban issues in the post disaster reconstruction phase. Birkland (2001) says that disasters can alter the environmental, socio-economic and political structures, institutions and organizations, by changing their interrelations and creating new conditions. The challenge is to either mould such changes to restore normalcy in citizen's lives or to embrace the changes in such a way that the citizens are better prepared to face the next disaster. This requires policies which will not only encourage the communities to adopt major roles.

Urban planning should never start after a disaster, instead should be taken up prior to a disaster. Only then, the spatial plans could be used as genuine recovery tools, else there are chances that the post disaster spatial plans would only help in infrastructure augmentation. The disasters are reminders to local authorities in urban areas to stop walking on the path of unsustainable development which has seen unprecedented land reclamation for urban and industrial expansion, and shrimp farming.

The policies viz-a-viz urban issues that came up after the disaster were impacted by a number of inherent factors such as extent of damage to urban areas, density and diversity of the population, multi-hazard threat to that urban area, resettlement demand, external and domestic budgets, and political leadership. The policies could have addressed the urban issues in a much better way as the daily issues faced by urban areas, compound the recovery and worsens the damage of both lives and infrastructure. Moreover, the urban issues like poverty, gender gap, and poor infrastructure pose as a hurdle in faster recovery of a city. The facilities provided for disaster management, as evacuation centers in the case of Aceh, highlights an important aspect. It is important to consider the construction as well as maintenance of such facilities, preferably by the community for whom it is built.

With little research on post disaster reconstruction phenomena and factors that affect the process, there is little understanding of the impacts of such policies on urban issues, and hence it will take a few years, to establish and implement such policies which will address both the disasters and the ills of urbanization.

Acknowledgement The first author acknowledges support from the Ministry of Education, Culture, Sports, Science, and Technology (MEXT) of Japan; and the Global Center of Excellence (GCOE) Program entitled "Global Center for Education and Research on Human Security Engineering (HSE) for Asian Megacities" of Kyoto University.

References

Active Learning Network for Accountability and Performance in Humanitarian Action (ALNAP) (2012) Meeting the Urban challenge, adapting humanitarian efforts to an urban world. Overseas Development Institute, London. http://www.alnap.org/pool/files/meeting-paper-2012.pdf. Accessed 12 Mar 2014

Adger N, Hughes T (2005) Social-ecological resilience to coastal disasters. Science, 309:1036–1039. http://www.sciencemag.org/content/309/5737/1036.full. Accessed 13 Dec 2013

Birkland TA (2001) Scientists and coastal hazards: opportunities for participation and policy change. Environ Geosci 8(1):61–67

Birkmann J, Buckle P, Jaeger J, Pelling M, Setiadi N, Garschagen M, Kropp J (2008) Extreme events and disasters: a window of opportunity for change? Analysis of organizational, institutional and political changes, formal and informal responses after mega-disasters. Nat Hazards 55(3):637–655

Census of India (2011) Housing stock, amenities & assets in slums—census 201. Ministry of Home Affairs, Government of India, New Delhi. http://censusindia.gov.in/2011. Accessed 31 Jan 2014

Climate Change Action (CCA) (2010) National policy dialogue on climate change actions: linking grassroots action to policy debate, up scaling knowledge sharing and science. Conference proceedings, New Delhi. www.climatechangeaction.in/workshop/delhi/chennai-action-plan. Accessed 12 Mar 2014

CNN-IBN (2011) Climate change action plan awaits implementation. http://ibnlive.in.com/news/climate-change-action-plan-awaits-implementation/198234-60-120.html. Accessed 13 Mar 2014

Department of Census and Statistics, Sri Lanka (2004) Census of buildings and persons affected by the Tsunami—2004. Department of Census and Statistics, Colombo

Dercon B, Kusumawijaya M (2007) Two years of settlement recovery in Aceh and Nias what should the planners have learned? http://unhabitat-indonesia.org/data-files/rep-696.pdf. Accessed 31 Jan 2014

Fitzpatrick K (2005) Land mafia the latest threat to Acehnese. The Canberra Times. http://www.unisdr.org/wcdr/intergover/official-doc/L-docs/Hyogo-framework-for-action-english.pdf. Accessed 25 Feb 2005

Jayawardane AKW (2006) Disaster mitigation initiatives in Sri Lanka. Faculty of Engineering, University of Moratuwa, Sri Lanka. http://management.kochi-tech.ac.jp/PDF/IWPM/IWPM_Jayawardane.pdf. Accessed 26 Jan 2014

Mannakkara S, Wilkinson S (2012) Building back better in Japan—lessons from the Indian Ocean Tsunami experience in Sri Lanka. http://iiirr.ucalgary.ca/files/iiirr/A2-2_.pdf. Accessed 01 Jan 2014

Matsumaru R, Nagami K, Takeya K (2012) Reconstruction of the Aceh region following the 2004 Indian Ocean tsunami disaster: a transportation perspective. IATSS Res 36(1):11–19

OHCHR (2013) Response: Malaysia. Office of the High Commissioner of Human Rights, United Nations Human Rights. http://www.ohchr.org/Documents/Issues/Housing/SecurityTenure/ResponseMaldives01072013.pdf. Accessed 15 Nov 2013

Sheth A, Sanyal S, Jaiswal A, Gandhi P (2006) Effects of the December 2004 Indian Ocean Tsunami on the Indian Mainland. Earthquake Spectra 22(S3):S435–S473

Sridhar A (2006) Pre-and post-tsunami coastal planning and land-use policies and issues in India. Draft Submitted to FAO. RAP, Bangkok

Sugiyasu K, Murao O, Nakazato H (2010) Trinational comparative study of emergency support for victims and framework of recovery planning in Indonesia, Thailand, and Sri Lanka after the 2004 Indian Ocean Tsunami. AIJ J Technol Design 16(33):811–816

Tsunami Global Lessons Learned Project (TGLLP) (2009) The Tsunami legacy—breakthroughs, innovations and change. http://www.recoveryplatform.org/assets/publication/the-tsunami-legacy.pdf. Accessed 12 Jan 2014

Twigg J (1999) The age of accountability?: future community involvement in disaster reduction. Aust J Emerg Manage 14(4):51

United Nations (UN) (2005) Hyogo framework for action 2005–2015. Building the resilience of nations and communities to disasters. United Nations, Geneva. http://www.unisdr.org/wcdr/. Accessed 11 May 2008

United Nations (UN)—India (2008) Disaster recovery: a focus on good practices. UN Tsunami Response—India, United Nations, India. http://nirapad.org/admin/soft_archive/1308458712_A%20Focus%20on%20Good%20Practices.pdf. Accessed 2 Feb 2014

United Nations Environment Programme (UNEP) (2013) Maldives: post-tsunami environmental assessment. http://www.unep.org/tsunami/reports/maldives.pdf. Accessed 25 Dec 2013

Chapter 3
Environmental Recovery and Mangrove Conservation: Post Indian Ocean Tsunami Policy Responses in South and Southeast Asia

Rajarshi DasGupta, Rajib Shaw, and Miwa Abe

Abstract The Indian Ocean Tsunami in 2004 was a mega-disaster that triggered massive disruption of ecological services across the South & Southeast Asian Coast. Particularly, loss of precious mangroves was some of the most discussed environmental consequence of the event. In many South & Southeast Asian countries, mangrove bio-shielded the coastal communities and saved human lives and properties, however in doing so, it also suffered irreversible damage. Importantly, the Indian Ocean Tsunami was also important event in the context of rejuvenation of an already degraded mangrove ecosystems surrounding the Indian Ocean as many countries reemphasized the protective role of mangroves and other coastal forests and developed integrated coastal zone management policies. This chapter examines the role of mangrove forests in disaster risk reduction in four most affected countries (Indonesia, Sri Lanka, India & Thailand) and analyzes the policy amendments in the respective countries pertaining to mangrove conservation & restoration in the post Tsunami recovery period. The analysis was conducted based on six policy principles of mangrove conservation & restoration. It was observed that Indonesia & India responded the Tsunami event with significant amendments of its existing policies; whereas the policy responses of Sri Lanka & Thailand were not adequate to the purpose.

Keywords Coastal Zone management policies • Environmental recovery • Indian Ocean tsunami • Mangrove conservation

R. DasGupta (✉) • R. Shaw
Graduate School of Global Environmental Studies, Kyoto University, Kyoto, Japan
e-mail: rajarshidg1@yahoo.co.in; shaw.rajib.5u@kyoto-u.ac.jp

M. Abe
Center for Policy Studies, Kumamoto University, Kumamoto, Japan
e-mail: m-abe@kumamoto-u.ac.jp

R. Shaw (ed.), *Recovery from the Indian Ocean Tsunami: A Ten-Year Journey*, Disaster Risk Reduction, DOI 10.1007/978-4-431-55117-1_3

3.1 Introduction

Impacts of mega-disasters are multi-dimensional. At one hand, it causes unprecedented loss of human lives and properties, while on the other; it triggers massive disruption of ecosystem and environmental services. Although, environmental loss to disasters receives good amount of attention; however, much lesser attention is generally given to poor environmental management practices and ecological degradation, which aggravates a disaster's impact (Srinivas and Nakagawa 2008). Undoubtedly, the huge loss of human lives serves as the most tragic consequence of any mega-disasters which generally receive extraordinary national and international humanitarian response; yet, such disasters also have critical environmental consequences that are often ignored against post disaster reconstruction and developmental priorities. Moreover, in recent years, disaster recovery is perceived as 'building back better' which essentially advocates to reduce the vulnerability components of the communities at risk. Needless to say, rejuvenating environment and ecological services formulate one of the key aspects of disaster recovery and therefore, requires special initiatives from the government, national & international NGOs, community based organizations and most importantly, donor agencies. Further, 'environmental recovery' also includes the idea to develop and sustain new environmental resources that can promote effective disaster risk reduction for the affected and vulnerable communities. However, to initiate such activities, the local government & the coastal managers, at first, should be equipped with adequate policy and legislations in order to avoid any kind of conflicts. It is rather apparent that although the inherent links between disaster risk reduction and environmental management are mostly recognized, the concept of using environmental tools for disaster reduction still needs to be further streamlined at policy level (ISDR 2004; Shaw 2006). It is, therefore, highly imperative that environmental recovery is given adequate priority during the post disaster reconstruction phases.

The Indian Ocean Tsunami was one of its own kind of 'never experienced before' disaster that left wide scale damage across 12 developing countries redefining the region's vulnerability against an apparent 'Tsunami-free' Indian Ocean. Out of these 12 countries, the most affected countries were Indonesia, Sri Lanka, Thailand & India (in terms of loss of human lives to the Tsunami). The Tsunami of 26 December 2004 left wide spread damage with a death toll that have crossed over a quarter of million and such damage was largely confined to coastal rural areas that hosts some of Asia's poor & vulnerable population (Shaw 2006). Yet, the 'boxing day Tsunami' was more than just a human disaster (Wong 2009). This is partly because of its place of occurrence and the trans-boundary impacts which coincide exactly with the Indo-Malayan mangrove region in South & Southeast Asia (IUCN 2005). The Indo-Malayan region host approximately 40.4 % of the existing mangrove forests and considered as most diverse mangrove region in the world (Spalding et al. 2010). Therefore, the environmental impact of the Tsunami was enormous in terms of loss of mangroves and its precious ecosystem services. Although, in many cases the mangroves shielded the Tsunami impacts and resulted in lesser damage of the communities; however, in doing so, it also underwent unprecedented damage especially along the coast of Indonesia and Andaman & Nicobar Islands of India. Further, other coastal ecosystems such coral reefs, coastal wetlands, sea grasses were irreversibly damaged and it may take long time and significant resources to restore and rejuvenate the lost ecosystem services. In places where vast tracts of pristine mangroves and other coastal have been shattered, it is believed that it may pose a long-term threat for the region not only for forest and biodiversity conservation, but also in terms of economic & livelihood security for the coastal communities (IUCN 2005). Indeed, the Indian Ocean Tsunami resulted in wide spread destruction of mangroves which has already been under the serious threat of extinction. Importantly however, all the affected countries have shown keen interest in recovering the environmental loss incurred during the Tsunami as a key learning from the Tsunami event; while some of them have come up with important policy responses such as an improved coastal zone management plan and protective legislations. As many consider that the Indian Ocean Tsunami triggered a paradigm shift in existing disaster management practices, particularly in South & Southeast Asian countries, this chapter, which is mostly an overview in nature, aims to provide an outline of the loss and recovery of mangrove forests and to give a detailed outline of the proactive policy responses by the affected countries in terms of conservation and restoration of mangroves forests.

3.2 Role of Coastal Forests in Hydrometeorological Disaster Risk Reduction

Coastal forests, such as mangroves represent strong natural capitals that are capable of producing extremely essential environmental & economic services. Over the previous years, mangroves have sustained more than 70 direct human activities and have been an enormous source of wood, wax, tannin, charcoal, traditional medicines and various other forest products. Yet, such tangible services are often outnumbered by the indirect benefits from the mangroves, particularly its impeccable environmental services ranging from arresting soil erosion, water purification, flood, tropical storm and tsunami risk reduction. Although, it can be scientifically argued the extent of mangrove's efficiency in arresting huge wave energies of Tsunami, yet, it is also extensively believed that the dense and mature mangrove forests growing along the shoreline can greatly help to reduce its impacts. It has been estimated that a mangrove density of 30 trees per 100 m^2 in a 100-mt wide belt may reduce the maximum tsunami flow pressure by more than 90 % (Danielsen et al. 2005). Importantly, the trend of devastating coastal disasters such as tropical cyclones and Tsunamis are increasing and in response to such rising disasters, several structural measures are being adopted to combat the woeful situation. However, structural resistance to Tsunamis & other coastal hazards are also cost intensive and at times can be environmentally destructive. Consequently, several developing

Table 3.1 Classification of conservation forest in Japan

Conservation forest	Disaster prevention forests
Headwaters Conservation Forest	
Erosion Control Forest (sediment discharge)	Categorized as Disaster Prevention Forests
Erosion Control Forest (mass failure)	Categorized as Disaster Prevention Forests
Blowing Sand Break Forest	Categorized as Disaster Prevention Forests
Wind Break Forest	Categorized as Disaster Prevention Forests
Flood Damage Prevention Forest	Categorized as Disaster Prevention Forests
Tidal Wave and Salty Wind Prevention Forest	Categorized as Disaster Prevention Forests
Drought Damage Prevention Forest	
Snow Break Forest	Categorized as Disaster Prevention Forests
Fog Break Forest	Categorized as Disaster Prevention Forests
Avalanche Prevention Forest	Categorized as Disaster Prevention Forests
Stone Fall Prevention Forest	Categorized as Disaster Prevention Forests
Belt of Trees Plantation as a Firebreak	Categorized as Disaster Prevention Forests
Fish Shelter Forest	
Protection Forest for Navigation Sign	
Protected Forest for Health	
Forest Grown for Scenic Beauty	

countries may find it extremely difficult to implement such hard engineering measures. For example, in case of Japan, artificial coastal barriers such as sea walls & breakwaters have played an important role in safeguarding the communities from Tsunamis and tidal waves, yet, it requires high cost of construction and maintenance (Harada and Imamura 2005). Therefore, coastal forests provide a cost effective, environmental friendly and ecologically benign solution particularly relevant to the developing countries. Even in the developed countries, such as Japan, which has a long history of adapting to extreme coastal hazards has prioritized its policies towards conservation of forests exclusively for disaster risk reduction purposes. For example, as per the forest act of Japan, there are total 17 kinds of conservation forests which include 11 kinds of disaster prevention forests. These forests are artificially grown for preventing the impacts of natural/human disasters and can be grown by individuals, local governments or even private sectors. Table 3.1 provides the list of the recommended disaster prevention forests as per the Article 25 of the Japanese forest act. In the past, a number of evidences are available from Japan which essentially highlights the protective role of mangrove & other coastal forests. For example, during the Shouwa Sanriku Tsunami in 1933, high density coastal forests (called as "Takada Matsubara") were responsible of protecting large residential area in Rikuzentakada city of Japan. Similarly, during the Chili earthquake & Tsunami (1960), 'Takada Matsubara' protected severe housing damage from the tide-water. At Akamae beach, six power-driven vessels which were washed away by the waves (each weighed about 10 t) were virtually screened by the coastal forests. The Nihonkai Chubu Earthquake and Tsunami in 1983 also revealed similar kind of experiences. The initial Tsunami waves of 5 m at Oohiraki, Noshiro city was reduced to mere 70 cm as it passed through a 150 m thick tide-water control black pine forest (Japan Forest Agency 2012). Most recently, during the Great East Japan Earthquake & Tsunami of 2011, it has been reported that coastal forest prevented floating objects such as ships, cars and rubbles of housing to crush into the densely populated habitat in Aomori, Iwate and Fukushima Prefecture although it could not sustain the high wave energy of the Tsunami at the coastal front. However, approximately 1,718 ha of coastal forests was completely damaged by the gigantic Tsunami waves.

3.3 Role of Coastal Forest in Mitigating the Indian Ocean Tsunami in 2004

The Indian Ocean Tsunami was in fact an eye-opener for many coastal managers in affected the region about the protective role of coastal forests. In the aftermath of the Tsunami, several reports and experiences of affected communities, national & international NGOs, individual scientists & research organizations revealed that the mangroves not only mitigate the impacts of the waves but also trapped huge amount of floating objects saving a precious human lives and properties. It also prevented the people from being washed out into the sea which has been a major cause of casualties (EJF 2006). Since the last decade, several empirical studies based on the evidences of Indian Ocean Tsunami established significant relationship between the existence of mangroves and relative lesser loss of lives and properties. The Table 3.2 summarizes the protective roles of mangroves during the Indian Ocean Tsunami across the severe impact zone as per the reported human causalities (Indonesia, Sri Lanka, India & Thailand).

3.4 Occurrence and Threats on Mangroves of South & Southeast Asia

As mentioned earlier, the affected countries in South & Southeast Asia host some of the world's dense mangrove forests accounting for 40.4 % of the total mangroves in the world. These mangroves are the part of Indo-Malayan mangrove that spreads over the sub-tropical and tropical coastlines of South and Southeast Asia encompassing approximately 6.13 million ha (Spalding et al. 2010). Out of its global share of 40.4 %, Indonesian mangrove alone accounts for 22.6 %, whereas Malaysia (3.7 %), Myanmar (Burma) (3.6 %), Bangladesh (3.2 %), India (2.7 %), contribute significantly to this highly fragile ecosystem (Giri et al. 2011). However, the region also suffered the highest loss of mangrove forest in the last three decades. The United Nation's Food and Agricultural Organization (FAO) reported that in South and

Table 3.2 Protective role of Mangroves in the Indian Ocean Tsunami, 2004

Name of the country	Affected areas	Causalities	Role of Mangroves and other coastal forests
Indonesia	Northern & Western Tip of Ache Province	More than 130,000	Since the area served as the epicenter of the Great Sumatra Earthquake, it underwent unprecedented damage. It is widely believed that coastal shrimp farming development in the mangrove areas largely aggravated the loss the human lives; although it is doubtful that even if the mangroves were present, perhaps it would not able to sustain the immediate impact of the Tsunami as in the case of Ulee Lhee (EJF 2006). However, Kathiresan and Rajendran 2005, EJF 2006 reported that in Simeuleu Islands which is merely 41 km away from the epicenter was saved partly due to substantive mangrove cover
Sri Lanka	Eastern and the Southern Coast	More than 35,000	Coastal forests proved effective in protecting some village in Sri Lanka .e.g. Kapuhenwala, a coastal hamlet with dense mangrove cover, only had two casualties because of the dense mangroves (EJF 2006). At Odu lagoon and Nasiva village (Valachenai), in Batticaloa district, the 6 m Tsunami waves penetrated about 1 km inside, however, a mixed landscape comprising beach, mangrove-fringed lagoon, coconut plantation, scrub forest and home gardens would able to absorb and dissipate much tsunami energy turning the waves to just 40 cm high. There was no loss of Life although there were severe damage to the mangrove and other coastal forests (UNEP 2005)
India	Eastern and Southern Coast	More than 12,500	In part of Tamil Nadu, mangroves proved effective in safeguarding the lives and properties of the coastal hamlets. Kathiresan and Rajendran (2005) mentioned that the loss of human lives had significant correlation with the absence of mangrove forests. Danielsen et al. (2005) mentioned that in Cuddalore district, there are strong evidences based on the pre and post Tsunami satellite images that coastal hamlets protected by mangroves suffered lesser damage
Thailand	The Andaman Coast	More than 5,400	In Phang Nga province of Thailand, large mangrove forests in the north and south of the province have significantly reduced the impact of the Tsunami where as the unprotected area suffered greater loss (UNEP 2005)

Southeast Asia, mangrove cover has decreased by 25 % from its original extent in 1980, while the Indo-Malayan mangroves suffered a net loss of nearly 1.9 million ha (FAO 2007). No wonder that the annihilation rate of mangroves is more threatening than the tropical rain forests or the adjacent coral reefs with an alarming disappearing rate of 1–2 % per year (Duke et al. 2007). Such extremely high disappearing rate is sustained through a combination of human and natural factors. However, it is rather evident that the human intervention into the mangrove ecosystems remains as the sole cause behind the degradation of mangrove forests whereas scientists predict climate change does not possess an immediate effect on mangroves; although it may result to 10–15 % reduction of mangroves in longer term scenario (Alongi 2008).

In case of South & Southeast Asia, the major two delineating factors behind the mangrove forest annihilation can be regarded to coastal agricultural development and forest diversion for shrimp/pond cultivation (DasGupta and Shaw 2013). The other associated causes involve development of tourism front, coastal urbanization and port & waterfront development. Historically due to the immense population pressure in the coastal zones, South & Southeast Asian countries has been generally poor in effectively develop and implement scientific coastal management plans. As a result, coastal development has been executed with tremendous disregard to environment. For example, conversion of mangrove forest for coastal agriculture purposes is typically prominent in countries like India, Bangladesh, Sri Lanka, Thailand, and Indonesia. A remote sensing based analysis by Giri et al. (2008) revealed that agricultural activities in the mentioned countries are responsible for nearly 82 % of the reported loss during 1975–2005. Importantly, agriculture and fishing still remains the main occupation of the majority of the coastal communities living in these countries therefore over the past large scale organized deforestation of mangroves has been a common trend in all these countries. On the other hand, among the global shrimp production, approximately 75 % of the shrimp is produced by South & Southeast Asian countries and Thailand has been the single largest exporter of shrimps. Shrimp farming or artificial cultivation of shrimps in shallow marshy lands along the coast provides an economically lucrative livelihood for the coastal communities. Unfortunately, the process of shrimp farming in artificial ponds is environmental destructive in nature as it involves massive deforestation of mangrove forests in the low lying coastal areas. As per Giri et al. (2008), approximately 41 % (18,816 ha) of Thailand mangroves, 63 % (2,0956 ha) of Indonesian mangrove and 22 % (7,554 ha) of Indian mangrove, were diverted to shrimp ponds during 1975–2005. Such large scale deforestation also brought massive alteration of coastal land use in the respective countries with associated adverse environmental consequences. Since, a huge amount of population depends on such activities; many of the governments of these countries have historically paid lesser attention on such unscientific & unplanned coastal development against short term developmental & economical priority. However, it's longer term consequences has been disastrous. Another important cause of mangrove destruction has been referred to rapid development of coastal tourism. Development of coastal tourism with gross disregard for environmental consequences has been extremely prominent in several countries such as Thailand. In short, strong disregard for coastal environment has been the triggering factor for the massive degradation of mangrove forests in this region.

3.5 Impacts of Indian Ocean Tsunami on the Mangrove Forests

Although this chapter has discussed in details the protective role of mangroves during the Indian Ocean Tsunami, it also important to mention that the Indian Ocean Tsunami itself served as one of the major factor behind natural degradation of mangrove forests in this region. The major affected countries were Indonesia, Thailand, Sri Lanka and India where substantial area of mangrove habitats were destroyed by the powerful waves of the Tsunami. However, it is sometimes difficult to assess the exact loss of mangrove forests due to the sheer unavailability of reliable data in pre and post Tsunami period. Also, on the other hand, Alongi 2008 mentioned that the human impact of the disaster were so severe that little to no attention were provided in the damage assessment of mangroves. However, following the Tsunami event, some countries as well as international organizations established mangrove damage assessment mechanisms. It was reported that all the countries suffered massive loss of mangrove habitats in different Tsunami hit locations. Indonesia, which hosts the majority of mangrove forests in the region, also suffered the highest loss of mangrove forests accounting for nearly 25,000 ha. EJF (2006) reported that along the Banda Ache & Northern Sumatra 30 % of 48,925 ha of coastal forests is believed to be severely damaged by the Tsunami. However, as per the Indonesia government's official report sever damage of mangroves in the country was restricted to 300 750 ha (EJF 2006; UNEP 2005).

In India, as per the rapid damage assessment report of Forest Survey of India, the devastation of mangroves was greatly limited to the Nicobar Islands in the Andaman Sea. On the Indian mainland, mangroves were not severely affected and rendered an excellent service of bio-shielding; however is has been estimated that 40 ha of forest cover in Cuddalore District of Tamil Nadu, 8 ha of mangrove cover in Ernakulam District of Kerala, were lost after the Tsunami. In Andaman group of Islands, loss of forest cover was concentrated in Little Andaman Island which was reported as 258 ha. However, the Nicobar group of Islands suffered the highest loss of forest cover. Total loss of forest cover in Nicobar group of islands was assessed to be 12,224 ha including the mangroves and other coastal forests. The mangroves of Thailand, on the other hand, were relatively undamaged due to the Indian Ocean Tsunami, with a reported loss of 306 ha, affecting only 0.2 % of the total mangrove areas (EJF 2006). In case of Sri Lanka, the reported loss of mangrove forests was insignificant as only the 2–3 m mangrove fringes were destroyed by the Tsunami whereas the matured mangroves could sustain the wave energy of the Tsunami.

3.6 Post Tsunami Mangrove Recovery: Policy Implication in South & Southeast Asia

It is widely believed that the Indian Ocean Tsunami event was actually a major triggering factor which brought a paradigm shift in the proactive risk reduction mechanism in South & Southeast Asian countries. Following the Indian Ocean Tsunami, as the governments settled down after high level relief and rehabilitation missions; a greater consensus about the integrated coastal zone management and need for environmental preservation strongly emerged in these countries. Such consensus was instrumental for initiation of massive restoration drive for mangroves and other coastal forests through the formation of new and improved policies. For example, the Government of Indonesia, pledged to restore about 600,000 ha of degraded & fragmented mangrove forests in the West coast of Ache & Northern coast of Java (ENN 2005). Similar responses were also recorded from the Government of Thailand, India and Sri Lanka. On the other hand, other countries such as Malaysia promised $25 million to restore and develop 4,000 ha of mangrove forests as it clearly mentioned the importance of having mangrove forests as coastal barriers (EJF 2006). However, conservation of mangroves is far more complicated issue than such commitments alone. It is important to mention that despite of the generous commitments made by the respective governments, the existing restoration of mangroves does not reach up to the satisfactory limits. As mentioned earlier, the gradual destruction of mangroves in this region has been triggered by several factors which determine the economic wellbeing of the communities and developmental priorities of the local government. Therefore, in order to have successful restoration and development of mangrove resources, it is extremely important to have adequate policy arrangements to back such initiatives. It is also very important to have proper scientific understanding of the plantation site, for example, choice of species, hydrological scenario and slope etc. However, prior to that adequate policy support needs to be in place. In order to analyze whether the affected countries are legally well equipped to control the major threats on mangroves, the chapter examines four country's policy arrangements & amendments after the Indian Ocean Tsunami in order to protect & restore the mangrove forests.

In general, DasGupta and Shaw 2013 mentioned that there are six policy aspects of mangrove conservation in the backdrop of the South & Southeast Asian countries based on the country experiences of this region. These six are basically a combination of preventive & restrictive policies and/or legislations that supports the growth of mangroves. The following are listed below:

Principle 1: Legal arrangements for in-situ conservation of mangroves (Statutory provisions)
Principle 2: Institutional Provision of Community involvement in mangrove restoration & conservation (Legal arrangements of participatory management)
Principle 3: Legal arrangements on Control of shrimp farming/aquaculture in mangrove habitats
Principle 4: Legal arrangements for Marine & Coastal Protection from Industrial & domestic pollution
Principle 5: Legal & institutional arrangements for Coastal Zoning
Principle 6: Legal arrangements for Coastal Greenbelt or Barrier Plantation development

Importantly, following the Indian Ocean Tsunami, some countries in this region adjusted or developed new legislations under the above mentioned dimensions. Therefore the objective of this section is to provide a detailed outline of the policies of undertaken by the respective governments in the post Tsunami regime. However, for the ease of discussion the chapter limits its discussion in the four most affected countries.

3.6.1 Indonesia

In Indonesia, the nodal agency for mangrove conservation is the Ministry of Forestry and conservation of mangroves is promoted under the Act No. 41/1999 (Forestry Law) & Act No. 5/1990 (Conservation of Biological Resources and their Ecosystems). These specific acts envisage the sustainable use of resources that has ecological importance with a motto to promote a balance between development & ecological conservation. These acts also define the concept of national parks for conservation of native species which satisfies the principle 1. However, the forest governance in Indonesia holds a typical 'top–down' hierarchical approach despite of some decentralization proposed in Law no 32/41 of 2004. Therefore, the extent of community participation in mainstream forest management seems to be limited. Following the Indian Ocean Tsunami, Indonesia has taken strong legislative measures in controlling the unplanned and destructive fishing in its shores. A desired update in Law number 45/2009 (Fisheries Law) provides a detailed outline for scientific harvesting of the fisheries resources while implementing adequate conservation of coastal and marine resources. In 2007, the Indonesian government passed another important piece of legislation; Law 27/2007 (Law on Coastal Management and Small Islands) which requires all the provincial governments in coastal regions to develop strategies, zoning, and management plans of coastal areas through the proper involvement of multiple stakeholders. The Law 27/2007 established a legal basis for long-term coastal water use rights involving major marine based industries such as tourism, aquaculture, pearl farming and fishing. Similarly, coastal pollution, degradation of coastal areas through minerals and coal mining is presently screened through the Act No. 32/2009 (Environmental Protection and Management) and Act No. 4/2009 (Minerals and Coal Mining). Another important policy evolution since the Indian Ocean Tsunami was the Act No. 26/2007 (Spatial Use Management) which also has significant contribution towards mangrove protection & rehabilitation. The Indonesian Government prepared a Master Plan for Aceh and Nias Rehabilitation and Reconstruction which was officially released as Presidential Regulation (PR) No. 30/2005. This important policy document proposed 100–300 green belt development along the Aceh coast; however, the implementation of such policies is restricted with many hindrances. In short, the Government of Indonesia came up with important policy responses although the status of its implementation is not as desired.

3.6.2 Sri Lanka

The occurrence of mangrove forests in Sri Lanka is highly scattered and the management framework of mangroves itself is not well established. In Sri Lanka, patches of mangroves are generally managed by three government agencies (i.e. the Forest Department, the Department of Wildlife Conservation and the State Timber Corporation) under the Ministry of Environment & Forests. In general, the mangrove protection framework is inadequate and the government does not have any specific policies except the coastal conservation act of 1981. One significant change in policies which may be attributed to conservation of coastal forests is that—during the post Tsunami period the cabinet decided to implement buffer zones of 100 and 200 m in the west and south coasts and north and east coasts respectively. The Forest Conservation Department commenced some coastal forest rehabilitation program with the help of international NGOs and funds. Consequently, the government of Sri Lanka developed some guidelines for coastal forest restoration and plantation. However, there was no legal or policy alteration for this.

3.6.3 India

India is one of the countries in this region which already had strong policies for mangrove conservation. In India, conservation of mangroves is promoted by the Ministry of Environment & Forests in the federal and provincial government. Conservation of mangroves is promoted through two major piece of legislations i.e. Forest Conservation Act of 1980 and Wildlife Protection Act of 1972. While the former establishes the legal protection categories of natural resources (National Park, Reserve Forests etc.) and restrict illegal feeling of forest resources, the later is involved in declaring Marine Protection Area which has direct consequence on mangrove conservation. On the other hand, the establishment of Joint Forest Management in 1990 has strong implication of community involvement in mainstream forest management. Earlier much of the mangroves have been converted to coastal aquaculture pond in India. One extremely important piece of legislation that evolved after the Tsunami was the Coastal Aquaculture Authority Act, 2005. The act is instrumental for the establishment of Coastal Aquaculture Authority which is a legislative body to make sure that no mangrove or ecologically sensitive area is converted to aquaculture pond in coastal areas. The Coastal Zone Management (CZM) notification of 1994 was also amended in 2011. This suggests mandatory preparation of a Coastal zone management plan of all the coastal provinces of India within a span of 2 years. This also suggests a number of restrictive measures to protect the coastal ecological diversity and restrict new industrial establishment in coastal areas. Apart from the coast it also envisages proper scientific management of 12 nautical miles in the sea. However, the main classification of 4 zones under the CZM notification (1994) remains unaltered. Despite of the fact that many local & provincial governments include greenbelt development as a strategy to mitigate coastal disasters and also the Environmental Protection Act of 1986 mandates industrial greenbelt development, plantation for coastal defence is yet to be institutionalized in India.

3.6.4 *Thailand*

In case of Thailand, management of mangroves in governed by the Royal Forestry department and Ministry of Natural Resources & Environment. The Thai mangroves are conserved under the National Forest Reserve Act of 1964. Over the years, Thailand has sustained high rate mangrove degradation due to establishment of tourism front and development of coastal aquaculture pond. These two activities supported the economic need of the coastal communities. In 2007, The Royal Government of Thailand introduced a Community Forestry Bill which essentially includes the mangrove dependent community to involve in forest conservation and sustainable utilization. Following the Indian Ocean Tsunami, the Government of Thailand proposed an umbrella act named as Marine and Coastal Resources Management Act. This proposed umbrella act includes a wide range of proactive policies including community participation in managing coastal resources, development of Guidelines for formulating marine and coastal resources policies, compensation for conducting ecological damage and further to monitor the post 2004 Tsunami ecological and community rehabilitation. However, on contrary, Thailand is yet to develop any legal arrangement to control the shrimp/pond farming which has been responsible of large scale degradation of Thai mangroves.

3.7 Conclusion & Way Forward

The Indian Ocean Tsunami was a unique event in the disaster history of the region and possibly chances of such extreme events is less in near future. However, the same region is equally or even more vulnerable to tropical cyclone, storm surges and sea level rise. The region also hosts some of the Asia's poor and vulnerable population. Therefore, conservation of mangroves has great ecological, social and environmental significance. As mentioned by several researchers that the Indian Ocean Tsunami was a painful eye opener and brought several changes in disaster management policies of the respective country. On an environmental dimension, it strongly established the disaster-environment linkages; thereby tossing an urgent call for restoration and rehabilitation of coastal forests. The chapter analyzed the policy perspective of coastal forest conservation and restoration aftermath the Indian Ocean Tsunami.

Furthermore, it tried to broadly analyze the institutional response to restoration & conservation of coastal forests in four most affected countries. In order to assess the policy alteration since the Indian Ocean Tsunami, the chapter used six policy principles (indicators) to analyze the existing management practices in the respective countries and emphasized the post Tsunami policy evolution for the management of coastal forests with special reference to mangrove forests. It is observed that Indonesia and India have responded with adequate policy arrangements (e.g. coastal green belt development and control of coastal aquaculture) and a great deal of rehabilitation work has been undertaken by the local and federal governments. On the contrary, Sri Lankan response to restoration of mangrove forests were largely driven and motivated by the International NGOs and other conservation bodies, whereas the government did not come up with much needed policies and action plans on coastal forest development. In case of Thailand, policy responses were mixed; while the Thai government identified the need of coastal conservation, however, failed to address extremely important issues such as propagation of aquaculture ponds in mangrove areas.

In general, the Indian Ocean Tsunami brought about extensive change of perception among the community members as well as the local and federal governments. A clear sense of urgency in restoring mangroves was observed in all the affected countries both at the community level and the local institutional level. Whereas, at an academic arena mangrove management evolved as a critical dimension of natural resource governance and disaster risk reduction research. The chapter therefore analyzed the changes of coastal forest management policies and identifies two major changes which were triggered by the Indian Ocean Tsunami. The first was an evolution of the legalization of coastal forest development, particularly the coastal green belt concept to promote the mangrove forests as the first line of defense. As discussed, such policy approach was undertaken by Indonesia, Malaysia and several other countries and local governments. India, on the other hand, came up with an extremely important regulation of coastal aquaculture with appropriate legislative arrangements in 2005. However, till date, the restoration of mangroves is perplexed with several critical issues such as lack of administrative framework at local level and high resource dependent communities. Therefore, this chapter argues that at first, country and local level policies needs to be further strengthened & coordinated with clear demarcation of responsibilities and the communities can be further involved in mainstream management.

Importantly, in order to 'build back better' through better environmental preparedness, policies need to translate into actions. Therefore, it is extremely important to have a strong monitoring mechanism both at the local and federal government level. The year 2014 marks the 10 years anniversary of the deadly Tsunami, therefore, it is extremely important to look back and find the difficulties observed over the last decade in restoring the mangroves and coastal forest as such. Of course there are many obstacles, such as loss of suitable land for rehabilitation, community rehabilitation issues, community's inclination to live by the sea, high prices & lack of participation in plantation, lack of technical know how etc. However, this can be overcome with proper planning and involving the communities at risk. A 'bottom up strategy' for community based coastal zone management and environmental recovery perhaps hold the clue.

Acknowledgement The first author greatly acknowledges MEXT (MONBUKAGAKUSHO) scholarship provided by the Japanese Government for conducting research in the Graduate School of Global Environmental Studies of Kyoto University. The authors also acknowledge the support of GOCE-ARS (Sustainability/survivability science for a resilience society to extreme weather conditions) and Studies on the Connectivity of Hilltop, Human and Ocean (CoHHO) program of Kyoto University.

References

Alongi DM (2008) Mangrove forests: resilience, protection from tsunamis, and responses to global climate change. Estuar Coast Shelf Sci 76(1):1–13

Danielsen F, Sørensen MK, Olwig MF, Selvam V, Parish F, Burgess ND, Suryadiputra N (2005) The Asian tsunami: a protective role for coastal vegetation. Science (Washington) 310(5748):643

DasGupta R, Shaw R (2013) Cumulative impacts of human interventions and climate change on mangrove ecosystems of South and Southeast Asia: an overview. J Ecosyst doi:10.1155/2013/379429

Duke NC, Meynecke JO, Dittmann S, Ellison AM, Anger K, Berger U, Cannicci S, Diele K, Ewel KC, Field CD, Koedam N, Diele K, Ewel KC, Field CD, Koedam N, Lee SY, Marchand C, Nordhaus I, Dahdouh-Guebas F (2007) Letter: a world without mangroves? Science 317(5834):41–42

EJF (2006) Nature's defence against Tsunamis: a report on the impact of mangrove loss and shrimp farm development on coastal defence. In: Mangroves. Environmental Justice Foundation, London

Environmental News Network (ENN) (2005) Indonesia to replant mangroves in Tsunami defense. http://www.enn.com/top_stories/article/799

FAO (2007) The World's Mangrove 1980–2005, food and agriculture organization, 2007

Giri C, Ochieng E, Tieszen LL et al (2011) Status and distribution of mangrove forests of the world using earth observation satellite data. Glob Ecol Biogeogr 20(1):154–159

Giri C, Zhu Z, Tieszen LL, Singh A, Gillette S, Kelmelis JA (2008) Mangrove forest distributions and dynamics (1975–2005) of the tsunami-affected region of Asia. J Biogeogr 35(3):519–528

Harada K, Imamura F (2005) Effects of coastal forest on tsunami hazard mitigation—a preliminary investigation. In: Tsunamis. Springer, Netherlands, pp 279–292

IUCN (2005) Early observations of tsunami effects on mangroves and coastal forests. http://www.iucn.pk/pdf/TsunamiForest.pdf

ISDR (2004) Living with risk: a global review of disaster reduction initiative. United Nations Publication, Geneva

Japan Forest Agency (2012) The role of coastal forests in future disaster prevention (Tentative translation of title (in Japanese)). http://www.rinya.maff.go.jp/j/press/tisan/pdf/120201-01.pdf

Kathiresan K, Rajendran N (2005) Coastal mangrove forests mitigated tsunami. Estuar Coast Shelf Sci 65(3):601–606

Shaw R (2006) Indian Ocean tsunami and aftermath: need for environment-disaster synergy in the reconstruction process. Disast Prev Manag 15(1):5–20

Spalding M, Kainuma M, Collins L (2010) World atlas of Mangroves. Earthscan, London

Srinivas H, Nakagawa Y (2008) Environmental implications for disaster preparedness: lessons learnt from the Indian Ocean Tsunami. J Environ Manage 89(1):4–13

UNEP (2005) After the tsunami: Rapid environmental assessment (Compiled Country Reports). http://www.unep.org/tsunami/reports/TSUNAMI_report_complete.pdf

Wong PP (2009) Rethinking post-tsunami integrated coastal management for Asia-Pacific. Ocean Coast Manag 52(7):405–410

Chapter 4
Lessons from the Recovery of the Education Sector After the Indian Ocean Tsunami

Glenn Fernandez, Rajib Shaw, and Miwa Abe

Abstract In the aftermath of a disaster, the repair and reconstruction of schools can symbolize community survival. The resumption of classes can be one of the most effective ways to demonstrate a return to normalcy to the local population. Schools become the glue that holds a fractured community together. This chapter reviews the recovery of the education sector after the Indian Ocean Tsunami, emphasizing the similarities and differences in the recovery experience across India, Indonesia, the Maldives, Sri Lanka, and Thailand. Good practices as well as unsuccessful attempts in several issues, such as education sector recovery coordination, resumption of classes, reconstruction and relocation of schools, school-community partnerships, integration of disaster risk reduction education into the school curriculum, and reaching out to the most disadvantaged children, are highlighted to provide valuable lessons for education sector recovery in future disasters.

Keywords Disaster recovery • Education sector • Indian Ocean Tsunami • Schools

4.1 Introduction

Disasters over the past 20 years have wiped away billions of development investments in education facilities (Bastidas and Petal 2012). Significant impacts of disasters on the education sector include the following: students and teachers are killed, injured, or displaced; school buildings are damaged; classes are disrupted for long periods; vital records are lost; students are forced to drop out; etc. (Shaw et al. 2012; Fernandez et al. 2012). Disasters disrupt children's academic progress and diminish their long-term educational outcomes (Peek 2008; Rognerud 2009).

G. Fernandez (✉) • R. Shaw
Graduate School of Global Environmental Studies, Kyoto University, Kyoto, Japan
e-mail: glenn.fernandez@gmail.com; shaw.rajib.5u@kyoto-u.ac.jp

M. Abe
Center for Policy Studies, Kumamoto University, Kumamoto, Japan
e-mail: m-abe@kumamoto-u.ac.jp

R. Shaw (ed.), *Recovery from the Indian Ocean Tsunami: A Ten-Year Journey*,
Disaster Risk Reduction, DOI 10.1007/978-4-431-55117-1_4

In terms of economic costs, disasters can have direct and indirect costs as well as secondary effects to the education sector. Direct costs to education relate to the physical damage to capital assets and include the cost of damage to classrooms, sporting facilities, books, computers, etc. and the cost of demolition and clearing of unsalvageable facilities. Indirect costs to education refer to damage to the flow of services and consist of the cost of education and sport facilities used as evacuation and relief centers; additional transportation costs due to damaged roads; loss of income to teachers; and other educational services disrupted. Secondary effects refer to both the short and long-term impacts on overall economic performance, such as lost contribution to the national economy or production generated by the education sector; variations in employment rates; impact on the public sector; etc.

Schools can get damaged beyond repair in a disaster and, as a result, students are left with nowhere to go. Devastated schools require a level of reinvestment many times higher than the initial small incremental cost of building safely, which can be a substantial burden on the economy (GFDRR 2009). The destruction of other infrastructure such as roads can lead to a worsening of learning conditions and, if vital education-related infrastructure inside and outside the school is permanently damaged and not restored causing children not to return to their studies, this results to a direct loss of human capital (Baez et al. 2010).

The 2004 Indian Ocean Tsunami affected the lives of thousands of students. Table 4.1 shows the impact of the disaster on the education sector in five of the most affected countries. It was fortunate that the tsunami occurred during the weekend; otherwise, the number of children directly affected, especially in coastal areas, would have been significantly higher (Jovel 2005).

The immediate recovery of schools constitutes a key element after a disaster because it allows children to return to a state of normality and to replace their emotional crisis with the joy of being around other children and have a space in which to learn and play at the same time (HHF 2005). Communities also benefit from school recovery as while children are attending classes, parents and guardians are able to focus on returning to their daily work, which is important in order to feed their families. The return to employment activities also helps communities to rise again and contributes to rebuilding the economy. Communities with strong social capital and a tradition of community activities can pro-actively participate in the process and thereby contribute to a successful and speedy recovery (Nakagawa and Shaw 2004).

4.2 Education Section Recovery

When schools are destroyed by a disaster, their reconstruction and the resumption of classes can be one of the most effective ways to demonstrate a return to normalcy to the local population. Schools become the glue that holds a fractured community together. When schools continue to operate, they signify that a community is functionally competent (Masten and Obradovic 2008; Kuo and Means 2012).

Table 4.1 Impact of the 2004 Indian Ocean Tsunami on the education sector

Country	Schools damaged or destroyed	Effects on students and teachers
India	In Kerala, one government school required complete reconstruction and five needed major repairs. In Tamil Nadu, 252 schools needed complete reconstruction, 19 major repairs, and 49 minor repairs. In Andaman and Nicobar Islands, 53 of the 208 primary schools, 29 of the 92 senior secondary schools, and 18 of the 55 upper primary schools were seriously damaged by the tsunami	Seventy eight teachers were reported to be dead or missing
Indonesia	2,135 schools were either completely destroyed or heavily damaged. Of these, 1,521 were primary schools	40,900 students and 2,500 teachers died or were missing
Maldives	About 100 of the country's 315 schools were affected, of which two-thirds were destroyed or suffered severe damage	There was a considerable loss of the country's teachers; 40 % of the expatriate teachers did not return to the Maldives after the tsunami
Sri Lanka	182 schools (5 % of the total number of schools) were severely damaged or destroyed, while 287 schools served as emergency shelters for thousands of people made homeless	Some 80,350 students and 3,300 teachers were displaced from schools in ten districts
Thailand	A total of five schools were fully destroyed and 51 were slightly damaged	

References: ADB, UN, and World Bank 2005; Bitter and Edirisinghe 2013; Jovel 2005; Shaw et al. 2012; UNICEF 2005a, b; UNICEF 2009a, b, c, d; UNICEF 2010

In the aftermath of a disaster, the resumption of schools can symbolize community survival. Failure to prioritize school re-opening after a disaster jeopardizes community recovery (Bastidas and Petal 2012).

In an effort to extract valuable lessons for future disasters, we have studied and analyzed the education sector recovery efforts implemented in India, Indonesia, the Maldives, Sri Lanka, and Thailand. It is hoped that the positive recovery experiences obtained from the five countries can be adapted and applied in other places. Knowledge transfer from the experiences of the featured five countries might be able to help in expediting recovery and building back better and in avoiding proven ineffective measures.

4.2.1 India

In India, more than 2.7 million people were affected by the tsunami in the coastal areas of Andaman and Nicobar Islands, Tamil Nadu, Puducherry, and Kerala (Nikku et al. 2006). In some communities, schools were destroyed or badly damaged

(Rodriguez et al. 2006). In Kerala, one government school in Kollam required complete reconstruction and five needed major repairs (ADB, UN, and World Bank 2005). In Tamil Nadu, 252 schools needed complete reconstruction, 19 required major repairs, and 49 were in need of minor repairs (ADB, UN, and World Bank 2005). On the Andaman and Nicobar Islands, 53 of the 208 primary schools, 29 of the 92 senior secondary schools, and 18 of the 55 upper primary schools were seriously damaged by the tsunami (UNICEF 2005a).

In the states of Kerala and Tamil Nadu, government authorities acted promptly in collaboration with NGOs in ensuring that most children returned to school, in replacing books and uniforms, in making minor repairs, in organizing alternative sites for holding classes, and in providing transport (UNICEF 2005a). NGO volunteers organized interactive sessions with children beyond school hours and assisted in cleaning and improving the appearance of classrooms (UNICEF 2005a). The United Nations Children's Fund (UNICEF) played a supporting role by delivering school supplies and sports equipment, rebuilding libraries, organizing play activities for children, and training teachers to recognize signs of distress and provide basic emotional counseling (UNICEF 2005b).

Some of the immediate recovery needs in the aftermath of the tsunami included finding alternative hosts for relief camps so that schools would no longer be used for that purpose and classes could resume; lack of potable water in many schools; ensuring communities were consulted and issues of access, especially of girls, were taken into consideration when relocating schools (ADB, UN, and World Bank 2005). The recovery experience of one affected school is shown in Box 4.1.

Box 4.1 School Recovery at P.S. High School (North) (Source: Shaw et al. 2012)

P.S. High School (North) is located in a coastal area in Mylapore, Chennai. It has about 450 students and 20 teachers. Although the school was not directly affected by the tsunami, the houses of several students were damaged. A large number of students were paralyzed by fear due to the massive devastation and chaos caused by the tsunami.

In the aftermath of the deadly tsunami, Disaster Management was mainstreamed as a subject to be taught at public schools in Tamil Nadu and this initiative took effect in 2005. At the P.S. High School (North), disaster education is now taught in standards 6–10 science classes. In the science exhibition organized by the Tamil Nadu State Government in 2008, P.S. High School (North) emerged as the winner in the Chennai District. Using a model, P.S. High School (North) made a presentation on how to explain to school children the impacts of a tsunami.

(continued)

Box 4.1 (continued)

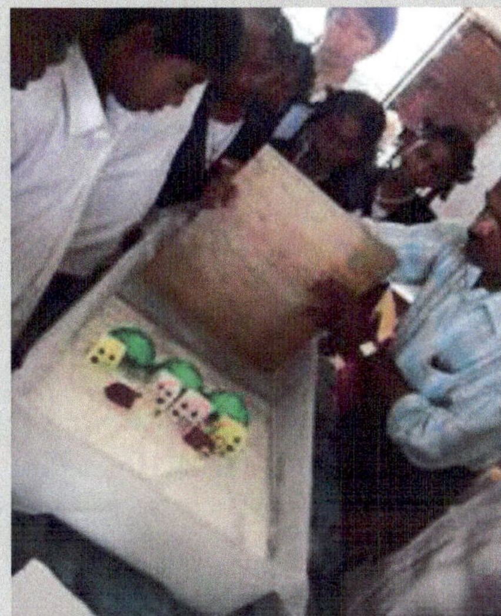

Demonstration of an occurring tsunami (model)

Further activities at the P.S. High School (North) included programs of tree planting and town watching supported by the United Nations International Strategy on Disaster Reduction (UNISDR), Madras University (local university), and Kyoto University (Japan). These programs took place in 2008. Today, disaster education is continuously taught at this school to sensitize students about potential disasters. This is still done in cooperation with Madras University. Disaster education is crucial to be taught as school children are valuable transmitters to their wider communities.

4.2.2 Indonesia

Indonesia suffered the most severe impacts of the 2004 Indian Ocean Tsunami due to its proximity to the epicenter of the earthquake. The tsunami caused the loss of thousands of lives and the displacement of a huge number of people. Three months after the tsunami, a strong earthquake on the island of Nias further added to the catastrophe. The two disasters caused overwhelming destruction to infrastructure, including education facilities. The tsunami damaged 2,135 schools in Indonesia leaving 150,000 students without schools (Shaw et al. 2012). In Aceh, about 41,000 students and 2,500 teaching and non-teaching staff were killed or missing, and about 150,000 children lost their schools (UNICEF 2009a). Substantial administrative and professional capacity of the education system was lost. In addition, damaged roads and broken telephone lines made communication among national, provincial, district, and sub-district government offices difficult (UNICEF 2009a).

Prior to the tsunami, the education sector in Aceh had been already reeling from the devastation of 30 years of armed conflict. Because of this post-tsunami responses in Aceh had to deal with social and systemic displacements that were significantly more complex than those produced by the tsunami alone (UNICEF 2009a).

Despite the tragedy, schools reopened 1 month after earthquake and tsunami (UNICEF 2010). Following its Core Commitments for Children in Emergencies (CCC), UNICEF and the Provincial Office of Education (POE) in Aceh led the first back-to-school campaign after the tsunami. More than 800,000 children (90 % of those at school age) were able to resume their schooling (UNICEF 2009a). Information materials, meals, stationery kits, sets of textbooks, school tents, School-in-a-Box kits, and recreation kits were distributed. The recovery experience of one affected school is shown in Box 4.2.

But the scale of destruction meant that permanent schools were not in place for several months. Major challenges in the recovery of the education sector included the following: many government employees died during the disaster; the presence of over 300 local and international NGOs completely overwhelmed government counterparts; and there was lack of clear directions of authorities of different government bodies, creating confusion and misunderstandings among actors (UNICEF 2010). There was also overlapping of school sites due to lack of communication among different actors and levels of education authorities. The communities also made several agreements with various humanitarian agencies as a ‘protection’ mechanism, because they had become used to ‘empty promises’ previously made to them (UNICEF 2010).

According to UNICEF (2009a), education policy changes that can be tied directly to the tsunami and its aftermath have been few and not especially significant. Financial and technical resources have been most influential in enabling the implementation of already existing policies, which had previously lacked support and attention due to the armed conflict in Aceh.

Box 4.2 School Recovery at SDN 81 Banda Aceh School (Source: Shaw et al. 2012)

SDN 81 Banda Aceh School is located two kilometers from the coast. It has 200 students and teachers. Since the school is located near the coast, the school building was completely destroyed and swept away by the big waves of the tsunami. It caused a large number of casualties among students and teachers, of which only 62 students survived. Classes were suddenly interrupted. The school children and the teachers were evacuated to a safer place. Some of the classes were conducted in an emergency shelter and some classes were conducted in open ground. The school children and the teachers of SDN 81 Banda Aceh School resumed classes in various means that they could think of.

In 2005, the Allianz Group in cooperation with the German Agency for Technical Cooperation (GTZ) constructed a school building for SDN 81

(continued)

Box 4.2 (continued)

Banda Aceh School. Allianz employees worldwide donated money amounting to 210,000 euros to fund the school construction project. Staff of Allianz Indonesia identified, coordinated, and monitored the project. The school rebuilding process commenced directly at the beginning of 2005 and it was accomplished 1 year later, in January 2006.

Aside from Allianz, the Indonesian National Commission for Child Protection and a bilateral Indonesian-GTZ technical cooperation were involved in the recovery and reconstruction process. With six classrooms, a library, a prayer house (*Musholla*), a ritual cleaning place, and a school guardian house, the school provides basic community support. Two of the classrooms are linked and can be used as a community assembly room, and in the event of extreme weather conditions, the second story can be used as a shelter for up to 900 people. In addition to the essential role as a project coordinator, employees of Allianz Indonesia have volunteered to become foster parents of schoolchildren. As foster parents, every year employees sponsor textbooks, uniforms, and school supplies, as an effort of supporting the continuation of students' education.

New school building donated by international donors

The recovery process cemented a bond between the students, teachers, and donors. Along with the newly built school, a new education perspective, especially for disasters, is strengthened. By the provision of a community assembly room in the school area, it allows mutual attachment of school and community education.

4.2.3 Maldives

The tsunami had a substantial impact on the education system in the Maldives. The tsunami flooded all but nine islands, causing significant damage to infrastructure, including educational facilities. Financial losses in the education sector was estimated at USD 21.1 million in rebuilding and construction costs, school supplies, uniforms, equipment, and teaching and learning materials (UNICEF 2009b). About 100 of the country's 315 schools were affected, of which two-thirds were destroyed or suffered severe damage (UNICEF 2009b). There was also a considerable loss of the country's teachers; 40 % of Maldives's 1,800 foreign contract teachers (35 % of the entire teaching force) who were on vacation when the tsunami happened did not return to the Maldives after the tsunami (UNICEF 2009b). To address the human resource gap immediately following the tsunami, UNICEF and the Maldivian Ministry of Education collaborated in a successful and innovative effort to recruit 180 final-year students from the College of Higher Education, Faculty of Education, to replace the large number of foreign teachers who did not return. Called the "tsunami teachers" by the schools, they were in place very quickly, brought improvised teaching aids with them, and had the advantage of speaking Dhivehi. This enabled schools to be available to children during the recovery period (UNICEF 2009b). The experience also helped to provide on-the-job training that would contribute to the teachers' development and teaching abilities.

Recognizing the need for trained teachers, particularly in the primary schools, the Ministry of Education also initiated a system of 'twinning' between schools in Malé and those on the islands, an arrangement through which Malé staff could visit affected schools, provide them with lesson plans, undertake informal training workshops, and demonstrate active learning methods (UNICEF 2009b). Overall, there was relatively little major shifting of students and schools after the tsunami. But the migration to the capital from the atolls put extra demand on the overcrowded schools in Malé (UNICEF 2009b).

4.2.4 Sri Lanka

The 2004 Indian Ocean Tsunami was the largest natural disaster to ever affect Sri Lanka. More than 35,000 people were killed, one third of whom were children. The country was already under great strain prior to the tsunami due to a lengthy armed conflict between the Liberation Tigers of Tamil Eelam (LTTE) and the Government of Sri Lanka, which had previously resulted in 65,000 deaths and displaced up to 700,000 people (UNICEF 2009c). Out of 25 districts in Sri Lanka, 14 were affected by the tsunami, with nine designated as those in need of the most recovery assistance. Six of these nine districts were simultaneously impacted by the conflict. The conflict created barriers in getting aid to LTTE-controlled areas, with issues of unequal distribution of resources between geographical areas and between tsunami-displaced and war-displaced persons (UNICEF 2009c).

The tsunami caused significant impact on the education sector, damaging or destroying 182 schools representing 5 % of all schools and displacing more than 80,000 students and 3,000 teachers (UNICEF 2009c). But despite considerable challenges, schools reopened in the third week of January 2005 to restore normality to children's lives and prevent a serious break in the regular learning cycle (UNICEF 2009c). There were public announcements informing families that children would be allowed to attend school without uniforms and birth certificates. But resumption of schooling did not mean all children immediately returned to school. In the few schools that reopened, attendance was about 50 %. Many children died or were injured during the tsunami and some parents hesitated to send their children back to school because of safety concern.

In Sri Lanka, especially significant in the recovery process and in fostering social inclusion and participation was the establishment of the Tsunami Education Rehabilitation Monitoring Trust (TERM). TERM is an NGO created by the Ministry of Education mandated with coordination and monitoring of the overall construction program and was responsible for implementing the requirements and standards for repair and reconstruction (UNICEF 2009c).

A critical factor to the success of education recovery was the overall strength of the education sector before the tsunami (UNICEF 2009c). A series of national plans provided the framework for restoring the education system in broad terms of equity in access, quality of teaching and learning, and efficiency in management and service delivery. The rapid response of the Government, the availability of funds and expertise from a wide spectrum of donors, and the resilience of communities themselves enabled the country to cope with the effects of the disaster with relative speed and success. Critically contributing to children's capacity to return to school were the following: health services were restored; food aid was provided to over 900,000 people; and repairs to basic infrastructure such as major pipelines and water sources, roads, bridges, electricity and telephone lines, were undertaken with speed (UNICEF 2009c).

4.2.5 Thailand

The 2004 Indian Ocean Tsunami caused widespread devastation along Thailand's 400-km southern coastline, directly affecting 407 villages and completely destroying 47 of them (UNICEF 2009d). Fortunately, the tsunami's impact on the education sector was minimal in Thailand. According to the Thai Ministry of Education, a total of five schools were fully destroyed and 51 more were slightly damaged (Jovel 2005). Most schools reopened on January 4, 2005 (UNICEF 2005b). Back-to-school campaigns and creation of secure temporary learning spaces enabled pre-tsunami enrollment levels to be attained before the end of 2005 (UNICEF 2009d).

4.3 Similarities and Differences in Education Sector Recovery

Education sector recovery may include the resumption of education activities under special conditions to compensate students for the loss of educational opportunity (World Bank 2010). Recovery activities may include intensive utilization of undamaged education facilities by establishing several daily shifts instead of the normal schedule; rental of alternative premises for holding classes; and setting up temporary classrooms by using tents, containers, or other similar facilities. Another common recovery activity is the immediate replacement of essential education materials. Special schemes for fast track the training of new teachers may be implemented in the case of massive disasters that have a heavy toll on human life (World Bank 2010). In the following sections, the experiences of the affected countries in addressing important recovery issues will be compared.

4.3.1 Education Sector Recovery Coordination

When the 2004 Indian Ocean Tsunami occurred, discussions to implement a cluster approach in emergencies to improve coordination among various actors were still in progress within the UN system (UNICEF 2009d). The cluster approach was not carried out in this disaster. Different coordination mechanisms were used.

In Indonesia, due to unusually large funding in the wake of the tsunami, there was a proliferation of agencies working in the education sector (Telford and Cosgrave 2006). From July 2005 onward, education sector coordination meetings were held every 2 weeks. Minutes of the meetings were distributed through a mailing list of over 150 organizations and individuals (UNICEF 2009a).

In the Maldives, over the first few days and weeks following the tsunami, the Ministry of Education and donors cooperated in assessing and acting on the situation of schools under the umbrella of a recovery task force (UNICEF 2009b). By the end of January, all affected schools had been visited and assessed and preparation of rehabilitation plans had begun.

4.3.2 Resumption of Classes

In all countries affected by the tsunami, children were back in school relatively quickly (Telford and Cosgrave 2006). Governments and partners cleaned, constructed, or made available existing buildings, temporary schools, and tents and provided learning materials, books, uniforms, school bags, and school furniture to hundreds of thousands of children. Some local governments, such as the Tamil Nadu State Government in India, provided midday meals in schools, supported by UNICEF, who provided plates, tumblers, and water containers for the schoolchildren

(Nwe 2005). However, when schools reopened in India, attendance ran as low as 30 % (UNICEF 2005b). In the Maldives, Sri Lanka, and Thailand, UNICEF carried out repairs and rehabilitation of schools where damage had been relatively light and where the buildings were still safe for children (UNICEF 2005b).

The recovery of education systems is essential to the protection of children in the affected areas (Lauten and Lietz 2008). When students feel that teachers and friends support them, they are more likely to engage in normal activities, even in the face of significant adversity (Wickrama and Kaspar 2007). Delayed resumption of classes could lead students to retreat to the past as an escape from present worries, spurring depressive feelings that might eventually result in the development of disorders. It is observed that for many people who are hurting, a busy schoolyard is the best medicine (UNICEF 2005b).

4.3.3 Repair, Reconstruction, and Relocation of School Buildings

Under a "building back better" strategy, reconstruction of destroyed schools aims to ensure future disaster resilience through strict enforcement of improved construction standards and improved quality of education facilities (World Bank 2010). Non-disaster resilient schools not only kill or injure students and teachers, the destruction of the physical educational infrastructure can be a great economic loss for a country. The flip side of the coin, however, is that the destruction of old and unsafe physical assets offers the opportunity to replace them with safer infrastructure (Baez et al. 2010). In Indonesia, school buildings that were no longer safe because of the continuing earthquakes had to be demolished (Nwe 2005).

Investing in strengthening school structures before a disaster occurs reduces long-term costs, protects generations of children, and ensures educational continuity after disaster events (UNISDR 2006). However, gaps in practice persist. There is evidence that in newly built schools in Aceh, builders may not have adequately addressed construction criteria for earthquake preparedness and education administrators in various schools were unaware of the current or planned state of safety and protection standards (UNICEF 2009a).

In Sri Lanka, progress in the restoration of damaged schools was neither linear nor smooth (UNICEF 2009c). There were delays in the rehabilitation and reconstruction of some damaged or destroyed schools. Students in several areas have had to continue in damaged buildings and sheds because no donor has assisted, the community has refused to relocate the school, or there was a resurgence of armed conflict.

4.3.4 School—Community Partnerships

In some cases, post-tsunami interventions appeared to have reinforced community links to schools (UNICEF 2009c). In Sri Lanka, there was a high degree of in-kind support from parents, many communities contributed to decisions on school relocation and back-to-school campaigns, and some supported extracurricular activities. There appeared to have been a shift in perception, away from the state as exclusively responsible for the welfare of schools to one in which parents and community-based organizations are more involved and have a sense of shared community ownership. For example, some School Development Societies (SDSs) were revived, allowing parents and communities to participate in school management activities. Mechanisms that link the school to its community remain necessary constituents of effective child-friendly learning, but are typically expected to act with resources that are technically too limited and short-term. In the Maldives, significant attention was given to making community mobilization the cornerstone of the recovery process, a point of departure that continued through much of the relief to recovery transition in the education sector (UNICEF 2009b). Teachers and community volunteers were trained to provide psychosocial support to children.

4.3.5 Integrating Disaster Risk Reduction into the Curriculum

Making disaster risk reduction education part of formal school curriculum fosters awareness and better understanding of the immediate environment in which children and their families live and work (UNISDR 2006). From past experience it is observed that children who are taught about natural hazard risks can play an important role in saving lives and protecting members of the community in times of crisis. One classic example of this is the story of British schoolgirl Tilly Smith, who saved many lives by urging people to flee from the shore when the 2004 Indian Ocean Tsunami struck a beach in Thailand (Peek 2008). Her Geography class in the UK had enabled her to recognize the first signs of a tsunami.

In Sri Lanka, with assistance from the German Society for International Cooperation (GIZ), disaster risk management (DRM) was integrated into the school curricula and in all pre-service and in-service teacher training curricula and respective training modules (Bitter and Edirisinghe 2013). There was actually a conceptual shift from DRM education to Disaster Safety Education (DSE). DSE aims to foster disaster preparedness, prevention, and mitigation among school children and in school communities.

On the tsunami-affected coast of Thailand, there are new curricula that focus exclusively on tsunami (Wisner 2006). As an example of bilateral exchange, the Asian Disaster Reduction Center in Japan developed tsunami teaching materials alongside Thai teachers for use in schools in the tsunami-affected coastal areas of Phang Nga and Phuket (Wisner 2006). A 40-page textbook was developed that covers not only tsunami but earthquake, flood, fire safety, and evacuation (Wisner 2006). In another example, Save the Children had supported the Thai Volunteer Association in developing a school curriculum on tsunami hazard in a participatory manner with children that were affected by the 2004 Indian Ocean tsunami (Wisner 2006).

In-school DRR programs designed for young people have the potential for significant impact of within communities (Peek 2008). When students are educated about hazards risk, they will share the knowledge with their households and the larger community, hopefully leading to positive changes. Several studies have confirmed that children who participate in school-based hazards education programs tend to have increased accurate knowledge of hazards, increased reports of hazards preparedness at home, reduced levels of fear, and more realistic risk perceptions than their peers who did not receive such kind of education (Peek 2008).

4.3.6 Recruitment and Training of Teachers

One critical outcome defining progress toward a better-built education sector would be the adequate number of qualified and effective teachers (UNICEF 2009a). In Indonesia, UNICEF recruited and trained more than 1,500 teachers to start teaching in July 2005 to address staff shortages and subsidized their salaries for the first 6 months, after which the Government of Indonesia covered their salaries (UNICEF 2009a). Persistent insufficiencies of qualified teachers have not yet been resolved. Based on surveys, teacher qualifications in Aceh's primary schools continue to be poor (UNICEF 2009a).

In the Maldives, low numbers of trained teachers and low retention rates at all grade levels among those trained specifically in child-friendly active learning and teaching persist (UNICEF 2009b). There is continued deployment of untrained teachers to fill short-term posts and reliance on foreign contract teachers. Preschools, which do not fall under the formal school system, continue to struggle with poor conditions of service for teachers (UNICEF 2009b).

4.3.7 Reaching Out to the Most Disadvantaged Children

As part of the recovery of the education sector, the issue of reaching out to the poorest and most disadvantaged children, especially those who had never been to school before, needs to be addressed (Nwe 2005). In the education sector evaluation in Indonesia, findings showed that there should be a proactive approach for seeking out children not in school, including those with physical disabilities or special needs (UNICEF 2009a). The need to improve inclusion was also not evident in the conflict areas in Aceh (UNICEF 2009a).

In the Maldives, although the Ministry of Education had begun to make progress on addressing the situation of children with special needs, the more vulnerable, hard-to-reach children were not actively sought out and the numbers of excluded children may be higher than what official numbers suggest (UNICEF 2009b).

In the past in India, marginalized children from the fishing and historically discriminated Dalit (untouchable) communities did not go to school. To prevent this trend from continuing, several actions have been taken (Nwe 2005). Some schools had been instructed to exempt tsunami-affected students from fees for a year.

Children from certain vulnerable groups were eligible for free education through government schemes. District education officials, UNICEF, and NGO partners began a post-tsunami campaign to identify children not in school and enroll them in a catch-up or bridging program, with a view to channeling them into the formal school system.

In Thailand, the Government provided education grants to all orphaned children, which covered the period until the child graduated from higher education (Nwe 2005).

4.4 Ongoing Challenges

Recent tsunami events have increased tsunami awareness throughout the world, leading to the emergence of a "tsunami culture" (Esteban et al. 2013). Such tsunami cultures already existed in a number of countries, such as Japan or Chile, due to the numerous events that took place in the past, and which continue to linger in people's memory either through oral transmission or education. In other places, the cultural link to past events is less clear. It is imperative that the awareness about disasters continues to be enhanced so that safety measures are promoted in all areas at risk of tsunamis.

Stronger participation of communities, community-based organizations, and NGOs in the education sector recovery should be supported to leverage their knowledge and encourage their commitment to sustaining and strengthening outcomes of tsunami interventions. Recovery investments should be protected by planning for sustainability, such as maintenance support and continuity of human resources capacity development (UNICEF 2009a). There should also be continued support for evidence-based systems of data collection and management to better inform policy, strategic planning, and allow for strengthened monitoring and evaluation (UNICEF 2009a). Formal assessments should link interventions such as curriculum infusion or teachers' training with the learning outcomes of students. The validity, reliability, and significance of education recovery statistics should be improved as a step toward creating a reflective and self-correcting education system (UNICEF 2009a).

4.5 Conclusion

It is hoped that the lessons and experiences from the education recovery from the 2004 Indian Ocean Tsunami briefly presented in this chapter will contribute to future efforts of the education sector in disaster-affected countries to achieve sustainable recovery. Recovery efforts are most effective when they are informed by the lessons learned from previous disasters as well as by knowledge of new disaster risk reduction measures. Although disasters destroy education systems, they also provide opportunities to "build back better." Disaster-affected areas can take these opportunities to introduce and strengthen disaster risk reduction education, improve the quality of education in general, offer child-centered teaching and learning methods, train competent education planners and managers and teachers, and produce disaster-resistant, safe, and accessible schools.

Acknowledgements We thank the Ministry of Education, Culture, Sports, Science, and Technology (MEXT) of Japan for the generous scholarship grant that made it possible for the first author to pursue his Master's and PhD degrees at Kyoto University. We also acknowledge the financial support provided by the Global Center of Excellence Program on Sustainability/Survivability Science for a Resilient Society Adaptable to Extreme Weather Conditions (GCOE-ARS) of Kyoto University for the first author's research activities.

References

ADB (Asian Development Bank), UN (United Nations), World Bank (2005) India post Tsunami recovery program: preliminary damage and needs assessment. http://www.tnrd.gov.in/externallyaidedprojects/Tsunami_rehabilation/india-assessment-full-report[1].pdf

Baez J, de la Fuente A, Santos I (2010) Do natural disasters affect human capital? An assessment based on existing empirical evidence. http://ftp.iza.org/dp5164.pdf

Bastidas P, Petal M (2012) Assessing school safety from disasters: a global baseline report. UNISDR, Geneva

Bitter P, Edirisinghe MNS (2013) Introduction of disaster safety in the Sri Lankan education system: factors for success. http://www.ukfiet.org/cop/wp-content/uploads/2013/11/Bitter-Edirisinghe_-Introduction-of-Disaster-Safety-in-the-Sri-Lankan-Education-system-Factors.pdf

Esteban M, Tsimopoulou V, Shibayama T, Mikami T, Ohira K (2013) Analysis of Tsunami culture in countries affected by recent Tsunamis. Proc Environ Sci 17:693–702

Fernandez G, Shaw R, Takeuchi Y (2012) School damages in Asian Countries and their implication to Tohoku recovery. In: Shaw R, Takeuchi Y (eds) East Japan earthquake and Tsunami: evacuation, communication, education, and volunteerism. Research Publishing Services, Singapore, pp 201–221

GFDRR (Global Facility for Disaster Reduction and Recovery) (2009) Guidance notes on safer school construction. GFDRR, Washington

HHF (Happy Hearts Fund) (2005) Educate yourself of disaster statistics. http://www.happyheartsfund.org/disstat.php

Jovel R (2005) The economic impact of the 26 December 2004 Indian Ocean Earthquake and Tsunami in Thailand. http://www.adpc.net/maininforesource/dms/thailand_assessmentreport.pdf

Kuo S, Means B (2012) Corporate social responsibility after disaster. Wash Univ Law Rev 89(5):973–1016

Lauten AW, Lietz K (2008) A look at the standards gap: comparing child protection responses in the aftermath of Hurricane Katrina and the Indian Ocean Tsunami. Children Youth Environ 18(1):158–201

Masten A, Obradovic J (2008) Disaster preparation and recovery: lessons from research on resilience in human development. Ecol Soc 13(1)

Nakagawa Y, Shaw R (2004) Social capital: a missing link to disaster recovery. Int J Mass Emerg Disasters 22(1):5–34

Nikku BR, Sah N, Karkara R, Ahmed S (2006) Child rights perspective in response to natural disasters in South Asia: a retrospective study. Save the Children Sweden, Kathmandu

Nwe YY (2005) Children and the Tsunami, a year on: a draft UNICEF summary of what worked. http://www.unicef.org/spanish/emerg/disasterinasia/files/WhatWorked.pdf

Peek L (2008) Children and disasters: understanding vulnerability, developing capacities, and promoting resilience—an introduction. Children Youth Environ 18(1):1–29

Rodriguez H, Wachtendorf T, Kendra J, Trainor J (2006) A snapshot of the 2004 Indian Ocean Tsunami: societal impacts and consequences. Disast Prev Manag 15(1):163–77

Rognerud EW (2009) Experiences in post-disaster education recovery and school reconstruction: a case for investment in 'Soft Components.' http://gfdrr.org/docs/Presentation_WB_21Jan_09_EWR.pdf

Shaw R, Takeuchi Y, Fernandez G (2012) School recovery: lessons from Asia. http://www.iedm.ges.kyoto-u.ac.jp/School%20recovery_low.pdf. Accessed 28 Mar 2014

Telford J, Cosgrave J (2006) Joint evaluation of the international response to the Indian Ocean Tsunami: synthesis report. http://www.sida.se/Publications/Import/pdf/sv/Joint-Evaluation-of-the-International-Response-to-the-Indian-Ocean-Tsunami_3141.pdf

UNICEF (United Nations Children's Fund) (2005a) Children eager to return to schools in Andaman and Nicobar Islands. UNICEF, New York. http://www.unicef.org/infobycountry/index_25059.html

UNICEF (2005b) Hundreds of schools destroyed by Tsunami. http://www.unicef.org/media/media_24847.html. Accessed 18 Apr 2014

UNICEF (2009a) Children and the 2004 Indian Ocean Tsunami: evaluation of UNICEF's response in Indonesia (2005–2008), Country Synthesis Report. UNICEF, New York

UNICEF (2009b) Children and the 2004 Indian Ocean Tsunami: evaluation of UNICEF's response in Maldives (2005–2008), Country Synthesis Report. UNICEF, New York

UNICEF (2009c) Children and the 2004 Indian Ocean Tsunami: evaluation of UNICEF's response in Sri Lanka (2005–2008), Country Synthesis Report. UNICEF, New York

UNICEF (2009d) Children and the 2004 Indian Ocean Tsunami: evaluation of UNICEF's response in Thailand (2005–2008). UNICEF, New York

UNICEF (2010) Education in emergencies training facilitators' guide. Asia-Pacific Shared Services Centre, UNICEF, Bangkok

UNISDR (United Nations International Strategy for Disaster Reduction) (2006) Disaster risk reduction begins at school: 2006–2007 World disaster reduction campaign. http://www.unisdr.org/files/2105_VL108003.pdf

Wickrama KAS, Kaspar V (2007) Family context of mental health risk in Tsunami-exposed adolescents: findings from a pilot study in Sri Lanka. Soc Sci Med 64(3):713–23

Wisner B (2006) Let our children teach us! A review of the role of education and knowledge in disaster risk reduction. Books for Change, Bangalore

World Bank (2010) Estimation of post-disaster needs for recovery and reconstruction. World Bank, Washington

Chapter 5
Indian Ocean Tsunami Warning System (IOTWS)

S.H.M. Fakhruddin

Abstract The tremendous loss of life resulting from 2004 tsunami emphasized the necessity of establishing and improving tsunami warning systems in countries of the Indian Ocean. For a rapid onset disaster (i.e. tsunami), a robust end to end early warning system is necessary to protect lives and livelihood. Early warning system (EWS) is an integral part of human to influence perceptions, decisions and behaviour in times of adverse conditions and crises. The Indian Ocean Tsunami Warning and mitigation System (IOTWS) has developed rapidly since its establishment after the Indian Ocean Tsunami of 2004. One of the major elements of the IOTWS is the concept of a Regional Tsunami Service Provider (RTSP). An RTSP is a centre that provides an advisory tsunami forecast service to one or more National Tsunami Warning Centres (NTWC). The RTSPs have a number of requirements that they need to meet (IOTWS, Report of intersessional meeting of working group 5, 2009). An early warning system need to integrated science, institutions and society for hazard detection to trigger warning, influences behavior for decision making and community response. The notion of the Last Mile originated from the need that EWS have to reach people at the local and community level with appropriate information in order to ensure that anticipated responses at the Last Mile can take place. The issues of the Last Mile underscore that EWS need to pay more attention to risk knowledge, response capabilities, and vulnerabilities of communities, including aspects of temporary and long-term migration. This chapter discuss about the evaluation of IOTWS, current status, risk assessment and major gaps still remains in the system. This chapter also presented some good practices and lessons learned based on recent incidents to strengthen IOTWS.

Keywords Community capacity • Early warning • Resilience • Tsunami preparedness

---

S.H.M. Fakhruddin (✉)
Politecnico Di Milano, Piacenza, Italy

Regional Integrated Multi-Hazard Early Warning System (RIMES), 58 Moo 9 Paholyothin Road, Klong Nung, Klong Luang, P.O. Box 4, Pathumthani 12120, Thailand
e-mail: smfwater@gmail.com

R. Shaw (ed.), *Recovery from the Indian Ocean Tsunami: A Ten-Year Journey*, Disaster Risk Reduction, DOI 10.1007/978-4-431-55117-1_5

5.1 Introduction

Coastal inundations are an increasing threat to the lives and livelihoods of people living in low-lying, highly-populated coastal areas. Of them, tsunami is one of the most devastating natural coastal disasters. Most of large tsunamis are generated by submarine earthquakes occurring in subduction zones. Tsunamis can also be triggered by volcano eruptions and large landslides. Although the skill for predicting earthquake is still in its infancy, a tsunami warning system is still possible for a distant tsunami if the tsunami can be detected in the open-ocean. Information on the arrival time and the height of leading tsunami wave in areas far away from the source region can be predicted with some confidence (Wei et al. 2003; Joseph 2011).

Early warning is a key element for disaster risk reduction. Early warning systems alone do not prevent hazards turning into disasters, early action is essential in order to mitigate potential damage (World Disasters Report 2009; Assilzadeh and Mansor 2011; Behrens et al. 2010; Borga et al. 2011; Diodato et al. 2011; Essink et al. 2010; Frédéric and Patrick 2008; Hapuarachchi et al. 2011). Early warning and early action together can save thousands of lives and livelihoods; reduce vulnerability and strengthen resilience. Nevertheless without lead time to react an early warning is almost ineffective (Babel et al. 2013; Fakhruddin 2013; Parker et al. 2005). The World Disaster Report 2009 said

> *Communicated the importance that understanding early warning as a system rather than a technology highlights the need to address risk assessment, communication and dissemination, and preparedness to act with the same level of commitment provided to the technological aspects of early warning* (WDR 2009).

Therefore a breakdown in any one of these elements of early warning can cause warning messages to fail to reach and motivate their intended recipients. The United National International Strategy for Disaster Risk Reduction (ISDR) provides a framework for an effective tsunami early warning system that is people-center and integrates four interrelated elements: (1) knowledge of risks faced, (2) technical monitoring and warning service, (3) dissemination of meaningful warnings to those at risk, and (4) public awareness and preparedness to act. While tsunami warning centers are focused primarily on monitoring and warning, they play a vital role in providing technical input and developing community partnerships for the other elements of risk reduction that are essential for an effective tsunami early warning system. For any natural hazard mitigation is a complex endeavor that requires direct links between natural and social sciences (Liu et al. 2009). For example, an effective early warning system must include not only the ocean technologies to accurately detect an emerging tsunami, but also a public notification system through which the population can be timely warned by the local government and other sources.

In the aftermath of the Great Sumatra Earthquake 2004, countries of the Indian Ocean basin formed an Intergovernmental Coordination Group (ICG) for the development of an Indian Ocean Tsunami Warning System (IOTWS). The ICG/IOTWS was formed under the auspices of the United Nations Educational, Scientific, and Cultural Organization (UNESCO) Intergovernmental Oceanographic Commission (IOC).

Subsequently, ICGs have also been established in the Caribbean, Atlantic, and Mediterranean basins to guide the development of tsunami warning centers in those areas. The ICG/IOTWS coordinates the implementation of an Indian Ocean wide tsunami warning system that is being established as a network of national systems. The network comprises of National Tsunami Warning Centres (NTWC) in each Member state receiving tsunami advisories from Regional Tsunami Advisory Services Providers (RTSPs) that are establishing capabilities to issue regional tsunami bulletins for the Indian Ocean. The National Tsunami Warning Center's (NTWCs) of individual countries, through bilateral agreements, will be able to choose the RTSP(s) from which they wish to access tsunami watch information from. Ultimately it is the NTWCs, operating within the legal framework of the sovereign nation in which they reside and serve, that provide warnings, watches, and advisories to their citizens, public and private agencies. The IOTWS will therefore operate as a "system-of-systems". This will require all tsunami watch information to be interoperable, i.e. use common and agreed formats for information exchange, address service requirements, follow agreed, high-level operating Standard Operating Procedures (SOPs), share information on procedures and processes. The IOTWS implementation plan outlines the RTSP service requirements and capabilities, Concept of Operations (ConOps), SOPs, performance indicators, and implementation schedule (IOTWS 2009). A tsunami warning system can only be successful if it spans the continuum of activities associated with an "end-to-end" approach. An end-to-end tsunami warning system begins with the rapid detection of a tsunami wave and ends with a well prepared community that is capable of responding appropriately to a warning.

5.2 The Regional Tsunami Advisory Services Providers

The Pacific Tsunami Warning Center (PTWC), based in Ewa Beach, Hawaii, serves as the operational headquarters for the Tsunami Warning System in the Pacific (TWSP). PTWC works closely with other regional and national centers in monitoring seismological and sea level stations and instruments around the Pacific Ocean to evaluate potentially tsunamigenic earthquakes. The TWSP disseminates tsunami information and warning messages to over 100 points across the Pacific. Regional tsunami warning centers operated by the U.S.A., France, Russia, and Japan provide regional warnings to the U.S. Alaska and west coast and Canada, French Polynesia, and the northwest Pacific, respectively. The tsunami warning system in the Pacific is one of the most successful international scientific programs with the direct humanitarian aim of mitigating the effects of tsunami to save lives and property. PTWC has been working as interim RTSP for the Indian Ocean till after tsunami 2004.

From 12 October 2011 three centers Commenced operations as RTSP of them are Australia (JATWC), India (ITEWC), Indonesia (InaTEWS) (IOC, http://ioc-unesco.org/). Roles and responsibilities of RTSP's are as follows:

- Monitor earthquakes and provide timely initial earthquake information for those that could generate a tsunami

- Generate specific coastal-zone threat information using output from scenario databases produced by tsunami models, using earthquake source information and verified by sea level observations
- Provide timely tsunami products for use by NTWCs in preparation and issuing of national tsunami warnings
- Monitor tsunami propagation and provide updated tsunami wave amplitude observations
- Receive timely National Warning Status Reports from NTWCs and show on RTSP Public Webpages
- Provide public bulletins of tsunami events, containing National Tsunami Warning Status information
- Serve as a backup centre to other RTSPs
- Serve as an NTWC for the country in which it resides

IOTWS has three defined RTSP Service Levels for tsunami events: Service Level 1—Earthquake Information (in operation); Service Level 2—Tsunami Threat Information (in operation) and Service Level 3—Inundation Mapping (not yet in operation).

- Service Level 1—(in operation) provides detection and reporting of earthquakes potentially capable of generating tsunamis (above magnitude 6.5 km and undersea, or near coast); optional qualitative threat assessment (e.g. "earthquakes of this size sometimes have the potential to generate ocean-wide tsunamis…") and report generated shortly after earthquake (target 10 min)
- Service Level 2—(in operation) provides assessment and reporting of tsunami threat, providing specific information for coastal zones, predicted maximum wave amplitudes; predicted times of arrival; predicted coastal zones above threat threshold and tsunami wave observations received
- Service Level 3—(not yet in operation) to provide NTWCs under bilateral agreements with RTSPs will develop enhanced national tsunami warnings using like inundation mapping, risk and hazard assessments

As risk reduction at the community level, people need very simple information about the local level risk and inundation scenarios to take appropriate response. For this a well established service level 3 is essential. Unless a well functioning service level 3 people at the coast are highly vulnerable.

RTSP services are Tsunami Advisory Information for NTWCs only and RTSP is not authorized to provide warnings. National Tsunami Warnings are the responsibility of the NTWCs (unless bilateral arrangements are established between an NTWC and an RTSP). RTSPs generate Service Level 1 Earthquake Bulletins for all global undersea earthquakes with magnitude>=6.5. RTSPs then generate Service Level 2 Threat Assessment Bulletins for all Indian Ocean earthquakes and earthquakes outside the Indian Ocean if the magnitude is>=8.0. RTSPs tsunami bulletins for NTWCs are located on password-protected websites, in the form of earthquake Bulletins, tsunami threat assessment bulletins: no threat, potential threat and confirmed threat; final bulletin (threat passed). RTSPs transmit Notification Messages to NTWCs, notifying that the bulletins have been generated (modes of dissemination: WMO GTS circuits, email, fax, SMS). RTSP password-protected websites contain all the generated tsunami bulletins; maps showing the tsunami source, tsunami energy dispersion, and threatened coastal zones; tables giving details of the predicted tsunami wave heights and arrival times in each Indian Ocean coastal zone. The NTWC Warning Status Reporting Form, for NTWCs to report their current national warning status back to the RTSPs. RTSPs are also required to provide public bulletins with information about the tsunami source, tsunami wave observations, and the national warning status of each Indian Ocean country.

The tsunami prediction information provides for each coastal zone containing maximum tsunami wave amplitude at the shore line (max_beach); maximum tsunami wave amplitude in deep water (max_deep) and depth of the water. The threat categories are Threat or No Threat, based on 0.5 m threshold at shore line (i.e. 1 m wave crest-to-trough)

- T1—Arrival time of the first detectable tsunami wave (2 cm positive amplitude wave)
- T2—Arrival time of first wave exceeding 0.5 m threat threshold
- T3—Arrival time of maximum amplitude wave
- T4—Arrival time of last wave exceeding 0.5 m threat threshold

The Table 5.1 shows the bulletin type and content of RTSP.

The heart of a tsunami warning system is its operations center. The primary mission of a full service NTWC or RTSP is to provide accurate and timely tsunami warnings and bulletins to coastal populations in its area of responsibility (AOR) 24 h per day, 7 days per week. To accomplish this mission a tsunami warning center detects and

Table 5.1 Bulletin type and Content of RTSP

Bulletin type		
Type 1 Earthquake bulletin	Earth quake information, plus optional qualitative threat assessment	Target within 10 min
Type 2 Threat assessment bulletin	No threat bulletin, based on assessment using model scenarios	Target within 20 min
	Potential Threat Bulletin, based on assessment using model scenarios. Contains specific threat information for each coastal zone	
Type 3 Threat confirmation bulletin	Confirmed threat bulletin based on real time sea level confirming a tsunami was generated. Also contains specific threat information for each coastal zone	When the first real time sea level observation confirming tsunami waves is available
Type 4 Updates	Confirmed threat bulletin containing real time sea level observations and specific threat information for each coastal zone. Also contains THREAT PASSED information for zones where all significant waves have passed	Hourly updates/as and when the subsequent real-time sea level observation are available
Type 5 Final bulletin	THREAT Passed all zone	120 min after the last exceedance of 0.05 M threshold at last Indian Ocean member state

analyzes earthquakes throughout its adjacent ocean basin. Earthquakes that are above a previously established threshold activate the center's alarm system and initiate an earthquake and tsunami investigation that includes the following four basic steps:

- Locate and characterize the earthquake's source and the probability of it creating a tsunami via the collection of data from seismic networks;
- Review automated earthquake analysis, and modification of the automated results by the duty scientist (watch stander) if necessary;
- Obtain sea level data from tide gage sites, and where available, data from Deep Ocean Assessment and Reporting of Tsunamis (DART) buoys, to verify the existence of a tsunami and to calibrate models; and
- Prepare and disseminate information to appropriate emergency management officials and others.

5.3 End to End Early Warning System

Early Warning Systems (EWSs) provide warning information that enable policy-makers, responders, and the community at large to plan and implement emergency response programs. The Hyogo Framework for Action 2005–2015 emphasized that early warning systems must be centered on the needs of people, implying that warnings must be timely and understandable to those at risk. Despite the great advances in technology and EWSs, most present systems have not completely served their intended purposes as a result of weakness in some parts of the relevant EWSs, especially in areas of public response to official emergency edicts and actions (Quansah et al. 2010). Problems were encountered at several levels, ranging from the insufficient application of existing technology, especially regarding the integration of various systems pertaining to different and sometimes conflicting agencies, to the almost complete absence of communities from the design process of alert messages and procedures (Menoni and Margottini 2011). Other reasons for poor performance of early warning systems are found in the areas of hazard detection failure, and inadequate message design and communication.

There are challenges regarding conveyance of early warnings, as well as the responsiveness of different communities following the provision of early warnings. Hallegatte (2012) noted that the benefits from forecasts depend largely not only on their accuracy, but also the trust of the population. A decision to evacuate people is often difficult if the probability of false alarm is too high (or if the warning area is too large): as, after a few unnecessary evacuations, the trust into the warning system disappears, and the warning system becomes useless (Hallegatte 2012). Hallegatte (2012) added that trust is often lacking due to poor communication from the specialists to the public, as well as the reluctance by specialists in explaining the limits of forecasts and warning. Users of information seldom know how to use the information, especially for forecasts that are always uncertain. Demeritt et al (2010) highlighted that concerns about the effects of false alarms on public confidence in and responsiveness to warnings reinforce an institutional tendency to avoid issuing early warnings.

This bias helps protect the public and forecasting agency against the costs of issuing false alarms, but it also increases the risk that warnings will not be issued early enough to allow precautionary action (Demeritt et al. 2010).

Therefore A breakdown in any one of these elements of early warning can cause warning messages to fail to reach and motivate their intended recipients. From the above statements and from other research of early warning, it's clear that early warning is not helpful unless its reach to the people who need to act. To response to the early earning the information need to understand and internalized by the people. Thus an interpretation and translation of the science information is essential. People do not immediately respond to early warnings because people worldwide first "search" for additional information to "confirm" that they are really at risk. This searching happens despite the technology used to give warnings. Searching is a social phenomenon. It involves talking things over with others and seeking to hear the same warning multiple times from different sources. Warned people turn to friends, relatives, and strangers to determine if they agree that risk is present and if protective actions are warranted. This process, constructing new perceptions of risk out of existing perceptions of safety adds time before protective actions are taken—it is fundamental to all human beings worldwide, and it is difficult to change. Early public warnings work best when they are under mandate from a government that is trusted as they can facilitate the process and speed it along (World Bank and UN 2010). Ignoring these basic human warning elements may continue to cost lives. Figure 5.1 shows an end to end early warning system structure.

A NTWC or RTSP might not be directly responsible for some of the links in the end-to-end chain. In many cases, other government agencies and private groups will be charged with taking Center products and notifying the populous. These authorities must provide understandable safety messages to the public to ensure that persons at risk move to safe areas. Even when not directly charged with notification and action planning, the Center can contribute to stronger links in the overall chain if the center understands and works with the other groups to insure the warnings get all the way to the beach and proper actions are taken.

Successful end-to-end systems require cooperation at all levels, a commitment of all stakeholders to work together during a tsunami warning, and over the long-term, a sustained effort of activity to keep awareness and preparedness at high levels. A mechanism to build organizational support that bring together stakeholders from warning centers, emergency management and first responders, tsunami scientists, and other government/non-government agencies, and the private sector. Due to the infrequent nature of tsunamis it is also strongly advised that a tsunami warning system be embedded in a multi hazards framework to ensure sustainability.

5.4 Risk Assessment and Communications

Risk communication has been understood as an interactive process of exchanging information, knowledge and opinions between stakeholders regarding the nature and associated risks of a hazard on the individual or community, and the appropriate

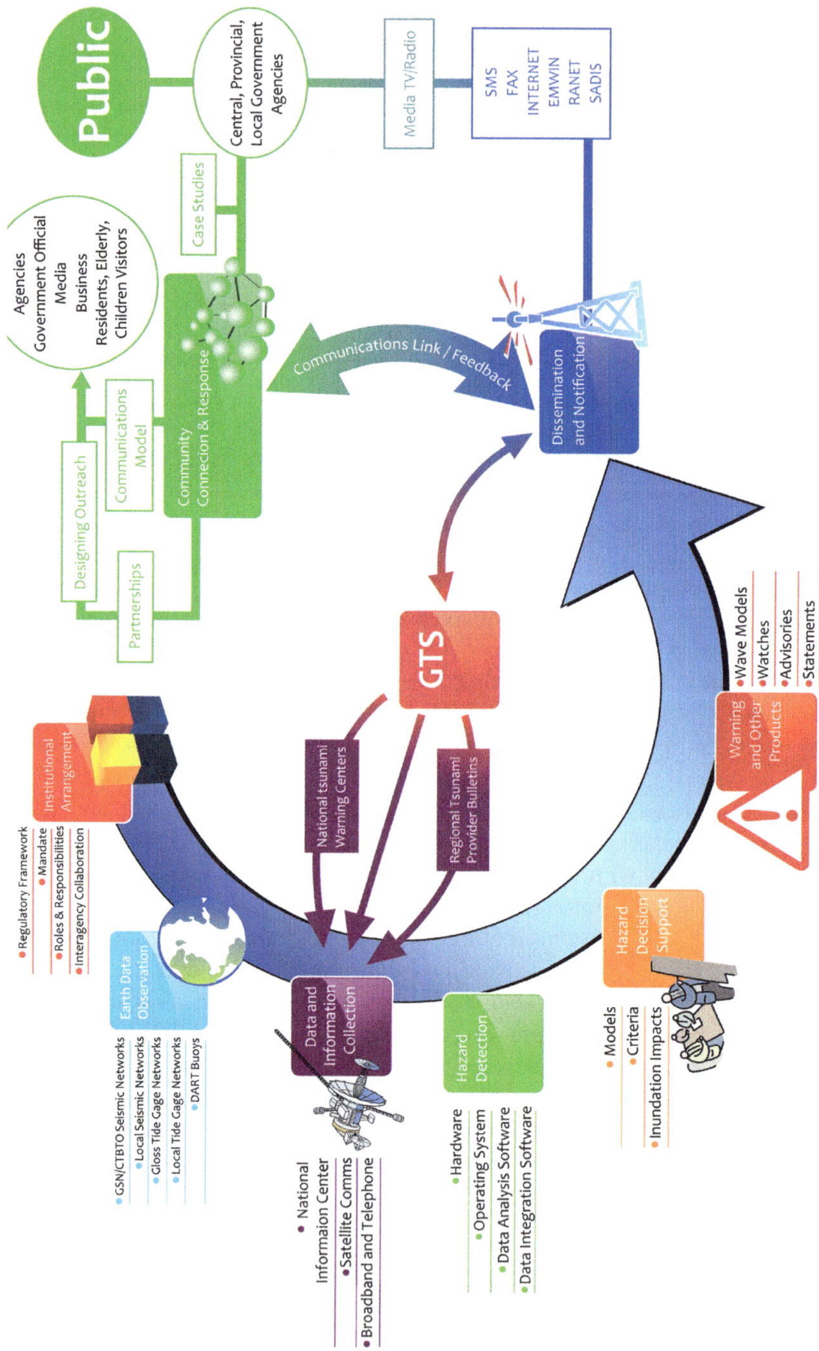

Fig. 5.1 End to end early warning system (modified from US IOTWS)

strategies or responses to minimise the risks (Alan et al. 2002; Höppner et al. 2010; Hurford et al. 2012; Klein and Methlie 1990; Mirza 2011; Rahmstorf 2010; Rossa et al. 2011; Tiranti and Rabuffetti 2010; Vogelbacher 2011). Overall, in Europe, risk communication with the general public is largely limited to one-way communication aiming at informing the public (on the hazard, on risks, on mitigation and prevention measures, on how to behave in the case of an event, and to promote acceptance), raising awareness, triggering protective action, and warning of upcoming events (Höppner et al. 2010). Among the various information sources, print and media approaches were the most frequently used methods of risk communication (Fitzpatrick-Lewis et al. 2010). Höppner et al. (2010) noted that too many sources of risk communication were found to be problematic as they led not only to confusion, but also reduced the credibility of both the message and the sender.

The observation has been that risk communications attempting to change people's attitude and behaviour (or to promote a particular behaviour) are still the most prominent kind of communication in relation to climate related hazards (Höppner et al. 2010). Between rational (e.g. probabilities) and emotional appeals (e.g. personal consequences of hazard events) in risk communication, Höppner et al. (2010) argued that emotional appeals seemed more effective than rational ones. Similarly, Fitzpatrick-Lewis et al. (2010) found no clear answer as to whether presenting information through statistics and numbers was more effective than through narrative styles such as personal accounts. According to Höppner et al. (2010), a perennial challenge in all risk communication is how to target the right safety message to the right audience. Höppner et al. (2010) noted that the public is not a single and uniform entity. Therefore risk communications must be tailored to their requirements taking into consideration the differences in their experiences, interests and needs. People generally paid attention to information delivered by sources that they trusted and considered credible (Fitzpatrick-Lewis et al. 2010; Höppner et al. 2010). The conclusion drawn by Höppner et al. (2010) was that risk communications alone do not necessarily lead the public to take the desired and expected actions as people are not simply passive recipients who receive risk communication messages and act on them.

5.5 Warning Dissemination

Warnings are the end result of a sequence of actions taken by forecasters to inform the public of a heightened probability of severe and/or hazardous weather that often begins days in advance of the event (Stensrud et al. 2012). Early warnings are disseminated in several ways. With respect to the communication mode, Moser (2010) concluded that 'face-to-face' communication in dissemination of warnings was more persuasive and impactful on personal behaviour than mass-media(ted) communication (Höppner et al. 2010). According to Höppner et al. (2010), 'one-way, written or verbal communications tended to enable learning and active engagement less well than dialogic and interactive forms of communication'. The RTSP product format and dissemination are conducted through notification messages are issued in

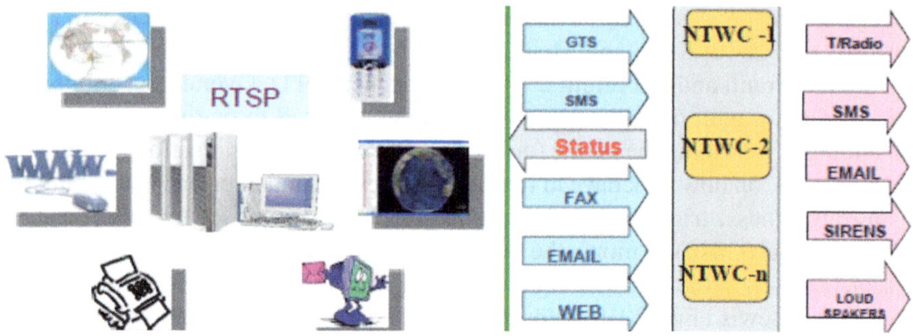

Fig. 5.2 RTSP's warning dissemination procedures (*source*: BoM)

Table 5.2 RTSP's performance indicators (*source*: IOC)

Indicators	Time
Elapsed time from earth quake to generation of earthquake information bulletin	10 min
Elapsed time from earth quake to generation of threat assessment bulletin	20 min
Percent of countries issued notifications of product generation	100 %
Probability of detection of earthquakes with magnitude>=6.5	100 %
Probability of detection of tsunami above threshold (0.05 m)	100 %
Accuracy of earthquake location	30 km
Accuracy of earthquake depth	25 km
Accuracy of earthquake magnitude	0.03
Accuracy of tsunami forecast amplitude/height	Factor of 2
Reliability of RTSP operations (power, computers, communications)	99.5 %

text format; bulletins are generated in both text and HTML formats on the websites; graphics are generated in .jpg or .png format on the websites and spatial data is also available in dbf format on the websites. Figure 5.2 shows the RTSP's warning dissemination procedures.

Notification Messages are transmitted to NTWCs via email, fax, SMS & GTS, when an RTSP bulletin is generated. RTSP bulletins are available to NTWCs through the RTSP password-protected websites. RTSP graphical products such as coastal zone threat maps and tsunami energy maps and coastal zone spatial data are also available to NTWCs on the RTSP websites. Table 5.2 shows the RTSP's performance indicator.

For a rapid onset disaster like tsunami, a robust and redundant tsunami alert rapid notification system (TARNS). TARNS refer specifically to the dissemination of the tsunami alert from the central focal point at the national level to the public at risk and all relevant entities within the country. Communication of a tsunami warning from a regional Indian Ocean warning center to the Thai national focal point is currently within the purview of other entities such as the Inter-governmental Oceanographic Commission. TARNS protocols are unique to a country, influenced by government structure, culture, available technology, and mechanisms for mass communication, existing infrastructure, and other factors. An effective TARNS consists of three components:

- A system design and plan that includes: regulatory framework, respective roles and responsibilities of government levels, role of the media, schematic of communication pathways for the warning, required interagency agreements, standard operating procedures that are updated annually, checklists and provisions for periodic testing and evaluating the system.
- Appropriate technology to facilitate rapid movement of warning information as per the TARNS plan and operating procedures.
- Procedures to test and evaluate all aspects of the system on a routine basis as identified in the TARNS plan and required by the characteristics of the technology utilized.

5.6 Community Interpretation and Response

Forecast interpretation and decision making processes are the major components for community level preparedness and response for any hazards. However, perceptions of risk vary not only across natural hazards, but also across individuals and groups of individuals, and this has an influence on forecast interpretation at community level (Höppner et al. 2010). Different perceptions of risk often lead to different judgments of a situation and of the need to act. A resulting challenge for communication is to foster mutual understanding and to mediate between different views (Höppner et al. 2010). Höppner et al. (2010) noted that the higher the perception of risk, the stronger the desire to receive information and to engage in discussions/decisions on the prevention and mitigation of hydro-meteorological hazards. There was a tendency to delegate the responsibility for safety to the agencies in charge, with a lack of personal involvement and urgency (Scolobig et al. 2012). Scolobig et al (2012) added that feeling more or less endangered was linked to the conceptual framing of the risk as the least concerned concentrated mainly on the presence of structural devices, whereas the most concerned focused on the unpredictability and exceptionality of the events, as well as on personal knowledge of the territory and its danger sources.

Disasters manifest pre-existing conditions within the social, economic, physical, and environmental fabrics of a society (Villagrán de León 2006). Thus, community studies are important in tsunami management. In summarizing lessons from Thailand and Indonesia in the 2004 tsunami, Atwood (2006) suggests that disasters lead to an exaggeration of previous inequities, enhancing the vulnerability of the most vulnerable. At the onset of a disaster, most families are equally needy, but not all are equally vulnerable. In planning a medium and long-term response it is important to identify those who are most vulnerable. Baseline data are also crucial for immediate assessment and response planning after a disaster event.

Implementing an emergency response plan to an issued forecast or early warning before the actual impact is a challenging task requiring collaborative efforts at all levels of government, including federal, state, county, and local authorities and significant prior planning and practice. In addition to forecast interpretation, failure to interpret early warnings by the communities at risk of hydro-meteorological disasters renders the EWSs useless. As such, a well-defined approach to data interpretation at community level is also required to improve the utility of an early warning system (Quansah et al. 2010). Höppner et al. (2010) identified six steps that are involved before a community can respond to an early warning. These include hearing the warning; understanding the contents of the warning message; believing the warning is credible and accurate; personalizing the warning to oneself; confirming that the warning is true and others are taking heed; and then responding by taking a protective action. According to Höppner et al. (2010), a warning by itself may not lead to action if the community at risk fails to interpret what actions they should take. Community response to risk communications is a function of personal risk perception, previous personal experience with risk, sources of information and trust in those sources, as well as preferences for information (Fitzpatrick-Lewis et al. 2010). Prior education is therefore critical in ensuring that the community is able to recognize and interpret environmental signals and respond appropriately, even in the absence of official warnings (Gissing et al. 2010).

5.7 Conclusion

The 2004 Indian Ocean tsunami was the first natural disaster in recent memory that affected many countries simultaneously, making it a truly international catastrophe. Because of its sheer scale of impact, this disaster broke new ground in many aspects of natural hazard management and response. How coastal communities manage risks associated with major tsunamis is an issue of global importance. The IOTWS now provides warnings to all Indian Ocean country members, reaching millions of people who had no warnings in 2004. Furthermore, through IOC and other donors initiatives tsunami hazard mapping and evacuation planning has been carried out for hundreds of coastal communities to enhance community capacity on tsunami hazards. The IOTWS is now fully operational, comprising a set of RTSPs issuing tsunami advisories to all NTWCs of the Indian Ocean rim countries. The IOTWS also developed the first international guidelines for tsunami hazard and risk assessment. As well, warning systems are also being established in the North East Atlantic, Mediterranean and Adjoining Seas, and the Caribbean. Protection is also being reinforced in the South West Pacific and the South China Sea. For effective tsunami warning system must include not only the ocean technologies for accurately detecting an emerging tsunami, but also a civil communication system through which the local government can effectively and timely warn the population. In fact, the evacuation time (how quickly a community can evacuate) is a key factor to make a warning system effective. Thus it is essential to invest in disaster education and training.

Investments leading to an increase in social capital will enable communities to cope with disasters of all kinds. In the last decade tsunami warning systems have shown a significant development but it may get feeding based on the occurrences. Sustaining RTSP is also essential to provide quality information with enhancement of service level 3. To provide a service level 3 and end to end mechanism is essential.

References

Alan FH, Daniel H, Dennis PL (2002) Economic value of long-lead streamflow forecasts for Columbia River hydropower. ASCE J Water Resour Plann Manage 128:91–101

Assilzadeh H, Mansor SB (2011) Natural disaster data and information management systems. Commission report. University Putra Malaysia, Serdang. http://unpan1.un.org/intradoc/groups/public/documents/apcity/unpan025913.pdf

Atwood SJ (2006) Shocks and aftershocks: lessons from Thailand and Indonesia. In: Distinguished lecture at the Morss colloquium: lessons from the 2004 Indian Ocean Tsunami. Woods Hole Oceanographic Institution, Woods Hole

Babel MS, Fakhruddin SHM, Kawasaki A (2013) A decision support framework for flood risk assessment: an application to the Brahmaputra River in Bangladesh, vol 357. IAHS Publication, Oxford

Behrens A et al (2010) Future impacts of climate change across Europe. Centre for European Policy Studies (CEPS) Working Document No. 324/February 2010. Available at SSRN 1604531

Borga M et al (2011) Flash flood forecasting, warning and risk management: the HYDRATE project. Environ Sci Policy 14:834–844

de León V (2006) Vulnerability: a conceptual and methodological review, vol 4, Studies of the University. United Nations University Institute for Environment and Human Security, Bonn

Demeritt D et al (2010) Challenges in communicating and using ensembles in operational flood forecasting. Meteorol Appl 17:209–222

Diodato N et al (2011) How the aggressiveness of rainfalls in the Mediterranean lands is enhanced by climate change. Climatic Change 108:591–599

Essink GHPO et al (2010) Effects of climate change on coastal groundwater systems: a modeling study in the Netherlands. Water Resour Res 46:W00F04

Fakhruddin SHM (2013) Flood early warning system in Bangladesh. In: Report on using science for disaster risk reduction. UNISDR technical advisory Group-2013. UNISDR, Geneva

Fitzpatrick-Lewis D et al (2010) Communication about environmental health risks: a systematic review. Environ Health 9:1–15

Frédéric A, Patrick H (2008) Encyclopedia of decision making and decision support technologies. Information Science Reference, Hershey

Gissing A et al (2010) Future challenges and directions of flood emergency management

Hallegatte S (2012) A cost effective solution to reduce disaster losses in developing countries: hydro-meteorological services, early warning, and evacuation. World Bank policy research working paper, vol 6058. Available at SSRN 2051341

Hapuarachchi H et al (2011) A review of advances in flash flood forecasting. Hydrol Process 25:2771–2784

Höppner C et al (2010) Risk communication and natural hazards. CapHaz Project, Birmensdorf

Hurford AP, Priest SJ, Parker DJ, Lumbroso DM (2012) The effectiveness of extreme rainfall alerts in predicting surface water flooding in England and Wales. Int J Climatol 32:1768–1774. doi:10.1002/joc.2391

Indian Ocean Tsunami Warning and Mitigation System (IOTWS) (2009) Report of intersessional meeting of working group 5. Intergovernmental Oceanographic Commission (of UNESCO). Hyderabad, India. 7–9 April 2009

Joseph A (2011) Chapter 14—IOC-UNESCO Tsunami early warning systems. In: Joseph A (ed) Tsunamis. Academic, Boston, pp 171–245. ISBN 9780123850539. http://dx.doi.org/10.1016/B978-0-12-385053-9.10014-6
Klein FH, Methlie L (1990) Knowledge based DSS. In: Information system and decision science. HEC School of Management, Jouy-En-Josas
Liu PL-F, Wang X, Salisbury AJ (2009) Tsunami hazard and early warning system in South China Sea. J Asian Earth Sci 36(1):2–12, ISSN 1367-9120
Menoni S, Margottini C (2011) Inside risk: a strategy for sustainable risk mitigation. Springer, New York
Mirza MMQ (2011) Climate change, flooding in South Asia and implications. Reg Environ Change 11:95–107
Moser SC (2010) Communicating climate change: history, challenges, process and future directions. WIREs Clim Chang 1(1):31–53
Parker DJ, Tunstall S, Wilson T (2005) Socio-economic benefits of flood forecasting and warning. In: International conference on innovation advances and implementation of flood forecasting technology, Tromso, Norway, 17–19 Oct 2005
Quansah JE et al (2010) Early warning systems: a review. J Terrestrial Observation 2:24–44
Rahmstorf S (2010) A new view on sea level rise. Nat Rep Climate Change 4:44–45
Rossa A et al (2011) The COST 731 action: a review on uncertainty propagation in advanced hydro-meteorological forecast systems. Atmos Res 100:150–167
Scolobig A et al (2012) The missing link between flood risk awareness and preparedness: findings from case studies in an Alpine Region. Nat Hazards 63:499–520
Stensrud DJ et al (2012) Progress and challenges with warn-on-forecast. Atmos Res 123:2–16
Tiranti D, Rabuffetti D (2010) Estimation of rainfall thresholds triggering shallow landslides for an operational warning system implementation. Landslides 7:471–481
Vogelbacher A (2011) Flood warning in Bavaria, Germany. In: Environmental knowledge for disaster risk management. ekDRM Secretariat (GIZ-NIDM), New Delhi, p 128
Wei Y, Cheung KF, Curtis GD, McCreery CS (2003) Inverse algorithm for tsunami forecasts. J Waterway Port Coastal Ocean Eng 129(2):60–69
World Bank, The United Nation (2010) Natural hazards, unnatural disasters: the economics of effective prevention. World Bank, Washington, ISBN: 978-0-8213-8050-5
World Disasters Report (2009) Early warning for early action. International Federation of Red Cross and Red Crescent Societies, Geneva

Chapter 6
Post-tsunami Urban Recovery Process and Current Conditions in Sri Lanka, Thailand, and Indonesia

Osamu Murao

Abstract Each country damaged by the 2004 Indian Ocean Tsunami adopted individual recovery planning strategies according to the need for resettling victims after the disaster. It is important to accumulate records of the different recovery processes with comparative studies, and to clarify the relationship between each post-disaster recovery process and strategy in discussions concerning future post-disaster recovery initiatives. However, it is difficult to compare recovery phenomena because of the different social, religious, political, or economic context in each country.

To remove the contextual bias from comparisons, a method for generating a recovery curve, which is developed using a building construction dataset, was proposed as a quantitative measure in previous studies by the author. This chapter first outlines the social contexts and the recovery policy and planning conducted in three countries, Sri Lanka, Thailand, and Indonesia, that were seriously damaged by the 2004 Indian Ocean Tsunami. Recovery curves are subsequently developed based on the building construction dataset of temporary and permanent housing, and then followed by a quantitative comparison. The result of the comparison indicates that the swiftest recovery was in Sri Lanka, and the slowest was in Indonesia. Finally, the physical conditions of areas affected by the tsunami as of 2012 are reported.

Keywords Comparative study • Housing reconstruction • Probability density function • Recovery curve • Urban recovery strategy

6.1 Significance of Comparative Study on Post-disaster Recovery

As of December 2013, 9 years have passed since the 2004 Indian Ocean Tsunami struck the coastal areas of several Indian ocean rim countries. The author followed post-tsunami urban recovery processes for several years in the three countries most

O. Murao (✉)
International Research Institute of Disaster Science, Tohoku University, Sendai, Japan
e-mail: murao@irides.tohoku.ac.jp

R. Shaw (ed.), *Recovery from the Indian Ocean Tsunami: A Ten-Year Journey*, Disaster Risk Reduction, DOI 10.1007/978-4-431-55117-1_6

devastated by the event: Indonesia, Sri Lanka, and Thailand. Each country adopted individual recovery planning strategies according to the damage sustained and social context, and most victims resettled themselves in permanent housing as a result. To make future recommendations for recovery strategies, comparative studies, both qualitative and quantitative, are indispensable. However, it is difficult to compare recovery phenomena because of the different social, religious, political, or economic context in each country. To remove the contextual bias from comparisons, measures to quantitatively evaluate the recovery processes are necessary.

It is important to clarify the relationship between the recovery process and policy after disasters in discussions concerning future post-disaster recovery initiatives. For example, Haas et al. (1977) suggest a model of disaster recovery activity after the 1906 San Francisco Earthquake. However, it is difficult to clarify the relationship because of several complicated factors, such as the difficulty of collecting data or diversity of social backgrounds.

Murao et al. (2007) proposed a method to construct a recovery curve for Chi–Chi, one of the areas seriously damaged due to the 1999 Taiwan Earthquake, as a tool to solve these problems. While there are many indicators for understanding the urban recovery conditions after disasters, the authors chose the construction ratio of necessary buildings in the affected areas, such as transitional/temporary housing or permanent housing, as an important physical indicator for most victims who lost houses. Following the research in Chi–Chi, recovery curves were developed for Thailand (Murao et al. 2008), Sri Lanka (Murao and Nakazato 2010), and Indonesia (Sugiyasu and Murao 2010) to assess those recovery processes after the 2004 Indian Ocean Tsunami. These recovery curves were developed using the rate of construction of both transitional and permanent housing from the viewpoint of the urban physical environment.

This chapter first outlines social contexts and the recovery policy and planning conducted in the three countries, Sri Lanka, Thailand, and Indonesia, that were seriously damaged by the 2004 Indian Ocean Tsunami. Recovery curves for these countries are subsequently presented based on a building construction dataset of temporary and permanent housing, and then followed by a quantitative comparison to easily understand each recovery process. Finally, the physical conditions of areas affected by the tsunami as of 2012 are reported.

6.2 Social Contexts and Outline of Policy and Planning Conducted by Sri Lanka, Thailand, and Indonesia

It is necessary to conduct a comparative study on post-disaster recovery with social contexts of the countries. Table 6.1 shows basic information of the three counties as of 2004 before the 2004 Tsunami. Each social context is explained in the following subsections.

The 2004 Indian Ocean Tsunami adversely affected several Indian Ocean Rim countries, as shown in Table 6.2. This chapter focuses on Sri Lanka, Thailand, and Indonesia for a comparative study.

Table 6.1 Comparison of social contexts of the three countries (as of 2004)

Official name	Democratic socialist republic of Sri Lanka	Kingdom of Thailand	Republic of Indonesia
Regime	Republic	Constitutional monarchy	Republic
Official language	Shinhala, Tamil	Thai	Indonesian
Religion	Buddhist 70 % Hindu 15 % Christian 8 % Muslim 7 %	Buddhist 95 % Muslim 4 % Followed by Christian, Hindu, Sikh, and Taoist	Muslim 77 % Christian 13 % Hindu 3 % And others
Administrative divisions	Nine provinces	Seventy-five provinces and the capital Bangkok	Thirty-three provinces
Area (km^2)	65,610 (119)	514,000 (49)	1,919,440 (15)
Population	19,905,165 (53)	64,865,523 (19)	238,452,952 (4)
Population density	303/km^2	126/km^2	124/km^2
GDP (MER)	$244 billion (79)	$1,764 billion (36)	$2,813 billion (26)
GDP per capita (PPP)	$3,555 (177)	$6,838 (84)	$3,217 (122)

The figures in brackets indicate ranking in the world
Data sources: Ministry of Foreign Affairs of Japan (2008)

Table 6.2 Casualties in the countries affected by the 2004 Indian Ocean Tsunami as of February 10, 2005

Country	Displaced	Missing	Death	Missing/death
India	112,588	5,640	10,749	16,389
Indonesia	417,438	127,749	114,573	242,322
Malaysia	8,000	6	68	74
Maldives	10,568	26	82	108
Myanmar	2,592	—	61	61
Seychelles	160	—	3	3
Sri Lanka	500,668	5,644	30,959	36,603
Thailand	—	3,062	5,392	8,454
Somalia	2,320	158	394	552

Data sources: Governments of affected countries, the International Federation of Red Cross and Red Crescent Societies (IFRC)

6.2.1 Social Context and Post-tsunami Recovery Strategy in Sri Lanka

Sri Lanka is a republic and a unitary state governed by a presidential system. The major religion is Buddhism, which accounts for 70 % of the total population, followed by Hinduism, Islam, and Christianity. Sinhalese constitute the largest ethnic group in the country, making up about 75 % of the total population. The second major ethnic group is Sri Lankan Tamils, making up about 11 % of the total population. After achieving national independence in 1948, the Sinhalese and Tamils were embroiled in an ethnic conflict, which erupted into the Sri Lankan Civil War between the government and the Liberation Tigers of Tamil Eelam (LTTE) in 1983.

Although it ended in 2009, the unstable political situation affected the recovery process in several regions.

In Sri Lanka, located to the west of the epicenter, tsunamis struck both the east and west coasts of the island. The damage covered thirteen districts in five provinces along 70 % of the coastal areas in the country. The number of people either missing or dead reached more than 36,000, and the number of damaged buildings was nearly 100,000.

The Coast Conservation Act (No. 57 of 1981) defines the Coastal Zone as "*the area lying within a limit of three hundred meters landwards of the mean high water line and two kilometers seawards of the mean low water line and in the case of rivers, streams, lagoons, or any other body of water connected to the sea permanently or periodically, the landward boundary shall extend to a limit of two kilometers measured perpendicular to the straight baseline drawn between the natural entrance points thereof and shall include the waters of such streams and lagoons or any other body of water so connected to the sea*" (Sect. 42). In December 1982, the Government announced in the Gazette (Extraordinary) No. 223/16 (Democratic Socialist Republic of Sri Lanka 1982) that the coastal area lying within 100 m from the coastline should be developed as an Urban Development Area. The Urban Development Area has been under control of Urban Development Authority (UDA) since then. However, some areas were privately owned, even though they were in the Urban Development Area, and private houses that had been constructed in the area were destroyed by the 2004 Indian Ocean Tsunami.

The Sri Lankan Government established the Task Force to Rebuild the Nation (TAFREN) just after the disaster to comprehensively coordinate its post-tsunami recovery planning and to organize related projects. The Tsunami Housing Reconstruction Unit (THRU) was then placed in charge of housing resettlement projects.

The recovery process for residents planned by Rebuilding and Development Agency (RADA) consisted of three stages: (a) emergency shelter, (b) transitional housing, and (c) permanent housing (Fig. 6.1). In the third stage, they provided two different resettlement programs. The Donor Driven Housing Program was for victims whose house was located in the Coastal Conservation Zone. These victims were able to relocate to permanent housing prepared by the Government and could do so without any payment. On the other hand, the Homeowner Driven Housing Reconstruction Program was for victims whose house was located outside the zone. The Government provided them some financial support to rebuild their houses based on the level of damage the house sustained.

Fig. 6.1 Three types of shelter/housing in the stages prepared by TAFREN. (**a**) Emergency shelter. (**b**) Transitional housing. (**c**) Permanent housing

6.2.2 Social Context and Post-tsunami Recovery Strategy in Thailand

Thailand is a constitutional monarchy, headed by King Rama IX, and Buddhists make up about 95 % of the total population. Because of its high economic status the GDP per capita of the country was almost double that of Sri Lanka and Indonesia, Thailand refused financial support from other countries after the disaster. Post-tsunami recovery support by politicians or political parties was prohibited just after the disaster because of the general election in February 2005, and this influenced the recovery activities to some degree. On the other hand, the royal families actively assisted in reconstruction of the affected areas.

In Thailand, the second closest country to the epicenter following Indonesia, the southern six provinces, including Phuket and Phang Nga along the coast of the Andaman Sea, were affected by the tsunamis. Alternatively, consider: "A large number of foreigners were among those killed in Thailand because the areas damaged by the disaster included world famous resorts, such as Phuket and Khao Lak. Approximately 3,900 of the 8,500 killed in the disaster were foreigners." Nearly 6,800 buildings were damaged by the tsunami.

The Thai Government did not establish any special horizontal organization for the recovery, and each existing ministry dealt with related matters respectively. The matters related to housing resettlements, including selection of construction sites, housing construction, and financial support to repair moderately damaged houses, were conducted by the Ministry of Defense, Ministry of Industry, Ministry of Social Development and Human Security, and Ministry of Finance. The victims were financially supported to build their own house, or were provided a house by the Government, though they had to rent the land. Permanent houses were constructed with application by the victims, and donors were given a construction subsidy by the Government. However, each local government developed its own recovery plans and led its own recovery process.

In Phuket, the affected residents living in popular resort areas had to relocate to other areas because of the regulation that residential buildings are not permitted in areas within 15–25 m of the coast. Instead of constructing temporary housing, the Local Government of Phuket, with support from donors such as NGO, World Bank, or UNESCO, provided permanent housing.

Khao Lak Province was the most severely damaged area in Thailand. After first providing temporary housing, the local government eventually provided lots of large-scale permanent housing throughout the province.

6.2.3 Social Context and Post-tsunami Recovery Strategy in Indonesia

The Republic of Indonesia is a country having the largest number of islets in the world. The major religion is Islam, which accounts for 77 % of the total population, followed by Christianity and Hinduism. The country experienced the 1992 Flores

Tsunami, but the 2004 Indian Ocean Tsunami was more devastating to Nanggröe Aceh Darussalam (Aceh). It has abundant natural resources of petroleum and natural gas, which spurred the Free Aceh Movement for independence in 1976 by Gerakan Aceh Merdeka (GAM). GAM fought against Indonesian government forces in the Aceh Insurgency from 1976 to 2005, with a peace agreement finally reached during the year following the tsunami.

The most devastated country was Indonesia, which was located near the epicenter. Buildings were strongly shaken by the earthquake and villages were washed away by huge tsunamis. The death toll topped 240,000, 81.7 % of the total number of those killed by the disaster, and at least 510,000 buildings were damaged, 79.6 % of the total number of buildings damaged. Banda Aceh was decimated by the tsunami because of its low-lying geographical location.

In Indonesia, Badan Perencanaan dan Pembangunan Nasional (BAPPENAS) was in charge of financial support, emergency supplies, and management for the forthcoming recovery phase, and Badan Perencanaan Pernbangunan Daerah (BAPPEDA) and local governments implemented substantial measures until April 2005. They subsequently established a special 4-year, time-limited agency, Badan Rehabilitasi dan Rekonstruksi NAD-Nias (BRR), to deal with the enormous number of tasks required for recovery. This agency functioned well in grasping the various recovery-related issues throughout the affected areas and in implementing the recovery planning.

The following procedures were used to develop the recovery plans in Indonesia. In January 2005, the National Development Planning Board produced a basic framework called a "blueprint" for the rebuilding of Aceh. The blueprint was then revised to concrete recovery master plans for each jurisdiction supported by Japan International Cooperation Agency (JICA) in September 2005. In contrast, Village Plans were arranged to cover smaller residents' communities in cooperation with the United States Agency for International Development (USAID) in October 2005. Unfortunately, the two different recovery plans led to confusion in some seriously damaged areas in Banda Aceh and Aceh Besar as to the proper execution of the envisioned recovery (BRR 2007).

6.3 Comparison of Recovery Process Using Recovery Curves for Housing Reconstruction

6.3.1 Method

Figure 6.2 demonstrates sequential changes in the general recovery process from the occurrence of a disaster event to resumption of the states quo. Our research is aiming to make this kind of recovery curve for the affected countries by the 2004 Indian Ocean Tsunami. A point of Y axis on the recovery curve can be various elements to indicate the ratio of recovery after the event for individual or regional situation.

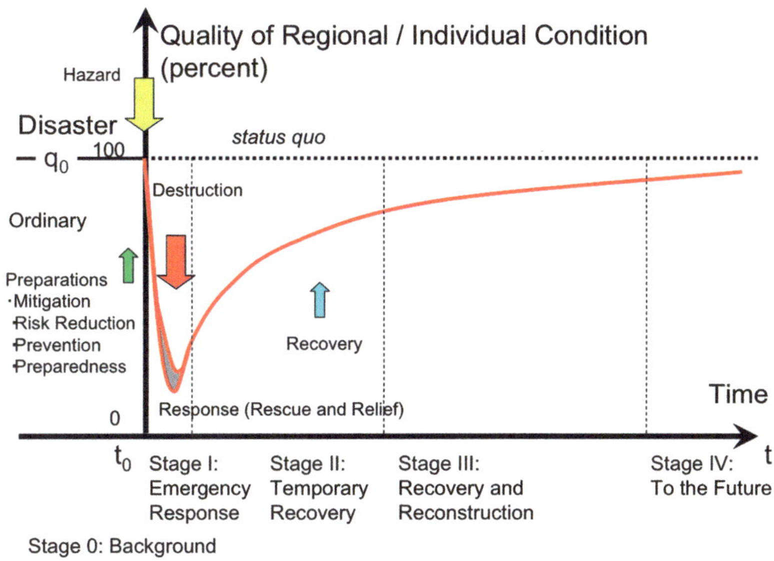

Fig. 6.2 Conceptual chronological model of social resilience after a disaster (Murao 2005)

For example, the number of population, regional income, or re-opened business ratio could be applied. Here, building construction ratio is used as the indicator because building recovery situation, namely transitional housing and permanent housing, is one of the most important factors for victims as a hub for household rehabilitation.

The following procedure is employed to construct the post-tsunami recovery curves and to compare the curves for the three countries.

Recovery functions for the three countries were derived from the building construction data of temporary/permanent housing provided by the respective governments through author's previous studies. The factors of time (months) and completion ratio of building construction were used to construct the recovery functions. The time period starts in December 2004, with January 2005 regarded as passing month "1". The time period extends 15–50 months respectively in each country until the time that most planned houses are completed. The completion ratio of building construction for a given time period is calculated based on the total amount of completed buildings.

At time T when the victims' lives have been sufficiently settled, for a time period of t (months), the cumulative completion ratio of building construction R (t) is assumed to have the cumulative normal distribution, as follows:

$$R(t)=\Phi\left(\left(t-\lambda\right)/\zeta\right) \tag{6.1}$$

where Φ represents the standard normal distribution, and λ and ζ are the mean and standard deviation of t, respectively. The two parameters for the distributions, λ and ζ, are determined using the least squares method on probability paper,

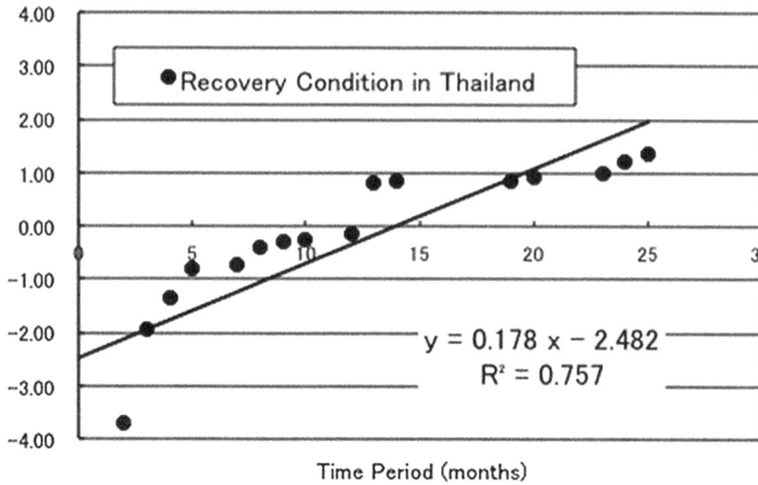

Fig. 6.3 Relationship between time and completion ratio of permanent housing construction in Thailand, as plotted on probability paper

Table 6.3 Construction status of permanent housing in Sri Lanka, Thailand, and Indonesia

	Sri Lanka[a]	Thailand[b]	Indonesia[c]
Planned number of permanent houses	29,971	2,251	132,928
Completion time of the first permanent housing site	Apr. 2005	Mar. 2005	Dec. 2006

Data sources:
[a]RADA (2006)
[b]Provided by Provincial Social Department and Human Security in Feb. 2007
[c]BRR (2005), BRR (2007), BRR (2006), and provided data by BRR in Feb. 2008

as shown in Fig. 6.3, for the case of Thailand. Recovery curves are then drawn, as shown in the following subsection.

The constructed recovery curves have two important parameters, mean (λ) and standard deviation (ζ). These parameters suggest the start time and average construction time period of the permanent housing built in each country, and are compared to quantitatively understand the national recovery tendency.

6.3.2 *Recovery Curves for Permanent Housing*

After the disaster, each government discussed strategies for urban recovery according to the damage sustained and social context. By arranging post-tsunami recovery plans, they estimated the number of required permanent houses for the victims before actual construction. Table 6.3 shows the number of planned permanent houses and the completion time of the first permanent housing site. The datasets of permanent housing information provided by the governments shown in the table were used to generate the recovery curves.

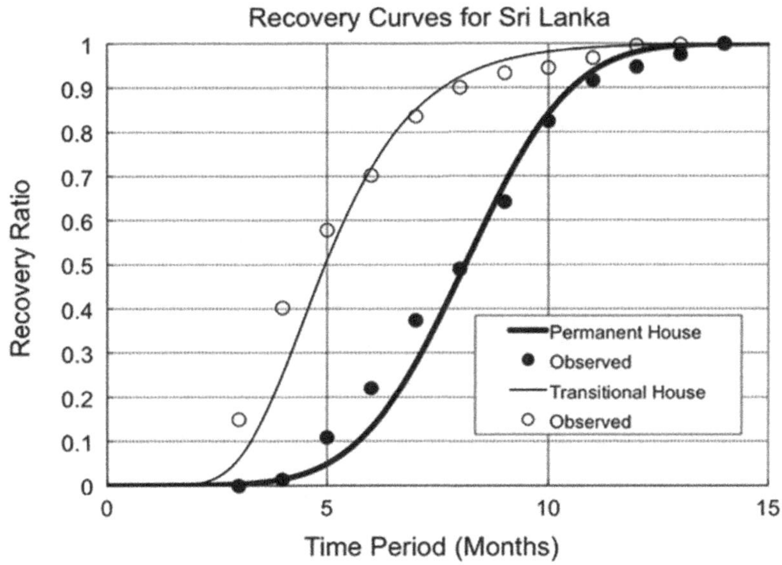

Fig. 6.4 Recovery curves for permanent housing and transitional housing in Sri Lanka (Murao and Nakazato 2010)

Figure 6.4 shows the recovery curves for Sri Lanka along with the observed recovery ratio calculated by Murao and Nakazato (2010). Although the curve for the permanent housing was fitted using the cumulative normal distribution as mentioned previously, the curve for the transitional housing was based on the Gompertz distribution.

Figure 6.5 shows the recovery curve for the permanent housing in Thailand along with the observed recovery ratio (Murao et al. 2008).

Finally, the recovery curves for permanent housing and transitional housing in Indonesia (Sugiyasu and Murao 2010) are shown in Fig. 6.6.

6.3.3 Comparison of the Recovery Curves Between Sri Lanka, Thailand, and Indonesia

One aspect of the comparative study is construction speed. Showing the two parameters of the recovery curves, mean (λ) and standard deviation (ζ), the recovery curves for the permanent housing indicate how each government dealt with resettlement of the victims. Table 6.4 shows the parameters of the curves for the three countries and the construction completion time obtained by the curves. Mean (λ) shows the number of months in which the construction completion ratio of the permanent housing is 50 %. The standard deviation (ζ) indicates that 68.27 % of the scheduled buildings are mostly constructed during the time period.

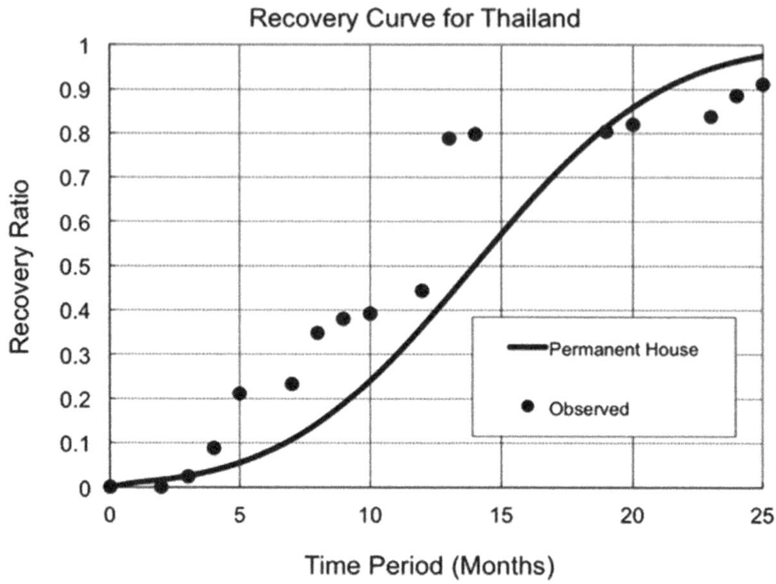

Fig. 6.5 Recovery curve for permanent housing in Thailand (Murao et al., 2008)

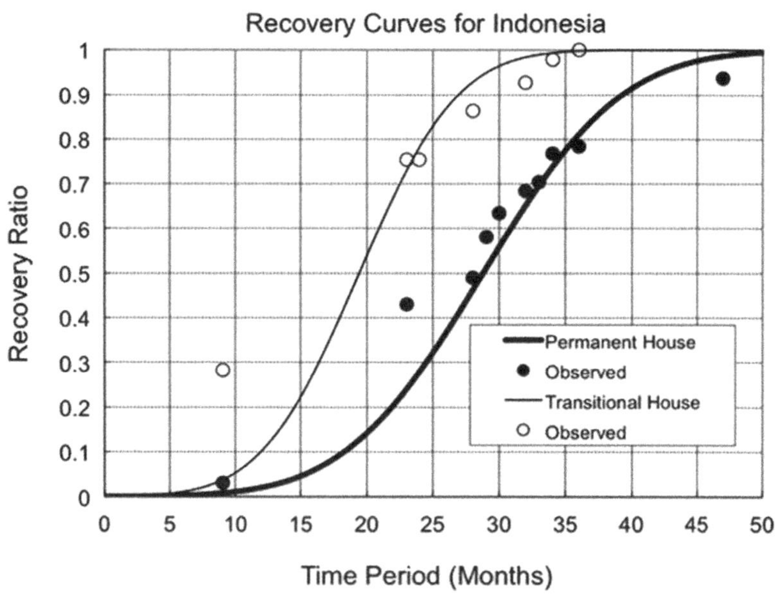

Fig. 6.6 Recovery curves for permanent housing and transitional housing in Indonesia (Sugiyasu and Murao 2010)

Table 6.4 Comparison of parameters for construction of permanent housing

	Sri Lanka	Thailand	Indonesia
Mean (λ)	8.2	14.0	28.8
Standard deviation (ζ)	1.9	5.6	8.2
Coefficient of determination (R^2)	0.96	0.76	0.92
Time period of most building construction (2ζ)	3.8	11.3	16.4
Completion time of the first permanent *housing* site	Apr. 2005	Mar. 2005	Dec. 2006
50 % completion time	Aug. 2005	Feb. 2006	Apr. 2007
90 % completion time	Oct. 2005	Sep. 2006	Mar. 2008

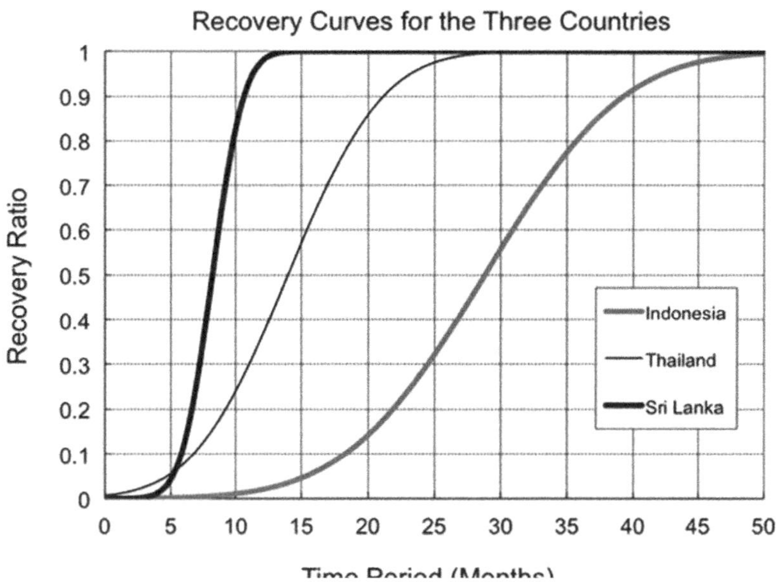

Fig. 6.7 Comparison of recovery curves for Sri Lanka, Thailand, and Indonesia

The recovery curves for permanent housing in the three countries are demonstrated in Fig. 6.7, and probability density functions are shown in Fig. 6.8. These figures and the above table clarify quantitative differences of the post-tsunami recovery process among the countries in terms of permanent housing construction.

The speediest recovery from the disaster was in Sri Lanka; the slowest was in Indonesia. The first permanent housing site in Thailand was finished in March 2005, followed by Sri Lanka in April 2005. Compared with these two countries, the completion of permanent housing sites in Indonesia was too late, in December 2006, 21 months after construction was finished in Thailand.

Figures 6.7 and 6.8 show the time of the peak (λ: 50 % completion time) for each country: 8.2 months (August 2005) in Sri Lanka, 14.0 months (February 2006) in Thailand, and 28.8 months (April 2007) in Indonesia. The gap between Sri Lanka

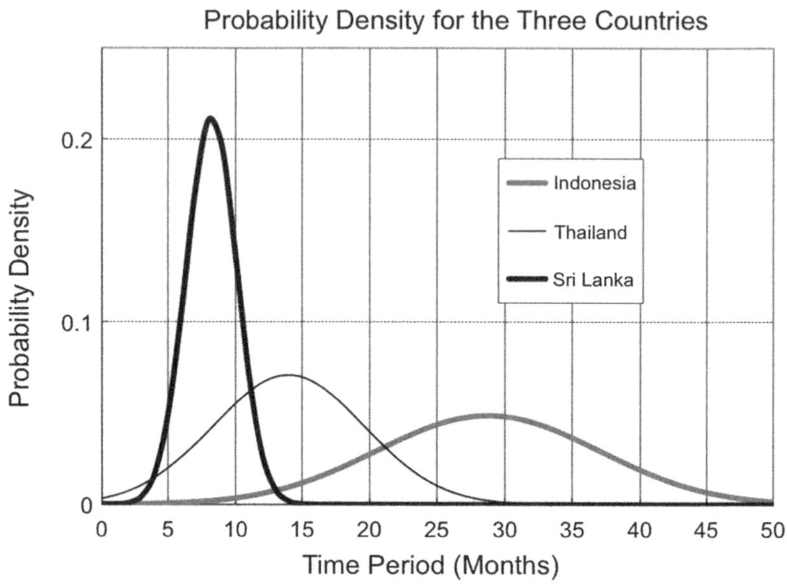

Fig. 6.8 Comparison of probability density functions for Sri Lanka, Thailand, and Indonesia

and Indonesia is 20.6 months, more than one and a half years. The 90 % completion time is October 2005 in Sri Lanka, September 2006 in Thailand, and March 2008 in Indonesia. The difference between Sri Lanka and Indonesia is about 30 months.

Time period in which 68.27 % of the buildings are mostly constructed (2ζ) is 3.8 months in Sri Lanka, 11.3 months in Thailand, and 16.4 months in Indonesia. The recovery in Indonesia took four times as long as that in Sri Lanka. This indicates that Sri Lanka was able to concentrate the construction effort over a shorter period than the other countries.

6.3.4 Future Discussion

Constructing those recovery curves, we are able to quantitatively compare the recovery processes of these countries. The question is, how we can utilize the recovery curves.

Each curve presents the national recovery situation of the countries, and it is also possible to deal with regional recovery situation in a country. Murao and Nakazato (2010) tried to construct regional recovery curves for Sri Lanka and presented the regional differences. We conducted interviews to stakeholders in a field survey in Sri Lanka to understand the differences shown by the recovery curves. Then we found some reasons; for example, the delay in Trincomalee was caused by construction strikes because of the civil war, and the swift recovery situation in Hanbantota was caused by the fact that the place is the hometown of powerful politicians in the country.

The time difference of the completion of construction between Trincomalee and Hambantota was 2.2 months.

Using recovery curves, we can quantitatively recognize the differences with some explanations as mentioned above. Even more importantly, the curves can be used as a tool to consider future post-disaster urban recovery strategies. If we can recognize dominant factors to determine the speed of recovery, we can control the factors. The factors might be the amount of money invested in recovery activities, the level of affected areas, or organization systems. In any case, it is necessary to collect and accumulate the data of post-disaster recovery cases.

Of course, not only speed but also quality should be considered when we discuss regional/individual recovery.

6.4 Current Urban Recovery Situation in Areas Affected by the 2004 Indian Ocean Tsunami

The Indian Ocean Tsunami that took place on December 26 in 2004 influenced the lives of more than 1.2 million people and dramatically changed the physical environment of the damaged coastal areas. The current situation of the coastal areas is a result of the governments' long-term efforts to create a better environment and reduce future tsunami disasters. Now that 9 years have passed since the catastrophic event, the recovery efforts implemented in each country need to be examined to better develop future post-disaster recovery strategies. This section briefly reports the current situation of urban recovery in Sri Lanka, Thailand, and Indonesia from field surveys conducted by the author in 2011 and 2012.

6.4.1 Post-tsunami Recovery Situation in Sri Lanka as of December 2012

Following the recovery plan, most of the damaged residential coastal areas were changed into public spaces, such as memorial parks, and the surviving residents were moved to permanent housing. A more remarkable result, however, is that the 30-year civil war between the Government and LTTE, which had influenced recovery activities since the tsunami disaster, ended in a military victory in 2009. President Mahinda Rajapaksa, who was born in Hambantota and won the presidential election in 2005, contributed to the peace settlement. Through his efforts, Hambantota, a district seriously damaged by the tsunami in the Southern Province, has undergone dramatic changes and is now an important international city in Sri Lanka. Several urban development projects have been completed or are underway (Fig. 6.9): (1) the first highway in the country was constructed between Colombo and Galle in 2011, which enabled easy access to Hambantota; (2) an international airport is under construction; (3) the ferry terminal was constructed with support from China; and (4) the government office plans to move from the coastal area to a newly developed inland district. Hambantota has been tremendously influenced by the 2011 Tsunami.

Fig. 6.9 Large projects in Hambantota, Sri Lanka. (**a**) International airport. (**b**) International ferry terminal

Fig. 6.10 Waterfront development in Phang Nga, Thailand. (**a**) Baan Nam Kem Tsunami Memorial Park. (**b**) Tsunami evacuation building

6.4.2 Post-tsunami Recovery Situation in Thailand as of October 2011

Coastal areas in Phuket and Khao-Lak, world-famous sightseeing locations, have recovered almost completely. As well as these places for visitors, other coastal areas enacted changes to reduce future tsunami damage, such as installing early-warning systems and building evacuation facilities (Fig. 6.10). Nam Kem Village, the northern district of Phang Nga Province, was severely damaged by the tsunami, and its devastated waterfront residential area was turned into a memorial park. Signboards depicting evacuation routes were placed within the surrounding areas and indicate how to evacuate from risky waterfront areas to safer inland locations in the event of a tsunami. Several tsunami evacuation facilities were also constructed along a systematic evacuation route.

Fig. 6.11 New facilities in Banda Aceh, Indonesia. (**a**) Community building. (**b**) Aceh Tsunami Museum

6.4.3 Post-tsunami Recovery Situation in Indonesia as of February 2012

In the decimated region of Aceh, each district was redeveloped according to the recovery plan. The change in Banda Aceh has been remarkable because of the global attention the city received following the tsunami (Fig. 6.11). During the recovery, several community buildings, both in Aceh and other provinces, were designated as evacuation facilities for vulnerable waterfront residents. Unfortunately, some of them are now occupied with bicycles or motorcycles and are unlikely to be used during an emergency evacuation. Other tsunami evacuation facilities have deteriorated because of management problems, even though only a few years have passed since those construction. In preparing these kinds of facilities in future disaster management, it is important to consider not only construction but also constant management.

One of the most meaningful facilities to come out of the recovery is the Aceh Tsunami Museum. It was established to educate people about tsunamis and to commemorate the victims. It became popular as an international landmark of the 2004 Indian Ocean Tsunami.

6.5 Conclusion

This chapter presented the recovery curves for permanent housing constructed in Sri Lanka, Thailand, and Indonesia to quantitatively understand the recovery processes after the 2004 Indian Ocean Tsunami. The constructed curves clarified differences in the recovery process among the three countries: (1) the swiftest recovery from the disaster was in Sri Lanka; (2) the slowest recovery was in Indonesia; (3) the time period in which most buildings were constructed was 3.8 months for Sri Lanka, 11.3 months for Thailand, and 16.4 months for Indonesia. These results indicate that Sri Lanka was able to concentrate the construction effort over a shorter period than the other countries.

The differences in the recovery process among the countries were a result of the expanse of regional damage, complexity of the society, governmental budgets, political leadership, and so on. Further discussions are required to clarify these relationships.

The comparative study of post-disaster urban recovery is crucial for developing effective recovery procedures and uncovering problems associated with the recovery processes. This, however, is easier said than done. Each recovery process has some bias according to its social context. To remove this bias, a quantitative measure to objectively compare recovery processes in different contexts is required. The recovery curve method proposed in this chapter is one such measure.

This chapter also briefly reported the current recovery situation of the three countries. The process of recovering from urban destruction is a significant concern for future cities. By monitoring individual post-disaster recovery, and understanding the successes and failures, we will be able to make future recommendations.

References

Badan Rehabilitasi dan Rekonstruksi NAD-Nias (BRR) (2005) Laying down the foundation for a better future. In: Six-monthly report of the executing agency for the rehabilitation and reconstruction of Aceh and Nias

Badan Rehabilitasi dan Rekonstruksi NAD-Nias (BRR) (2007) Tsunami recovery indicators package for Aceh and Nias english edition. BRR Information analysis Section

Badan Rehabilitasi dan Rekonstruksi NAD-Nias (BRR) (2009) http://e-aceh-nias.org/home/. Accessed 31 Jan 2009

Democratic Socialist Republic of Sri Lanka (1982) The Coast Conservation Act No. 57 of 1981, Section 14 and Section 42, Gazette (Extraordinary) No. 223/16. Accessed 17 Dec 1982

Haas JE, Kates RW, Bowden MJ (1977) Reconstruction following disaster. MIT, Cambridge, ISBN-13: 978-0262080941

Ministry of Foreign Affairs of Japan (2008) Regional index of Asia, http://www.mofa.go.jp/mofaj/area/asia.html. Accessed 17 June 2009

Murao O (2005) Description of building reconstruction process of Chi–Chi area using image archives after the 1999 Chi–Chi Earthquake, Taiwan. In: Proceedings of the 1st international conference on urban disaster reduction, Kobe, Japan

Murao O, Nakazato H (2010) Recovery curves for housing reconstruction in Sri Lanka after the 2004 Indian Ocean Tsunami. J Earthquake Tsunami 4(2):51–60

Murao O, Mitsuda Y, Miyamoto A, Sasaki T, Nakazato H, Hayashi T (2007) Recovery curves and digital city of Chi–Chi as urban recovery digital archives. In: Proceedings of the 2nd international conference on urban disaster reduction (CD-ROM), Taipei, Taiwan

Murao O, Sugiyasu K, Nakazato H (2008) Consideration of recovery process and development of recovery curves for buildings in Thailand after the 2004 Indian Ocean Tsunami. J City Plann Inst Jpn 43(3):745–750 (in Japanese)

Rebuilding and Development Agency in Sri Lanka (RADA) (2006) Progress report

Sugiyasu K, Murao O (2010) Comparative analysis of the reconstruction process of urban facilities in Indonesia based on recovery curves after the 2004 Indian Ocean Tsunami. In: Proceedings of the international conference on sustainable built environment 2010, vol 2, pp 363–370, Kandy, Sri Lanka

Chapter 7
Critical Factors for Sustainable Post-tsunami Resettlement: Cases from India and Sri Lanka

Miwa Abe and Rajib Shaw

Abstract After the Indian Ocean Tsunami, affected people were forced to leave their original place either due to the government policy or physical issues such as eroded shoreline caused by the waves. However, organizations that had supported the affected areas did not evaluate their resettlement projects nor the lives of the people who were relocated. Resettlement issues have been discussed by many organizations and researchers, however they were mainly case studies with non-specific conclusions. This study aims to: (1) identify the critical factors required to achieve sustainable settlement, and (2) identify problems arising from displacement. To obtain the relevant information, questionnaire survey was conducted in Tamil Nadu, India and Kalutara and Galle district in Sri Lanka.

Keywords Disaster recovery • Resettlement • Tsunami

7.1 Introduction

In 2004, the Indian Ocean Tsunami struck the coast along seven nations, resulting in approximately 280,000 deaths. The tsunami, with waves up to 30 m, was caused by an earthquake originating in the Indian Ocean on the northern Sumatra Island, Indonesia, and hit the shores of Indonesia, Sri Lanka, India, Thailand and other countries, causing serious damages and deaths with impacts affecting nations as far as the eastern coast of Africa. After the Indian Ocean Tsunami, affected people were forced to leave their original place either due to the government policy or physical issues such as eroded shoreline caused by the waves. However, organizations that had supported the affected areas did not evaluate their relocation projects

M. Abe (✉)
Center for Policy Studies, Kumamoto University, 2-39-1 Kurokami, Chuo-ku, Kumamoto 860-8555, Japan
e-mail: m-abe@kumamoto-u.ac.jp

R. Shaw (✉)
Graduate School of Global Environmental Studies (GSGES), Kyoto University, Yoshida-Honmachi, Sakyo-ku, Kyoto 606-8501, Japan
e-mail: shaw.rajib.5u@kyoto-u.ac.jp

R. Shaw (ed.), *Recovery from the Indian Ocean Tsunami: A Ten-Year Journey*, Disaster Risk Reduction, DOI 10.1007/978-4-431-55117-1_7

nor the lives of the people who were relocated. Researchers pointed out the importance of considering socio-culture with relocation in target communities (Boen and Jigyasu 2005; Chen et al. 2005). However, it remains unclear about the kind of influence that 'socio-cultural' consideration has on disaster recovery support.

Resettlement has been studied for years but inquiry into the individual's life after relocation is lacking. Relocation remains unpopular as plans to relocate entire communities are costly, mainly because of the need to acquire alternative locations and the position of housing for those resettled (Ngai 1995). Ngai pointed out that "relocation is, therefore, seldom a popular solution in hazard and disaster management." Blaikie et al. (1994) classify relocation as the worst option as it gives the impression that nothing else can be done about hazards and hazard-zone inhabitants have to be moved.

Disasters are increasing year by year (CRED 2012) and it is important for decision-makers and local people to consider not only risk reduction, but also disaster recovery. People cannot escape relocation after natural disasters, and can be considered as an important disaster recovery strategy. Therefore, one needs to understand positive and negative impacts by relocation for making sustainable community life in disaster affected areas. Relocation issues have been discussed by many organizations and researchers, however they were mainly case studies with non-specific conclusions. This study aims to: (1) identify problems arising from relocation and (2) identify the critical factors required to achieve sustainable settlement. To obtain the relevant information, questionnaire surveys were conducted in Sri Lanka and India during 2006–2010.

7.2 Relocation Study and Stream

In the previous post-disaster relocation study, two aspects, namely (1) positive impacts from perspectives of international organizations and (2) negative impacts from academic perspectives were discussed (Imura and Shaw 2009). International organizations, which provides support to disaster affected victims (World Bank 2010; UNDP 2004; ADB 1995; USAID 2007) and local government, pointed out relocation should need to manage, but it takes relocation as one of the development strategies because victims can live in new settlement area without disaster risk, in addition, people can receive new facilities and education opportunities such as environment, disaster management, livelihoods support and so on. Certainly, it may be said that relocation has some positive effects such as loss vulnerability to hazards, opportunity of getting new life and education.

From academic perspectives, Rodriguez et al. (2006) pointed out the extent to which relocation of communities and industries was possible when land acquisition, community acceptance, and industry feasibility were in question. Relocation initiatives must take into consideration the impact such movements will have on the communities' social, economic, cultural, and political activities. Especially, social and cultural impact

Table 7.1 Relocation issues from previous studies

1. Relocation distance and effect to occupation by distance, changes in life style, kind of job and mental problems (Sonogashira and Murosaki 2000; Yamori et al. 2006; Maki et al. 2003; Higashi 2000; Chiba 1995; Suzuki et al. 2007; Rodriguez et al 2006)
2. Land use in relocation area by government or support organization and relocation planning (Ando et al. 2000; Maki et al. 1999, 2003; Ishikawa et al. 2006; Suzuki et al. 2007; Rodriguez et al 2006; Ragsdale et al. 2009)
3. Infrastructure problems such as water supply, electricity in new land and access to town and job place etc. (Ye and Okada 2002; Dikmen 2010; Nakagawa et al. 2008; Ishikawa et al. 2004)
4. Loss of familiar environment and memories such as previous house, land, family and neighbors (Miura et al. 1996; Badri et al. 2006; Chiba 1995; Menoni and Pesaro 2008)
5. Political problems such as government and support organization's strategy, policy and implementation about relocation/resettlement (Kimura and Takahashi 2004; Ishikawa et al. 2007)
6. Human relationships with same community members, relocation members, neighboring people and host community people (Ninomiya et al. 2000; Sonogashira and Murosaki 2000; Perry and Lindell 1997; Maruyama 2003; Aota et al. 2006; Badri et al. 2006)

by relocation was pointed out from several case studies (Oliver-Smith 1991; Boen and Jigyasu 2005). Recently, the necessity of victims joining in the decision making process has become increasingly critical. Academic studies provided some more detail findings on relocation after disaster. Findings from previous relocation studies identified issues, as described in Table 7.1.

In post-disaster recovery, relocation does not only involve housing and land support but also includes area development. It also includes human relationships, political issues, and so on. Therefore, relocation can say that is complex issues when policy maker try to do relocation for housing recovery after disaster. Hence, it is very difficult to evaluate relocation. For a reason of this, usually relocation will be implemented in same community members. Imura et al. (2011) showed those same and different community members' relocation cases after Indian Ocean Tsunami in 2004. It have shown that setting a common purpose in resettled community, it will be a possibility of making a new relationships.

7.3 Impact of Indian Ocean Tsunami, 2004

7.3.1 Cases of Sri Lanka

The tsunami swept over Sri Lanka on 26th December in the day between 9:15 and 14:30 local time. Sri Lanka is located 1,700 km away from the epicenter of the earthquake therefore nobody felt the ground shake. The tsunami hit 60 % of the coastal area in Sri Lanka 2 h after the earthquake. Tsunami struck about two-thirds of Sri Lanka's 1,600 km coast, destroying the shoreline and an inland strip of lowland that varied, according to the configuration of the coastal lowlands between 200 and 400 m. The reported number of damages from the tidal waves, resulting in

Fig. 7.1 Tsunami affected housing and relocated site in Kalutara District, Sri Lanka. (**a**) Tsunami affected house (kitchen). (**b**) Tsunami affected house. (**c**) Relocated site (recovery housing). (**d**) Relocated site (recovery housing as an apartment)

30,196 deaths and 15,683 injured, 3,792 mission, and 838,088 was displaced in the county (Fig. 7.1; Grant-in-Aid for Special Purposes Research Report 2005). The people of Sri Lanka do not know about tsunami and had never experienced such a phenomenon. There were no shaking experience, no tsunami information and no safe evacuation area against tidal waves, which were the main reasons why the casualties exceeded 40,000 in Sri Lanka. Among the dead, there were more women and children than men. The total number of persons affected was about one million and two-thirds of the victims were in the northeast regions. Other damages caused include 29,700 fishing vessels lost and 92 destroyed, as well as 90 damaged schools. The economic loss, estimated by the government of Sri Lanka, was approximately 2.2 billion USD, which is significant considering the status of Sri Lankan economy.

After the tsunami, the government decided to give compensation from the Presidential Fund but it did not continue for long. Ten months later, the International Labor Organization (ILO) pointed out that more than 90 % of the tsunami victims were unemployed. The fisher folks were the hardest hit by the tidal waves. Fishermen lost their family members, fishing vessels and fishing tools including motors and nets. The others with some livelihood such as running a small shop or as a farmer also had their shops or land destroyed by the tsunami. The damage caused by the tsunami affected not only private land but also public institutions. People lost not only their own land and houses but also public facilities such as school, hospital, government offices and post office and so on. The tsunami also damaged the physical infrastructure in the coastal regions including roads, railway lines and the telecommunications network, water and electricity supply and drainage systems, etc.

7.3.2 *Cases of India*

In India, the tsunami struck hard in Tamil Nadu across the 2,260 km long coastline, and caused severe damages in Andhra Pradesh and Kerala states. In particular, three districts in Tamil Nadu namely Nagapattinam, Cuddalore and Kanniyakumari suffered severe damages from the tidal waves, resulting in 12,405 deaths and 6,913 injuries, 157,393 dwellings destroyed, and 31,755 livestock killed (Cuddalore Collecterate 2005). After the tsunami, the Tamil Nadu government and many organizations such as NGOs addressed tsunami recovery issues, including relocation. Majority of the severely affected communities tried to recover and adapt to a new location with other communities (Fig. 7.2). Tamil Nadu located on south India, has 32 districts, of which 13 districts borders on Bengal bay. The total population of Tamil Nadu is approximately 62 million.

Fig. 7.2 Relocated site and recovery housing in Tamil Nadu, India. (**a**) Relocation site. (**b**) Recovery housing. (**c**) Relocated site (recovery housing as an apartment). (**d**) Sharing water resources in relocated area

After the Indian Ocean Tsunami, Tamil Nadu government established the Tamil Nadu Tsunami Resource Center (TNTRC) in Chennai. They received many support from other countries and international/local organizations and NGOs. Recovery and reconstruction projects including relocation projects began after disaster. In projects that focused on the relocation, government prepares the new land, and supporting organizations sponsored the construction of the new houses. However land preparation was difficult, and in some cases, people were resettled on flood prone areas or near the crematory and etc.

7.4 Methods of Survey

Most of the tsunami affected people were fisherman who lived near the shoreline and they were given priority when receiving support from the government. This study focuses on seven communities. In India case, there are four villages which are named 'Thazhanguda', 'Devanampattinam", 'Tsunami Nagar' and 'HUL Nagar' in Tamil Nadu. And in Sri Lanka, there are three villages which are named 'Fukuoka village', 'Lagoswatta II' and 'Vaddavan'. Each village was affected by tsunami in 2004 and people had to relocate to other places because of government policy. Table 7.2 shows the details of the target communities. These seven communities were relocated based on the different relocation indicators such as main occupation, relocation distance, and relocation style.

Data for this study were collected from a randomly chosen sample of 2,544 total residents aged over 15 from all communities. These communities were selected based on their relocation style, occupation, area (urban/rural) and support organization so as to analyze the relocation impact at coastal zone. Sample size for each community was proportional to its population. In total, 712 usable responses were received (adjust response rate = 28.0 %). The results of the questionnaire survey were analyzed by SPSS (Statistical Package for the Social Sciences).

7.5 Results of Data Analysis

Questionnaire survey was conducted to understand the living condition following relocation. Table 7.3 shows the percentage distribution between those who preferred their previous life and those who preferred life after relocation between the relocated people in India and Sri Lanka. Findings showed that a higher percentage of people in Sri Lanka preferred 'previous life' compared to those in India (Table 7.3). This may suggest that people in Sri Lanka faced issues, complaints or are unsatisfied in the relocated area, and as a result, responded that they preferred their previous life.

Regarding the reason of decision making of relocation site showed on Table 7.4. In India, many people did not consider relocation site when people had to decide their new housing area. Therefore 210 people responded that they could settle anywhere.

Table 7.2 General information of target community

Community name	Devanam Patthinam	Thazhanguda	Tsunami nagar	HUL nagar	Fukuoka village	Lagoswatta II	Vaddavan
Total household	648	235	384	963	47	67	200
Collected samples	131 (20.2 %)	49 (20.9 %)	78 (20.3 %)	260 (27.0 %)	38 (80.9 %)	55 (82.1 %)	101 (50.5 %)
National	India	India	India	India	Sri Lanka	Sri Lanka	Sri Lanka
District	Cuddalore	Cuddalore	Cuddalore	Chennai	Kaltara	Kaltara	Batticalor
Support organization	Local NGO	International/local NGO, private company	International/ local NGO	Local government	UN Org.	International NGO	Local NGO
Relocation distance[a]	650 m	550 m	1 km (5 km)	3 km	1–1.5 km	5 km	1–1.5 km
Relocated style[b]	Same	Same	Mixture	Mixture	Same	Same	Mixture
Main occupation	Fishermen	Fishermen	Fishermen	Non-fishermen	Non-fishermen	Non-fishermen	Fishermen

[a]It is direct distance on the map. At Tsunami Nagar, there are two distance '1 km (5 km)'. It means on the map, the distance is only 1 km, but if people access to previous/post relocation site, actually they have to use a bridge for reach their destination through a roundabout path

[b]Same' means relocation with same community members, and 'Mixture' means relocation with mixture community members

Table 7.3 Comparison between relocated people who preferred their previous life and those who preferred their present life in India and Sri Lanka (frequency and percentage)

	Previous life	Present life	Total
India	236 (56.3 %)	183 (43.7 %)	419
Sri Lanka	118 (69.4 %)	52 (30.6 %)	170
Total	354 (60.1 %)	235 (39.9 %)	589

Table 7.4 Reasons of decision making of relocation site (number of answers; people)

	India	Sri Lanka	Total
Good natural environment	130	4	134
Good income	44	1	45
Good relationships with neighborhoods	57	2	59
I want to settle anywhere	210	32	242
NGOs told me	105	22	127
Government told me	152	112	264
I don't know	30	1	31
No other option	96	89	185

Table 7.5 Considering relocation styles

	India	Sri Lanka	Total
With same community members	237 (63.0 %)	120 (68.2 %)	357 (64.7 %)
With different community members	54 (14.4 %)	18 (10.2 %)	72 (13.0 %)
Nothing particular	85 (22.6 %)	38 (21.6 %)	123 (22.3 %)

On the other hand, Sri Lanka people decided new housing area based on what the government had suggested. This suggests that Sri Lanka case had a higher impact from the government policy than India case, and also had a more systematical relocation process. Other significant findings is that approximately 200 people in both countries answered—'No other option'. This number suggests the number of 'forced' relocation. The foregoing data shows that people are less picky about relocation site in India case because people can get a new house and new land with relocation. In addition, people have a chance to consider a suitable relocation site in Sri Lanka case. However, 200 people who answered—'No other option' did not have the option to decide their own new housing area in their relocated site.

Data showed that over 60 % people hope to relocation with same community members in both countries. Majority preferred that relocation should have the same community members so as to retain the original community networks. However, approximately 20 % people showed little concern on the relocation style in both countries (Table 7.5). Approximately 35 % people had not specific comments about relocating with same village members.

Data was analysed to identify the relocation issues and important factors of relocation. Ten key factors were hypothesised for relocation issues and important factors for relocation (Table 7.6). Relocation issues were based on earlier study and

Table 7.6 Result of logistic regression analysis (relationships with present life)

Variable	Definition	β	Significance probability	Exp (β)
1. Relocation issues				
Lost previous housing and land***	Yes=1, No=0	−0.911	0.000	0.402
Lost our history		0.128	0.602	1.137
Occupation/school**		−0.689	0.015	0.502
Aid, support from Gov. or NGOs		0.268	0.488	1.307
Access, transportation		−0.398	0.148	0.672
Culture/religion		1.095	0.076	2.988
Human relationships		0.026	0.924	1.027
Women's life**		1.546	0.018	4.694
Infrastructure***		−1.001	0.000	0.367
Community activities***		2.137	0.000	8.471
2. Important factors for relocation				
Relocate with same community members***	Yes=1, No=0	−0.655	0.009	0.519
Gov's management		0.253	0.389	1.288
NGO's management		0.455	0.225	1.576
Money		−0.019	0.955	0.981
Women's activity		1.002	0.061	2.724
Community activity***		1.767	0.001	5.854
Infrastructure**		0.457	0.031	1.580
Culture, religion		0.464	0.243	1.591
Increasing life level, life condition		0.290	0.388	1.336
Return previous life condition***		−2.368	0.000	0.094

(df=1)
Source: Field survey by the authors
Note: *** and ** represent statistical significance in the level of 1 % and 5 %, respectively
Software: SPSS version 21

results from field survey. It has been noted that in addition to physical factors such as ownership of housing and land, social factors such as human relationships, culture and religion as well as mental factors such as memories were significant factors in relocation. Similarly, impact factors of relocation were analyzed by logistic regression analysis.

This study selected two indicators to study relocation issues. The first indicator is the comparison of the previous and present life among the relocated people. Relocated people were to answer which life they prefer after relocation (previous life=1, present life=3). The other indicator is problem raised after relocation, as listed in Table 7.6. The relocated people were asked to answer yes, it is problem=1 or no, it is not problem=0. Other important factors for relocation also were analyzed. Dependent variable suggests that relocated people prefer to previous life even after relocation to the new site.

Results of logistic regression analysis suggests that 'Lost previous housing and land (β=−0.911, p<0.01)', 'Occupation/School (β=−0.689, p<0.05)', 'Women's life (β=1.546, p<0.05)', 'Infrastructure (β=−1.001, p<0.01)' and 'Community activities (β=2.137, p<0.01)' are significant among those that preferred the present life. High scores on 'Lost previous housing and land', 'Occupation/School', and 'Infrastructure' were seen among those who preferred previous life. Meanwhile, high score on 'Women's life' and 'Community activities' suggests to present life. These five factors are related with relocation issues for relocated people. In order that relocated people prefer the relocated area, these five factors are important. Low quality infrastructure, distance from work place or school and lost of ownerships is directly connected to economic well being, therefore, considered a relocation 'issue'. These three factors have changed and introduction of activities within the new relocation site helps improve the communication within the community.

'Relocate with same community members (β=−0.655, p<0.01)', 'Community activities (β=1.767, p<0.01)', 'Infrastructure (β=0.457, p<0.05)' and 'Return previous life condition (β=−2.368, p<0.01)' were found to be significant factors in relocation. 'Community activities' and 'Infrastructure' had high score suggesting its significance in influencing people to prefer the present life. For sustainable resettlement, it is important that it reflect people's evaluation. Through this data analysis, 'Infrastructure' and 'Community activities' are important factors. People do not only hope to get good infrastructure in relocated site. They also expect to start some of activities with other members. Relocation and resettlement after disaster is one way of starting life anew for them. Therefore, relocated people have expectations on their new life condition, which includes good infrastructure, new relationships with other members. In addition, minimal economic impact is another factor that is considered by the relocated people. Previous study suggested that 'Human relationship' is an important factor in relocation, however, in this survey, this factor was not found to be a significant relocation issue nor an important factor in relocation. Moving forward, if support groups continue to focus on 'Human relationships' as an important in relocation projects, it may be a possible hinderance for the relocated community in starting their life anew in the relocated area.

7.6 Conclusion

This study purpose was to identify the relocation problems and identify the critical factors required to achieve sustainable settlement. In the previous study of relocation after disaster, issues that were pointed out were relocation distance, land use with relocation planning by government and support organizations, infrastructure, loss of environment and memories, political problems and human relationships and so on. Prior to considering the 'sustainability' of relocation, support organization and local government officers, who are the decision makers, should consider what sustainable relocation is. Sustainable resettlement should be been determined by how long

Table 7.7 Critical factors for sustainable resettlement

Positive	Culture/religion, women's life, community activities, infrastructure
Negative	Lost previous housing and land, occupation/school, infrastructure
	Relocate with same community members, return previous life condition

relocated people are staying in the relocated land, but rather, how satisfied the relocated people feels living in the relocated area. To achieve sustainable resettlement, the affected people and support organizations needs to consider what are the most important relocation issues as well as the appropriate factors to support the community.

Result of data analysis showed that positive and negative impacts. Four factors as 'Culture/Religion', 'Woman's life', 'Community activities' and 'Infrastructure' have positive impact to relocated life. It means that if policymaker or support staffs consider relocation plans with these factors, there is lower tendency of the people wanting to return to their previous land. On the other hand, five factors namely 'Lost previous housing/land', 'Occupation/School', 'Infrastructure', 'Relocate with same community members' and 'Return previous life condition' have negative impact, which is connected to people's tendency of returning to previous land. In addition, relocation with same community members and policy of returning previous life condition will influence the wish of returning to previous land. Result of logistic regression analysis showed that relocation issues and important factors in India and Sri Lanka (Table 7.7).

'Infrastructure' has both impacts to relocated life. It is an important factor to achieve sustainable settlement. This factor is related to the positive/negative impacts in relocated life. Road construction is a transportation factor, that helps connect people in the previous or neighboring communities. Water and electricity will impact the continuity of daily life before and after relocation. Following relocation, it is inevitable for the relocated people to compare their present life with their previous life. Therefore, if the relocated people experience inconvenience in transportation to work place or school, poor infrastructure, concerned in the previous housing and land, they will not be satisfied with their present life.

It is important that the relocated people hold hope in the continuation of their daily lives in the relocated area. Following a disaster, people creates new network and participates in community activities so as to continue their daily livelihood. 'Human relationship' has long been seen as one of the main relocation issue. However, in this study, findings showed that this issue was not significant. Instead, 'Infrastructure' and factors causing inconvenience in their daily lives were of greater importance.

These critical factors can separate two groups (Fig. 7.3) such as to promoting with reconstruction work in the relocation site, and to promoting implementation after relocation. It should be addressed at different phase. For viability of sustainable resettlement, policy maker should consider these factors, stage of starting recovery support to relocated people along with phase of needs, and both groups should be include in recovery plan.

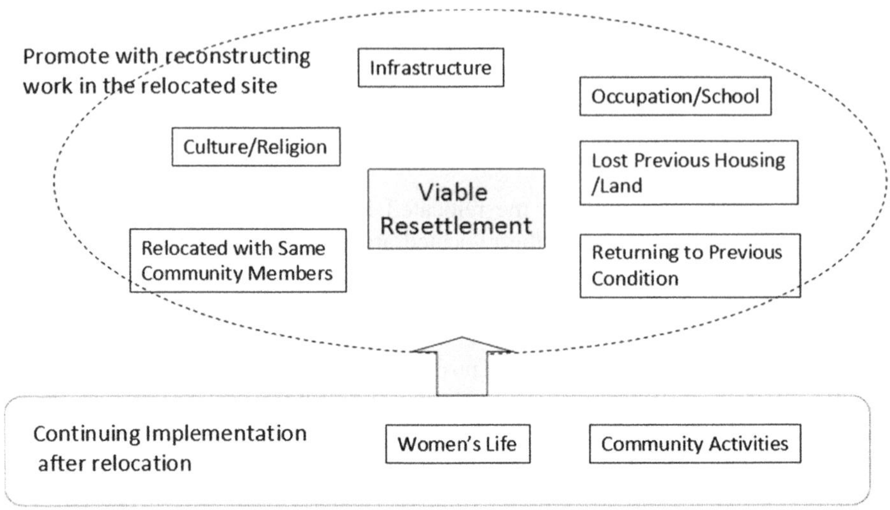

Fig. 7.3 Critical factors for resettlement in post-tsunami disaster

Acknowledgement The first author acknowledges the support of JSPS (Japan Society for Promotion of Science) for the fellowship to conduct this research. The authors are thankful to Prof. R. R. Krishnamuty of University of Madras, Mr. Shanmugan and Mr. Ilayaraja of the University of Madras and Ms. S. Kabita for their assistance and support in the field work, and to the people of the villages for their spontaneous response to the questionnaires. And thankful to Dr. Deepthi Wickramasinghe, in the University of Colombo, and Dr. Vinya Aryaaratne and Mr. Nandana in Sarvodaya which one of the local NGO in Sri Lanka for their assistance and support in the field work, and to the people of the villages for their spontaneous response to the questionnaires.

References

Ando A, Sasaki Y, Akatani R, Miura T (2000) A historical study on tsunami disaster prevention of the coastal area in Iwate. J Jpn Soc Civil Eng 693:1–11

Aota R, Hokugo A, Matsushita T (2006) A study of support of rebuilding houses for victims after the Sumatra tsunami disaster in Sri Lanka. Rep Res Center Urban Saf Secur Kobe Univ 10:177–195

Asian Development Bank (1995) Involuntary resettlement http://www.adb.org/sites/default/files/pub/1995/involuntary_resettlement.pdf. Accessed 8 June 2012

Badri SA, Asgary A, Eftekhari AR, Levy J (2006) Post-disaster resettlement, development and change: a case study of the 1990 Manjil earthquake in Iran. Disasters 30(4):451–468

Blaikie P, Cannon T, Davis I, Wisner B (1994) At risk: natural hazards, people's vulnerability and disasters. Routledge, London

Boen T, Jigyasu R (2005) Cultural considerations for post disaster reconstruction post-tsunami challenges. In: UNDP Conference. www.adpc.net (http://www.adpc.net/irc06/2005/4-6/TBindo1.pdf). Accessed 8 June 2012

Center for Research on the Epidemiology of Disasters (CRED) (2012) The International Disaster Database (EM-DAT). http://www.emdat.be/. Accessed 02 Nov 2013

Chiba T (1995) A Study on rearrangement of residential zone at villages in Hokkaido: Part 1: settlement relocation at Kitahiyama. In: Summaries of technical papers of annual meeting architectural

Institute of Japan. E-2, architectural planning and design II, dwelling houses and housing sites, rural planning, education, pp 479–480

Cuddalore Collecterate (2005) http://www.cuddalore.tn.nic.in/. Accessed 20 May 2010

Chen E, Ho E, Jallad N, Lam R, Lee J, Zhou Y, Del Re D, Berrios L, Nicolino W, Ratti C (2005) Resettlement or resilience? The Tsunami safe(r) project. In: International symposium disaster reduction on coasts scientific sustainable holistic accessible. Monash University, Melbourne

Dikmen N (2010) Relocation or rebuilding in the same area: an important factor for decision making for post-disaster housing projects. In: Proceedings of international conference. http://www.grif.umontreal.ca/pages/DIKMEN_Nese.pdf. Accessed 5 Sept 2010

Grant-in-Aid for Special Purpose Research Report (2005) Ministry of Education, Culture, Sports, Science and Technology, Japan. (Grant Number 16800055, Leader Yoshiaki Kawata, Kyoto University)

Higashi M (2000) The analysis of the case of planning for moved village which suffered from flood of Changjiang. Trans Rural Plan 2:121–126

Imura M, Shaw R (2009) Challenges and potentials of post-disaster relocation. Asian J Environ Disast Manag 1(2):197–219

Imura M, Shaw R, Takeuchi Y (2011) Adaptive and sustainable post tsunami human resettlement in Sri Lanka and India. Research Publishing, Singapore

Ishikawa E, Fukutome K, Minai N, Ikeda H, Nakabayashi K (2004) A study about housing relocation policy for earthquake recovery and living environment in Turkey: comparative studies on Urban reconstruction process after earthquake among Turkey, Taiwan and Japan: Part 17. In: Summaries of technical papers of annual meeting architectural Institute of Japan. F-1, Urban planning, building economics and housing problems, pp 345–346

Ishikawa E, Ikeda H, Nakabayashi K (2006) Conditions on reaching consensus on recovery project with housing relocation in depopulated mountainous area and issues for sustainable village: focusing on various attributes of suffered families on past examples in Nigata prefecture. In: Summaries of technical papers of annual meeting architectural Institute of Japan. E-2, architectural planning and design II, dwelling houses and housing sites, rural planning, education, pp 451–452

Ishikawa E, Ikeda H, Nakagayashi K, Minai N (2007) Recovery process of housing and suffers' lives in landslide areas: recovery project with group relocation and landslide predictions after the Niigata-ken Chuetsu earthquake. J Hous Res Found 34:303–314

Kimura T, Takahashi K (2004) A study on the reconstruction process of dwellings after a volcanic disaster: a case study in Kamikoba district, Shimabara City. J Nat Disast Sci 23(2):229–244

Maki N, Hayashi I, Hayashi H (1999) Disaster management in the Papua New Guinea Tsunami disaster on 17 July, 1998: recovery process of the social flows and stocks from disaster. J Soc Saf Sci 1:195–200

Maki N, Miura K, Kobayashi M, Hayashi H (2003) Relocation process at resettlement site after 1992 flores earthquake and tsunami disaster. J Archit Plan Environ Eng (566):1–8

Maruyama J (2003) The impacts of resettlement on livelihood and social relationships among the Central Kalahari San. Afr Stud Monogr 24(4):223–245

Menoni S, Pesaro G (2008) Is relocation a good answer to prevent risk? Criteria to help decision makers choose candidates for relocation in areas exposed to high hydrogeological hazards. Disast Prev Manag 17(1):33–53

Miura K, Maki N, Kobayashi M (1996) Relocation to restoration housing due to the eruption of Mt. Unzen-Fugendake: natural disaster and environmental transition, Part I—transactions of AIJ. J Archit Plan Environ Eng (485):87–96

Nakagawa Y, Maeda M, Kanki K, Takada M (2008) Consideration on issues of the relocation program through typology of resettlement sites: analysis on location and surrounding environment of resettlement sites constructed in Sri Lanka after the Indian Ocean Tsunami 2004 with GIS-Part 2. In: Summaries of technical papers of annual meeting architectural Institute of Japan. F-1, Urban planning, building economics and housing problems, pp 327–328

Ngai WC (1995) Flood disaster management in Malaysia: an evaluation of the effectiveness of government relocation schemes. Disast Prev Manag 4(4):22–29

Ninomiya K, Murosaki M, Hokugo A, Koshiyama K (2000) A report about the housing reconstruction after Unzen-Fugendake eruption. In: Summaries of technical papers of annual meeting architectural Institute of Japan. F-1, Urban planning, building economics and housing problems, pp 1247–1248

Oliver-Smith A (1991) Successes and failures in post-disaster resettlement. Disasters 15(1):12–23

Perry RW, Lindell MK (1997) Principles for managing community relocation as a hazard mitigation measure. J Contingencies Crisis Manag 5(1):49–59

Ragsdale T, Oberhagemann K, Faisal MA (2009) Involuntary meets disaster mitigation resettlement: a case study from erosion mitigation on the lower Brahmaputra/Jamuna River in Bangladesh. http://www.his.com/~mesas/articles/Involuntary-mitigation.pdf

Rodriguez H, Wachtendorf T, Kandra J, Trainor H (2006) A snapshot of the 2004 Indian Ocean tsunami: societal impacts and consequences. Disast Prev Manag 15(1):163–177

Sonogashira S, Murosaki M (2000) The influence on situation of stress comes from residential removing: #3. The relation between the elements of residential removing and stress. J Asian Architect Build Eng (Kinki) 40:281–284

Suzuki S, Maki N, Furusawa T, Hayashi H, Kawata Y (2007) Social responses and recovery processes in the April 2, 2007 Earthquake Tsunami disaster of the Solomon Islands. J Jpn Soc Nat Disast Sci 26(2):203–214

The World Bank (2010) To relocate or not to relocate. In: Jha AK (ed) Safer homes, stronger communities, a handbook for reconstructing after natural disasters. The International Bank for Reconstruction and Development/The world Bank with support of the Global Facility for Disaster Reduction and Recovery, 370 pp

United Nations Development Programme (2004) Reducing disaster risk: a challenge for development. United Nations Development Programme Bureau for Crisis Prevention and Recovery. www.undp.org/bcpr. 146 pp

United States Agency for International Development (2007) How resilient is your coastal community? A guide for evaluating coastal community resilience to tsunamis and other hazards, U.S. Indian Ocean Tsunami Warning System Program, Bangkok. (http://www.crc.uri.edu/download/CCRGuide_lowres.pdf)

Yamori K, Yokomatsu M, Okumura Y, Shiroshita H, Kawata Y (2006) Social responses and recovery processes in the 2006 landslide disaster in Southern Leyte, Philippines [in Japanese]. J Jpn Soc Nat Disas Sci 25(1):99–111. http://ci.nii.ac.jp/naid/110004745191

Ye Y, Okada N (2002) Integrated relief and reconstruction management flowing a natural disaster. In: Second annual IIASA-DPRI meeting, integrated disaster risk management: megacity vulnerability and resilience. IIASA, Luxemburg

Chapter 8
Institutional Arrangements for Managing Large-Scale Recovery: Key Lessons from 2004 Tsunami

Sudhir Kumar

Abstract Disaster recovery programs, especially large-scale recovery programs, throw a number of challenges from planning to monitoring & evaluation and from implementation to exit. These challenges call for a robust but adaptive, central but localized and swift but consultative institutional arrangements, which is key for undertaking effective and efficient recovery program. The institutional arrangements need to be informed of scale of recovery, governance of the affected region, exiting capacity, internal and external stakeholders of recovery, socio-economic condition of the affected region, timeline, transition to development phase, etc. In order to undertake recovery programs in the aftermath of the Indian Ocean Tsunami 2004, different types of institutions were set-up by the five worst affected countries namely India, Indonesia, Maldives, Sri Lanka and Thailand to meet their objective of recovery. As the socio-economic and geographical condition, governance, impact of tsunami, country capacity, engagement of stakeholders, etc of these worst affected countries varied, so was their institutional arrangements for recovery. This chapter analyzes these institutions and draws key lessons for setting up institutional arrangements in different context. The chapter provides guidance, along with examples, on lead institution for undertaking recovery, approach for recovery, different models of recovery institution, key features of these institutions such as adaptibility, human resource, etc, leveraging local institutions and exist strategy for recovery institution. These lessons will help policy makers and managers of disaster recovery in setting up efficient and effective institution for managing recovery programs in the aftermath of large-scale disasters.

Keywords Climate change • Exit strategy • Institution • Leadership • Recovery • Tsunami

S. Kumar (✉)
United Nations Development Programme, Tacloban, Philippines
e-mail: sudhir.ise@gmail.com

R. Shaw (ed.), *Recovery from the Indian Ocean Tsunami: A Ten-Year Journey*,
Disaster Risk Reduction, DOI 10.1007/978-4-431-55117-1_8

8.1 Introduction

The world has witnessed some of the worst disasters in the recent times, which includes the Haiyan Cyclone of 2013 in Philippines, the Canterbury, New Zealand Earthquake of 2010–2012 the Great East Japan Earthquake of 2011, the Earthquake of 2010 in Haiti, the Sichuan Earthqauke of 2008 in China, the Cyclone Nargis of 2008 in Myanmar, the Pakistan Earthqauke of 2005, the Cyclone Katrina of 2005 in United States of America, the Indian Ocean Tsunami of 2004 and the Gujarat, India Earthquake of 2001.

8.1.1 Importance of Institution in Recovery and Reconstruction

In the aftermath of disasters, including mega disasters, massive recovery and reconstruction interventions are undertaken and some act decisively to reduce future losses, the others just return to status *quo ante* (National Research Council 2006). There are many examples of realized and missed opportunities. Recovery and reconstruction of Lisbon after the earthquake, fire and tsunami in 1755 by Marquis do Pombal is an example of realization of some of the post-disaster opportunities not only for development and future disaster mitigation (Alexander 2004) but also for political and economic consolidation (Dynes 2000, 2005). Recovery and reconstruction programs failing to build back better, despite aiming to do so, is evident in the case of 1976 Guatemalan earthquake (Bates 1982), Hurricane Mitch (Wisner 2004) etc. Similarly, Haiti and New Zealand experienced earthquake of a similar magnitude within a year of each other, yet due to broader institutional reasons investing in Port-au-Prince, Haiti is far riskier than investing in Christchurch, New Zealand (Smart 2012).

The success or failure of a recovery and reconstruction largely depends on the ability of the State to tackle the governance issues in reconstruction (Harvey 2009), as testified by the study of post-disaster reconstruction programs in different countries. States which were pro-active towards tackling governance issues through appropriate mechanisms could build-back-better and the others missed the window of opportunity provided by the disaster (Thiruppugazh 2014). The institutional arrangement for recovery is very important as recovery calls for massive development and reconstruction compressed in time and space (Olshansky et al. 2012). Also in recovery and reconstruction a range of actors from government departments to UN Agencies and from donors to NGOs/CBOs are involved, hence coordination is a major challenge. For example, in Indonesia in the aftermath of 2004 Tsunami, 56 countries and 800 organizations were engaged (Fan 2013) and it calls for cooperation and collaboration among stakeholders.

Another key challenge in recovery is to strike a balance between speed and quality and consultative process and quick decision making in a compressed time frame

(Olshansky et al. 2012), which requires robust governance mechanism for recovery. Also, moblization of resources especially external resources is highly dependent on transparency and accountability. For example, donors appreciated the Government of Indonesia's move to create a separate government agency expressly to handle the reconstruction, and were reassured by its dual-role function. The Economist reported on May 26, 2005, that BRR was "a promising new government body like no other i.e., clean, efficient, well managed and results-oriented" (Subekti 2009).

Hence, the enormous challenge of implementing an emergency reconstruction program requires an efficient and responsive institutioal mechansism (GSDMA 2001). It has more relevance in the era of climate change as it is most likley that some of the future recovery and reconstruction interventions will be of unprecedented scale and may throw many challenges to future recovery and reconstruction managers as climate change is reconfiguring the hydro-meteorological hazards and thus increasing disaster risk. In case of cylcone, global warming causes sea level rise, which amplifies storm surges and flooding associated with cyclone. Also, global warming has increased the amount of moisture in the air, causing more rainfall and amplifying flooding during cyclone, and lastly, warmer oceans are fuel for cyclone (The Guardian 2013). There is an evidance, based on data since 1950, that climate change has changed the magnitude and frequency of some climate extremes in some global regions already. As climate change becomes more dramatic, its effect on a range of climate extremes will become increasingly important and will play a more significant role in disaster impact (Climate and Development Knowledge Network 2012).

The 'recovery and reconstruction', often mentioned as 'recovery', is the critical link between response and relief and long-term development, and considered to be the most important as well as challenging phase as in this phase a series of decisions are made with limited available information and time, which have long-term impacts and requires complex coordination and deep local understanding (BRR 2009). Also, it offers a window of opportunity to 'build back better and safer'. In the past, some of the recovery interventions also helped in initiating peace process in conflict affected zone including the case of Aceh, Indonesia (Fan 2013). Thus, it is very important to harness the opportunity offered by recovery and governance mechanism or the institutional arrangements for undertaking recovery is of utmost importance as it is linked from planning to implementation and from engagement of community in recovery process to linking recovery process to long-term development process.

8.2 The 2004 Indian Ocean Tsunami

On 26th December 2004, the earthquake induced tsunami occurred in the Indian Ocean which left countless communities—from Banda Aceh to Batticalao, Puntland to Phang Nga and Noonu to Nagapattinam—without homes or livelihoods, swept way critical infrastructure and damaged large swaths of coastal area. In all, more than 228,000 people in 14 countries died and it is considered to be one of the

Table 8.1 Impact of Tsunami 2004 in Maldives, Sri Lanka and Thailand

Impact	Maldives	Sri Lanka	Thailand
Number of deaths	102	35,399	8,345
Injured	2,214	23,176	
Displaced	13,000	480,00	
Total damage and loss in million USD	470.1	1,316.5	405.2
Houses affected	NA	114,069	4,806

Source: International Recovery Platform (2014)

worst disasters in the recent history. The five worst affected countries were India, Indonesia, Maldives, Sri Lanka and Thailand. A snapshot of impact in Maldives, Sri Lanka and Thailand is at Table 8.1.

8.3 Approach and Institutional Arrangements for Recovery by Countries

These five worst affected countries adopted their own approach for recovery including institutional arrangements for managing recovery as impact of disaster, governance, socio-economic context, etc. varied. An overview of the approach to institutional arrangements in the five worst affected countries is as follows:

8.3.1 India

The Government of India set up a core group for tsunami recovery and rehabilitation at the Planning Commission, a high-level committee, and an inter-ministerial central team to facilitate assessment, development, implementation, and monitoring of the immediate, short, and long-term programs. The Tsunami Rehabilitation Program Report finalized by the Planning Commission, in consultation with the states (provinces), broadly outlined the impact of the tsunami on the Indian subcontinent; defined the guiding principles for the program; summarized information about damage assessment provided by the provinces (states/union territories) and the assistance sought for relief, reconstruction and rehabilitation and provided assistance. It highlighted the needs of areas where financial assistance was required for reconstruction and rehabilitation, and provided a Program Implementation Plan along with an organizational structure for the program (Fig. 8.1). The program sought to apply the following guiding principles in its implementation—environment friendly, egalitarian approach; private sector and nonprofit sector participation; gender empowerment; child welfare; structural and nonstructural rehabilitation measures; and transparency and accountability (ADB 2008).

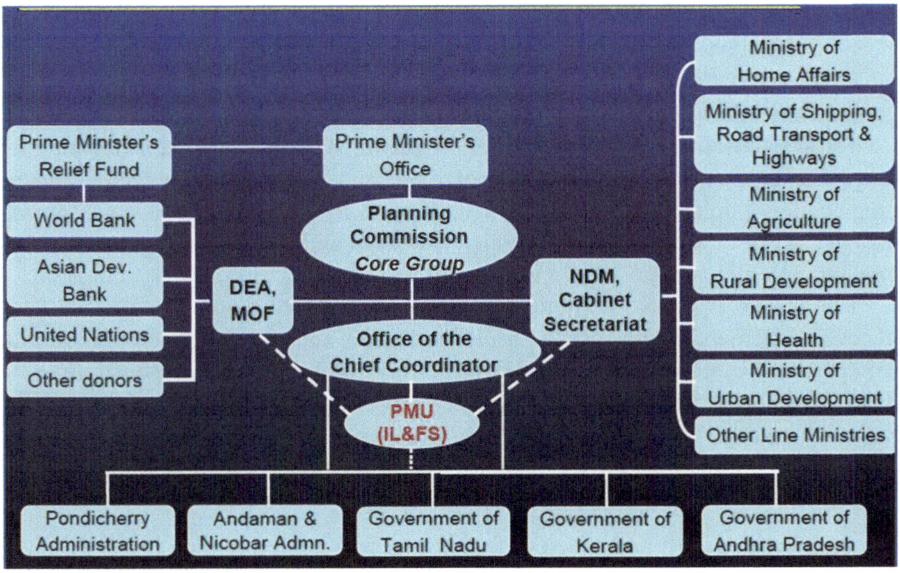

Fig. 8.1 Institutional structure for implementation of recovery (2004 Tsunami), India. *Source*: Planning Commission, India (2005)

The program implementation responsibility were at two levels, namely:

- National level: It was responsible for fund allocation, part implementation and monitoring.
- Provinces (States/Union Territories) were responsible for implementation of the program.

Institutional Arrangements at Province Level: In order to implement the recovery program, Tamil Nadu government (provincial government) established the Project Management Unit (PMU) through Government Order, under the existing government mechanism i.e. under the Revenue Administration, Disaster Management and Mitigation Department. The PMU is headed by Project Director/Officer on Special Duty (RR) and is supported by staff with required expertise (Government of Tamil Nadu 2013). The existing district mechanism was used for implementation of the program. Supporting staffs along with the specialists were also appointed on contract basis to assist the Project Director.

8.3.2 Indonesia

The Government of Indonesia through its Ministry of National Development Planning prepared the 'Master Plan for the Rehabilitation and Reconstruction for the Region and People of the Provinces of Nanggroe Aceh Darussalam and Nias Islands' for medium-term rehabilitation and reconstruction efforts for Aceh and

Nias islands and released in April 2005. The Master Plan was prepared taking into account the 2004–2009 National as well as Regional Medium-Term Development Plan. The plan covered the 5-year period from 2004 to 2009 in line with the medium-term development plan.

The key objectives were to draft the post-earthquake-and-tsunami action plan for the rehabilitation and reconstruction of Aceh and Nias Islands, which could be immediately implemented by concerned agencies in coordinated and synchronized manner. Also, objective was to mobilize resources and dissemniate recovery information in public domain.

The Master Plan also laid out the outline of the institutional arrangements for recovery and mentioned to establish a rehabilitation and reconstruction agency with comprehensive, centralized, coordinated duties, authority and responsibilities to plan, implement, control and evaluate such process based on the principle of good, effective, transparent and accountable governance.

It also spelt that the recovery agency would directly report to the President and would include:

- *Advisory Board*: It had advisory function of providing direction in the formulation, planning and implementation of the rehabilitation and reconstruction process. Members of the Advisory Board are the representatives of various stakeholders, such as the elements of the communities concerned, academicians / universities, government—both at the central and regional levels, and other elements supporting the advisory function.
- *Supervisory Board*: It was entrusted with the supervisory function and members comprised individuals with sufficient supervisory skills including national figures and independent Acehnese opinion leaders. Moreover, it was possible for the representatives of donor countries/agencies to become members of the Supervisory Board.
- *Implementing Agency*: It was tasked with an implementation function to, among other things: (1) formulate operational strategy and policy; (2) prepare an action plan and conduct activities, (3) conduct rehabilitation and reconstruction activities for certain projects based on agreements with other agencies/institutions; and (4) ensure that the rehabilitation and reconstruction funds are utilized in a manner that highly regards integrity and is free from criminal acts of corruption (Republic of Indonesia 2005).

It also mentioned that the Implementing Agency can be expanded based on the needs, with the aim of optimizing performance in implementation. The structure of BRR (implementing agency) evolved overtime and organogram of BRR of 2005 is at Fig. 8.2.

8.3.3 Maldives

The Government of Maldives set up a Ministerial Committee and Task Force. It also established the National Disaster Management Center to facilitate response and coordination. The Ministry of Defence coordinated the overall relief effort and the

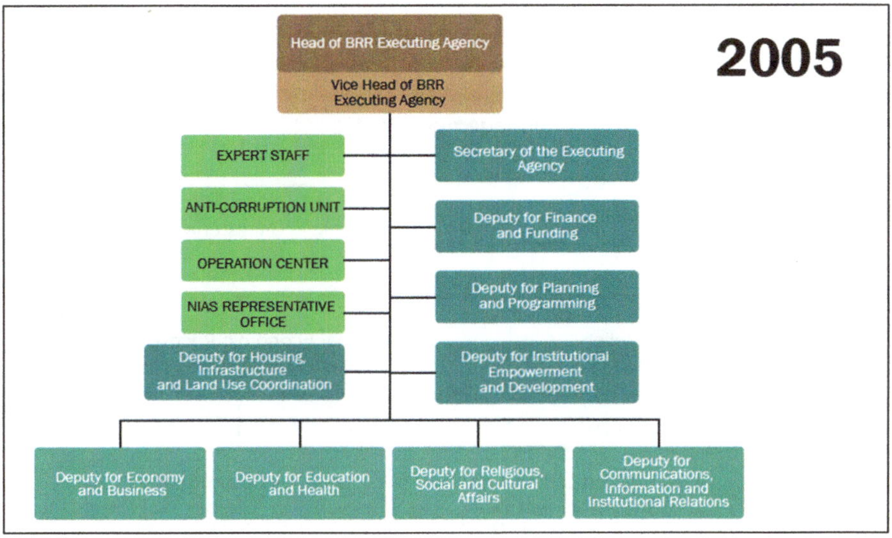

Fig. 8.2 Organizational structure: BRR (Agency for R&R, NAD), Indonesia. *Source*: BRR (2009)

Ministry of Finance and Treasury coordinated the donor assistance while the Ministry of Planning and Development coordinated data gathering and long term response. The Center was the focal point for all response, relief and recovery activities (Maldives 2007).

The organizational structure of National Disaster Management Center is at Fig. 8.3.

8.3.4 Sri Lanka

The Reconstruction and Development Agency (RADA) was established in November 2005 (UNDP/RADA 2006) with the vision to create a single government agency to focus on reconstruction and development issues across all sectors and stakeholders in affected areas. It was under the Presidential Secretariat. It aimed to *Build Back Better* the affected communities with mission to accelerate and coordinate the reconstruction and development of areas affected by manmade and natural disasters, and ensure sustainable recovery of properties, livelihoods, industries and infrastructure to mitigate the effect of such disasters in the future.

In order to achieve its mission, RADA also set objectives, which included optimization of resource usage, protection of vulnerable groups, compliance of core principles and accelerate sustainable reconstruction and development.

It is important to mention that RADA was established, functionally replacing the tsunami organizations such as The Task Force for Rebuilding the Nation (TAFREN), Task Force for Relief (TAFOR) and Tsunami Housing Reconstruction Unit (THRU) as well as a significant part of the former Ministry of Relief, Rehabilitation and Reconstruction (Tsunami Information Center 2006).

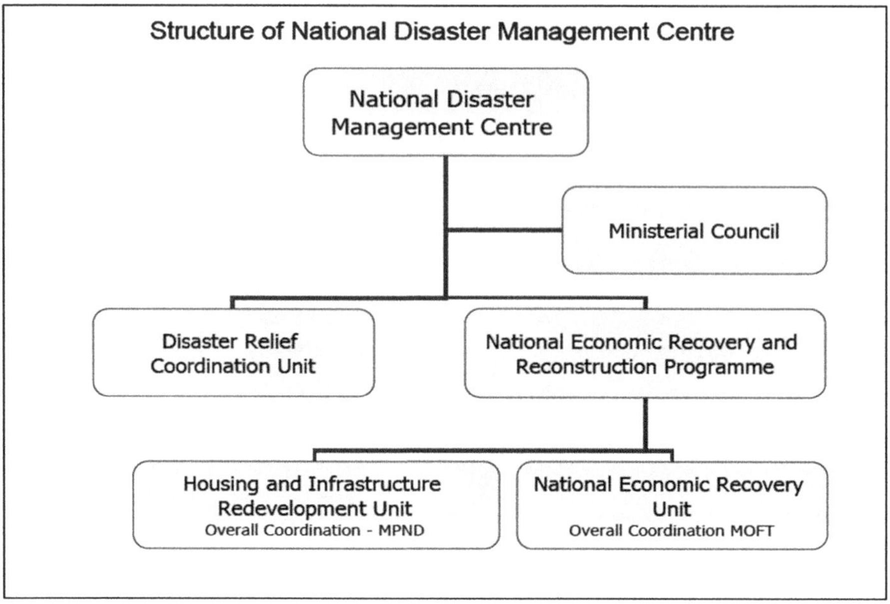

Fig. 8.3 Organizational structure of NDMC, Maldives. *Source*: Maldives (2007)

8.3.5 Thailand

The recovery in the aftermath of 2004 Indian Ocean tsunami was primarily undertaken by the Government. The legal framework and resulting structural arrangements for disaster management were relatively clear, national disaster management was under the direct authority of the Prime Minister, who directed various bodies of administration through the Ministry of Interior and the National Defense Council. This structure was replicated at provincial and district levels under the authority of the Governor and of the District Civil Defense Director. The Department of Disaster Prevention and Mitigation (DDPM) under the Ministry of Interior, being the principal government agency for disaster management coordinated the recovery.

In January 2005, the government agencies were given the responsibility for recovery work. Three Deputy Prime Ministers were assigned responsibilities for long-term coordination of the recovery work- overseeing the recovery of the environmental sector, the recovery of livelihoods and the establishment of a tsunami early warning system (Office of UNRC, Thailand). The arrangement is listed in Table 8.2.

Table 8.2 Agency-wise recovery responsibilities, Thailand

Agency	Responsibility
Provincial governments	Cleaning up of affected beaches and streets
Public Works Department	Infrastructure repair and maintenance
Ministry of Natural Resources	Clean water sources and monitoring water quality and supply
Agricultural Ministry and Thai Military	Repair and building of fishing boats
Finance Ministry	Loans for business recovery initiatives
Education Ministry	Rehabilitating schools and free board and education to all orphans
Interior Ministry with the Ministry of Social Development and Human Security and the Thai Military	Temporary and permanent housing

Source: Office of UNRC, Thailand

8.4 Key Lessons for Institutional Arrangements for Recovery

The five worst affected countries of 2004 Indian Ocean tsunami adopted their own approaches to manage recovery including different institutional arrangements for managing recovery. An analysis of the institutional arrangements is at Table 8.3

These processes have generated a plethora of learning, which can benefit the future recovery policy workers and managers for setting-up institutional mechanism for recovery. Some of the key lessons are following:

8.4.1 Government Led Institutional Mechanism for Recovery

In all five worst affected countries, the institutional arrangements for recovery were led by the respective government: Badan Rehabilitasi dan Rekonstruksi (BRR) led the recovery in Indonesia, Provincial governments led recovery in India, Provincial governors along with Department for Disaster Prevention and Mitigation (DDPM) led recovery in Thailand, etc. The government leadership helps in implementation of recovery program by all actors, external as well as internal, within the framework of the national recovery framework. It also ensures engagement of the government from the start and helps in ensuring accountability thus, sustainabilty of the recovery outcomes increases many fold. For example in Indonesia, the Government ensured the transition of unfinished projects, before the closure of BRR in April 2009, to line ministries. These projects were funded through various sources such as the Indonesian government, the Asian Development Bank, the World Bank, Japan Bank for International Cooperation, etc. (Subekti 2009).

Table 8.3 Analysis of recovery and reconstruction institutional arrangements

Key feature	Indonesia	India	Maldives	Sri Lanka	Thailand
New institution constituted for managing recovery	Yes	No	Yes	Yes	No
New recovery institution was permanent in nature	No	–	Yes	No	–
Organizational structure went through major change	Yes	No	Yes	Yes	No
Legal status and political support	Yes	Yes	Yes	Yes	Yes
Lead agency for managing recovery	Central and Provincial government led recovery	Provincial government led recovery	Central government led recovery	Central government led recovery	Provincial government led recovery

Source: Author (based on Sect. 8.3)

Also, the leadership role of host government in recovery has been well articulated in various resolutions or declarations such as the UN Humanitarian Resolution which states that "the affected State has the primary role in the initiation, organization, coordination, and implementation of humanitarian assistance within its territory" (United Nations 1991). The ASEAN Agreement on Disaster Management and Emergency Reponse has also vested the primary responsibility with the affected country (ASEAN 2009). Even in the case of countries with limted capacity and fragile and uncertain socio-economic context, country's ownership to recovery process should be respected (UNCTAD 2010).

8.4.2 'Single-Window' Approach

Most of the disasters affect across sectors and hence recovery calls for multi-sectoral approach and which in turns leads to multiple set of actors engaged in recovery. For example, in Aceh–Nias, Indonesia, there were more than 800 reconstruction actors, so the chances of geographical and sectoral overlaps are high in the absence of coordinated approach. And, more than 20 national Red Cross and Red Crescent societies were active in recovery in Indonesia. BRR was the nodal point of coordination for recovery in Indonesia. Also, the UN Office of the Recovery Coordinator was established to coordinate UN activities and was single point of contact for the BRR to the UN. Similarly, IFRC had a staff member at the BRR to coordinate its own member country activities.

Similarly, in Tamil Nadu, India the district governments were given responsibility for coordination of reconstruction as single focal point.

8.4.3 Evolutionary Institutional Mechanism with In-Built Flexibility

The institutional mechanism for recovery should have in-built flexibility and should evolve as recovery program generally happens in complex, challenging and dynamic environment.

For example, in the case of Sri Lanka, several response mechanisms were enacted to expedite the relief effort immediately after the tsunami. A Center for National Operations was formed under the president to liaise with all involved with the emergency response and coordinate the relief effort. Three task forces were also formed under the president: Task force for rescue and relief, Task force to rebuild the nation and Task force for logistics, law and order, while disaster management authorities were appointed at district levels to coordinate local relief efforts. The structure of the coordination mechanism changed subsequently with the disbanding and/or merger of different agencies during the recovery and reconstruction period. In November 2005, all government agencies involved with tsunami recovery and reconstruction were merged to form a single agency, the Reconstruction and Development Agency (Harsha et al. 2007). In Indonesia, BRR structure changed every 6–12 months to meet the changing needs and the organisational struture was evolutionary to meet the needs (Kumar 2013).

8.4.4 Selecting Suitable Institutional Model

In order to undertake recovery, the five worst affected countries of 2004 tsunami set-up institution or designated existing institution as per the need. However, these institutions can be broadly clubbed under two categories:

- Through existing government institutional set-up
  Under this model, the recovery and reconstruction is coordinated and implemented by the existing line ministries and departments. For example, in Thailand, the Tsunami 2004 recovery was managed by the existing government departments and the provincial authorities. Similarly, the Tsunami 2004 recovery, in Tamil Nadu, India was coordinated and implemented by existing administrative arrangements by devolving significant authority to local administrators.

  This model suits to the situations where significant prior experience of recovery management exists or where the strong disaster management system is in place or impact of disaster is limited and thus recovery intervention is limited. It is also important to note that its success is very much dependent upon there being a high level of preliminary planning or well-practiced administrative and operational procedures. And administrative planning includes planning for additional capacity in government departments to cope with the surged demands, which may stretch over a considerable time period.
- Through setting up new temporary or permanent agency
  Under this model, a new agency is established, which may be temporary or permanent in nature, and is tasked to undertake recovery. For example, in Indonesia, BRR was set-up for 4 years to complete the recovery and was closed after 4 years. Similarly, in case of Maldives, National Center for Disaster Management was established and which became permanent body. In case of Sri Lanka, RADA was constituted for recovery as temporary institution.

  The permanent agency has advantage that the capacity building of individual and agency on recovery is preserved and learning of past recovery can improve future recovery. As once temporary agency is dissolved, too often the experience and learning is lost, forcing governments to reinvent the wheel with each recovery. In case of temporary recovery agency it is a good option if few resources can be drawn from on-going government operations. It also presents a unified continuity particularly if it can draw non-partisan support.

8.4.5 Putting the Best Team as Recovery Is Not Business as Usual

The chaos of recovery, especially initial phase, requires reliance on formal as well as informal networks to win over key decision-makers and recovery actors, hence leadership role is very important. Also, the strong leadership set the tone for recovery. For example, BRR, Indonesia was led by Dr. Kuntoro Mangkusubroto, who served as Minister as well as Academia, had wealth of public and private sector experience and he in turn selected highly qualified deputies from private and public sector drawn from Aceh and other provinces. Similarly, in Tamil Nadu, India recovery, specialists were appointed on contract basis to assist the Project Director.

8.4.6 Political Support/Legal Authority for Recovery Agency

The agency managing recovery requires political support/legal authority to take key decisions for speedy implementation and has to work with the sectoral agencies, which has almost same *locus standi*. For example, in Sri Lanka, RADA was placed under the Presidential Secretariat. In Indonesia, BRR was established through Government Regulation and ratified by Law No. 10/2005 by Government of Indonesia, and thus BRR was at equal footing with line ministries. In Tamil Nadu, India, Government sanctioned the establishment of Project Management Unit headed by the Project Director for overall implementation of recovery through Government Order.

8.4.7 Exit Strategy for Recovery Institution

The recovery process creates a number of assets, institutions including the recovery agency. It is important to plan exit strategy for recovery institution at the start. For example, in case of Maldives, the National Disaster Management Center created during 2004 tsunami relief and recovery has continued and Disaster Management Bill is under discussion, which includes strengthening this institution. In case of Tamil Nadu, India, the Project Management Unit created under the existing department has been engaged in similar other interventions to harness its expertise. In Indonesia, BRR was closed as originally planned after 4 years in 2009 but to preserve its knowledge and expertise of R&R, BRR Institute has been created.

8.4.8 Leveraging Local Institutions

The marginalized groups which are mainly women, children, the urban and rural poor, ethnic minorities, etc. are sometimes left out in the recovery process even unintentionally. Under the recovery policy boundary or rigid nature programming, sometimes it is difficult to meet the special needs of the marginalized groups. The local institutions are better placed to address these issues in partnership with the authorities. For example, in 2004 tsunami, Sri Lanka Women's Coalition for Disaster Management (WCDM) played important role on gender issues in Tsunami recovery (Tsunami Evaluation Coalition 2006). In Batticaloa and Ampara districts in Sri Lanka, WCDM, which was initiated by Suriya, a local women's organisation in Batticaloa, in mid-January, played an important role in post-Tsunami relief and reconstruction work. The WCDM initially lobbied for a women's committee to be set up in every camp. The committees then identified the basic needs of women, such as private space, appropriate facilities (such as private bathing and toilet facilities) and access to supplies.

The WCDM also formed an action group called Gender Watch, involving local and international non-government organisations. The initiative enabled women to report domestic violence, sexual harassment and discrimination to the group. Gender Watch documented violations in the camps and distributed the information to international agencies and the government. Remedial action taken included: suspending a government officer for violations; protecting five orphaned children; ensuring women have access to oral contraceptives; facilitating access to the police in the case of domestic violence; providing temporary shelters to single women who were originally excluded because they did not possess the right papers; and registering women for the provision of ration cards to provide them with access to goods.

Similarly, in Tamil Nadu, India, South Indian Federation of Fishermen Societies, a non-governmental organization working in the marine fisheries with over 6,000 member fishermen, played a major role in relief and rehabilitation. It worked on restoring and further enhancing fisheries livelihoods in many areas.

8.4.9 *Institutional Mechanism at Local Level/Ground*

The field level institutions execute the recovery program and thus play critical role in its success. The activities which local government can undertake in recovery should be undertaken by it (local government) and central mechanism should create the enabling environment. The local government might have limited capacity compared to central government and the capacity of the local government might have been further eroded due to the disaster, however, it is important to use its existing capacity and identify the means to reinvigorate and strengthen local capacity through international assistance and the recovery process.

The local level institutional mechanisms are important in recovery as

- They are more familiar with local social, economic, environmental and political systems, hence better placed to assess the needs of the affected communities and devise more relevant recovery programs.
- The local government leaders (particularly when elected) are often more accountable to their constituencies than central governments. This provides them with an incentive to ensure greater quality of recovery services.

- They are typically responsible for development planning of their respective constituencies.
- Being on-site makes it better placed to coordinate recovery efforts.

In case of Indonesia Tsunami 2004 recovery, BRR was established in Aceh, but after a year BRR realized the need to open district level offices. As part of its initial strategy the BRR focused on developing the capacity of local governments to manage their affairs and deliver effective services. In addition to this, the BRR also focused on enhancing the effectiveness of the relationship between central government agencies and local governments.

8.5 Conclusion

The 'Handbook for Reconstructing after Natural Disasters' of the World Bank mentioned that 'Every reconstruction project is unique' as the last word (Jha et al 2010). However, learnings from the past recovery can provide guidance to future recovery. The recovery program undertaken in the worst affected countries of 2004 Indian Ocean Tsunami provided a number of lessons including valuable lessons on institutional arrangements for recovery. The key lessons for institutional arrangements for recovery are reconstruction are following:

- Institutional arrangements or governance of recovery and reconstruction is the most important determinant of a successful recovery program.
- The recovery process including the institutional arrangements should be led by the host/disaster impacted government.
- The selection of institutional model for recovery needs to be determined by a number of factors such as scale of recovery, governance of the affected region, exiting capacity, stakeholders of recovery, socio-economic condition of the affected region, timeline and transition to development phase.
- The institution mandated for recovery should be constituted through legal provision and it should have the highest level of political backing.
- Some of the inherent features of a successful recovery institution includes leadership, professional team, flexibility and single-window approach.
- The recovery institution should leverage the network, rapport and goodwill of the local institution.
- The exit strategy should not be developed at the closing phase but at initial phase of recovery.
- Lastly, as each recovery is unique so is the institutional arrangements, thus recovery institution should be adaptive and evolutionary.

It is important to consider these lessons while setting-up institutions for managing recovery in future as these valuable lessons were learnt at a very high price and most importantly, governance or institutional arrangements is key to success of any recovery.

References

Alexander D (2004) Planning for post-disaster reconstruction. Paper presented at 2004 international conference and student competition on post-disaster reconstruction "planning for reconstruction." Coventry, 22–23 Apr 2004

ASEAN (2009) Asean agreement on disaster management and emergency response, Vientiane, Article 3, 4 pp

Asian Development Bank (ADB) (2008) Tsunami recovery in India: 3 years on. p. 5, ISBN 978-971-561-719-2

Bates FL (ed) (1982) Recovery, change and development: a longitudinal study of the 1976 Guatemalan earthquake. University of Georgia, Athens

BRR (2009) Ten management lessons for host governments coordinating post-disaster reconstruction. Executing agency for rehabilitation and reconstruction of Aceh-Nias 2005–2009, 11 pp

Climate and Development Knowledge Network (2012) Managing climate extremes and disasters in the agriculture sector: lessons from the IPCC SREX report. p. 1, ISBN:978-1-907288-90-6. Available online at www.cdkn.org/srex

Dynes RR (2000) The Lisbon earthquake in 1755: contested meanings in the first modern disaster. TsuInfo Alert 2(4):10–18

Dynes RR (2005) The Lisbon earthquake in 1755: the first modern disaster. In: Braun TEB, Radner JB (eds) The Lisbon earthquake of 1755: representations and reflections, Studies on Voltaire and the eighteenth century. Voltaire Foundation, Oxford, pp 34–49

Fan L (2013) Disaster as opportunity? Building back better in Aceh, Myanmar and Haiti. HPG working paper. Overseas Development Institute, 6 pp

Government of Tamil Nadu, India (2013) Tsunami rehabilitation programme. http://www.tn.gov.in/tsunami/. Accessed 24 Nov 2013

GSDMA (2001) Gujarat earthquake reconstruction and rehabilitation policy. J Nat Disast Sci 29(1):27

Harsha ARR, Saman PS, Fumihiko I (2007) Post Tsunami recovery process in Sri Lanka, 1 July 2007, 23 pp

Harvey P (2009) Towards a humanitarian government: the role of the affected state in disaster response. Humanitarian Practice Network, Overseas Development Institute, London

International Recovery Platform (IRP) (2014) Indian Ocean Tsunami, 2004. Brief overview of the disaster by countries. http://www.recoveryplatform.org/countries_and_disasters/disaster/15/indian_ocean_tsunami_2004. Accessed 11 Jul 2014

Jha AK, Baresnstein JD, Phelps PM, Pittet D, Sena S (2010) Safer homes, stronger communities: a handbook for reconstructing after natural disaster. The World Bank, Washington, IX pp

Kumar S (2013) Experience sharing: developing guidelines on recovery (ppt). ASEAN workshop on recovery planning, 5–6 Nov 2013

Maldives (2007) Recovery status report. International Forum on Tsunami and Earthquake, Japan. Accessed 15–16 Jan 2007

National Research Council (2006) Facing hazards and disasters: understanding human dimensions. The National Academies Press, Washington

Office of the UN Resident Coordinator, Tsunami Thailand, One year later: national response and contribution of International partners. http://www.preventionweb.net/files/2203_VL323153.pdf

Olshansky RB, Hopkins LD, Johnson LA (2012) Disaster recovery: processes compressed in time. Nat Hazards Rev 13(3):173–178

Planning Commission, Government of India (2005) K. S. Sandhu, Chief Coordinator, Tsunami rehabilitation program. Presentation at Asian Development Bank (ADB) on 18th March 2005, Manila. http://old.icsf.net/icsf2006/jspFiles/tsunami/docs/rehabDocs/ind0236.pdf. Accessed 11 July 2014

Republic of Indonesia (2005) Master plan for the rehabilitation and reconstruction of the regions and communities of the province of Nanggroe Aceh Darussalam and the islands of Nias, province of north Sumatera

Smart J (2012) The role of post-disaster institutions in recovery and resilience: a comparative study of three recent disasters—Victorian Bushfires (2009), Queensland Floods (2010–2011), and Canterbury Earthquakes (2010–2012). Working Paper 12/02, 6 pp

Subekti A (2009) Finance: the seven keys to effective aid management, BRR. ISBN 978-602-8199-48-3. Multi Donor Fund and UNDP, pp 113–114

The Guardian (2013) Will extreme weather like super typhoon Haiyan become the new norm?, 20 Nov 2013. http://www.theguardian.com/environment/climate-consensus-97-per-cent/2013/nov/20/climate-change-haiyan-hurricanes. Accessed 7 Jan 2014

Thiruppugazh V (2014) Post-disaster reconstruction and institutional mechanisms for risk reduction: a comparative study of three disasters in India. In: Shaw R (ed) Disaster recovery: used or misused development opportunity. Springer, Tokyo, pp 19–20, http://www.springer.com/cda/content/document/cda_downloaddocument/9784431542544c1.pdf?SGWID=0-0-45-1423216-p175147203. Accessed 7 Jan 2014

Tsunami Evaluation Coalition (2006) Impact of the Tsunami response on local and national capacities. Sri Lanka country report, 44 pp

Tsunami Information Center (2006) Sri Lanka Tsunami situation report, report number 6, 15 Mar 2006, 4 pp

UNCTAD (2010) Rebuilding Haiti: a new approach to international cooperation, 13 Mar 2010

UNDP/RADA (2006) Good practice note: Sri Lanka Tsunami aid coordination tools, 2 pp

United Nations (1991) United Nations general assembly, A/RES/46/182, strengthening of the coordination of humanitarian emergency assistance of the United Nations, 19 Dec 1991

Wisner B (2004) Risk and the neoliberal state: why post-mitch lessons didn't reduce El Salvador's earthquake losses. Disasters 25(3):251–268

Chapter 9
When Is Too Much Money Worse Than Too Little? Giving, Aid, and Impact After the Indian Ocean Tsunami of 2004

Malka Older

Abstract This chapter examines the problematic of donations and humanitarian aid, a context already fraught with power imbalance, when donors give so much that it is arguably more than what is needed. Although similar situations have occurred over the past decade, the devastating Indian Ocean tsunami of 2004 was one of the first examples of this in modern aid work, inspiring a response from donors that was both quantitatively and qualitatively unprecedented. Both the size of the response and the structural differences because of the large percentage of non-institutional donors led to a number of problems that highlight issues in the aid paradigm. The large amount of funding did not lead to improved performance; rather, it raised expectations to an unrealistic level and led to a focus on media-friendly projects rather than need-based programming. Despite the surplus in funding, agencies continued to compete, either for grants or for beneficiaries, which became in shorter supply than money; and this competition, rather than improving performance, tended to worsen it.

Keywords Aid • Disaster • Donors • Funding • Tsunami

9.1 Introduction

The devastating Indian Ocean tsunami of 2004 inspired a response from donors that was both quantitatively and qualitatively unprecedented. It was "the largest international response to a natural disaster on record" in absolute terms and also in relative terms; the per-person amounts of aid were not only the largest ever documented, but "of a completely different order of magnitude from those of previous disasters" (Flint and Goyder 2006). Qualitatively, this funding was different because of the extraordinarily high percentage that came from the general public, rather than governments or

M. Older (✉)
Centre de Sociologie des Organisations, Institut d'Études Politiques (Sciences Po), Paris, France
e-mail: malka.older@sciencespo.fr

R. Shaw (ed.), *Recovery from the Indian Ocean Tsunami: A Ten-Year Journey*, Disaster Risk Reduction, DOI 10.1007/978-4-431-55117-1_9

other institutional donors. In other words, much of the funding came in small amounts from a broad base of private citizens; the Tsunami Evaluation Coalition (TEC) report on funding found that "[p]olls in Spain, France and the US suggest that around one third of the population gave a donation" (Flint and Goyder 2006).

It would seem that such generosity would be enough, and it was enough money; the TEC found that "international donations and pledges have, very unusually, been at least sufficient for both emergency relief and reconstruction" and indeed that "[e] stimates suggest that the international pledges and donations exceed the total economic impact of the tsunami (US$9.3 billion)" (Flint and Goyder 2006). However, the extraordinary circumstance of having enough money for the work at hand does not seem to have translated into a response that was sufficient for the needs caused by the disaster; the massive cluster of studies commissioned and consolidated by the TEC came to the conclusion "that the response did not achieve the potential offered by the generous funding" (Cosgrave 2007). Why does more funding not lead to improved quality of response? Is it possible that the vast amounts of money donated may have not only failed to achieve their objective, but actually caused more harm than good?

Why might additional aid funding be more harmful than helpful? This chapter will review the impacts of this extraordinary outpouring of aid to trace the ways in which it might have had negative effects. This includes both practical consideration of the structural elements of the aid process that make such successful fundraising problematic, as well as the contradictions in the socio-symbolic construction of aid work highlighted by such extreme amounts of money. In other words, do extremely large amounts of money fail to improve performance and possibly worsen it because of practical concerns, because they lead to shifts in the relationships, expectations, and understandings involved, or both?

These issues came into a certain degree of focus after the 2004 tsunami. Although there was some reluctance to pose the question in a way that might make donors hesitate about future giving, the event (in part because of the abundance of funding available) spawned evaluations of a scale and quality not previously possible. It also, in part because of the gap between expectations raised by those funds and performance, contributed to significant reforms in the humanitarian system for disaster response. However, more recent disasters such as the 2010 Haiti earthquake suggest that the difficulty in managing highly mediatized, highly funded disasters has not been resolved. Openly considering the implications of what some reports refer to in coy quotes as "'too much' money" (Flint and Goyder 2006) remains a pressing concern for the humanitarian industry.

9.2 Context

While the particular characteristics of the "second tsunami" of response were enough to make the 2004 disaster an important event in humanitarian aid, it also occurred at a particular moment in the development of the humanitarian system, the characteristics of which contributed to its watershed effects. The 1990s saw rapid and broad changes in the practice of aid, enough to suggest the transformation from a fairly idiosyncratic, ad hoc activity into an industry. Barnett (2005) writes that "in the 1990s humanitarianism became a field, with regular interactions among the members, an increase in the information and knowledge that members had to consider, a greater reliance on specialized knowledge, and a collective awareness that they were involved in a common enterprise." The idealistic, unregulated, largely individualistic practice of humanitarian work was becoming more formalized. Barnett sees this as largely driven by external issues: "The institutionalization of humanitarianism was largely driven by challenges to the emerging field's legitimacy and effectiveness-challenges that emanated from donors that paid the bills and members who were experiencing a crisis of confidence in reaction to new circumstances and short-comings." It was also related to the growth of the endeavor, in terms of individual and organizational headcount, financial resources, and visibility.

Not all of the shifts were entirely the result of external forces. Of particular interest to us is the question of accountability. While much of the emphasis on accountability, results, and measurement arose as a means of control by donors (Stein 2008), Barnett notes that "those within the sector increasingly sought greater accountability—to recipients. It was not enough to be accountable to donors for how their money was spent; it shows it also was important to be accountable to the supposed beneficiaries of their activities." As professionalization and principles of new public management diffused through the field, some of these concepts were, to a limited extent, co-opted by the sector for use within its own moral framework.

Nonetheless, this growth and systemization often raised conflicts for humanitarian actors who were, for the most part, rooted in idealistic principles and a culture of independence and neutrality. As the changes took root, there was increased questioning of the new approaches, and whether they were compatible with the principles of humanitarian aid. At the same time, there were more practical concerns that despite these and other efforts at reform, accountability, and coordination, effective performance still seemed elusive. Shortly before the Indian Ocean tsunami the UN Emergency Relief Coordinator commissioned the Humanitarian Response Review setting into process what would become the wide-ranging and influential Humanitarian Reform Agenda. The immediate trigger for it was the response in Darfur, which was a complex, highly political emergency addressed with a humanitarian effort generally seen as slow, underfunded, and inadequate.

The response to the Indian Ocean tsunami was almost the polar opposite to that of Sudan: rapid, overfunded, and dealing with a "natural" emergency that many saw as pure and unencumbered with politics. Where the Sudan crisis required celebrities to gain any fundraising traction, the tsunami became a *cause célèbre* that burnished stars rather than the other way around. Despite all these differences and the seemingly highly favorable circumstances, the response in Southeast Asia was still for the most part considered inadequate. This disillusion influenced the subsequent reform process and continues to have echoes to this day. Without the excuse of not enough money, evaluations and internal reviews had to consider other reasons for a less than ideal performance. On the other hand, the extreme amount of money exacerbated other, previously unimportant problems. While the "too much" situation is unusual, it has continued to occur, making an understanding of its dynamics important; moreover, studying this exceptional case offers insights into the more general functioning of the humanitarian system.

9.3 Questions of Quality and Quantity

Humanitarian work remains challenging to evaluate and compare, particularly in major emergencies. The reforms and changes of the last two decades include substantial steps have been taken towards standardization and accountability, most significantly the Sphere Project (www.sphereproject.org), which offers minimum standards across a number of sectors, indicators for assessing success in meeting them, and a humanitarian charter that addresses more high-level issues. Nonetheless, given both the practical difficulties in measurement and the huge variety of contexts and baselines, quantitative evaluations are rarely considered sufficient or even useful. Mixed methods, especially with "participatory evaluation" are almost *de rigeur*. Academic studies on paradigms of aid work tend to use qualitative or mixed methodology (see for example Daly and Brassard 2011; Henderson and Lee 2011).

Beyond the challenges of measurement are broader questions of what should be measured. Barnett (2005) suggests that the structural changes in the industry have contributed to "a shift from deontological, or duty-based, ethics to consequentialist ethics", or from trying to do the right thing to trying to effect positive change. Output indicators were no longer enough; outcomes had to be tracked. Even the latter remains fraught, however; as Barnett writes: "The issue for humanitarian organizations is becoming not whether aid has negative and unintended consequences—for it almost always does—but whether, on balance, it does more harm than good"—something that is almost impossible to measure, and certainly to calculate beforehand. Henderson and Lee (2011) frame a different type of question: "How do we evaluate the efficacy of aid delivery—what aid agencies do a better job in delivering higher quality housing or boats, or in attainment of certain social objectives?" In other words, should evaluation of aid prioritize outputs or large-scale goals? Others question even the premises of evaluation and accountability (see Dufour et al. 2004; Stein 2008).

Despite all this difficulty, some methods of evaluation have been developed, and even agreed upon. The Sphere guidelines are widely used. The TEC makes use of the Good Humanitarian Donorship principles (informed, sufficient, timely, flexible, coordinated, impartial, accountable, effective, and efficient). An in-depth review of evaluation methods is beyond the scope of this chapter, which does not attempt to assess specific interventions or practices; rather it seeks to trace the relationship between the amounts and types of funding with the evaluation results as presented in the existing reports.

9.4 Structural Elements and Practical Considerations

9.4.1 Asset Replacement and SWEDOW

Despite the (repeatedly acknowledged) difficulty with evaluations, the reports of the unprecedented Tsunami Evaluation Coalition (TEC) are not shy about critiquing the humanitarian community's response to the disaster (although this is possibly also somewhat strategic, given the huge media coverage of the aid effort). The synthesis report suggests that this is in part because of the large amount of money raised: "There was early recognition that the exceptional response to the tsunami disaster, including the amount of money given, [...] demanded a high standard of accountability to a generous public" (Cosgrave 2007). Whether or not this report satisfied the "generous public" (or whether any of them actually read it), it does identify a number of issues with the response, none of which result from lack of funds and many of which were caused or exacerbated by the surplus.

It may seem counterintuitive to suppose that too much money can make output worse. More money can, after all, buy more things. In fact, buying too many things was one of the TEC criticisms of the response: "A focus on asset replacement has meant that poor labourers, who had no assets to replace, got left out" (Flint and Goyder 2006) and "[b]ecause most international agencies concentrated on the replacement of physical assets, this may have overshadowed the broader economic development of communities, and livelihoods within them" (Bennett et al. 2006). The choice of assets as a focus, itself, may be related to the levels of funding, but even if other types of programming are preferable, more money should enable the hiring of better staff, faster transport, additional resources of all different types. Money is fungible; enough money can buy anything.

However, as many developed, technologically countries have found to their chagrin, this is not always the case in a post-catastrophe setting. When infrastructure and stocks have been devastated, there is a limit to what can be acquired and transported within the brief timeframe of relevance in a severe disaster. The TEC synthesis report offers the example that "One INGO had planned to hire a fleet of civilian helicopters, but found that there were simply not enough available on the market for short-term contracts at short notice" (Cosgrave 2007). Even for replacing assets, money can only reduce the time needed for production or transport to a certain point. Holguín-Veras et al. (2012a) describe the difficulties of logistics after a catastrophic event based on case studies in Haiti and Japan:

> it requires: (1) considerable resources e.g., manpower, (2) distribution efforts spanning a large area, (3) travel over relatively long distances where the transportation infrastructure has been devastated, and (4) logistical operations of great complexity that can involve tens of thousands of individuals who must be organized, trained and deployed across the geography of the disaster area.

This is further complicated by a phenomenon that Holguín-Veras et al. (2012a, b) refer to as "material convergence", also known in aid worker parlance as Stuff We

Don't Want (SWEDOW).[1] In the more technical description of Holguín-Veras et al. (2012a), "a significant portion of useless unsolicited donations creates major complications for the disaster response." This is not a minor problem. Figuring out what is useless or searching for appropriate projects for unwanted donations is time-consuming, and such materials clog warehouses and supply chains. In interviews of professional logisticians involved in the responses in Haiti and northeastern Japan, Holguín-Veras et al. (2012a) found that when researchers asked "what [the logisticians] considered to be the most significant obstacle to the response, the consensus was 'unsolicited donations.'" Despite the "overwhelming" evidence that these donations worsen responses, "humanitarian organizations remain reluctant to refuse shipments for fear of upsetting donors which might negatively impact future funding" (Holguín-Veras et al. 2012a), a pattern which offers an exaggerated reflection of the slightly more subtle problem of too much general funding.

Although the examples cited above are from later disasters, SWEDOW was clearly a problem after the tsunami as well. The TEC Synthesis report uses this as an example of "poor quality relief work" and specifically cites "inappropriate clothing; expired or culturally inappropriate food; and inappropriate drugs or medicines" (Cosgrave 2007). The problem was serious enough for "the Secretary General of the Indonesian Red Cross to send a harsh letter to other RC Societies complaining of the failure to adhere to guidelines on such donations" (Telford et al. 2006).

Even appropriate and useful goods donated in another country soak up money in transport and warehousing.[2] Rapidly changing demands can mean that what is needed one day is not the next. Holguín-Veras et al. (2012b) offer the example of blankets in northeastern Japan, urgently needed during the first frigid week after the disaster, but quickly received far in excess of needs; just one prefecture, Iwate, "had 100,000 blankets in excess of the 100,000 that were actually needed. Thus, at least 50 % of the blankets that arrived at Iwate Prefecture were not used at all."

While these examples demonstrate why so many aid commentators now urge donors to give cash rather than goods, they also show how difficult it can be to get even the most needed items into an emergency area rapidly. The delivery of goods in a humanitarian setting is a non-trivial task, no matter how much money is thrown at it; at some point, roads and bridges can not be repaired any faster, qualified logisticians can not be found or flown in any more quickly, and trucks are no longer available for rent at any price. Holguín-Veras et al. (2012a) estimate, for example, that "delivering and distributing supplies to the two million Port-au-Prince residents after the 2010 earthquake required 20,000 staff members (comparable in size to a US Army division, which takes 3–4 weeks to fully deploy)." That gives a sense of the challenges of distributing goods that would normally be commonly available. Sourcing qualified human resources or building relationships are even less responsive to funding increases.

---

[1] See http://talesfromethehood.com/2010/04/20/swedow/.

[2] They may also impair local economic recovery.

9.4.2 Scaling Up and Elasticity

One of the issues that Holguín-Veras et al. identify in post-disaster humanitarian logistics is the difference between disasters, which they consider relatively localized, and more widespread catastrophes. Although this use of the terminology is not widely recognized, the distinction is useful, and indeed one of the widely noted learning from the 2004 tsunami was the difficulty for the humanitarian community in scaling up from a more normal level of operations to the one required after a disaster that affected hundreds of thousands over large stretches of coastline in multiple countries.

The TEC reports attempt to make the case that the problem with scaling up to meet the needs of an enormous catastrophe is not entirely due to the amount of money, or the size of the disaster; but rather to the amount of money in every other disaster. The report on coordination notes that the vast difference between "normal" emergency operations—averaging 5.4 billion USD for all emergencies in a year during the 5 years before the tsunami—and a monster disaster that garnered 14 billion was too great for agencies to quickly meet the challenge: "The normal level of funding available for emergencies is much lower than that available here, so the 'baseline' of staff resources is low" (Bennett et al. 2006). This suggests an understandable delay in up-scaling operations; a problem more with the size of the emergency needs than with the amount of funding. The synthesis report elaborates further: "the TEC reports found that agency managers both overestimated what they could do with the available funds and underestimated how long it would take. They did not allow for the personnel, procurement and other factors that constrained their efforts to scale up" (Cosgrave 2007). The problem was not only the time constraint in hiring and sourcing material; it was a lack of familiarity with the scale of operations, leading to errors in estimations.

Parts of the evaluations, however, suggest that this is an overly generous interpretation. The TEC describes "agencies moving into sectors outside their competence and reduced information sharing between agencies" (Cosgrave 2007) as well as "extend[ing] their mandates beyond areas of traditional competency and [making] promises that in subsequent months had to be retracted" (Bennett et al. 2006). This suggests that competition between agencies worsened coordination and therefore the response, as organizations stopped sharing information in an attempt to get a competitive edge and inflated their claims of capacity to donors but ended up being unable to comply. Again we find contradictions: Why should large amounts of money, more money than anyone can spend, lead to increased competition? Beyond that, shouldn't competition make the response better, as incompetent agencies are weeded out?

9.4.3 Competition

The answers come back to the peculiar structures and strange double accountability of the aid industry. One explanation returns to the TEC point that it was not the absolute scale, but the relative scale that was problematic. As noted above, the large amount of funding in the case of the Indian Ocean tsunami was extremely unusual. For organizations used to "scrambling" for funds, as described in Cooley and Ron's 2002 article about the market and incentive forces working on NGOs, the shift to a surplus may have been too shocking to adjust to immediately. There are more rational explanations as well. Despite the restrictions on relief funding which prevent it from being completely fungible to another, there is enough slippage, particularly in hard-to-fund administrative and overhead budget lines, that would lead agencies to try to "stock up" or, as Cooley and Ron write, "to raise funds for future work by raising their flag in media-saturated humanitarian 'hot spots.'" Working a disaster response can also help mitigate barriers to entry; Indonesia and Sri Lanka temporarily loosened controls on visas and organizational registration, and as in many disasters NGOs used the catastrophe as "foot in the door" programming that eventually extended into, or was leveraged into, long-term non-emergency projects in the country.

There were also reputational issues at stake. To not compete for money to respond to the tsunami would have been, or at seemed, tantamount to abstaining from participation in humanitarian aid altogether. Redfield (2008) writes that "In a surprisingly short period of time, tsunami response defined the very currency of humanitarianism." As the disaster raised NGOs and the work they did to new highs of popular awareness, few agencies felt confident enough not to compete (Médecins Sans Frontières was the highly visible exception). Beyond simple entry into the country or validation by basic participation in the response, there were also incentives even for large and credible agencies to continue to seek additional financing. Aid work is an iterative game with a growing number of NGOs or other implementers and a limited number of donors. Disasters with large amounts of funding offer an opportunity for a "foot in the door" with new donors as well as in new countries, as well as for extending the "organizational experience" CV of the agency.

Despite all these incentives, competition for funds did fade relatively quickly. Competition, however, did not. High levels of funding meant an urgent need to spend that funding. The focus shifted to "[i]ncreased competition between agencies for beneficiaries" (Cosgrave 2007) to comply with donor timeframes and "burn rates." This strange shift inverted the normal structure of the industry from the more usual situation, in which agencies compete for small amounts of funding and must select among more needs than they can possibly serve, to one in which agencies have more money than there are needs (at least the needs that money is designated to address). Stirrat (2006), writing about the response from an anthropological perspective, explains that "[g]iven the number of NGOs and the amount of money they had at their disposal this [competition over beneficiaries] was not surprising: after all, there were only a limited number of 'beneficiaries' and only a certain area of land which had been affected by the tsunami." Obligated to spend money, the agencies had to find someone and something to spend it on.

This is where the money begins to be, plausibly, too much, in the sense of worsening the response. Stirrat describes how "representatives of two agencies almost came to blows over which one had the 'right' to assist elderly female coir workers affected by the cyclone, the argument being couched in terms of the respective agencies' particular skills." The supply of aid may be exceeding the demand, or at least the demands of the donors who are "buying" it; the demand for people to aid is exceeding the supply. This situation is not contradicted by the previously discussed difficulties with actually delivering the aid in a timely fashion, because the competition is not over results, but over activities; over any way to spend the money that can plausibly be described as aid work. At least two factors contribute to this: the need for agencies to respond to the disaster of the hour, and the requirement of donors for budgets to be spent during the planned timeframe. After all, for funding to be considered part of the disaster response, it should be spent with some urgency. Less semantically, most institutional donors also work via yearly budgets, which they prefer to spend out as planned.

9.5 Non-institutional Donors

Most of the donors in this case, however, were not institutional, but private citizens. This brings us back to the second unusual fact about the tsunami money, beyond the sheer amount. The TEC report on funding notes that "the financial response of the general public was the defining characteristic" (Flint and Goyder 2006). Although not quite enough to reach half of the total international funding (the TEC report finds it at 39 %), it was far larger than the usual percentage, which the TEC puts at around 15 % (Cosgrave 2007). The quantity was decisive—the TEC states that "It was the private response that meant that the international response was, for once, sufficient"—and the different parameters of private funding had a significant impact on the results of the response. This was also one of the key elements that, within the development of the aid industry, made the tsunami such an important moment.

Throughout the development of the humanitarian field in the 1990s, donors and implementers had a more or less symbiotic relationship. Despite disagreements about emphasis and "direction" of accountability, both NGOs and donors had incentives to professionalize the industry. For NGOs this meant not only improving performance, but also raising the barriers to entry for new actors—including, it should be noted, local organizations without the wherewithal to fulfill complex donor requirements for financial accounting, English-language reporting, and previous, demonstrable experience with both. For donors it meant more control in a relationship that many have characterized as principal-agent (for example, Stein 2008). As the rules of the games were negotiated (with a balance of power tilted heavily towards donors) and solidified, and the activity became a field, donors and large NGOs learned to play by them together.

The response to the tsunami threatened this narrative of professionalization. The vast and unprecedented response from private donors—huge numbers of people that gave small amounts of money or raised slightly larger amounts of money—not only changed the operating environment from insufficiently funded to sufficiently or even excessively funded, but also changed the rules of the game.

This did have benefits for the large NGOs. With most of the private money going to them (and the Red Cross Movement), it gave them both a degree of freedom from institutional donors, and more clout in a system that had been dominated by the UN. However, it also had another effect, less welcome from the perspective of the established implementers: a rapid expansion of actors, including many agencies not familiar with the system that had become more or less accepted. The TEC report on funding points out that "it was the private response that made NGOs and the Red Cross Movement such important (and numerous) actors" (Flint and Goyder 2006). Although the TEC notes that it was unable to find exact figure for the number of implementing agencies,[3] it finds that "[a]t least 202 NGOs are recorded internationally as receiving private donations" and that although "[i]n most countries private donations were heavily concentrated on a small number of international NGOs," some of the additional implementers benefiting from private donations "did not have previous experience of emergency or reconstruction operations" (Flint and Goyder 2006).

New players not dependent on institutional donors—and some of them without the intent to work in future disasters—had little incentive to work within the system. Indeed, some of these actively positioned themselves as outside of or against the system, working based on what Stirrat describes as "a strong humanitarian impulse that 'something had to be done'"; this makes the conflict between them and the institutionalized NGOs, most if not all of which originated in the same way, all the more ironic. Stirrat describes the competition between "large international NGOs with a long track record in relief or development or both" and "a horde of small, often newly formed, foreign organizations with little if any experience in disaster relief" as a contest between slow dinosaurs and small furry mammals, although the metaphor may not hold since the dinosaurs seem, at least in the short term, to have survived.

9.5.1 Refusing Funding

In any case the degree of private funding had a direct impact on the large NGOs as well even when they also had institutional grants; 59 % of the private funds went to NGOs, and "on average the general public donated 77 % of NGO income" (Flint and Goyder 2006). Despite the fact that many of these donations were small, dispersed, and received via mail or web donations rather than as part of a contracting process the way institutional funds are allocated, large NGOs still professed, and seemed to feel, a sense of accountability to these donors. This was expressed mainly in a sense that the money they donated need to be used in accordance with their wishes. When

[3] One reason for this is because of the number of informal implementers, such as private citizens who collected donations and flew to an affected country to hand over money or carry out some project. For an example and some discussion of this phenomenon, see Fernando and Hilhorst (2006).

Médecins Sans Frontières (MSF) took the decision not to accept further funding for the tsunami, it was a public decision rather than a quiet reassignment of funds. Additionally, when MSF decided the funding they had already received was in excess of the needs, or at least the needs within their narrow mandate, it did not merely reprogram that funding, but "sought permission from its donors to divert almost 60 % of the funds it collected to the victims of other emergencies" (Flint and Goyder 2006). This negotiation of the implicit contract between the NGO and its donors was then criticized "for reallocating funds that 'belonged' to the Indonesian people" (Flint and Goyder 2006), going to the heart of the problematic of double accountability in the complex system of giving and receiving. Whom did the money belong to and who could decide what to do with it? Who gets to determine "need?' What exactly are donors "buying" with their money?

The refusal of private donor money was a challenging idea for many actors, and continues to be a question of debate within many NGOs.[4] The flood of private donor funds after the tsunami was seen by many NGOs as a gateway, in the same way that a disaster grant from a large institutional donor might be; if many of these donors were giving for the first time, there was hope that they would continue to do so, and perhaps to the same agency. The relative flexibility of these privately donated funds makes them more appealing to agencies, and for some of NGOs the response to the tsunami may have raised the (ultimately illusory) hope of becoming independent of controlling institutional donors (as, for example, MSF largely is).

Beyond these self-interested considerations, however, there were some more ideological concerns. As Redfield (2008) paraphrases some MSF contributors confused by their decision, "How could a humanitarian organization refuse donations for a worthy cause?" After all, most NGOs, and most NGO workers, were or at least had once been, driven by the same "humanitarian impulse" that drove people to send them a hundred dollars; why should they refuse that money? How could they decide when the amount of funding was "enough" or "too much"? There are of course practical difficulties in answering this question: one NGO receives only a small part of the total funding; needs are generally not clear in the first few weeks of a major disaster, and much less the division of those needs among the various actors. But it is the more abstract difficulties that prevent it from being answered. What is enough?

9.5.2 Proportionality

The TEC found that "If all the tsunami commitments were shared out equally between the 2 million people directly affected, each person would receive roughly US$5,500" and that even "If all the tsunami commitments were shared out between

[4] The 2011 tsunami in Japan is one example where the discussion of whether to raise money, and if so when to stop, occurred within several large NGOs. The evolving results were reflected on agency websites.

the 3.7 million people in areas with major impacts, each person would receive about US$3,000" (Flint and Goyder 2006). By comparison, per person amounts distributed in the subsequent 2005 Pakistan earthquake were between $250 and $1000, in Somalia between $114 and $178, and in Eritrea around $50 (Flint and Goyder 2006). While the difference in magnitude is striking, it does not answer the question of what is enough. Nor does it answer the slightly different question, perhaps politically easier to pose but no easier to resolve, presented in the TEC reports: what is proportional?

The TEC attempts to answer this in a way similar to its answer to the problem of capacity. The grossly different amounts given for different disasters "does not mean there was too much funding for the tsunami, but rather that most emergencies have funding that is far inferior to need" (Cosgrave 2007). While the latter point is undoubtedly true, it is also a somewhat disingenuous framing which avoids the difficult questions of how to determine that level of proportionality or fairness as well as the problem of the division between emergency and development aid. Is need in an emergency replacing what has been lost? The TEC's concern over asset replacement as a strategy would seem to contradict this, and the finding that the total funding for the tsunami response exceeded the estimate of economic damages under that interpretation would then definitely indicate too much funding. (It is perhaps in part in response to this problem, as well as because of the frustration of repeated emergency responses, that the slogan for the response became "building back better." Rather than giving back the money that exceeds the damages caused, or reprogramming it for less telegenic disasters, the answer was to do something more than return to the day before the tsunami hit.)

It is also striking that while the synthesis and funding reports from the TEC complain about the "arbitrary nature of the current funding system for humanitarian emergencies" which "produces an uneven and unfair flow of funds for emergencies" (Cosgrave 2007), the TEC report on the local response asks instead "What can be done with the amounts that seem to exceed 'relief' requirements and do these resources remain available in a form for use that supports longer term development issues related to disaster mitigation?" (Kessler 2006). There were certainly needs in most of the tsunami-hit areas before the tsunami hit; as the concept vulnerability has extended the "responsibility" for disasters to underdevelopment, the question of why disaster funding can not be used for building "resilience" through development has become more difficult. Donations can be seen as out of proportion to one emergency as relates to another; they can also be seen as out of proportion to an emergency in relation to the long-term ravages of poverty, preventable disease, or repeated, less-mediatized disasters, such as regular crop flooding. At least one group, however, saw the level of funding as proportional: the donors. What is less certain (although there are theories) is why; as the TEC notes, "it is not clear what triggers this impulse as some emergencies, such as the Bangladesh floods in 2004, get relatively little funding from the public" (Telford et al. 2006).

9.5.3 Unforced Accountability; Unclear Communications

Despite all these concerns about proportionality based on various different benchmarks, however, most agencies still felt a strong compulsion to comply with the desires of these non-institutional donors. The TEC Initial Findings report points out that while "[a]ccountability to donors is normally ensured by the rules of formal institutional donors," in the case of the tsunami "[t]he prevalence of individual private donors meant that this mechanism did not apply" (Telford et al. 2006). Theoretically, there is nothing compelling agencies to follow the desires of private donors, or keep them appropriately apprised of how their funds are being used. Indeed, the TEC finds that the voluntary reports some NGOs made to the general public, "are generally uncritical summaries of what agencies have done, aimed more at promoting their "brand" than presenting a balanced view of their performance" (Telford et al. 2006).

Yet the sense of accountability to private donors, despite not being described in a detailed legal document, prevented agencies from reprogramming funding without permission, either to other countries or to longer-term programming. As the most immediate relief needs were addressed and the sheer amount of funding that had been collected became clear, the question of what "counted" as an emergency program became a key topic of debate within NGOs. (Given that programming continued in many of the tsunami-affected locations almost until this tenth anniversary, some very fine lines were drawn about what could still be considered "responding to the emergency.")

So despite the surplus of money available, even donors of relatively small (individual) amounts were collectively able to have a large impact on how the response was carried out. This was problematic in part because of the weak links of communication between this highly dispersed group of donors and the implementing partners. Even in rapid, short-term (3-to-6 month) emergency programming, an institutional donor will require a full monitoring and evaluation plan complete with indicators and targets and often with an interim as well as a final report. The implementer may not have much leverage to negotiate with the donor, but they should have a fairly good idea of what the donor wants to see. This may under some circumstances have the effect that the evaluations are further from the "truth," and certainly there are troubling incentives on donors as well as implementers for ignoring or downplaying problems, but there is a certain amount of clarity.

With private donors, on the other hand, communication and approval are much more tenuous. There is no proposal describing the programming, but more usually a description of existing programming or a suggestion of what a certain amount of money (a "dollar handle") "might" mean for an affected person. Even if reports are made publicly available on websites or through mailing lists, there is an assumption that most of these private donors will not read them, will not care, or perhaps will not understand the system of standards and indicators constructed over a decade or two of interaction between governments and agencies. This perhaps contributed to the sway of the media. The TEC found that "huge amounts of funding encouraged a virtual obsession with 'upward' accountability to donors, the media and the public in donor countries," an obsession which not only "discouraged accountability to disaster-affected populations" but also "resulted in competition, duplication and waste" (Cosgrave 2007).

Stirrat describes how agencies found themselves

> faced with a vast number of 'stake-holders' who to a greater or lesser extent felt a sense of 'ownership' in the relief effort. From the beginning the pressure was on the agencies not only to be effective but to be seen to be effective. They had to try to justify the vast investment that millions of stakeholders had made in the relief effort. And they had to intervene in ways which these 'stakeholders' would recognize as being 'relief'.

The issue here is not that these donors were engaged in the response or questioned how their money was used; the problem was, in Stirrat's terms, that they would not necessarily recognize all forms of relief as 'relief'—or at least that implementers believed they wouldn't. In this dynamic, the needs of the people actually affected by the tsunami became invisible beside the interpreted demands of the people virtually affected by it.

This is not to criticize the people who gave generously in response to the plight of those on the other side of the world. Stirrat finds the problem in "the structural nature of the relief business"; he believes that the moral impulse itself, siphoned through this structure, leads to a problematic outcome: "For the ethically motivated donor the issue is always how their donation can have its maximum effect and this inevitably generates competition between the donors to prove that they and not other organizations have the greatest impact in alleviating suffering and achieving humanitarian objectives."

This misses at least some of the point, however. For example, a purely ethically motivated donor, looking for the greatest impact, would not be tied to one disaster over another. More importantly, competition could be a healthy force in the aid industry, as long as that competition is tied as closely as possible to actually impacts, rather than perceived impacts. Stirrat writes that "[w]hat starts out as a gift becomes a commodity", but in fact the gift itself has very little value by itself (or, in the cases of some material aid, negative value). Rather, from the donor's perspective the "gift" is bartered for—something. Some form of satisfaction, some feeling of self-worth, a form of self-definition as virtuous, a story to tell at parties—something that it is important to the giver. Perhaps that satisfaction comes from competition over having made the "best" gift, or perhaps it really is based on the actual impact the gift has had on people who need it. Without clear information one way or the other, NGOs, led by the media, seem to have assumed the lowest common denominator.

Much of the blame, of course, is put upon the media. The TEC finds that "[i]t was television coverage of the disaster that provided the basis on which funding decisions were based rather than any more formal assessment of needs" (Telford et al. 2006). Stirrat observed that

> NGO representatives were only too aware of the presence of TV teams and reporters, and many organizations had their own film crews to record their activities [...] The result was to privilege certain sorts of activities, such as distributing new fishing craft or constructing housing, rather than less visible or more indirect forms of disaster relief, for instance rehabilitation of government offices destroyed by the tsunami.

This brings us back to the problematic of "buying too many things"; whether in the form of material goods or in fungible cash, large amounts of donations seems to have encouraged physical, visible, photographable types of programming.

9.6 Discussion: Double Accountability

The aid and humanitarian industries claim to function on the basis of double accountability: responsibility both to beneficiaries and to donors. In underfunded disasters where the main donors are institutional, it is easy for NGOs to position themselves as relatively weak players, advocating for the unheard to the best of their ability in the face of a huge power imbalance and stretching the meager funds they receive. When the money is suddenly "enough," it puts all this into question.

From the study of the 2004 tsunami it is clear that money—even extraordinary amounts of money—and good intentions are not enough to guarantee good or even competent, timely programming. Understanding the non-fiscal constraints of humanitarian work, whether in terms of setting realistic expectations for logistics or for situational awareness, can attenuate the problematic link between financial generosity and misplaced priorities.

Because, despite all protestations to the contrary, the money, and the mediatization that was both cause and effect of the massive outpouring of donations, do seem to have affected planning and prioritization during the response to the tsunami. For all the efforts towards professionalization and standardization that had taken place over the previous decade and a half, the intense attention and large-scale funding brought an almost immediate shift to a situation in which the response was led by the interpreted priorities of non-professionals and driven by image rather than indicators.

Although aid agencies continued to talk about double accountability, the weakening of formal, explicit, legal accountability to donors with the shift towards non-institutional, private funders did not strengthen accountability to beneficiaries. Instead of improved performance based on competition, this weak accountability and tenuous communications meant that the terms of the competition shifted, worsening performance.

In a piece on generosity, Korf (2006) argues that unlike the unconditional acts of hospitality or kindness practiced at the local level immediately after the tsunami, the international mustering of aid, which "became almost like a sports competition over who was the most generous" is in fact conditional: "since the consumer of generosity wants to see his or her money materialize in a consumption good - possibly the feeling of doing something good. Private donors want to see their donations flourish - they want to see it materialized in the smile of children in Sri Lanka who can go back to school, for example." This may be true, or not; or it may be true in some cases and not in others. Certainly many NGOs acted as though it were true, and unchangeable.

9.7 Conclusion

The years since the 2004 tsunami have seen many reforms and initiatives, many of them directly inspired by the experience, aimed at improving performance, coordination, and accountability in major disasters, and many of these have had effect. However, more recent highly publicized catastrophes, like the 2010 Haiti earthquake, have demonstrated that high levels of funding continue to bring their own challenges to the aid enterprise. As the industry continues to standardize and professionalize, efforts towards common goals and understanding of aid priorities need to be expanded: to beneficiaries, to the "amateur" groups that flood to a major disaster, and to private donors. As long as double accountability is as unbalanced as it was in 2004, there is the risk that humanitarian work will be driven by the generosity of givers as opposed to the needs of the affected.

References

Barnett M (2005) Humanitarianism transformed. Perspect Polit 3(4):723–740

Bennett J, Bertrand W, Harkin C, Samarasinghe S, Wickramatillake H (2006) Coordination of international humanitarian assistance in tsunami-affected countries. Tsunami Evaluation Coalition, London

Cooley A, Ron J (2002) The NGO scramble. Int Secur 27(1):5–39

Cosgrave J (2007) Synthesis report: expanded summary. Joint evaluation of the international response to the Indian Ocean tsunami. Tsunami Evaluation Coalition, London

Daly P, Brassard C (2011) Aid accountability and participatory approaches in post-disaster housing reconstruction. Asian J Soc Sci 39:508–533

Dufour C, Geoffroy V, Maury H, Grünewald F (2004) Rights, standards and quality in a complex humanitarian space: is sphere the right tool? Disasters 28(2):124–141

Fernando U, Hilhorst D (2006) Everyday practices of humanitarian aid: tsunami response in Sri Lanka. Dev Pract 16(03–04):292–302

Flint M, Goyder H (2006) Funding the tsunami response. Tsunami Evaluation Coalition, London

Henderson JV, Lee YS (2011) Organization of disaster aid delivery: spending your donations. NBER Working Paper No. 17707. http://www.nber.org/papers/w17707

Holguín-Veras J, Jaller M, Van Wassenhove LN, Pérez N, Wachtendorf T (2012a) On the unique features of post-disaster humanitarian logistics. J Oper Manag 30:494–506

Holguín-Veras J, Taniguchi E, Ferreira F, Jaller M, Aros-Vera F, Thompson RG (2012b) The Tohoku disasters: chief lessons concerning the post-disaster humanitarian logistics response and policy implications. Provided by the first author via email. http://docs.trb.org/prp/12-1162.pdf

Kessler E (Asian Disaster Preparedness Center) (2006) Local response: overview. The international community's funding of the tsunami emergency and relief. Tsunami Evaluation Coalition (TEC), London

Korf B (2006) Commentary on the special section on the Indian Ocean tsunami: disasters, generosity and the other. Geogr J 172(3):245–247

Redfield P (2008) Sacrifice, triage, and global humanitarianism. In: Barnett M, Weiss TG (eds) Humanitarianism in question: politics, power, ethics. Cornell University Press, Ithaca/London, pp 196–214

Stein JG (2008) Humanitarian organizations: accountable—why, to whom, for what, and how? In: Barnett M, Weiss TG (eds) Humanitarianism in question: politics, power, ethics. Cornell University Press, Ithaca/London, pp 124–142

Stirrat J (2006) Competitive humanitarianism: relief and the tsunami in Sri Lanka. Anthropol Today 22(5):11–16

Telford J, Cosgrave J, Houghton R (2006) Joint evaluation of the international response to the Indian Ocean tsunami: synthesis report. Tsunami Evaluation Coalition, London

Chapter 10
Social Protection Ten-Years After the Tsunami: The Case of Indonesia and Thailand

Benigno Balgos and Jesus Dominic Dizon

Abstract An earthquake with an epicenter in the west coast of Sumatra, Indonesia sparked a chain of injurious tsunamis along the shoreline bordering the Indian Ocean, killing thousands of people in eleven countries. Aside from damage to properties and lives claimed, it resulted to exacerbating the dire conditions of the poor and vulnerable communities. The sources of livelihood of the people were destroyed. That said, rebuilding their lives of those who survived the catastrophe became a critical challenge. This chapter examines the social protection programs focusing on the livelihood of affected communities in Indonesia and Thailand after the tsunami. From community-level examples, the chapter will draw issues, challenges, and impacts in social protection in relation to building resiliency in a post-disaster context. The author argue that social protection is not only a means to aid survivors in recovery, but it also reduces poverty, improves the lives of survivors, and facilitates communities' adaptation and mitigation to future disaster risks toward resiliency.

Keywords Community resiliency • Disaster recovery • Livelihood • Social protection

10.1 Introduction

Recovery is one of the important thematic areas in disaster risk reduction. The United Nations International Strategy for Disaster Reduction (UN ISDR) defines recovery as the effort to improve the dire conditions of the communities affected by

B. Balgos (✉)
Center for Disaster Preparedness – Philippines, Block 25 Lot 3, JP Rizal,
New Capitol Estates 1, Commonwealth Avenue Brgy. Batasan Hills,
Quezon City 1126, Philippines
e-mail: cdp.phil@gmail.com

J.D. Dizon
Earthquakes and Megacities Initiatives, 3/F Puno Annex Building
47 Kalayaan Avenue—Diliman, Quezon City, Philippines
e-mail: jdvdizon@gmail.com

R. Shaw (ed.), *Recovery from the Indian Ocean Tsunami: A Ten-Year Journey*,
Disaster Risk Reduction, DOI 10.1007/978-4-431-55117-1_10

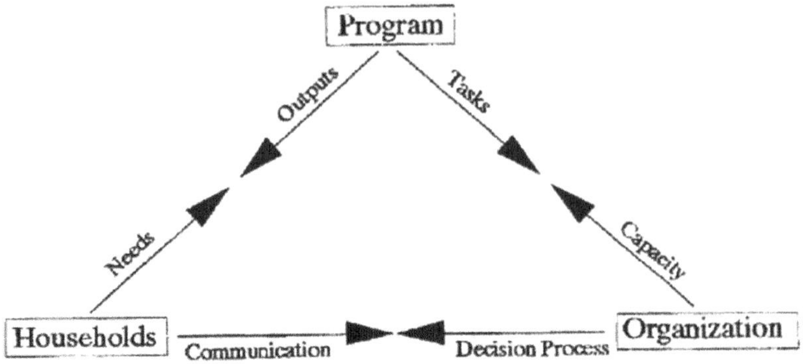

Fig. 10.1 Fit requirement for disaster recovery planning

disasters in relation to facilities, livelihoods, and risk reduction among others. In recovery, efforts to disaster-stricken communities should enable the survivors to bounce back from the stresses and shocks brought about by the catastrophe. That said, it is not about bringing the people back to the same old vulnerable and helpless condition. Recovery is looked at as a struggled against Mother Nature, so that the natural order of things in the community would be reestablished.

Recovering from disasters entails many things. The Asian Disaster Reduction and Response Network (ADRRN), a coalition of disaster reduction organisations in the Asia Pacific region that aims to stimulate coordination and collaboration work on risk reduction in the region, works on the following in the area of recovery: strengthen the ability of communities to combat disasters, protect critical facilities such as schools and hospitals, provide humanitarian assistance such as food, water, shelter and health care, create awareness on disasters, disaster response and disaster risk reduction, advocate for policy changes, and improve the capacity of members-organisations and their respective community-based partners. Recovery cannot be done overnight even though there is an overflow of resources. The publication entitled, *Holistic Disaster Recovery: Ideas for Building Local Sustainability After a Natural Disaster*, highlights that recovery takes years. The book underscores that, short-term recovery is usually performed during the first weeks, and the activities are mainly looking into responding to the emergency. Restoring the critical services for the people in the affected communities then follows this. Concurrently, the final phase is the rebuilding of the community, which entails addressing the political, economic, and social vulnerabilities that make the community at-risk.

Figure 10.1 provides the fit requirements for disaster recovery planning developed by Korten. The model underscores in order for the recovery efforts to be truly effective, intervention should address the very needs of the affected communities. Thus, the household needs are paramount in order for recovery programs to be successful. Also, the recovery programs should be undertaken by a strong and capable organisation with a clear development programs.

This chapter looks into community level livelihood social protection recovery programs in the tsunami-affected communities in Indonesia and Thailand. From the community-level examples, the chapter draws issues, challenges, and impacts in social protection in relation to building resiliency and adaptive capacity in a post-disaster context. The 2011 Global Assessment Report on Disaster Risk Reduction highlights that social protection can protect vulnerable communities before, during, and after disasters. Also, the report emphasizes that social protection has the ability to increase the resilience of communities and improve human capital. Social protection in the context of recovery takes many forms. It includes, but not limited to, conditional cash transfers, micro-insurance, and temporary employment programs. In this chapter, social protection in the form of livelihood intervention and initiatives to survivors of the tsunami in Indonesia and Thailand will be highlighted. The chapter is divided into three sections. *First*, the theoretical underpinning of social protection as a development strategy will be provided. Specifically, a discussion on how social protection, in the recovery context, in the form of livelihood programs can address disaster vulnerability by increasing people's adaptive capacity will be highlighted. *Second*, the social protection focusing on livelihood programs initiated in Indonesia and Thailand in the aftermath of the tsunami by various organisations will be presented. *Finally*, lessons and challenges for social protection in the recovery context will be discussed.

10.2 Theoretical Overview of Social Protection

Social protection is a core development strategy addressing poverty. It aims to reduce the underlying risk factor and exclusion of many poor and communities vulnerable to disaster. Nishino and Koehler (2011) note that it is a development tool, which responds to human needs and rights issues. Development institutions and think tanks consider social protection programs as imperative to concretize an inclusive development. On the other hand, Midgley (2013) held that the concept of social protection is not really new. In fact, the emergence of social development as a subfield of development studies looks into this "social" aspect of development. That said, the growing interest on social protection surprises him. He argues that there are three (3) the factors that led to the new interest in social protection, namely: (1) the creation of the conditional cash transfers in Latin American countries such as Brazil, Mexico; redesigning of social assistance in South Africa; (2) the introduction of old-age pension schemes in Botswana, Lesotho, and Namibia; and, (3) the launching of the National Rural Employment Guarantee Scheme in India. Midgley (2013) further notes that social protection is the umbrella term being utilized to cover comprehensive programs pertaining to social security, social insurance, social assistance, and universal social programs. Also, he held that most of these programs are initiated by non-government organizations with funding support form international donors.

As observed in the biblical proportions of the recent disasters such as Indian Ocean Tsunami (2004), the Sichuan earthquake in China (2008), Cyclone Nargis in Myanmar (2009) the flooding in Pakistan (2010), the eruption of Mount Merapi in Indonesia (2010), and the super tyhpoon Haiyan in the Philippines (2013), doubtless, disaster events have the propensity of jeopardizing the development initiatives and targets of a particular country. Evidently, the abovementioned both have economic and social ramifications that have endangered development. The poor and vulnerable section of the population such as women, children, and the elderly bear much of the brunt of disaster events and economic crises because they lack the needed resources and capacity to withstand its impacts. That said, long-term recovery efforts should take into consideration the immediate and foreseeable needs of the people.

In the overall framework of social protection, Nishino and Koehler (2011) note that rights-based approach is central to social protection programs. By capitalizing on human rights, they deem that social protection implies equity and universal coverage supplemented by interventions that address the specific needs of the poor and the marginalized sectors of the society. Concurrently, a policy paper published by the Center for Social Protection (CSP) and the Institute of Development Studies (IDS) remarked that: *'social protection enables people to survive and build resiliency to shocks (i.e. droughts and illness). Also, it facilitates people to evade harmful mechanisms such as selling their productive assets and allows people to take productive risks and invest in their livelihood. That said, low income nations that have pursued the path of economic growth without capitalizing on building a robust social protection mechanisms failed to create opportunities for the poor and vulnerable.'*

Samson (2009) observes that, traditionally, social protection is seen as a mechanism to reduce socioeconomic vulnerabilities with the aim to promote equitable human development. Nonetheless, as stated by Nishino and Koehler (2011), social protection can likewise be viewed as an imperative development strategy with three crucial tasks, namely to: (1) safeguard against income shocks offering minimal socioeconomic security to economically vulnerable groups, particularly those living under the poverty line; (2) support inclusive growth or contributing to social inclusion; and, (3) address disparities among groups, be it gender, age-based discrimination or exclusion vis-à-vis language, religious affiliation or ethnicity. In hindsight, given that many communities are vulnerable to disasters because they are poor, social protection programs address the very core of people's vulnerability.

Across literature, the general framework of social protection has four (4) components, which include: *protective, preventive, promotive*, and *transformative*. Figure 10.2 provides a summary of each of this component vis-à-vis time frame, specific instruments, and roles in crises. On the other hand, Nishino and Koehler (2011) held that in its radical form, social protection contributes in fulfilling social justice and transformation, which facilitate the underprivileged communities to move out of their dire state. The transformative approach to social protection, according to the two scholars, is an enhancement both in theory and policy of social protection. They argue that rights-based approach, both in the design and delivery

Timeframe	Type of Social Protection	Social Protection instruments	Role in this crisis
Short-term	*Protective* (social assistance)	- social transfers -disability benefit -pension schemes -social services	**Immediate protection and relief from poverty and deprivation**
	Preventive (Insurance and diversification mechanisms)	-social transfers - social insurance (pensions, health insurance, unemployment benefit) -livelihood diversification - savings clubs; funeral societies	**prevents damage to coping strategies**
	Promotive (economic opportunities)	-social transfers -access to credit -asset transfers/protection - school feeding -starter packs -access to common property resources	**promotes resilience through livelihood diversification and improves security**
Long-term		- public works programmes	
	***Transformative* (addressing underlying social vulnerabilities)**	**-promotion of minority rights -anti-discrimination campaigns -social funds**	**transforms social relations to reduce exclusion**

Fig. 10.2 The role of social protection in crises

of policy actions is being integrated in the transformative element. Concurrently, Devereux and Sabates-Wheeler (2004) held that social protection includes all development initiatives, whether formal and informal, providing: *social assistance* to extremely poor individuals, *social services* to groups who need special care or who would otherwise be denied access to basic services, *social insurance* to protect people against the risks and consequences of livelihood shocks; and *social equity* to protect people against social risks such as discrimination and abuse.

10.3 Social Protections After Tsunami: The Case of Indonesia and Thailand

The 26 December 2004 tsunami is considered as one of the catastrophic disasters in contemporary history. It started with an earthquake that sparked a chain of injurious tsunamis along the shoreline bordering the Indian Ocean, killing thousands of people in eleven (11) countries and territories. The earthquake had an epicenter in the west coast of Sumatra, Indonesia. According to report of the International Federation of Red Cross and Crescent Societies (IFRCCS), the total number of victims, both dead and missing is approximately 230,000. Among the country that were struck, Indonesia had the highest casualty (163,795), followed by Sri Lanka (35,399), India (16,389), Thailand (8,345), and Somalia (298) (Satake 2007).

At that time, never has the world seen a catastrophe like what occurred that fateful day. After the tsunami, the communities ravaged by the disaster became ghost towns. People who were able to survive the tsunami were searching for their loved-ones. Also, survivors tried recovering whatever salvageable properties can be used. Emergency response was the immediate action of the national governments of affected states. Furthermore, there was an influx of relief aids from humanitarian agencies, particularly in Aceh from all over the world. The urgent actions of humanitarian agencies were aimed toward providing the immediate needs of the victims, which include medicines, foods, potable water, clothes, tents, and burial for the dead bodies among others. In Aceh alone, it was estimated that over 350 aid organizations responded right after the tsunami (Vebry et al. 2007).

Aside from damage to properties and lives claimed, the tsunami has resulted to various economic woes as well. In fact, the major sources of livelihood of the people were destroyed. That said, it was a challenge to those who survived where they can start rebuilding their lives (Christoplos et al. 2007). In Aceh as well as the six (6) provinces in Thailand (Krabi, Phang Nga, Phuket, Ranong, Satun, and Trang) that were hit by the tsunami, economy is very potent. Aceh, for instance is one of Indonesia's rich provinces due to oil and gases. Concurrently, the areas hit in Thailand contribute to the country's tourism industry. However, the tsunami had turned around the situation for the Acehnese and the Thais. That said, the tsunami had placed those who survived in abject misery. As a result of the magnitude of the disaster, the survivors became homeless and helpless. On the other hand, the influx of relief aids and humanitarian assistance at the wake of the disaster did not make the survivors helpless. The outpouring of assistance was encouraging. In the case of Aceh, there was an apparent re-opening Aceh to the Indonesian society and to the rest of the world (Reid 2005).

After the short-term emergency relief phase, the focus of aid agencies and other development organisations shifted to long-term recovery plans. According to the long-tem recovery commences whenever the affected community starts to reconstruct or rebuild the damaged properties such as roads and bridges among others. It is also the time when enhancements of other non-physical properties are being undertaken. In hindsight, it is in this phase that social protection comes in the recovery process of the affected community. The following highlights the social protection programs carried out in the disaster-stricken areas of Thailand and Indonesia.

10.3.1 After Tsunami Social Protection Programs in Thailand

When the tsunami struck the southern Thailand, six (6) of its coastal provinces namely Krabi, Phang Nga, Phuket, Ranong, Satun, and Trang received the severity of impacts and damages brought by the hazard. The waves of the tsunami devastated communities and livelihoods, destroyed buildings and infrastructures, damaged properties and utility systems and worst of all, claimed thousands of life.

The Department of Disaster Prevention and Mitigation (DDPM) estimated the number of casualties at 8,457[1] while 12,815 households (58,550 persons) were displaced and directly affected by the disaster. Estimated amount of damages and losses for the affected areas was US $2.09 billion.[2] The estimates included both direct damages and indirect losses as consequence of the disaster.

Thailand was the second worst affected country in the Southeast Asian region, after Indonesia. Although the number of casualties was relatively lower than in Sri Lanka and Indonesia, the international attention focused on Thailand due to the affected areas, which are popular tourist destinations.

This disaster brought an extraordinary response from the Royal Thai Government (RTG) with support from the non-government organizations (NGOs) and private sector, and the international community. The RTG adopted a three-phase response strategy to this disaster. Phase I focuses on search and rescue operations for the survivors as well as the recovery and identification of the dead bodies. Phase II focuses on prevention of infectious diseases and secondary losses among those who survived the tragedy and Phase III on reconstruction and long-term rehabilitation.[3]

Despite the effective response and efforts of the RTG, the global community composed of bilateral agencies and international NGOs significantly contributed in providing financial and technical support in various sectors including child protection, psychosocial support, provision of livelihood programs, environmental rehabilitation, and capacity building and training.

While there was no appeal for financial assistance, the RTG welcomed support from UN County Team in a form of technical assistance to local governments and NGOs working in the affected areas. Through the UN Flash Appeal of January 6, 2005, the government was able to access the worldwide funding for tsunami in the form of project proposals. UN Flash Appeal was the primary mechanism for UN funding of tsunami relief and rebuilding projects in Thailand.

The United Nations Environment Programme (UNEP) reported that the tsunami hit the main economic sectors of Thailand including tourism, agriculture and fisheries. According to Coate et al. (2006), in order to recover from the tsunami there is a need to involve the community at the grass-root level through local participation in decision-making.[4] Recovery process involves the participation of various stakeholders such as the government, NGOs and civil society organizations and international institutions.

As part of the recovery initiatives, the USAID Regional Development Mission/Asia created the ECO-ASIA: Post Tsunami Sustainable Coastal Livelihoods

---

[1] See DDPM, Loss Assessment by Tsunami and Earthquake from Provinces (Data as of June 10, 2005)

[2] Regional Analysis of the Socio Economic Impacts of the Indian Ocean Tsunami, August 2005

[3] Tsunami Thailand: One Year Later: National Response and Contribution of International Partners. United Nations Country Team in Thailand, Office of the UN Resident Coordinator

[4] Saarni (2007).

Program in partnership with the University of Rhode Island, Asian Institute of Technology (AIT) and University of Hawaii-Hilo which focused on restarting and diversifying livelihoods and small enterprises, sustainable use and management of natural resources, infrastructure rehabilitation, and tsunami preparedness. The goal of this program is to rebuild livelihoods and rehabilitate coastal communities affected by the tsunami through participatory capacity building. Under this program, ECO-ASIA assisted five-tsunami affected villages achieve self-sufficiency, preparedness in the face of natural hazards and protection of the natural heritage. The program has five (5) major program components: build a common vision for action, reestablish and diversify livelihoods, enhance community readiness and resilience to coastal hazards, build capacity for planning and decision-making in the coastal zone, and share experience and best practices.

Another recovery project supported by the United Nations Development Programme (UNDP), the Chulalongkorn University and UNESCO entitled "Rehabilitation of Traditional Communities and Municipalities in Tsunami Affected areas in Thailand". The project is implemented in Phang Nga, Ranong, Phuket and Satun. To promote cultural heritage and build capacity of indigenous people to actively participate in the post-tsunami development of conservation regulation, various activities will be carried out. These include: (1) constructing of traditional boats and nature trails to transfer sea gypsy knowledge and skills and for community use; (2) rebuilding huts exhibiting on traditional culture of the people; (3) workshops with relevant stakeholders to explore options and provide recommendations for co-management in protected areas; (4) developing booklets for tourists on indigenous people and their natural resources; (5) developing local school curricula with an emphasis on sea gypsy traditional cultures; (6) GIS mapping; and, (7) collecting household data of prepare personal documents for the development of human security.

The project documented the best practices and lessons learnt by both communities and local government organizations and disseminated the information to enhance and build support networks for communities and relevant local government institutions. Several recovery and livelihood rehabilitation programs were conducted by international funding organizations in partnership with local government agencies and non-government organizations.

10.3.2 After Tsunami Social Protection Programs in Indonesia

Indonesia comprises of five (5) major island groups namely *Java, Sumatra, Sulawesi, Irian Jaya,* and *Kalimantan.* Aceh, the epicenter of the 9.1–9.3 magnitude earthquake, is situated strategically in the northwestern tip of Sumatra. Historical documents describe Aceh as a place rich in gas, oil, and potent commercial products during the pre-colonial period. In addition, historians strongly agree that Aceh plays an indispensable role in Indonesia's development both then and now (Gin 2004). In 1953, shortly after Indonesia gained its independence from the Dutch, a 10-year rebellion against the central bureaucracy took place. As a consequence, a *dearah istemewa* or a *special region* status was granted to Aceh whereby autonomy in the aspects of law, religion and education was permitted (Oey 1996).

The 2004 tsunami became the tipping point as it led to what Reid (2005) argues as the re-cosmopolitanisation of Aceh due to the influx of international donors, which put pressure again to the Indonesian government to re-think things. Truth be told, a meaningful end to one of the region's long standing conflict could have been done prior the tsunami. Then again it did not happen. Indubitably, even though majority of the international donors were not tide to any political organizations, there presence in Aceh after the tsunami fostered an atmosphere of transparency in governance of the province (Kivimaki 2008). In the same way, due to the tsunami incident in Aceh, much research has been written about the province. Doubtless, the increased interest in Aceh generated by the tsunami gave the province and its issues further room within academic discourse.

The impacts of the tsunami in Aceh in incomparable. Held that almost every sector in Aceh was affected by the disaster. It is estimated that US$ 4.45 billion or 80 % of Southeast Asia's gross domestic product was the total damage and losses incurred. Figure 10.3 highlights the impact of the disaster in Aceh as reported by the Badan Perencanaan Pembangunan Nasional (BAPPENAS), Indonesia's ministry on development planning.

The Indonesian President Susilo Bambang Yudhonoyo announced in March 2005 the end of the short-term emergency response in Aceh. As a result, the long-term recovery programs commenced. Figure 10.4 provides the reconstruction allocation in Aceh from various organisations such as the Government of Indonesia, international donors, and non-government organisations. As observed, the comprehensive recovery program in Aceh can be considered as one of the historic recovery and reconstruction experience in contemporary history.

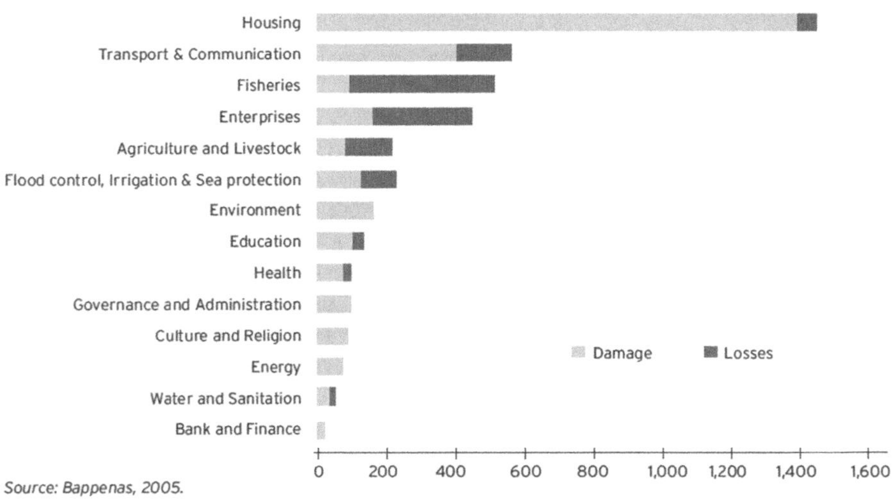

Fig. 10.3 Damage and losses of Aceh's tsunami

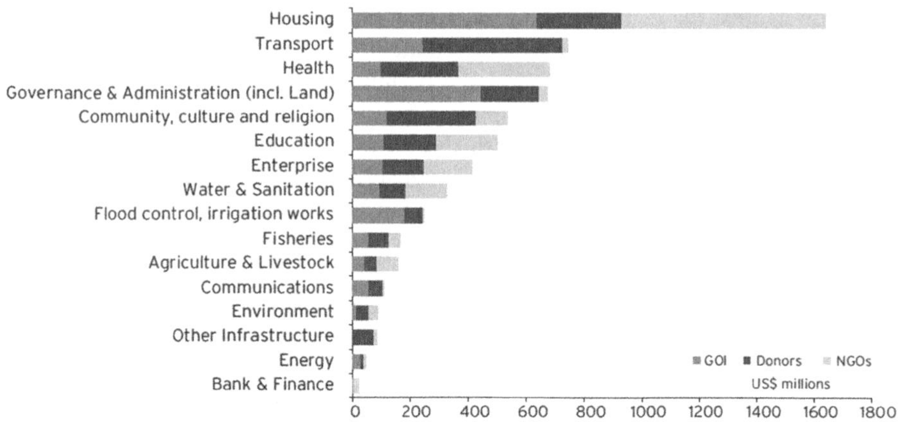

Fig. 10.4 Reconstruction allocation in Aceh

Table 10.1 Livelihood program of Save the Children. *Source*: Save the Children 2008b)

Employed 19,700 people who worked 108,000 working days at 191 sites—providing much-needed income for families.
Provided small grant disbursements to 9,613 people.
Established 20 local farmer groups engaged in agriculture and livestock activities, in partnership with a local nongovernmental organization (NGO) in Nias.
Supported the expansion of Afdhal, a local microfinance institution that provides financial services to women. Their three branches serve some 2,500 clients who maintain an average 99 % repayment rate.
Helped over 450 women pandan (dried leaf) weavers increase sales and income, improve product quality, develop new links to markets and diversify their range of products. We also helped to facilitate the establishment of a pandan weavers' cooperative.
Distributed improved rice seeds, chili, plastic mulch, watermelon seeds, hoes and machetes to 1,460 farmers.
Helped 47 farmers established a community cocoa nursery that is expected to benefit more than 1,000 local farmers.

Doubtless, the response for recovery in Aceh was staggering. For instance, in a report published by Save the Children four (4) years after the tsunami reveals that the allocation for recovery in Indonesia alone is estimated to $169,254,280 (Save the Children 2008b). The recovery programs for Aceh covers the following: child protection, education, livelihood and food security, and shelter among others. Save the Children, a UK based organisation, is one of the hundreds of groups that carried our recovery programs in Aceh. Table 10.1 highlights the support of Save the Children to the survivors of the tsunami in relation to livelihood and food security.

Another organisation that provided livelihood intervention as a means to recovery from the disaster is the Food and Agriculture Organization (FAO) of the United Nations. In terms of livelihood, FAO's focused their funding in assisting fishermen in Malahayati in replacing their boats/fishing vessels. In the same way, FAO also provided tools, seeds, and fertilizers for residents in Meulaboh. Also, the organisation helped the fishermen in Pidie toward sustainable fishing through the conduct of capacity building activities (Food and Agriculture Organization (FAO) 2005).

On the other hand, the International Youth Foundation (IYF) partnered with a local NGO, the Center for Community Development and Education (CCDE), in carrying out livelihood programs in tsunami-hit areas in Banda Aceh, Great Aceh, South Aceh, Southwest Aceh, and Nagan Raya. More pointedly, they enhanced the capacity of the youth with needed life and computer skills. The main stakeholders in the initiatives were the women to facilitate their empowerment (International Youth Foundation (IYF) u.d). As a result of the program, the IYF was able to train 320 youth, while 155 of them received loans to start up their businesses. Also, 216 youth took part in finance trainings. More importantly, among the trained youths, 71 were capacitated to lead Self Help Groups (SHG) (IYF u.d.). Finally, Help Age International Asia Pacific provided livelihood and cash grants to the older people in Banda, Aceh to help them recover from the disaster. The organisation also partnered with a local institution called Older People's Association (OPAs). The assistance to the elder was categorized to three (3) things, namely: cash for work, vouchers, and cash grants. The cash for work scheme were mainly for road and house construction. Concurrently, the vouchers were aimed toward accessing the basic needs of the elderly. Finally, the cash grants include distribution of cash to identified population as well as intended for the re-establishment of businesses, but in this case, a business plan in needed (HelpAge International u.d).

10.4 Conclusion

The Asia Pacific Region is becoming more and more vulnerable to the impacts of catastrophes than any other regions in the world. The crucial factors that led to this includes: geographical location, lack of resources to withstand shock and stresses, and effective system and structure are not well in place for proactive disaster risk reduction and management. As such, policies and programs on disaster risk reduction and climate change adaptation have to be crafted and effectively implemented at the local level.

The presence of the Hyogo Framework for Action (HFA) provides a general framework on how to combat vulnerability through the five (5) priority actions. Almost 10 years after its ratification in Japan in 2005, there were observable gains, milestones, and significant achievement at the regional, national, and even at the community levels. The signatories to the HFA were able to formulate national policies and framework on disaster risk reduction and management. Also, the regional integration and collaboration on risk reduction have been promoted. Moreover, public-private partnerships have been forged and community-level risk reduction initiatives have been intensified. These measures, among others, helped reduce the risk and vulnerability that countries in the Asia Pacific Region are facing.

Now more than ever, it can be said that there is conscious, concerted, and collaborative effort toward integrating disaster risk reduction and management in sustainable development framework. There is strong evidence of recognition among stakeholders (i.e. governments, non-government organization, and communities) that development is being jeopardized by disaster events, and that to be able to achieve sustainable development, disaster risk reduction and management should be given utmost prioritization.

It is likewise evident that people are resilient and able to overcome the effects and impact of disasters if policies take paramount consideration of their needs and contributions. The root causes of disasters mirror the structural constraints, which hamper fair and sustainable development for all. People are as vulnerable as their livelihoods are fragile in facing the complex hazards. Their ability to cope reflects the complexity of resources needed to facilitate such coping. Disaster policies whether for hazard prevention, for vulnerability reduction or for fostering resilience require the participation of the people whose life and livelihoods are at stake. No one knows the local contexts, issues and needs better than the members of local communities facing natural hazards (Gaillard 2010).

The paper discussed the components of social resilience coupled with illustrations from countries in the region such as Thailand and Indonesia. Nonetheless, there is still a crucial challenge to beef up and intensify the efforts of stakeholders to ensure that vulnerable groups are not just prepared for disasters rather they are able to mitigate and prevent these catastrophes from happening and affecting them. It is worth appreciating that policymakers are providing measures and facilities, so that the vulnerable groups could stand the impacts of catastrophes. Nevertheless, the even greater challenge is how to find ways on what could effectively be done by policymakers and government leaders to lessen the occurrence of disaster or at least minimize its devastating impacts. The following are the more specific recommendations to ensure that social resilience factors are profoundly considered in building resilient communities, particularly in the context of disaster recovery.

First, expanding and refining social protection mechanisms in the recovery phase. Social protection programs have proven to be an effective mechanism toward reducing poverty as it directly provides assistance to the poor. Also, the cases from Indonesia and Thailand show that social protection are crucial in the long term recovery of the survivors of disasters. The success of social protection programs in Latin America and several countries in Southeast Asia prove that it is a potent development strategy to alleviate the world's poor. Given that the targets of this program are the poor and the vulnerable, which are also the high-risk population in the context of disasters and economic shocks, social protection is a powerful measure to increase social resilience vis-à-vis catastrophes. Nonetheless, there are four (4) things that could be done to ensure inclusive growth and social inclusion; namely: (1) Expand the coverage. The reach and beneficiaries of social protection programs need to be expanded from its current implementation scheme. The coverage of this program includes: unemployment and income support, universal basic old-age and disability pensions, basic child benefits, universal access to essential health care.

It could also take into account other crucial areas depending on the specific needs of the vulnerable groups in a particular locality; (2) Increase the number of beneficiaries. One of the critiques to social protection program is the seemingly small number of beneficiaries' vis-à-vis the total number of people vulnerable in a particular sector. That said, the recipients of the program could be increased to make it more all-inclusive; (3) Refine and sharpen the targeting system. To address the issue on beneficiary-selection, there is a need to sharpen and refine the program mechanism, particularly the targeting system to ensure that the beneficiaries are really deserving and the most vulnerable population; and, (4) Simplify claiming mechanisms. Many beneficiaries of this program have not undergone higher education. Also, a lot of beneficiaries reside in far-flung areas. That said, there is an imperative to make the claims of benefits easier for the recipients.

Second, reinforce the accountability and transparency of local government authorities. In the area of governance, reinforcing its initiatives, accountability, and transparency is needed. There is also a need for a mechanism to monitor the delivery of social services by the local government. The mechanism should: evaluate whether local governments allow its people to participate in decision-making and they are consulted; appraise whether local governments allocate adequate resources to effectively deal with shocks; and, whether they have effective functioning structures and systems at local levels, with proven good coordination between national, provincial and local governments. People outside the local government unit should do the monitoring mechanism. For instance, the academe and NGOs.

Third, increase and strengthen community-based disaster risk reduction and management initiatives. There is a need to beef up local level projects to assist the communities, particularly those vulnerable to disasters, to become more proactive rather than passive. This would mean investing more in community-based projects that would mitigate, prevent, and prepare them for future risk and shocks.

Finally, promote and encourage private-public partnership. Public-Private Partnership (PPPs) needs to be strengthened. The private sector, having sufficient resources, could augment what is lacking on the part of the state. That said, they are promoting social responsibility by engaging in social development-related endeavors. On the part of the state, it is necessary to encourage the involvement of private sector in development initiatives.

References

Christoplos I et al (2007) Linking Relief, Rehabilitation and Development (LRRD) & social protection—lessons from the early tsunami response in Aceh. First international conference of Aceh and Indian Ocean studies, Aceh

Devereux S, Sabates-Wheeler R (2004) Transformative social protection in IDS working paper 232. IDS, Sussex

Food and Agriculture Organization (FAO) (2005) Tsunami communities reborn: rebuilding livelihoods better than before. FAO, Rome

Gaillard JC (2010) People's response to disasters: vulnerability, capacities and resilience in Philippine context. Center for Kapampangan Studies, Philippines
Gin OK (2004) Southeast Asia: a historical encyclopedia from Angkor Wat to East Timor. ABCCLIO, California
International Youth Foundation (u.d.) Rebuilding young lives and livelihoods: the tsunami reconstruction initiative. IYF, Baltimore
Kivimaki T (2008) Case study June 2008: non-government actors in peace process—the case of Aceh. Center for Humanitarian Dialogue. Geneva, Switzerland
Midgley J (2013) Social development and social protection: new opportunities and challenges. Dev South Af 30(1):2–12
Nishino Y, Koehler G (2011) Social protection in Myanmar: making the case of holistic reform. IDS working paper, vol 11, No 386. IDS, Sussex
Oey E (1996) Sumatra: Island of adventure. Passport Books, Lincolnwood
Reid (2005) Is a speech delivered by Reid in the opening of the Aceh Research Institute in 2005. http://www.ari.nus.edu.sg/docs/Aceh-project/AnthonyReidLecture-Jun2005-Aceh.pdf. Accessed: 20 December 2013
Saarni M et al (2007). And Thailand still smiles: a study of post-tsunami livelihood recovery of Phuket and Khao Lak. Unpublished. https://jyx.jyu.fi/dspace/bitstream/handle/123456789/38335/URN:NBN:fi:jyu-201208212204.pdf?sequence=1. Accessed: 20 December 2013
Samson M (2009) Social protection approach to Myanmar, unpublished draft. UNICEF, Myanmar
Satake K (2007) Preparation for future earthquake and tsunami hazards: lessons learned from 2004 Sumatra-Andaman earthquake and the Asian tsunami. International conference on Aceh and Indian Ocean studies, Aceh
Save the Children (2008b) Four years later, Rebuilding lives after the tsunami: the children's road to recovery. Save the Children, England. http://www.preventionweb.net/files/8692_4YearTsunamiReportPDFFinal.pdf
Vebry M et al (2007) Community development approach in Aceh reconstruction, reflecting on lessons learned for Yogyakarta. Yogyakarta. In: Amaratunga and Haigh (eds) Post-disaster reconstruction of the built environment: rebuilding for resilience. Wiley-Blackwell, London

Part II
Indonesia

Chapter 11
Reform Amidst the Rubble

Kuntoro Mangkusubroto

Abstract The epicenter of the earthquake and tsunami was centered in Indonesia's westernmost province of Aceh, where over 180,000 people were killed and missing; over 5 % of the total population. The losses in in several of Aceh's west coast cities and districts reached as high as 35 % of their populations. In addition to the loss of life was the loss of livelihoods with one eighth of the population losing accommodation. Schools, hospitals, roads, ports, airports and other critical infrastructure were destroyed. Even the software of modern society, such as the provincial lands registry was destroyed and land ownership records lost.

The Government of Indonesia viewed the unprecedented global outpouring of support to victims of the Indian Ocean Tsunami in Aceh and North Sumatera as a major responsibility to be executed with competence and integrity. Support was provided from 50 countries and from over 900 organisations. It would be fair to say that the Government saw the potential for corruption as a grave threat to the whole reconstruction effort. To prevent this, new systems for accountability and transparency needed to be applied to support effective and timely implementation. These systems had to be applied in communities already suffering the debilitating impact of a long term insurgency and whose governor had just been arrested for corruption. This, in a country, then rated among the bottom ten countries in terms of corruption perception (Transparency International: Corruption Perception Index for 2004. Indonesia ranked 137 out 146 countries with a rating of only 2 out of 10). Business-as-usual and standard operating procedures had to be shelved.

Keywords Accountability • Coordination • Corruption • Integrity • Participation • Transparency

K. Mangkusubroto (✉)
President's Delivery Unit for Development Monitoring and Oversight, Jakarta, Indonesia

Rehabilitation and Reconstruction Agency (BRR) for Aceh and Nias, Banda Aceh, Indonesia
e-mail: hanief.arie@ukp.go.id

R. Shaw (ed.), *Recovery from the Indian Ocean Tsunami: A Ten-Year Journey*, Disaster Risk Reduction, DOI 10.1007/978-4-431-55117-1_11

11.1 Introduction

On top of the losses caused by the earthquake and tsunami, the people of Aceh had also suffered from the effects of a violent secessionist insurgency that had been ongoing for a generation. Many benefits from Indonesia's political reform movement starting at the end of the 1990s, such as openness, active democratic participation, defense of civil, political and wider human rights, freedom of expression and organization, had yet to take root in Aceh. At the time of the tsunami Aceh was governed under Martial Law. State capacity and public trust in the Government was therefore poor while the social fabric of Acehnese society itself was deeply frayed. To make matters worse on the eve of the tsunami the then governor of Aceh was arrested on charges of corruption, demonstrating that there were already pre-existing problems with integrity in public administration in this region.

Considerations about the need for accountable and transparent management of the reconstruction program were compounded by the unprecedented scale of international support that was being pledged. For the Government of Indonesia this tsunami of goodwill represented not only a major source of support for the reconstruction effort. It also represented a major threat to the nation's standing should these funds not be managed with integrity and competence.

The second region most affected by the earthquake and tsunami was the Islands of Nias, part of Aceh's neighbouring province of North Sumatra. The communities on Nias have long been the"distant cousins" from the economic, cultural and socio-political mainstream of North Sumatra. Geographic separation from the mainland and distance from established trade routes have combined to create this historic isolation. Administratively separate from Aceh, yet in great need of special reconstruction assistance, meant that the Indonesian Government needed to create a special arrangement to ensure support reached the affected communities on Nias.

In short, responding to all these complex challenges meant a total rethink of the way the country could manage such a disaster. Central to considerations would be a need to ensure demonstrable transparency, substantive accountability including active engagement with communities together with streamlined management arrangements capable of removing bottlenecks and other obstacles to quick progress. Achieving this meant designing, establishing and empowering a new agency that was structured in a way that would permit it to achieve the objectives above.

The agency designed to achieve each of the above was the Agency for the Rehabilitation and Reconstruction of Aceh and Nias, BRR. This chapter reviews the array of policies, arrangements and systems that were established for, and later by, BRR that allowed it to operate so effectively. The most powerful indicator of the effectiveness of the approaches used was demonstrated by the fact that of the USD 7.2 billion generously pledged in support of reconstruction efforts in early 2005, some 93 % had been transformed into programs, projects and initiatives on the ground by the time BRR concluded its mandate in April 2009. Almost all of the remaining 7 % consisted of large multi-year reconstruction projects that would be completed a couple of years later.

11.2 Establishing the Agency with an Appropriate Legal Instrument

The place to begin, in terms of building an effective agency, was through establishing an appropriate legal instrument to equip the agency with authorities and powers necessary to lead the national effort unencumbered by business-as-usual administrative and politico-bureaucratic obstacles. President Susilo Bambang Yudhoyono (popularly known as President SBY) enacted Government Decree in Lieu of Law No. 2/2005 regarding rehabilitation and reconstruction of Aceh and Nias on April 16, 2005. The legal instrument chosen was a Government Decree in Lieu of a Law. Such decrees are passed much faster into law than regular legislation but still require ultimate parliamentary endorsement. Failure to endorse would have meant the Decree lapsed. The Parliament did indeed endorse this Decree during its next session.

11.3 Institutional Design

Having agreed to the appropriate an legal instrument to establish the BRR the next critical set of issues covered an agreed institutional design. Special features here included the following:

1. The agency was to be led by a Ministerial level appointee. The purpose of this was to ensure that the policy proposals from the agency would not get lost in thickets of administrative, bureaucratic and other interagency "turf battles". As a ministerial level appointee the head of the agency would enjoy direct access to the President and Cabinet.
2. With a mandate to coordinate all aspects of rehabilitation and reconstruction the agency could operate as the one-stop go-to place for all partners in reconstruction from citizens through to multilateral agencies.
3. The agency was headquartered in Aceh. Agreement to locate this national level agency outside the capital city was based on the view that the center of action in reconstruction would be in Aceh and in Nias, not Jakarta. Being close to the center of action would ensure the agency could be accessed easily by affected communities as well as to be responsive and better informed of any emerging issues.
4. The head of the agency had authority to hire and fire as well as on issues of staff remuneration. This included flexibility to source staff from all sectors of society, not just the civil service. On remuneration it was critical for the agency to recruit the best possible staff. This meant providing salary packages commensurate with the private sector and donors. It also meant ensuring that these staff could be retained and not "headhunted" by other agencies, a process that ironically leads to de-capacitation of an agency; a problem noted in other programs around the world in post-conflict or post-disaster efforts.

5. The head of the agency retained authority over the organizational structure of the agency. It was necessary for the agency to evolve quickly as it moved through each phase of reconstruction without being overly burdened by issues such as administrative precedent, nor for staff to become bureaucratically complacent. The key principle in this point and in the point above was to ensure the accountability and responsibility for reconstruction efforts that were carried by the head of the agency were properly balanced with the necessary authorities to be able to command these internal management structures.
6. The treasury payments system for vendors and contractors was reformed. The purpose of this was to ensure quick, efficient and fully transparent processing of all payments made through Indonesian Government Budget to all payees. This in turn was to ensure integrity and full accountability on payments and restrict the potential for abuse or misuse of these public funds. In keeping with high standards of financial integrity, responsibility for managing the treasury operations governing payments for BRR related payments remained with the Ministry of Finance.
7. Each of these reforms, which were enshrined in legal instruments to establish BRR, should be seen as providing the agency with a fighting chance of being able to operate with integrity, competence, coherence and to perform with high levels of accountability and transparency. In this regard these reforms may be seen as providing the enabling conditions that would permit BRR to be functional in the complex circumstances in which it had to operate.

11.4 Taking Charge

With a strong legal basis and a wide ranging mandate, it was now for BRR, once established, to demonstrate it could live up to the expectations of the diverse communities it would serve.

As noted above, one of the key challenges to confront the agency would be to gain the trust of the communities it served. The approach adopted to face this challenge was engagement, openness and accountability. Each of these factors is indeed closely related in terms of practical engagement in the field.

While accepting that the BRR Executing Agency was a technical agency, there was also a wide awareness that the communities in which the agency was to work were deeply divided politically. This meant it would be dangerously naive to pretend political dynamics could be ignored.

For the Head of the Agency it was deemed critically important for the agency to establish and maintain the trust of the varied constituencies to the reconstruction program for it to succeed in its ambitious program. This called for an open engagement with the communities. It meant listening carefully to their concerns, being open to adjusting approaches. On issues of fighting corruption and any perception of partisan bias, it was deemed critical to be totally up front.

In the interview with journalists in Aceh after the BRR was established one journalist about whether the BRR would be helping people from GAM (the insurgents, with whom at that stage no peace agreement had been agreed). The answer was that the Agency did not care what colour uniform one wore. If you were a victim of the tsunami the Agency would be there to help. Some gasping in the audience suggests this was not the kind of non-partisan answer they has presumed they would receive. In this the Agency's perspective was clear. There would be discrimination in terms of partisan position for staffing. There were no tests to vet out people with affiliations with GAM or any other group. Jobs were open to any person of good will and competence and appropriate skills needed for the job.

On building and maintaining trust, critical considerations included establishing great transparency on how the Agency operated. It means bridging potential cultural divides with communities. This meant no neck ties for staff and removal of shoes at the door in accord with local customs. This final principle was enacted at the start when the agency operated from a rented house with no furniture. Key early meetings took place on the floor with lap tops literally on laps! But having established this no shoes principle as a basic sanitary requirement, it seemed fine to maintain it even as the Agency moved into final and properly equipped facilities.

11.5 Coordination Is More Than Meetings

Key to the BRR approach was to apply principles of transparency and participation as part of the overall mandate of coordinating the rehabilitation and reconstruction effort. Too often coordination is reduced to a series of endless meetings where some things are shared, others discussed but little is actually coordinated.

BRR took the view that, as the agency invested with the task of coordinating all aspects of rehabilitation and reconstruction, it needed firstly to know just what was being rehabilitated and reconstructed, by whom and where and with what community support.

One problem faced in this regard was that the BRR had actually been established over 3 months after the Tsunami. The Government's emergency relief agency operated for the first 3 months. Once it wound up its emergency relief activities it was the turn of BRR to step up. As a result many other agencies, bilateral, multilateral and NGO were already well established before the BRR was established. As a start-up operation facing those already established it was up to BRR to work carefully with these agencies all the while establishing its own mechanisms for coordinating activities.

The approach to building coordination consisted of several components.

11.5.1 Concept Note Endorsement

The first stage called for all projects to be submitted to the BRR for a transparent review and approval/revision/rejection. As the "new kids on the block" the demand for this kind of review was not always met with enthusiasm by agencies wishing to simply get on with the job. Indeed in practical terms ongoing projects could hardly be stopped! Key to the process was of course to build up a clear picture of the who, what, when, where, and how for all project initiatives. In the end this meant BRR had to coordinate and implement over 12,500 projects.

The information format to be filled in by project proponents was designed quite deliberately to be simple and not too long. Critical needs were just to know the basics so as to begin to build a clear picture of the emerging reconstruction situation. Fortnightly, and at time more frequently, review panels were convened at which proposals were discussed, not just by BRR staff but with external community and peer reviews. This provided a basis for transparency and for people to raise a red flag if they were aware of similar initiatives taking place.

11.5.2 Building the Map of the Who, What, When, Where, and How

Visualising the emerging patterns of rehabilitation and reconstruction activity was also a critical tool for coordination. In this regard BRR decided to make use of the Donor Assistance Database that was being applied by other tsunami affected countries and then grow it. The resulting program came to be known as the Recovery Aceh Nias Database or RANDatabase.

In doing so BRR also decided to integrate this web based platform more carefully into the overall coordination effort. In this regard the system was adjusted so that each agreed project would manage its own access to the site covering their project. They would be responsible for updating data and providing information on progress juxtaposed with the agreed performance as endorsed between them and BRR as part of the concept note endorsement process.

The transparent presentation of status and project progress was open both for donors and communities to review. In this regard it was seen as a key tool for holding project managers to account. It was also a means of identifying progress by regions. This was a critical management tool for BRR as it revealed emerging needs, shortages or capacity to shift resources to better ensure needs were met with minimal overlap.

11.5.3 The Integrated Team: Facilitation and Control

At the first meeting between BRR and the NGOs engaged in reconstruction one issue raised by several people was the nuisance of the short term visas that their international staff had to use. This meant much time was wasted entering and exiting Indonesia to update their visas. Others expressed concerns that, as they were new to Indonesia, and not yet incorporated under Indonesia law, they were very restricted in what they could do administratively and financially.

On return to the office the Agency decided to look at how to help these organizations. As the Agency had argued very forcefully in favour of organizations operating with integrity in conducting their activities, it clearly behooved BRR to ensure these same organizations could actually do so by removing barriers to them undertaking basic activities like opening bank accounts, renting equipment or issues staff contracts without having to engage in "gray area" fixes. While cutting through the necessary red tape to facilitate this kind of support like an easy thing, it still took 4 months to achieve. But the result was actually better than initially planned, as BRR was able to extend considerably the kinds of support provided to BRR's recovery partners to cover the issuance of clearances and permits in areas of varied as forestry, industry, trade, customs, police registration, immigration and agency endorsement.

The interagency one stop shop or Integrated Team was established and staffed by officials from each respective agency but with each working together collectively to facilitate all necessary permits and clearance within their agencies' remit. The office operated in a large room without internal walls in order to facilitate a stronger integration of effort and collective spirit to assist those who were assisting us in Aceh and Nias.

The results were frankly impressive. In practical terms it meant the time required, for example, to issue an extension on a visa was reduced to a couple of days instead of several weeks and with the duration of a year, not a month.

The work of the Integrated Team also provided a critical additional service to coordinating efforts at rehabilitation and reconstruction. Through aligning closely with the Concept Note process, the only permits processing that the Integrated Team could undertake was for activities and projects endorsed through the Concept Note. Agencies that did not wish to operate through the Concept Note process found they could not get extensions or permits. The value added provided by the Integrated Team was invariably superior to any costs in terms of time required to go through the Concept Note endorsement process.

11.5.4 BRR Operations Support

While processing the concept notes for project endorsement was a key start for agencies, as was the need to manage their status through the RANDatabase and securing permits and clearance support through the Integrated Team, many also required further direct support to deal with the range of on-the-ground coordination and implementation challenges they faced. This came through working with the various technical Deputies of the BRR and especially towards the end of the program working with BRR's Deputy for Operations.

Support could be required, for example, along the logistics supply chain or through resolving a regulatory log jam that was inhibiting implementation. On other occasions there were conflicts over access to land or disputes over who was doing what where and how. Many of these last disputes were resolved as one of the parties to the conflict had never processed its activities through the Concept Note process. This essentially meant BRR could do nothing much to help them, but could certainly assist those groups which were included as part of the reconstruction process. It did not take long for these other groups to realize the obvious benefits of joining the overall team for reconstruction.

For many others the convening power of the Deputies notably the Deputy for Operations allowed for discussions with various stakeholders to take place that led to agreements that resolved disputes. This helped reduce conflict, open up new areas for effective collaboration and integrate the results transparently for all to review through the RANDatabase.

The key purpose of these engagements was problem solving, debottlenecking and, where necessary, dispute resolution. Where problems could not be resolved at these levels they could be brought to the Head of Agency or if necessary where it could not be resolved on the ground in Aceh or Nias, issues could be taken further up the national policy making chain. In essence through these structures stakeholders and partners to reconstruction had a single door through which they could seek solutions to their problems, namely through the BRR.

11.5.5 Networks of Coordination

Beyond the four forms of internal coordination outlined above, there were several other modes of coordination for working closely with various groups of development partners.

11.5.5.1 Coordinating with the United Nations

One of the most important initiatives for coordinating the work of one large section of the reconstruction effort was the establishment of the Office of the United Nations Recovery Coordinator for Aceh and Nias, UNORC. This was a very creative application of the One UN idea in practice. This agency operated as a central system for bringing together the work of various UN agencies and provided valuable additional work in keeping BRR well connected to developments taking place and trends emerging with other recovery efforts around the region. Such an agency reduced considerably the amount of one-on-one work which BRR would have had to undertake without the UNORC.

11.5.5.2 The Red Cross and Red Crescent Societies

Unlike so many disaster recovery missions of the past, in the case of the program in Aceh and Nias the collective resources of the non-government sector led by the national societies federated through the International Federation of Red Cross and

Red Crescent Societies, IFRC, mobilized more resources than all the bilateral donors. These valuable resources needed to be integrated well within the overall reconstruction effort. To do so required lots of mutual engagement and exchanges as well as linking program operations.

11.5.5.3 Multi Donor Fund and RAN Trust Fund

The BRR also established, with the World Bank, a collective pot into which donors could contribute. This pooling of resources did allow for the alignment of major projects with the Government's own national budgetary cycle. Beyond the Multi Donor Fund, BRR itself also established its own Trust Fund known as the Recovery Aceh Nias Trust Fund, RANTF. This pooled resources mostly from non-traditional donors including from other developing countries that were happy for their contributions to be managed directly through the BRR rather than through a multilateral agency.

11.5.5.4 Regularised Stakeholder Engagement

One further initiative undertaken by BRR in close collaboration with its many partners was the Consultative Forum for Aceh and Nias (CFAN) and a dedicated Nias Islands Stakeholder Meeting (NISM). These were conducted usually twice a year. They provided a forum where all groups working on all aspects of reconstruction and in all regions from government, parliament, ambassadors, business, donors, communities, NGO and media could meet. The purpose was for all to see the larger picture of overall progress and to highlight challenges.

One possible danger that can emerge as each groups works in its particular part of such a huge reconstruction program is to lose sight of the wider picture. These meetings offered an opportunity for updates and to provide forewarning of likely future developments. The meetings allowed communities to raise their concerns to a wider audience and more generally as an opportunityfor various constituents to the program to engage in exchanges that busy work schedules usually do not permit. In general these meetings proved a great way for all stakeholders to see how they were contributing to a major exercise.

One very valuable piece of feedback received early was concerns that women were losing access to land and property rights. In response BRR initiated a wide ranging process of public consultation to redress this threat to property rights. The result was the implementation of a joint land titling system under which wives are now able to secure and preserve equal legal rights to homes and property with their husbands. This breakthrough policy is now being applied across the nation.

11.6 Testing the System

None of the above should lead anyone to believe it was all smooth sailing. Far from it! At one point BRR staff were locked onto the premises as demonstrators besieged the agency to demand more be done for victims still being housed in emergency conditions. In a related issue some activist NGOs complained that the data provided by BRR on housing completions were false or even fabricated. In response BRR beefed up the capacity of the RAN Data base to demonstrate literally who built it each house and who now owns it, each with a picture of the house and owners concerned and its geographic coordinates. By the end of the program the location of each of the 140,309 houses built could be seen by anyone with an internet connection.

Complaints about corruption, as noted before, represented one core threat to the potential success of the whole program. In response BRR established a dedicated Anti Corruption Unit that was free to respond to any and every complaint made by any party openly or anonymously. Such an internal control system had never been attempted in an Indonesian government agency. Equally the head of the Unit was free to pass any complaint which his team had investigated and had concluded could be potential evidence of corruption and dispatch the case material directly, and untouched by any other part of the BRR including the Head of the Agency, to Indonesia's Corruption Eradication Commission (KPK).

The Unit was also aggressive in publicizing and in communicating with various stakeholders including the press on corruption developments. The Unit provided advice to other offices within the BRR on approaches to better corruption proof their programs and indeed provided similar such advice to BRR's development partners be they the multilateral agencies, donors or NGOs.

To maintain an atmosphere in which both staff and BRR's various partners were made well aware of the consequences of deviant behavior, BRR maintained an active dialogue with key agencies notably the KPK and also the State Audit Agency, BPK. Indeed BRR lobbied hard for both to establish active operations in Aceh. In regards to the BPK BRR faced some eight different audits in its first year. This was a welcome development as it helped to both ensuring BRR built the best systems possible and also as a reminder to all stakeholders concerned that it was not worth the effort to play games with public funds, given the breadth and frequency of review by competent state agencies.

In terms of outcomes while it would be bombastic and absurd to suggest that no money went missing through corrupt practices, the problems certainly never became systemic, but rather episodic events undertaken by gamblers and high risk takers. It was also noteworthy that many cases that did lead to subsequent convictions were actually brought forward from within BRR's own internal control and whistleblower systems. At the level of audit, the agency was the first ministry to manage large amounts of funds to receive a no reservations audit clearance from the BPK.

More significantly as a sign of integrity and efficacy of funds managed and overseen is the fact that by the time the BRR closed its doors a total of 93 % of all the very generous funds pledged soon after the tsunami had been disbursed. This level of success in converting pledge to completed programs in the field remains a record. The trust built and sustained between BRR and its partners, and also with the communities serviced, represented the basis for this collective success story.

Among the key factors, which made this possible in the face of numerous potential upsets and backslides, were that the basis for this trust was earned and demonstrated through the various special and way beyond business-as-usual arrangements established. None of these should be seen as a cost. Rather they should be seen as critical and very cost effective investments in ensuring that trust could be sustained in the face of the many stresses and strains that would inevitably emerge in the implementation of such a complex program particularly in regions with pre-existing social, political, economic and logistical challenges.

11.7 Epilogue

Indonesia's experience in managing its mega disaster offers many valuable lessons that have been adopted by the country itself and are also available for other countries as they face their own mega disasters.

One key consideration is that in circumstances of disasters there is a critical need to consider corruption as a potential threat that needs to be faced up front. To do so requires the Government creating the regulatory and management framework that will allow reconstruction actors and communities to operate with integrity. This means ensuring smooth, efficient administrative procedures that encourage compliance, rather than encouraging an incentive to "get around the system".

Another key lesson is that systems of accountability and transparency should be seen as investments and not costs. The value served by these systems are revealed as the inevitable conflict emerge as to which groups are perceived to be getting more or less than others etc. These systems also allow for means to identify potential problems quickly before they became deleterious to the overall program.

One further lesson is that seeking reforms to systems in the middle of all other stresses and strains of the immediate disaster situation is actually possible. Most of the key reforms initiated were set in place in the lead up to establishing the agency and during the first few months once the agency was established and operational.

A related measure is that a number of the reform related efficiency and integrity related initiatives can be applied to other parts of public administration. This argues for a need to be able to "plug" these initiatives into wider public sector reform streams to assist ease their capacity to be applied.

There is also great merit in trying to capture the lessons of the disaster response as an ongoing process, rather than as an "after-thought" at the end of the reconstruction program. This ensures that people actually involved at various stages of the program can be sourced at the time they are active and thus with memories fresh and before the scatter to other programs.

References

BRR Book Series (2009a) Story: feat of the daunting launch. BRR, Banda Aceh
BRR Book Series (2009b) Finance: the seven keys to effective aid management. BRR, Banda Aceh
BRR Book Series (2009c) Breakthrough: thousands of paths toward resolution. BRR, Banda Aceh

Chapter 12
Reconstruction Through External Support: Key Observation in Aceh

Satoru Mimura

Abstract As UNISDR defines Disaster as "a serious disruption of the functioning of a community or a society involving widespread human, material, economic or environmental losses and impacts, which exceeds the ability of the affected community or society to cope using its own resources", intervention by international society is expected after a large disaster. Though the governments of disaster affected countries are primarily responsible for the event and have obligations to protect their citizens and liaise with international society as duty bearer, international donors are also expected play essential role on the response. In the case of Indian Ocean Tsunami, the affected countries received the largest ever interventions for emergency response and reconstruction from international society. Because of the hugeness and variety of support programs in comparison with capacity of recipient governments and communities, inefficiency and inadequateness were observed. There were some other reasons for the problems beside the lack of capacity such as lack of credible coordination system and gaps between donors and recipients. Meanwhile, many trials were made to overcome such constraints. This chapter presents a summary of lessons learned from experience of reconstruction with external supports for Indian Ocean Tsunami, especially on reconstruction stage.

Keywords Aceh tsunami • Emergency response • External support • Reconstruction • Recovery

12.1 Introduction

Indian Ocean Tsunami was the disaster for which world society provided the largest ever humanitarian support. However, it is a complicated work to estimate the accurate amount of foreign aid for the tsunami, for a large number of aid providers as well as the complex channel of huge money flow. Funds were provided not only through official

S. Mimura (✉)
Fukushima University, Fukushima, Japan
e-mail: r649@ipc.fukushima-u.ac.jp

R. Shaw (ed.), *Recovery from the Indian Ocean Tsunami: A Ten-Year Journey*,
Disaster Risk Reduction, DOI 10.1007/978-4-431-55117-1_12

Table 12.1 International source of tsunami funding

Source of funds	US$ million
Governments	6,481
Private Donation	5,490
International Development Banks	2,095
Total International Funds	14,067

Source: Flint and Goyder (2006)

and diplomatic channel, but also private and unofficial channels. Such supports took many forms such as cash, in kind, donations and loans. Also many donors provided technical assistance by deploying technical personnel and experts in the affected areas. In addition, it is difficult to identify the fund was additional for the tsunami relief or should be defined as aid which was already scheduled before the disaster.

According to the evaluation by the Tsunami Evaluation Coalition (TEC), which was established in February 2005 as an independent learning and accountability initiative in the humanitarian response, the total amount of donated funds for the tsunami response by the international society amounted to US$14 billion, and was the largest international response to a natural disaster in the history. Out of US$14 billion, 46 % came from governments, 39 % from private and 15 % from multilateral (shown in Table 12.1).

The largest bilateral donors were Australia, Japan, United States, Germany and EU, as their funding amounted to 60 % of support from foreign governments. However, scale of amount was not the only appraisal. Many of the contributions were provided with various conditions set down by donors, while type of funding could be classified into grants and loans. Also time periods for which the funds were offered varied (Jayasuriya and McCawley 2010).

As shown in Fig. 12.1, the majority of the funds were allocated to the worst affected countries namely Indonesia, Sri Lanka, India and Maldives. Thailand and India accepted emergency support from international society, but they did not officially requested international funding for reconstruction.

12.2 Coordination of External Supports

The coordination of large and complex efforts to response the tsunami was extremely challenging task, as a large number of agencies that had different purposes and methodologies was involved. Especially for the first few months, inefficiency was observed due to the lack of capacity to coordinate and absorb external supports by recipients. In order to improve the situation, various attempts were made. In Indonesia, the World Bank established the Multi Donor Fund (MDF), while the national government established Badan Rekonstruksi dan Rehabilitasi (BRR) or Agency for the Rehabilitation and Reconstruction of Aceh and Nias in April 2005. Such mechanisms were effective for coordination with the national government and

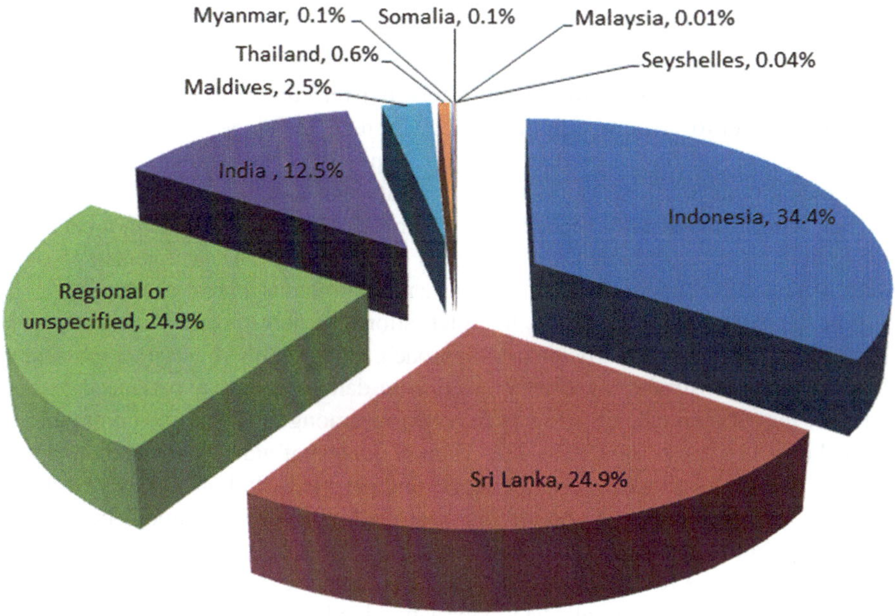

Fig. 12.1 Share of funding by recipient. *Source*: Funding the Tsunami Response (2006)

among donors (BRR 2009). Establishment of multilateral coordination body as well as credible national agency is the key to improve the coordination, however if incentives for donors to collaboration are weak, coordination would become difficult.

12.3 Supply Driven Support

TEC noted that much of the response by international donors was driven by the availability of funds, or by contextual opportunism, rather than by needs (Telford et al 2006). Inefficiencies in aid programs reflected the problems of such supply driven assistance. Also it is criticized that poor coordination among aid donors hampered effective delivery and distribution. Donors are often tending to be supply oriented. Of course their motives to provide humanitarian aid were to help affected people, however, their strategy and activity were tend to be influenced by many factors such as diplomatic and domestic policy.

In case of mega disasters in developing countries, capacity of recipient governments are always bottleneck for accepting international supports, and it was also observed in the case of Indian Ocean Tsunami. Needs for support of disaster affected countries are changing as time goes on, from emergency response immediately after the disaster to early recovery and reconstruction stages. On each stage, it is necessary for outside donors having adequate opportunities to grasp priority needs of affected people.

12.4 Quality or Speed

At the early response stage after the disasters, delivery of support without delay is a priority in order to save lives. In the subsequent recovery stage, restoration of housing and livelihood has to be done as soon as possible, but it is also important to secure quality as it will be the foundation of future of affected people.

Survivors of the disaster needed support for restoration of daily life in great haste, while donors were accountable to taxpayers or patrons, media, recipients and other donors for quality support. Meeting all those requirement is a great challenge.

For the disaster reconstruction, the international society professed to take "build back better" and "public participation" approach. However, there wasn't a moment to lose for the reconstruction and restoration from damages, they experienced friction whether they should take time to make consensus among residents and communities, or be hurry to recover their daily life. Though quality of aid for reconstruction was crucial, speed was also key issue for supporting programs. In housing sector support in Aceh, reconstruction on site or migration to higher ground should be decided through public participation taking enough time for discussions. However, it was urgent issue for the residents to secure their life, only the residents whose land were submerged or eroded were allocated alternative land on higher ground, while other houses rebuild on site though it had been inundated by the tsunami.

As large flow of support came into relatively small domestic markets in short period of time, shortage of materials and services in the local markets was invited in line with construction boom during reconstruction stage. Consequently it caused inflation of construction cost. The situation got worse when pressure of Dutch disease arouse in the country's economy.

12.5 Reconstruction Support of Aceh

As the disaster gave the largest ever impact to the world, international society including donor countries, international organizations, NGOs, private companies and individuals offered various supports not only for the emergency support but middle and long term reconstruction. Many discussions and evaluation were done on the response in the emergency stage but the consequent reconstruction stage.

As the disaster affected region had been suffering from long term conflict and less developed in the country, restoration of Aceh shouldn't just aim to recover damages by the disaster but rebuild the region better than before (International Cooperation and Mega Disasters Research Team 2013). Agency for the Rehabilitation and Reconstruction of Aceh and Nias—BRR followed the sequence of recovery activities as the diagram image illustrates in Fig. 12.2. The priorities in the early stage were on emergency response, consequently sector as the basic of restoration of daily life. After housing sector peaked off in 2006, reconstruction of infrastructures and business were priorities in order to revitalize livelihood and regional economy in the affected area.

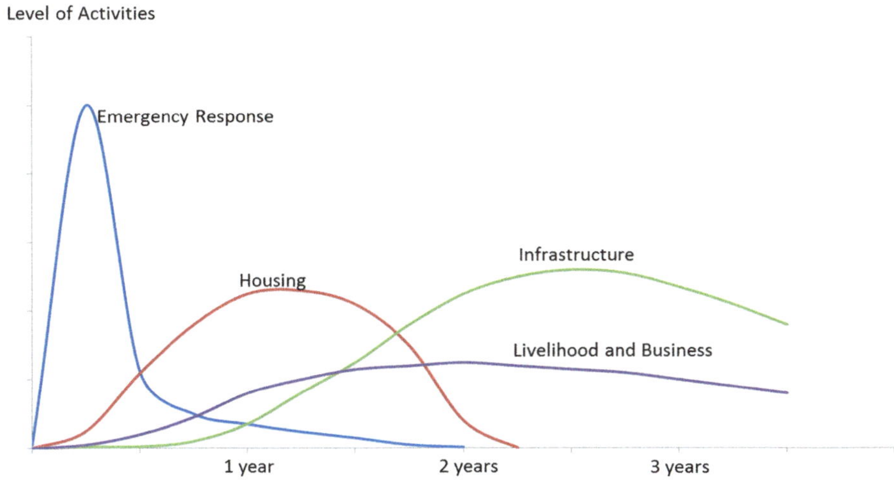

Fig. 12.2 Sequence of recovery activities. *Source*: Aceh and Nias one year after the Tsunami (2005)

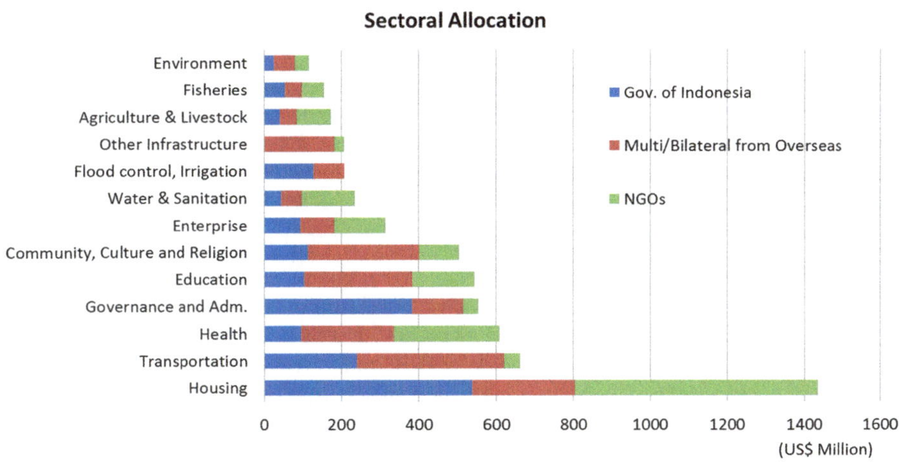

Fig. 12.3 Allocation by sector. *Source*: Aceh and Nias two year after the Tsunami (2006)

Expected amount for reconstruction and restoration of Aceh had amounted to $61 billion (BRR and Partners 2006), while 60 % of them had been expected to be provided by international donors and NGOs. Supports were provided in various sectors as shown in Fig. 12.3. Bilateral donors tended to support reconstruction of social infrastructures, while international organization and NGOs focused on housing, health and education sectors.

The selection of supporting sectors were decided in consideration with recipients' needs as well as interests of donors themselves, so needs in some sectors were not filled e.g. Environment. On the stage of reconstruction, restoration of livelihoods of affected people was crucial, but scarce donors and NGOs were aware, or could afford their support activities besides housing sector. Of course various programs of cash for work were planed and implemented, as well as job opportunities were provided related to project done by donors. However, those programs were ad hoc and not provided equally, as it was not directly intend to enhance capacity of people affected, nor to stabilize their livelihoods in longer term.

Following the various programs for restoration of the region, BRR, local governments and Japan International Cooperation Agency—JICA started collaborative project in order to support communities in Aceh Province. "Project on Self-sustainable Community Empowerment Network Formulation in Nanggroe Aceh Darussalam Province" aimed to establish self-sustain communities by strengthening livelihood of residents and capacity of local governments and their officers. The project commenced February 2007, and concluded March 2009, along with the operation of BRR.

The project followed the preceding study on urgent rehabilitation and reconstruction plan for Banda Aceh City by JICA, as they realized the importance to materialize community empowerment effort which was recommended in the plan. The project covers Banda Aceh and 6 provinces as shown in Fig. 12.4, while its budget was amounting to US$2.5Million including the cost for six technical experts from Japan.

One of the main components of the project was a model program to support livelihood of tsunami affected communities in Banda Aceh. They were supported to start small business such as small aquaculture, fish processing, sawing and cake making. As the autonomy of residents was respected, some of the groups could sustain their activities with minimal support, while some could not. By the end of the project, 34 community empowerment activities were organized. In 2012, JICA conducted ex-post evaluation on the project and confirmed 24 groups were still working.

Other components were replication of the model program to surrounding regions of Banda Aceh, and networking of stakeholders of the community empowerment program. It also contained disaster awareness and preparedness activities, as preparation for the next disaster was crucial point for "Build Back Better" approach.

The obvious feature of the project was that it also focused on the residents and community of Aceh Tengah, where was not directly affected by the tsunami but had been suffering from long term conflict in the region. It was because the Government of Indonesia and the Free Aceh Movement (GAM) signed a peace agreement on August 15, 2005 after 30 years conflict. So it was also the important mission for BRR to restore impoverished inland region due to the conflict. If the people in the inland area had felt inequity to people lived in coastal area, it might invite further confusion in the region.

12.6 Conclusion

Aid agencies are required to be accountable to tax payers and the public. In addition, they have to be accountable to the affected people and other aid agencies (HAP 2010). Thus the process of selections of sectors and target groups has to be transparent

Fig. 12.4 Project sites. *Source*: JICA (2013)

paying attention to support the most affected and in urgent need. Donors usually tend to be supply oriented and needs by affected people are changing continuously for the time being. Donors have to pay attention to grasp their priority needs.

Importance of coordination has been emphasized repeatedly. In developing countries, the coordination of responses against mega disasters is likely to be difficult due to the lack of coordination capacity in contrast to a large number of donors. Establishment of national and international mechanisms for coordination is urgent need immediately after the occurrence of disaster. It is crucial for recipient countries to clarify the responsible body of coordination and give them authority. In case of Indonesia, the Government established BRR in April 2005 as a responsible agency for reconstruction effort including coordination of donors.

Mega disasters attract a great deal of attention in the world, but it doesn't last long. Aid agencies tend to be disposed programs which can obtain obvious result in shorter term. In the case of Indian Ocean Tsunami, most of reconstruction supports were poured in housing sector. Though it was essential support to help the affected people, livelihood support in order to sustain their future should have more attention. There were many trials of cash for work programs, but it can't be sustainable without respecting initiative and autonomy of residents.

References

BRR (2009) Ten management lessons for host governments coordinating post-disaster reconstruction. Executing Agency for Rehabilitation and Reconstruction of Aceh-Nias, Banda Aceh

BRR and International Partners (2005) Aceh and Nias one year after the tsunami the recovery effort and way forward. BRR, Jakarta

BRR and Partners (2006) Aceh and Nias: two years after the tsunami—2006 progress report. BRR, Banda Aceh

Flint M, Goyder H (2006) Funding the tsunami response: a synthesis of findings. Tsunami Evaluation Coalition, London

HAP International (2010) The 2010 HAP standard in accountability and quality management. HAP, Geneva

International Cooperation and Mega Disasters Research Team (2013) Tackling mega disasters Japan and the world. Saeki Printing, Japan

Japan International Cooperation Agency (2013) The study of reconstruction processes from large-scale disasters: JICA's support for reconstruction—final report

Jayasuriya S, McCawley P (2010) The Asian tsunami aid and reconstruction after a disaster. ADB Institute, Cheltenham

Telford J, Cosgrave J, Houghton R (2006) Joint evaluation of the international response to the Indian Ocean Tsunami: synthesis report. Tsunami Evaluation Coalition, London

Chapter 13
Progress of Coastal Line Rehabilitation After the Indian Ocean Tsunami Around Banda Aceh Coasts

Syamsidik, Adhitya Iskandar, and Teuku Muhammad Rasyif

Abstract Coastal recovery at tsunami affected areas is another challenging work for practitioners. Pressures from demand to acquire ample and suitable land to restart economy activities and from harmonizing nature with the development are among the reasons why the coastal recovery may enter into complicated decision process. This chapter elaborates lessons learned from coastal recovery process around Banda Aceh to restore coastal lines around this area after severely damaged by the Indian Ocean Tsunami waves in 2004. A time series of maps to show the coastal morphological dynamics around Banda Aceh was used to compare the coastal features between 2004 and 2011 conditions. Other similar studies focusing on the coastal recovery around this area were also included to strengthen the lessons of the coastal recovery process orchestrated after the tsunami. Four coastal locations around Banda Aceh were selected to show the different approaches used to recover the coasts and impacts on the area after the efforts.

Keywords Coastal features • Coastal morphology • Hard structures • Sediment transport

13.1 Introduction

Coastal area and coastal lagoons around Banda Aceh have been long functioned not only for brackish water environment buffer zone, but also for fishery area. The coastal areas were used for location of shrimp and fish ponds. However, due to erosion problem caused by the Indian Ocean tsunami event on December 26, 2004, the coastal area were severely eroded. Land barriers situated between the coastal lagoons and the coastal water were washed away creating new forms of the coastal line.

Syamsidik (✉) • A. Iskandar • T.M. Rasyif
Tsunami and Disaster Mitigation Research Center (TDMRC), Civil Engineering Department of Syiah Kuala University, Jl. Prof. Ibrahim Hasan, Gampong Pie, Banda Aceh 23233, Indonesia
e-mail: syamsidik@tdmrc.org

R. Shaw (ed.), *Recovery from the Indian Ocean Tsunami: A Ten-Year Journey*, Disaster Risk Reduction, DOI 10.1007/978-4-431-55117-1_13

Wave hydraulic energy created large dragging force that demolished the coastal barriers. The damages led to highly saline intrusion to the fish and shrimp ponds around the area. This was followed by difficulties to the local fishermen to run the ponds again after several years of the tsunami event. To accelerate the fishery activity, it was considered important to recover the coastal barriers immediately. This was also to enable to re-create the brackish water environment again to the surrounding coastal zone. The brackish water salinity level that is suitable for the fish and shrimp ponds is between 0.5 and 17.0 ppt. The condition inspired approaches used in coastal rehabilitation process. Some of them were rehabilitated by mean of hard structures, and some parts of the lagoon were left to natural process for their recovery process.

During Aceh-Nias earthquake and tsunami rehabilitation and reconstruction process (BRR Aceh-Nias), conducted between 2005 and 2009, the coastal lines were rehabilitated by constructing about 6.5 km of revetment structures along the Banda Aceh coast. The structures were made from rubble mound materials where quarries of the material are easily found around the Banda Aceh city. The revetment structure did help the Banda Aceh to rehabilitate the coastal line quickly. However, the structure hindered long shore drift to supply sediment to down drift part of the revetment structure (Syamsidik et al. 2011). On the other hand, the coastal lagoons which were not protected by any revetment structures were successfully recover its coastal barriers in a longer period. Impacts of tsunami on coastal morphology have been studied at several occasions of tsunamis. Tanaka et al. (2012) showed examples of tsunami wave interactions with the coastal morphological changes, including the damages of coastal forest, around Tohoku area due to the Great East Japan Earthquake and Tsunami (GEJET) in 2011. Observations on coastal morphology recovery around the GEJET affected area was also observed by Liu (2013). The study revealed that coastal restoration process was hindered by the lack of natural sediment supply as it was shown in the case of Rikuzentakata coast. On the other hand, the study also discovered that after 1 year of the disaster event, the coast only could recover about 40 % of the coastal area compared to the condition before the tsunami event. Among the most important factors that determined the coastal recovery speed were hydrodynamic regimes and sediment supply availability (Tinh et al. 2012). Another coastal recovery process after the Indian Ocean Tsunami was also studied in Phang-Nga of Thailand (Choowong et al. 2009). The impacts of the tsunami waves on the coastal morphology further affected water bio-diversity around a lagoon or a river mouth. This was found in the case of South Western Thailand river mouth. The disappearance of sea grass vegetation at almost all area around the Southwestern coasts of Thailand, that were affected by the tsunami, were confirmed (Nakaoka et al. 2006).

The two approaches of the coastal rehabilitation around Banda Aceh post the Indian Ocean tsunami give important lessons learned for other similar rehabilitation around coastal zone. The coastal lagoons around the city of Banda Aceh have been shaping new morphology that further contributed to the changes of environment around the lagoons (Fig. 13.1). Prior to the Indian Ocean Tsunami, the lagoons were

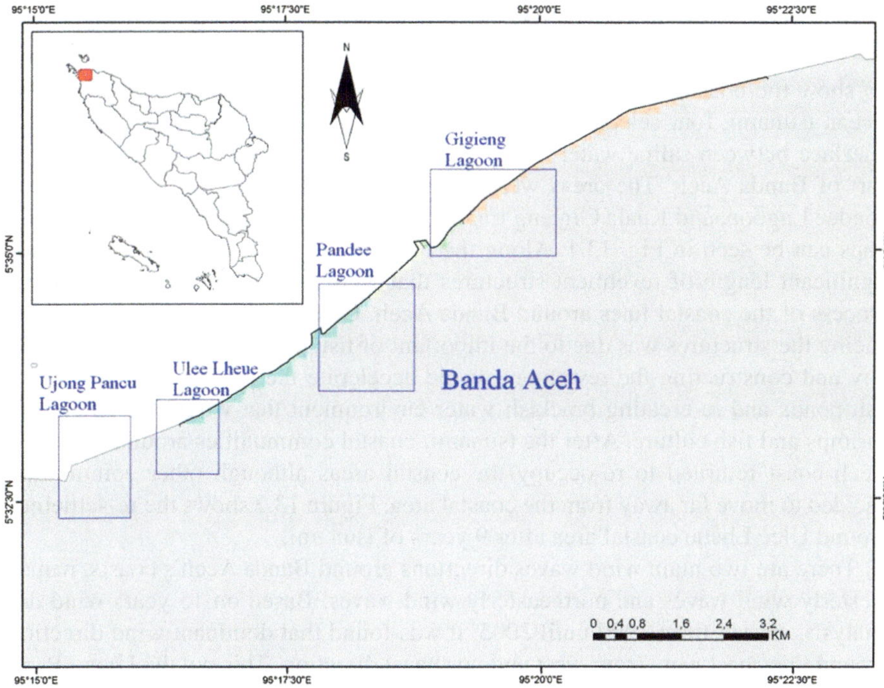

Fig. 13.1 Coastal areas for showing the coastal morphological changes around Banda Aceh after 8 years of Indian Ocean Tsunami

estimated to have brackish water. However, 8 years of the coastal lagoon process have changed the water environment. The changes of the water environment may also lead to the changes of coastal vegetation around the lagoons. Observations of several parts of the coastal lagoons have been done previously to observe the development of aquaculture environment before and after the Indian Ocean Tsunami by Griffin et al. (2013). They focused on incorporating information of the destruction of the fish/shrimp ponds around the coastal lagoons, development of mangroves forest around the area, and the livelihood of the coastal community. This study extended the investigation to relate the morphology dynamics at four coastal lagoons (in terms of the morphology features, i.e. land, water, ponds, and coastal vegetation) to hydrodynamic situation around the lagoon, and time-series salinity measurements at selected coastal lagoons around Banda Aceh.

This article is aimed at elaborating the coastal lagoons' morphology dynamic that are expected to recover the coastal water environment after the Indian Ocean Tsunami. This study was done by means of spatial analysis and field measurements at selected coastal lagoons around Banda Aceh.

13.2 Area of Observations

To show the coastal morphological changes around Banda Aceh after the Indian Ocean tsunami, four selected areas were studied here. The areas were selected as interface between saline water of sea water and brackish water around the inland part of Banda Aceh. The areas were Ujong Pancu Lagoon, Ulee Lheue Lagoon, Pandee Lagoon, and Kuala Gigieng Lagoon. A map of the coastal areas for observations can be seen in Fig. 13.1. Along the coastal line of Banda Aceh, we found a significant length of revetment structures that were built to accelerate restoration process of the coastal lines around Banda Aceh. One of the main reasons of introducing the structures was due to the important of fish ponds to the coastal community and constructing the revetment would accelerate the process of restoring the fish ponds and re-creating brackish water environment that would be suitable for shrimps and fish culture. After the tsunami, coastal communities around the Banda Aceh coast returned to re-occupy the coastal areas although other communities decided to move far away from the coastal area. Figure 13.2 shows the re-settlement around Ulee Lheue coastal area after 9 years of tsunami.

There are two main wind waves directions around Banda Aceh's coasts, namely westerly wind waves and northeasterly wind waves. Based on 15 years wind data analysis, started from 1990 until 2005, it was found that dominant wind directions around this coast were from west and northeast direction. This put the Ujong Pancu and Ulee Lheue Lagoons, those are located at the southwestern coast of Banda Aceh

Fig. 13.2 The Ulee Lheue Coast located at the Southwestern coast of Banda Aceh city taken in March 2013

Fig. 13.3 The Pandee Lagoon (*upper part*) and the Gigieng Lagoon (*middle part of the picture*). The picture shows the coastal lagoons condition in February 2013. Floodway jetty construction and the Lampulo fishery port structures are assumed to block the longshore sediment transport along the coastal area

as updrift part of a series of coastal structures (namely Lampulo Fishery Port Structures and Floodway Jetty Construction). On the other hand Gigieng lagoon is situated at the downdrift part of the structures. Therefore, in this paper, Ujong Pancu, Ulee Lheue and Pande Lagoons were classified as updrift lagoons. Meanwhile, Gigieng Lagoon is called as a downdrift lagoon. The two classifications were used for further analysis of the coastal lagoons morphology dynamics. Aerial photo for the locations and important coastal structures around this area can be seen in Fig. 13.3.

13.3 Morphology Dynamic Based on Spatial Analysis

Observation of the coastal morphology dynamic was done by comparing time series aerial images of the lagoons. The images were dated in 2004 (prior tsunami), 2005, 2007, 2009, and 2011. All images were obtained from the Google Earths' time series images those are also part of Quickbird satellite images. Spatial analysis was made to compare portion of land, coastal vegetation, and ponds (fish/shrimp ponds) that were found based on the image interpretation. The comparison of the spatial unit was aimed at providing information of the coastal lagoons' morphology dynamic in terms of land portion, development of coastal vegetation (either made from mangrove or *casuarinas* sp.), and fishery development by the coastal community from time to time. It was expected that all those three units of comparison could explain interaction between the coastal morphology dynamic and the coastal community during the observation. GIS technique was used to digitize the images and to measure the area of each of the spatial unit analysis.

Ujong Pancu coast was a place for early settlement in Aceh. It was told that this area was a place where around tenth century a large Hindu settlement was there prior to Islamic Kingdom era in Aceh. The settlement was known as Indrapurwa settlement. It was believed that the settlement was among three other Hindu settlements at that time in Aceh. Several tsunamis have impacted a large area of this coastal zone to be eroded and being submerged. Land-subsidence due to tectonic activities contribute to subsiding this area about 2 cm per year (McKinnon, 2013). Due to the Indian Ocean Tsunami in 2004, about 100 m of the coastal line around this area was eroded to inlandward.

In 2005, at the observation area of Ujong Pancu coast, there was about 50 % of the area was land and leaving no coastal vegetation and ponds at all at that time (Fig. 13.4; Table 13.1). After several attempts by constructing revetment to accelerate the coastal line recovery and to facilitate fishermen to start their economy activities, about 0.2 km^2 of fish ponds can be found around this area.

The fish ponds were made by the coastal communities after coastal lines being rebuilt by natural process. No revetments or other hard structures were made to recover the fish ponds around this area until 2009. Coastal vegetations were started to grow after 2009. The Ujong Pancu is one of best examples of letting natural process to recover the coastal area after a severe erosion due to tsunami. This was not seen at other parts of Banda Aceh coasts as they will be elaborated further in next section.

The second observation area is Ulee Lheue coastal lagoon. Coastal morphological dynamic around this area since 2004 until year 2011 can be seen in Fig. 13.5. Before the Indian Ocean tsunami, this area has about 8.6 km^2 of land area. Due to the tsunami, about 26 % of the land area was eroded by the tsunami waves. About 0.34 km^2 of coastal vegetation were also totally damaged. Recovery of the coastal lines around this area was started to show positive progress after 5 years of the tsunami. In 2009, about 95 % of the land area was recovered naturally. No hard structure was introduced to restore the coastal land. However, slow progress of coastal vegetation recovery was seen until 2011. Recently, a number of efforts to restore the coastal vegetation was done by various parties, including NGOs and local people. Comparison of the coastal features around Ulee Lheue coastal lagoon can be seen in Table 13.2.

Pandee Coastal Lagoon was also severely damaged by the tsunami waves. About 20 % of the land area was eroded and it recovers totally after 7 years later. Time series of the coastal morphological changes around Pandee Coastal area can be seen in Fig. 13.6. The process of the coastal recovery around this area was rather fast if it is compared to Ulee Lheue and Ujong Pancu coasts. This was because a set of revetment structures were introduced to speed up the coastal lines recovery. Before the tsunami there was 1.88 $km^2$of the areas were fish or shrimp ponds. Due to large

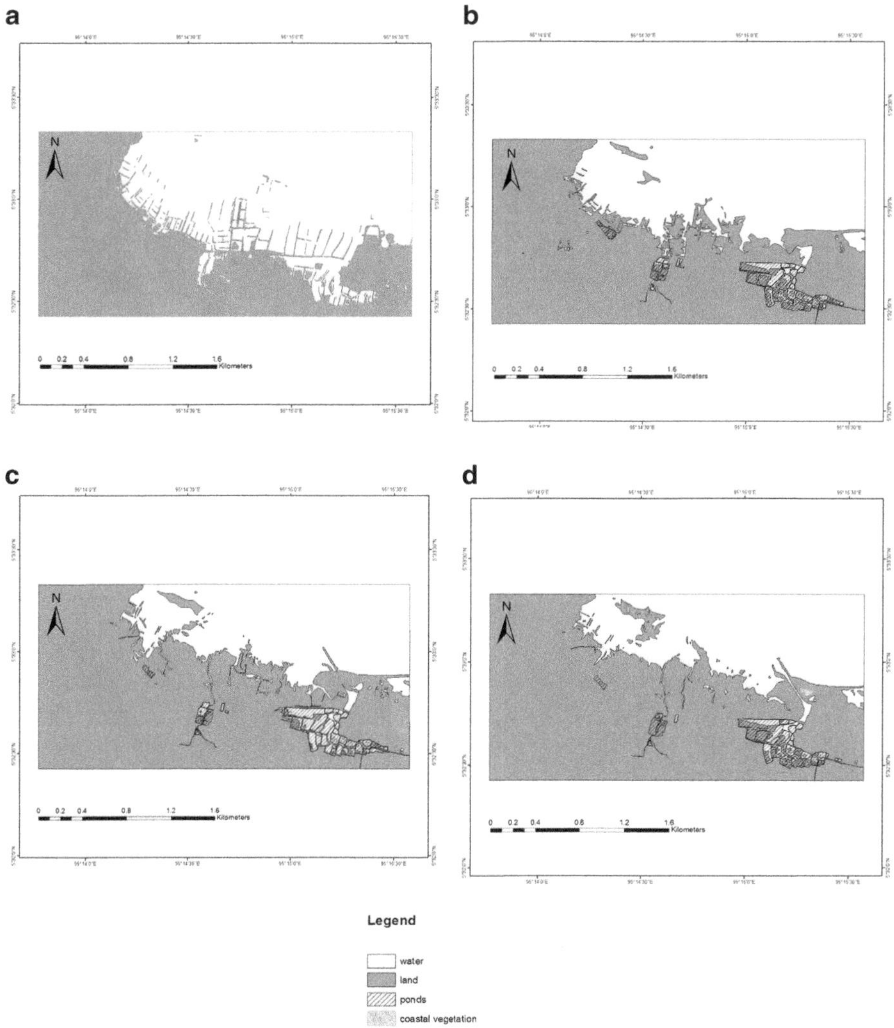

Fig. 13.4 Ujong Pancu Lagoon morphology dynamic. (**a**) Year 2005 (after Tsunami). (**b**) Year 2009. (**c**) Year 2010. (**d**) Year 2011

Table 13.1 Ujong Pancu coastal morphology dynamics

	Area (km^2) for year of image			
The coastal morphology features	2005	2009	2010	2011
Land	2.86	3.39	3.60	3.76
Water in the lagoon	2.71	2.17	1.95	1.81
Ponds	0.00	0.23	0.23	0.21
Vegetation (*Casuarina* sp./*Rhizopora* sp.)	0.00	0.01	0.01	0.01

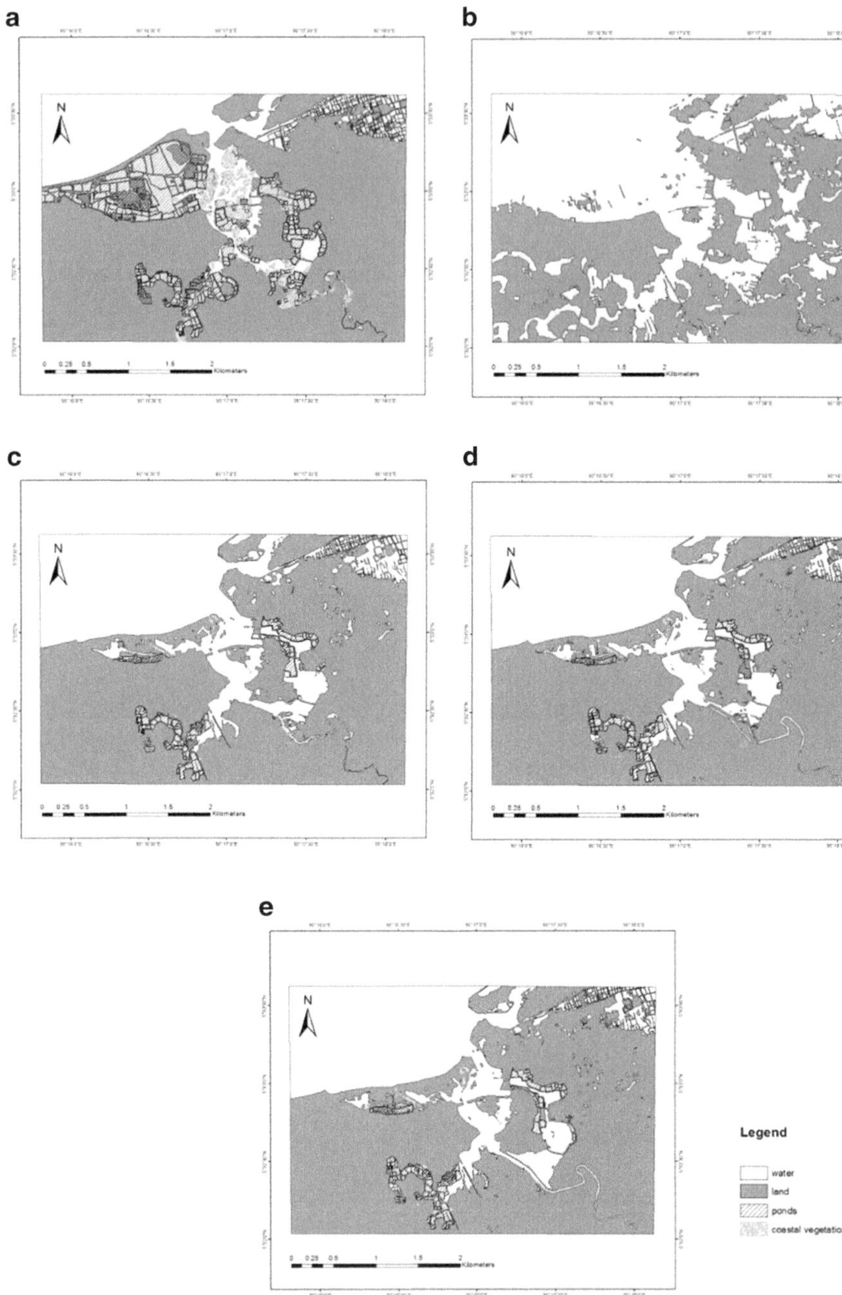

Fig. 13.5 Ulee Lheue Coastal morphological changes before and after the Indian Ocean Tsunami 2004. (**a**) Year 2004 (before Tsunami). (**b**) Year 2005 (after Tsunami). (**c**) Year 2009. (**d**) Year 2010. (**e**) Year 2011

Table 13.2 Coastal features area around Ulee Lheue coastal Lagoon

	Area (km^2) for year of image				
The coastal morphology features	2004	2005	2009	2010	2011
Land	8.60	6.35	8.13	8.26	8.26
Water in the lagoon	3.97	6.23	4.45	4.31	4.31
Ponds	1.90	0.00	0.40	0.39	0.39
Coastal vegetation (*Casuarina* sp./*Rhizopora* sp.)	0.34	0.00	0.02	0.03	0.04

economy activity around this area, a quick measure to restore the coastal line was considered important. This inspired the construction of the revetment structures around this area. The attempts were seen successful to accelerate the process. Coastal vegetation around this area was only 50 % recovered compared to the condition between 2011 and the before the tsunami. At the present, most of the coastal vegetation around this area were *Rhizopora* sp. Ponds were also not successfully recovered after the tsunami. A large portion of the community was relocated to other areas, which is far away from the coastal area. This forced some of the people to change their occupation from fishery jobs to other types of occupation. This also contributed to the low rate of the ponds' area recovery. A large portion of this area was still submerged by water and no significant activities around this area was found. On the left hand side of the observation area, there is a Ulee Lheue Ferry port. The port was also reconstructed after the tsunami. It is one of important ports in Aceh for sea-transportation from Banda Aceh area to other small islands around Banda Aceh, such as Weh Island and Pulo Aceh Islands. Comparison of the coastal features around this area can be seen in Table 13.3.

The next observation area is Kuala Gigieng Coastal Area. This area is located at the right hand side of flood canal of Krueng Aceh river. It is also named as Alue Naga floodway canal. Prior the Indian Ocean tsunami, this area was a coastal lagoon with active fishery activities around this area including fish and shrimp ponds. Due to the tsunami about, all the ponds were severely damaged and only 66 % of the ponds' area can be recovered until now. Outlet of the Kuala Gigieng coastal lagoon had been also shifted to landward direction. Coastal morphological changes around this area can be seen in Fig. 13.7. Comparison of the coastal features area of Gigieng coastal area can be seen in Table 13.4.

Progressive coastal erosion around Kuala Gigieng Coastal area was indicated by the recession of coastal line around this area. The coastal lines around this area are eroded due to the blocking of longshore sediment transport around the Alue Naga floodway Jetty construction. As this area is a downdrift part of the Banda Aceh coast, the Kuala Gigieng coastal area was expected to experience coastal lines' recession due to the disturbance of the Jetty construction. This area has been studied by Syamsidik et al. (2011) showing the progressive coastal lines recession after the completion of jetty construction after tsunami rehabilitation and reconstruction process.

Fig. 13.6 Pandee Coastal morphological changes before and after tsunami. *Dark grey* area is land area, *green color* is to represent coastal vegetation. (**a**) Year 2004 (before Tsunami). (**b**) Year 2005 (after Tsunami). (**c**) Year 2009. (**d**) Year 2010. (**e**) Year 2011

Table 13.3 Coastal features area around Ulee Lheue coastal lagoon

	Area (km²) for year of image				
The coastal morphology features	2004	2005	2009	2010	2011
Land	3.50	2.81	3.04	3.34	3.55
Water in the lagoon	4.93	5.61	5.38	5.08	4.87
Ponds	1.88	0.00	0.19	0.18	0.19
Coastal vegetation (*Casuarina* sp./*Rhizopora* sp.)	0.63	0.00	0.23	0.28	0.30

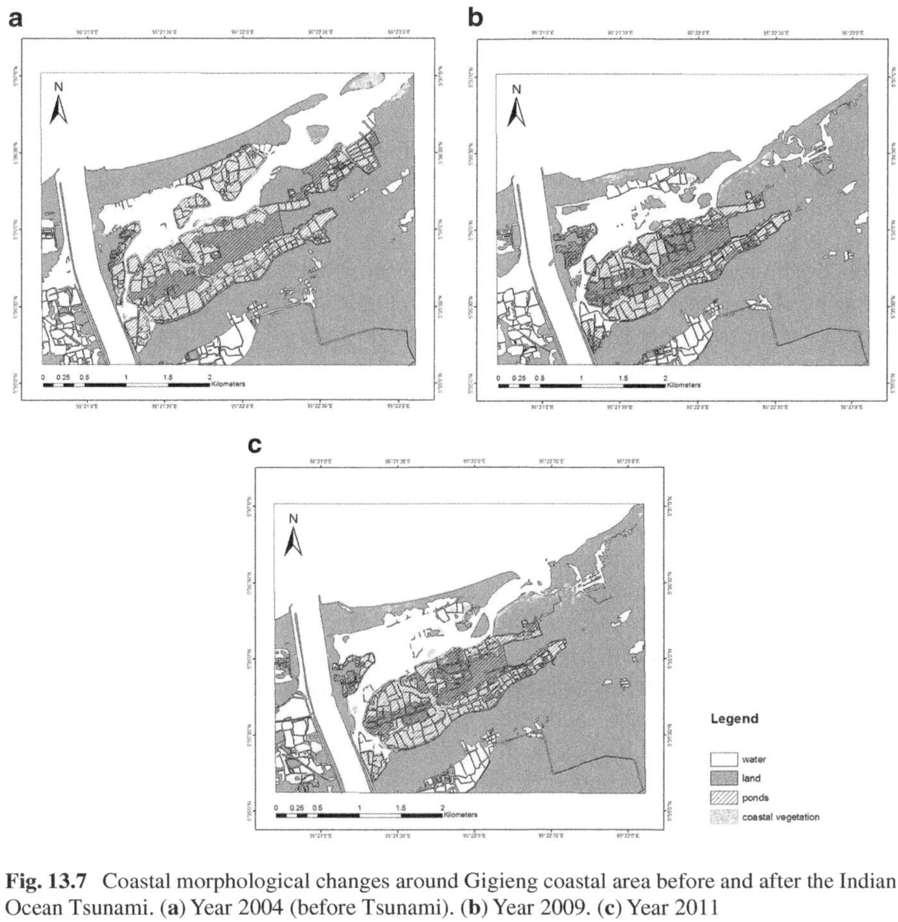

Fig. 13.7 Coastal morphological changes around Gigieng coastal area before and after the Indian Ocean Tsunami. (**a**) Year 2004 (before Tsunami). (**b**) Year 2009. (**c**) Year 2011

Table 13.4 Coastal features area around Gigieng coastal area

The coastal morphology features	Area (km^2) for year of image		
	2004	2009	2011
Land	7.39	7.46	7.35
Water in the lagoon	8.15	8.07	8.18
Ponds	2.73	1.81	1.79
Coastal Vegetation (*Casuarina* sp./*Rhizopora* sp.)	0.18	0.09	0.15

13.4 Lessons Learned from the Coastal Recovery

Coastal morphological changes around Banda Aceh give several lessons to be learned for recovery process at similar tsunami affected areas. The case of Ujong Pancu Coast and Ulee Lheue Coastal Lagoon are two important examples of how the recovery process was helped by natural process measures without any significant hard structure intervention. Although, it took longer time, these two locations were seem successful to re-create new land barriers around the area and shaping back the coastal lagoons. The sediment transport cells around these areas were rather well managed and no large sediment deficit was found after 8 years of the Indian Ocean tsunami. These areas could not be fully recovered to the condition prior to the Indian Ocean tsunami due to several reasons. The areas were not recipients of sediment budget discharged from Krueng Aceh River as the main river crossing Banda Aceh city (Meilianda et al. 2010). The contribution of the sediment from the Krueng Aceh river was important to the morphology dynamic around eastern part of the river mouth (Syamsidik et al. 2012). In this case, the Pande coastal area should be very much affected by the Krueng Aceh river's sediment supplies. However, a large harbor structure at the right hand side of the river mouth halted the movement of the sediment from the river leaving minimum impacts too on the Pandee coastal area. The study of the coastal morphology around this area was also been proven by Rasyif (2012).

Pressures to recover the coastal areas from the tsunami were often generated from community economy activities. One of them was to accelerate the coastal recovery for fishery activities such as for fish/shrimp ponds. To accelerate the coastal recovery, hard coastal structures were among preferable measures taken by the reconstruction and rehabilitation authorities. In the case of Banda Aceh, revetment and breakwaters were selected to be the structures for accelerating the recovery process. The hard structures create adverse impacts on the downdrift area of the coasts, as can be found around Kuala Gigieng coastal area. The longshore sediment transport was blocked by a set of jetty constructions around Alue Naga floodway. Furthermore, the salinity level of coastal lagoons' water around the area could not be recovered to become brackish water again as the salinity around these area were higher than 25ppt (Iskandar 2012). The changes of the freshwater discharges around the coastal lagoons also contributed to the change of the salinity level of the lagoons'

Table 13.5 Examples of type of coasts and key factors for morphology recovery

The coast/lagoon function	Sediment transport characteristics	Factors for recovery
Fishery area	Erosive coast	Hard Structures (revetment or seawall)
	Depositional coast	Coastal vegetation
Settlement area	Erosive coast	The settlement should be relocated
	Depositional coast	No specific intervention is needed
Not essential function	Erosive coast	No specific intervention is needed
	Depositional coast	No specific intervention is needed

water. In the case of Ulee Lheue lagoon, only a small tributary contributed freshwater inside the lagoon. As the landscape at the upstream part of the tributary was also changed during the rehabilitation and reconstruction process, the tributary could not provide ample freshwater to re-create brackish water inside the Ulee Lheue lagoon. Another example taken from New Zealand coastal lagoons' water quality changes after tsunami was linked to the geomorphic changes of the coastal barriers (Nichol et al. 2007).

Changes in term of outlets of the lagoons were found around Kuala Gigieng and Ulee Lheue coastal areas. In the case of Ulee Lheue coastal lagoon, the outlet was predicted as contribution of progressive land-subsidence around this area. The land-subsidence was already reported by McKinnon (2013) and Subarya et al. (2006). For the Kuala Gigieng coastal area, the outlet was shifted due to the coastal erosion around this area.

Learning from the elaborated findings after almost 10 year of the Indian Ocean tsunami, coastal morphology recovery policy should be done by means of hard and soft measures, or combination between the two approaches. Not all tsunami eroded coasts should be treated by constructing hard structures, such as revetment, break-water, and seawalls. The natural process of the coast could help the recovery process to take place. If the coast or a lagoon was used as important economy driving sector for the coastal community, hard structures could accelerate the recovery process. However, it should be anticipated carefully that erosion will take place at the down-drift part of the structures. When demand to recover the coastal line or coastal lagoons are not essential, the coast can be left to the nature to recover its morphology. This has been seen effective for some areas in Banda Aceh. Summary of the key characters and contributing factors of the coastal morphology recovery can be seen in Table 13.5.

13.5 Conclusions

Banda Aceh coastal areas have been recovered partially after almost 10 years of the Indian Ocean Tsunamis. Although, they were not fully recovered, but most of the areas have been restored, either by natural processes or by coastal hard structures interventions. Brackish water fishery activity, such as fish and shrimp ponds could not be restored to the condition prior to tsunami for several reasons. Among the reasons were the changes of salinity levels inside the coastal lagoons around Banda Aceh, due to the village relocation to other areas, and due to the change of people's occupations to other types of job. Coastal forests around Banda Aceh were slowly recovered. A promising coastal forest recovery was seen around Ujong Pancu and Ulee Lheue coastal areas. The efforts to bring back the coastal forests were performed by NGOs and local communities around the coasts. *Rhizopora* sp. and *Casuarina* sp. dominated the types of the vegetation introduced in the coastal forest recovery process. Sediment management around downdrift part of Banda Aceh coasts should be considered important due to the progressive erosion process around this area as it can be seen around Kuala Gigieng Coast. The coastal recovery process around Banda Aceh coasts can be a good examples for other similar tsunami affected areas, such as to the GEJET affected coast around Tohoku area of Japan.

References

Choowong M, Phatuwongraj S, Charoentitirat T, Chutakositkanon V, Yumuang S, Charusiri P (2009) Beach recovery after 2004 Indian Ocean tsunami from Phang-nga, Thailand. Geomorphology 104:134–142

Griffin C, Ellis D, Beavis S, Z-Nantes D (2013) Coastal resources, livelihoods and the 2004 Indian Ocean Tsunami in Aceh, Indonesia. Ocean Coast Manage 71:176

Iskandar A (2012) Influences of tidal forces to salinity level inside Ulee Lheue Coastal Lagoon (in Bahasa Indonesia). Bachelor Degree's Thesis, Civil Engineering Department, Syiah Kuala University, Banda Aceh

Liu H (2013) Dynamic changes of Coastal morphology following the 2011 Tohoku Tsunami. In: Proceedings of the 7th international conference on Asian and Pacific Coasts (APAC 2013), Bali-Indonesia, pp 594–601

McKinnon E (2013) Ancient Fansur, Aceh's 'Atlantis': the case for Lhok Pancu/Indrapurwa. NSC AU lecture series. National University of Singapore, Singapore. http://nscarchaeologyunit.wordpress.com/2013/04/24/nsc-au-lecture-series-ancient-fansur-acehs-atlantis-the-case-for-lhok-pancu-indrapurwa/. Accessed 20 Dec 2013

Meilianda E, Dohmen-Janssen CM, Maathuis BHP, Hulsche SJMH, Mulder JPM (2010) Short-term morphological responses and developments of Banda Aceh coast, Sumatra Island, Indonesia after the tsunami on 26 December 2004. Mar Geol 275:96–109

Nakaoka M, Tanaka Y, Mukai H, Suzuki T, Aryuthaka C (2006) Tsunami impacts on biodiversity of Seagrass communities in the Andaman Sea, Thailand: (1) Seagrass abundance and diversity. In: The Nagisa World congress, pp 49–56

Nichol SL, Goff JR, Devoy RJN, Chaque-Goff C, Hayward B, James I (2007) Lagoon subsidence and tsunami on the West Coast of New Zealand. Sediment Geol 200:248–262

Rasyif TM (2012) Numerical modeling of coastal morphology dynamic around Krueng Aceh River Mouth (in Bahasa Indonesia). Bachelor Degree's Thesis, Civil Engineering Department, Syiah Kuala University, Banda Aceh

Subarya C, Chlieh M, Prawirodirdjo L, Avouac J, Bock Y, Sieh K, Meltzner AJ, Natawidjaja DH, McCaffrey R (2006) Plate-boundary deformation associated with the great Sumatra-Andaman earthquake. Nature 440:46–51

Syamsidik, Muzailin, Bahri S (2011) Observations on beach deformation before and after the Indian Ocean Tsunami around Banda Aceh using satellite images. In: Suntoyo (eds) Proceedings of Japan Indonesia workshop on estuary and climate change. Institut Teknologi Sepuluh Nopember, Surabaya, p 85

Syamsidik, Arizal, Dewi S (2012) Applications of numerical simulation on coastal morphological dynamic studies. In: Azhari et al (eds) Proceedings of 5th international symposium of computation science. Gadjah Mada University, Yogyakarta, p 285

Tanaka H, Tinh NX, Umeda M, Hirao R, Pradjoko E, Mano A, Udo K (2012) Impact of the 2011 Tohoku Earthquake and Tsunami on beach morphology along the Northern Sendai Coast. Coast Eng J 54:1250009

Tinh NX, Tanaka H, Hirao R, Pradjoko E, Mano A, Udo K (2012) Morphology change by the 2011 Great Tohoku Tsunami wave mechanism of recent closure and breaching of the Nanakita River Mouth. Coast Eng 2012:1–10

Chapter 14
Disaster Waste Management: Lessons Learnt from Banda Aceh, Indonesia

Hari Srinivas and Farhan Helmy

Abstract Much attention has been paid to the human tragedy resulting from the 2005 Indian Ocean Tsunami. But the environmental impacts of the disaster, particularly in terms of the challenges faced as a result of the enormous amount of debris generated, provide important lessons for disaster waste "*preparedness*" for similar vulnerable cities and regions.

This chapter explores the challenges faced by the city of Banda Aceh, Indonesia in clearing the earthquake and tsunami debris. It looks at the actions taken by both the Government of Indonesia and international agencies/NGOs, and proposes a number of recommendations on the proper handling and management of disaster debris.

Lessons learnt from Banda Aceh's response call for the setting up of a long-term debris management plan, that include arranging mutual aid arrangements between neighbouring cities, updating the city's overall waste management, a community communication strategy creating an inventory of heavy equipment available for use, identifying a temporary collection site for debris, and the proper handling of hazardous and toxic wastes.

Keywords Environmental impacts • Indian Ocean Tsunami • Lessons learnt • Tsunami debris management

---

This chapter is based on the experiences of the authors in implementing an EU-funded project on debris management in Banda Aceh during 2004–2009, when they were working at the United Nations Environment Programme. The views expressed here, however, are those of the authors alone and should not be construed as official policy of the United Nations.

H. Srinivas (✉)
School of Policy Studies, Kwansei Gakuin University, Nishinomiya, Japan
e-mail: hari.srinivas@kwansei.ac.jp

F. Helmy
National Council on Climate Change (DNPI), Jakarta, Indonesia
e-mail: farhan.helmy@gmail.com

R. Shaw (ed.), *Recovery from the Indian Ocean Tsunami: A Ten-Year Journey*, Disaster Risk Reduction, DOI 10.1007/978-4-431-55117-1_14

14.1 Understanding Disaster Waste Characteristics

The impacts of disasters, whether natural or man-made, not only have human dimensions, but environmental ones as well. Environmental conditions may exacerbate the impact of a disaster, and vice versa, disasters have an impact on the environment (UNDAC 2005). Deforestation, forest management practices, agriculture systems etc. can exacerbate the negative environmental impacts of a storm or typhoon, leading to landslides, flooding, silting and ground/surface water contamination. A disaster event is, after all, an indicator that something is not right with our environment (UNEP IETC 2009).

Within the various impacts of a disaster on the local environment, the high volume of wastes from disasters, from households and debris from the disaster's impacts, constitute a particularly critical environmental aspect of disaster management (Baycan and Peterson n.d.). As highlighted by Tsunami in Banda Aceh, disasters can generate enormous volumes of debris, including soil and sediments, building rubble such as brick, concrete and timber, vegetation such as leaves, branches and trees,, hazardous and toxic materials such as oil drums, asbestos or batteries, mixed-up domestic and clinical wastes and, all too often, human and animal remains (UNEP 2005). In many cases, this waste presents a risk to human health from biological sources, for example, flies, rodents, rotting carcasses; chemical sources such as asbestos, oils, solvents; and physical sources such as cuts, abrasions, collapse. Debris also impedes pedestrian and vehicle access and blocks urban services such as drains, sewers (EPA 1999).

Disaster debris need to be handled in an environmentally sound manner including proper handling of scrap metals, for example copper, steel, aluminium, timber for reconstruction and cooking, demolition waste from buildings/structures for reuse, re-working as an aggregate or infilling/protection material, and uncontaminated soil/sediment for restoration or in-filling (UNEP 2005). In many cases, however, disaster debris place an additional burden on a nation or community already struggling to cope from the humanitarian aspects of the disaster.

14.2 Disasters and Debris

The amount and type of debris generated from a typical disaster varies from situation to situation. Table 14.1 outlines many of the types of debris that are created by different disasters (ACWMA 2005; EPA 1999; OES 2005; UNOCHA 2008)

In most cases, the debris is of mixed type—solid wastes are mixed with vegetative and other materials, making their processing, recycling and reuse more difficult.

In complex situations where a cascading set of disasters occur, such as the situation that Banda Aceh faced, there is a broader range of disaster debris deposited in a larger area that need to be properly cleared and disposed.

Table 14.1 Disasters and disaster types

Natural disasters	Man-made disasters
• *Tsunamis*: Widespread deposition of wastes on coastal fringes, and potentially pan-oceanic, including sub-sea deposition	• *Industrial Accidents*: Generally localized, waste types dependent on chemical release (or combustion by-product)
	• *Dam Breaches*: Similar to flood above
• *Earthquakes*: Localized generation of building material waste (and sediment from landslides) from seismic activity	• *Conflicts*: Bomb-damaged buildings (domestic and industrial) potentially impacted by depleted uranium (DU) and unexploded ordinance (UXO)
• *Floods*: Generally localized generation of soil, sediment and building material waste	
• *Hurricanes*: High-velocity winds and storm surge generally impact region of first landfall with high volumes of building material and vegetation waste being generated	
• *Forest Fires*: Although low volumes of waste are generated, includes building material waste; de-vegetated slopes are more vulnerable to mud-slides/landslides	

14.3 Tsunami-Generated Debris in Banda Aceh

It was estimated that 4.7 million tonnes of recyclable debris resulted from the damaged buildings and infrastructure in the Aceh Province. These wastes included brick and concrete blocks that were deposited around the buildings destroyed and damaged by the tsunami, and wastes generated prior to reconstruction works (UNEP IETC 2009). They included those derived from damaged buildings such as domestic housing and commercial buildings, and infrastructure including roads and bridges.[1]

Efforts were made by international organizations such as UNDP and UNEP, and the local government to collect, separate and recycle/reuse the debris (UNDAC 2005). The recycling of debris provided many benefits, including reduction of environmental risks, support and accelerate reconstruction works, sustainable employment generation, and demonstration and awareness-raising in waste recycling.

[1] In comparison, the total volume of debris generated from the 2011 Tohoku earthquake and tsunami in Japan has been estimated to be 25 million tons (This does not include vehicles and ships that were destroyed). Miyagi Prefecture generated 16 million tons of debris, Iwate Prefecture—6 million tons, and Fukushima Prefecture—2.9 million tons. Similarly, the 1995 Great Hanshin Earthquake that struck Kobe and surrounding areas generated 14 million tons and took more than 3 years to be cleared and processed.

14.3.1 Number of Buildings Destroyed

Comprehensive data on destroyed buildings/houses in Banda Aceh was not identified and analyzed; however, the Dept of Urban Planning and Housing (Dinas Perkotaan dan Permukiman, DPP) Banda Aceh estimated that 17,286 new houses were required to be built to replace the destroyed houses. JICA URPP team and Dinas Kebersihan dan Pertamanan (DKP), the Cleansing Department that is responsible for managing municipal wastes, estimated that Banda Aceh population in 2006 was 212,893, and the number of houses approximately 43,000 units (UNEP IETC 2009). This means that more than 40 % of the houses/building in Banda Aceh were destroyed during the tsunami disaster.

In comparison, Banda Aceh's population in 2005 was 177,881, and comprised of 35.557 households. This dramatic increase was caused by urbanization and overall economic development. Furthermore DPP Banda Aceh also estimated that out of the required new houses, 3,630 (8.4 %) units were reserved for low-income households.

14.3.2 Estimating the Volume of Debris Generated by the Tsunami and Earthquake

Most of the damaged and destroyed houses were either traditional wood houses (containing about 70 % wood, 15 % roof tiles and the remaining being mixed materials); or modern brick and concrete houses (containing 65 % brick/concrete, 15 % wood 10 % roof tiles/sheets, and remaining being mixed materials). Table 14.2 presents the summary of calculated debris' weight and volume generated during the tsunami disaster in Banda Aceh.

The estimated volume of waste generated from damaged buildings was 725,840 m^3. This amount did not include foundations of the buildings. Assuming that the foundation volume is about 15 % of the total mass volume of a building, then the volume of the foundation alone was about 128,090 m^3. The total volume of tsunami waste generated from destroyed building was approximately 853,930 m^3.

Table 14.2 Estimated debris generated during earthquake and tsunami in Banda Aceh, 2004

	Weight of debris generated		Volume	
Source	Ton	%	M^3	%
Single storied wooden house	29,503	4.85	95,722	13.2
Single storied brick and concrete house	290,492	47.77	315,005	43.4
Single storied commercial establishment	60,963	10.03	75,484	10.4
Multi storied commercial establishment	227,146	37.35	239,629	33.0
Total	608,104	100	725,840	100

The tsunami waves also scoured the bottom of the near shore, roads, bridges, vegetation and soil/mud from land when it moved inland, and swept away the debris when it moved back into the ocean. The international NGO, Oxfam GB, estimated that approximately 50 % of the tsunami waste consisted of soil/mud/solids (30 %) and vegetation (21 %). This approximation was likely based on the estimate of waste already dumped in temporary dump sites. Using these figures, the total estimated volume of tsunami waste generated in Banda Aceh then becomes 1,742,714 m^3. It was also estimated that parts of the tsunami waste were swept away by the backward wave to the sea, leaving approximately 35 % on the land, or 603,495 m^3 of mixed waste.

The above calculations illustrate the huge challenge faced by the Banda Aceh municipality to properly dispose the debris from the disaster. This presented a scenario that was well beyond the existing capacities of the waste management agency. It also presents a picture of the potential for recycling and reuse of the debris for various recovery and reconstruction purposes.

Along with the humanitarian effort put in place to help the residents of Banda Aceh, parallel efforts by the United Nations and international NGOs was mounted not only to clear this debris, but also generate livelihood options for the affected population (as cash-for-work programmes) and generate raw materials for reconstruction of infrastructure in the area.

14.4 Problems and Barriers Encountered in Banda Aceh

Besides UN agencies such as UNDP, UNEP and others, a number of other organizations, including GTZ, Oxfam, and local governments such as Rotterdam, were involved in debris management efforts either directly, or as parts of larger programmes in recovery and reconstruction.

Most worked with Banda Aceh's DKP in various aspects of disaster debris management, such as identifying and managing temporary dump sites, designating debris collection points, renewing the existing dumpsite of Banda Aceh to receive tsunami debris, identifying new dumpsites, and recycling of wastes for livelihood (for example, recycling wood to build furniture, reusing recovered bricks for construction, and crushing concrete blocks to make river embankments)

As these efforts were put in place over the weeks and months after the disaster, it enabled Banda Aceh's local government and DKP to get a better understanding of the various aspects of comprehensive debris management. In retrospect, local officials shared their experiences on the issues and challenges that they faced, which could have been avoided if there was better awareness and debris "preparedness."

Over a series of interviews and brain-storming meetings, a number of issues were raised by the officials, This section is based on interviews and brain storming meetings organized in 2007 (April and September) 2008 (March, June and October), 2009 (July and September) and 2010 (March) with local government officials and staff members of the local branch of DKP in Banda Aceh. Individuals interviewed include the mayor of Banda Aceh, Mr. Mawardi Nurdin and vice-mayor Mr. Illiza Saaduddin.

DKP staff members Mr. Teuku Zul Akhyar and Mr. Mirza Yanto were also interviewed for their views and insights. Besides these interviews, a total of six public workshops and seminars were organized in 2007, 2008 and 2009 with the Ministry of Environment of the Indonesian Government and the local government of Banda Aceh.

The above interactions provided highlights of the problems that local officials had faced and ways it could have been avoided:

1. There was no effort to estimate the magnitude and composition of likely wastes from a range of disaster scenarios. Attention should have been given to components with potential health and environmental impacts such as chemical contamination from commercial premises, wastes from health care institutions, asbestos and other substances commonly included in buildings, spilt fuels and oils, agricultural chemicals from stores and farms. Also included was damage to waste disposal facilities themselves, such as floods sweeping away a garbage dumpsite. Wastes arising from chemical or fuel spills should also have been considered.
2. Despite awareness in the region to earthquakes and related disasters, the vulnerability of municipal and commercial waste facilities, including landfills and dump sites, to natural disasters by themselves was not considered. Landfills on low-lying ground, transfer stations on the shoreline, wastewater treatment plants were just some examples cited.
3. Potential temporary storage or disposal areas for large volumes of inert solid debris close to where such waste might be generated—i.e. towns and industrial zones, were not identified. Figure 14.1 shows the temporary site that was setup

Fig. 14.1 Temporary site for collected debris

Fig. 14.2 Workshop recycling wood debris to make furniture

in Banda Aceh to collect and process debris. It is unlikely that debris will be carried very far during a time of crisis, and therefore such sites would have been critical to collect debris in an organized manner.

4. Additional removal, transport and handling personnel and heavy equipment that might be needed were not requisitioned. In effect, emergency services should have had a stand-by list of resources able to address immediately the waste management functions, and know the land areas available for storage/disposal.
5. Possible separation and recovery of potentially valuable waste components are important aspects that were not considered. As well as providing secondary materials for reconstruction, such operations would have provided some employment relief for households who have otherwise lost their livelihood. Figure 14.2 shows a workshop set up by international organizations for recycling wood debris to make furniture. Ownership, resale or donation conditions of such recovered waste should also ideally be pre-determined during the planning process to avoid unpleasant situations during the crisis.
6. Considering the potentially huge volume of debris to be cleared, there was no effort to explore the possibility of training non-waste personnel in the local government to assume waste management functions during a crisis.
7. Besides debris directly resulting from the disaster, anticipation and identifying wastes that might arise from a large-scale disaster relief operation, especially medical, health-care wastes and any equipment that would be eventually be discarded, was not considered.
8. The Banda Aceh's DKP realized early on that large industrial sites cause especially difficult circumstances as regular municipal authorities have little

Fig. 14.3 Uncollected municipal wastes during the recovery phases

experience in handling the debris from industrial plants. While the expertise in the industry can assist, there were legal liability questions, and a lack of facilities that could accommodate such special wastes, that were not explicitly planned for.

9. Integration of normal municipal waste generated during the emergency period into the waste management efforts was not carried out, creating additional problems for recovery and reconstruction. Figure 14.3 illustrates the problem of uncollected municipal wastes during the recovery phases. This had had to be managed at the same time as the debris.
10. Banda Aceh's garbage dump is located very close to the sea shore, and the initial back waves of the tsunami carried away much of this waste and debris into the ocean. Very little effort was made both by the government agencies and international organizations to clear this debris, except those that were re-deposited on the city's shores (UNDAC 2005). The debris seriously affected marine ecosystems and aquatic life, also resulting in disruption to livelihoods that depended on them.

In summary, the following ideas were raised during the interviews and workshops/seminars that ought to have been incorporated in plans for disaster debris management:

- Maintaining close links with disaster management agencies, and ensuring that waste management is incorporated into overall emergency plans
- Nominating stand-by waste personnel and equipment and ensuring training and practice, as in normal emergency management

- Identifying temporary waste handling locations
- Incorporating disaster mitigation measures in the design and operation of waste management facilities
- Elaborating special emergency waste systems for sensitive installations such as hospitals
- Incorporating disaster wastes into the scenario for overall waste planning at national and local levels

14.5 Lessons Learnt in Managing Disaster Wastes

A detailed evaluation of the debris management efforts carried out by the local government and international organizations highlights a number of lessons that can be effectively used by similar vulnerable cities and communities in managing disaster debris.

The strategies and approaches presented below were developed during interactive planning sessions between international waste experts and local government staff members for future waste preparedness and planning action for recovery. These lessons are useful for local and national governments that are developing action plans for disaster preparedness and recovery.

14.5.1 Making a Long-Term Debris Management Plan

There is a need for increasing existing emergency planning to include long-term debris management. Because natural disasters can generate tremendous quantities of debris, cities and communities should plan for the worst case (UNEP-IETC 1995). As illustrated by the situation that Banda Aceh faced, any such plan should include a detailed strategy for debris collection, temporary storage and staging areas, recycling, disposal, hazardous waste identification and handling, administration, and dissemination of information to the public.

The plan and work involved needs to be distributed to personnel from all concerned local agencies to ensure that it is implemented quickly and smoothly. The plan also needs to be reviewed and revised at least once a year as needed, in consultation and concert with larger efforts on disaster preparedness.

14.5.2 Considering Mutual Aid Arrangements

Mutual aid arrangements between nearby cities and communities are an important strategy to allow quick access to specialized personnel or equipment on a short-term basis. This allows sparse and expensive equipment to be shared between cities and regions. Usually the host city pays for the expenses of personnel as well as any maintenance or repair costs for equipment. These agreements can be developed for a local geographic area or can extend to cities in other provinces as well.

14.5.3 Implementing Recycling Programmes

Implementing a plan for recycling disaster debris is much easier if a city or community already has a recycling program in place. As a result, permits, enforcement, collection, processing, and marketing issues will already have been largely resolved.[2] After a disaster, the city will be faced with expanding current recycling practices rather than designing and implementing new practices. It is much easier to expand existing capacities and markets than to start these endeavors in the wake of a disaster.

14.5.4 Updating the City's Solid Waste Management Plan

It is important that a city's overall solid waste management plan reflect current practices and policies, especially those that apply in disaster situations. The plan is an official document that often is filed with the city office. Efforts to develop Banda Aceh's waste management plan was still being designed when the disaster struck the city.

It can also be beneficial to share the plan with private contractors and other community agencies (e.g., fire and police) that in the event of a disaster could potentially be involved in solid waste management services. Should a disaster occur, supporting agencies would find the plan useful because it describes established practices and policies, as well as the types, locations, and capacities of existing solid waste recycling and disposal facilities.

14.5.5 Developing a Communication Strategy

A communication strategy needs to be prepared ahead of time. Government officials will need to tell the community when, where, and how waste collection will resume, as well as provide special instructions for reporting and sorting disaster debris. For example, cities and communities can prepare radio announcements and flyers as part of their emergency plan.

[2] Most of the recycling efforts in Banda Aceh were carried out by the informal sector, which was also involved in recycling building waste. Particularly metals, bricks and other easily recyclable wastes were collected and processed by informal enterprises in the city.

As was seen in Banda Aceh, a city or community may however, lose electricity, telephone service, radio broadcasting capability, or newspaper service. Therefore, there should be more than one method of communication. Local media also need to be involved in communicating instructions in the event of a disaster.

14.5.6 *Preparing for Increased Outreach and Enforcement Staffing Needs*

In the aftermath of the disaster, waste management staff in Banda Aceh were overwhelmed by the demands placed on them to clear the debris. Much of their own equipment and facilities were destroyed by the disaster, further reducing their capacities. They realized that they must handle an increased number of requests concerning waste removal, staff to train and monitor debris collection contractors, enforce disposal restrictions, and help solve implementation problems. The use of members of the community itself as a temporary source of labour was considered, especially for low-income households.

14.5.7 *Obtaining Equipment and Supplies*

Officials in Banda Aceh mentioned that there was a need to identify in advance the types of equipment and supplies that waste teams will need to implement the plan. Quick procurement of these items through mutual aid agreements or standing contracts with the private sector, or stockpiling such equipment should be considered and planned for. If stockpiling is too expensive for one city alone, perhaps a group of neghbouring cities could stockpile the equipment.

Types of equipment that a community might need include chain saws, portable generators, cellular phones, flashlights, batteries, vehicle repair equipment, and extra work clothing. For example, a local government that routinely stores drinking water (e.g., for its solid waste collection crews) might want to make sure that water supplies are well-stocked.

14.5.8 *Selecting Collection and Storage Sites*

The most common suggestion made was to pre-select debris collection sites that could be used for temporary storage and processing of debris. Convenient local sites allow collection crews to reduce travel time when transferring debris to processing or disposal facilities and result in faster street clearing.

Sites are selected based on planned activities, such as collection, storage, sorting, recycling, landfilling, and burning of debris where feasible. Pre-selection of sites speeds the implementation of the debris management plan. It was clear from the experiences of Banda Aceh that access to heavy equipment, lack of impact on environmentally sensitive areas, and convenience to collection routes, need to be considered.

If residents are to be asked to bring disaster debris to collection sites, the city should include these locations in its disaster communication strategy, so that information is immediately available to the public in the event of a disaster.

14.5.9 Determining Management Options and Goals

After the initial response phases in Banda Aceh, collection and storage of debris in temporary sites was carried out quickly, but a proper disposal strategy was not put in place. Clearly, any disaster debris management plan should also include a final disposal strategy, including priorities for recycling wastes and determining the desired disposal options for the remaining waste.

14.5.10 Segregating Hazardous and Toxic Waste

In Banda Aceh, a number of hazardous and toxic wastes such as asbestos, paints and oils, petrol, chemicals etc. were mixed up with ordinary debris. Plans were put in place to manage these only after it was pointed out by international experts. The lesson learnt here was that segregation of hazardous from non-hazardous disaster debris should be carried out right at the beginning in order to avoid contamination of mixed wastes. Waste handlers need to understand these requirements as well as have a plan for controlling and diverting hazardous waste from the debris stream.

Taken together, these lessons learnt present important building blocks for an effective waste management mechanism that is able to respond to and manage disaster wastes effectively and efficiently.

Ultimately, preparing a disaster debris management plan in advance can pay off in the event of a natural disaster. Such planning can help a city or community identify its debris collection, recycling, and disposal options. Although the recovery process can take a long time, perhaps even years, careful planning will prevent costly mistakes, speed recovery, and avoid creation of more waste. A plan also can save money by identifying cost-effective debris management options and sources of help, increasing control over debris management in the city or community, and improving administrative efficiency.

References

ACWMA (2005) Alameda County disaster waste management plan. Alameda, Calif: Alameda County Waste Management Authority

Baycan F, Peterson M (n.d.) Disaster waste management—C&D waste. Ankara: Ministry of Environment, Turkey

EPA (1999) Planning for disaster debris. Washington D.C.: US Environmental Protection Bureau

OES (2005) Debris management in disaster recovery. Los Angeles: Response and Recovery Division, California Governor's Office of Emergency Services, January 2005

UNDAC (2005) Indian Ocean tsunami disaster of December 2004: UNDAC rapid environmental assessment of Aceh, Indonesia. United Nations Disaster Assistance Coordination, Geneva

UNEP (2005) United Nations post-tsunami waste management plan. United Nations Environment Programme, Geneva

UNEP IETC (1995) Proceedings of the earthquake waste symposium. United Nations Environment Programme, Osaka, 12–13 June 1995

UNEP IETC (2009) Final report: demonstrating ESTs for building waste reduction in Indonesia. United Nations Environment Programme, Osaka

UNOCHA (2008) Disaster waste management guidelines. UNOCHA-UNEP Joint Environment Unit, Geneva

Chapter 15
Community-Based Housing Reconstruction in Aceh, Indonesia

Chiho Ochiai

Abstract Housing reconstruction is one of the major tasks in post disaster recovery. Contrary to the conventional approach which is top-down decision making process, the participatory approach has gained the spotlight to improve the situation in housing reconstruction. In this chapter, the issues and specific difficulties in community-based housing reconstruction in Aceh are briefly described. Most housing reconstruction efforts have been completed and the lives of the people are getting back to normal. Their participation in the reconstruction process has given rise to increased community ties and a new found pride in the ownership of their homes. Modification of houses and development of landscaping after several years are a proof of people's adaptation to their living environment. However these participatory decision making processes were time consuming and it should be carefully evaluated. Further observation and extensive study are needed for better understandings of these complicated reconstruction processes.

Keywords Adaptation • Housing reconstruction • Participatory approach

15.1 Introduction

Post Housing reconstruction is one of the most important and critical issues in disaster recovery and reconstruction. The people's strong demands in housing reconstruction were observed and discussed after 2004 Indian Ocean Tsunami cases. "Housing" is one of the basic human needs and it is most important to realize the normalization of life after a disaster. Therefore housing and urban infrastructure has always been

---

Note: BRR (the Aceh and Nias rehabilitation and reconstruction agency): BRR was a temporary body created by the Indonesian government on 16 April 2005. Its mandate was to coordinate rehabilitation and reconstruction efforts following the tsunami in Aceh and Nias. In the last 4 years, the BRR coordinated more than 500 organisations, which ran more than 12,000 projects. The activities of the BRR officially ended on 16 April 2009.

C. Ochiai (✉)
Graduate School of Global Environmental Studies, Kyoto University, Kyoto, Japan
e-mail: Ochiai.chiho.2x@kyoto-u.ac.jp

R. Shaw (ed.), *Recovery from the Indian Ocean Tsunami: A Ten-Year Journey*,
Disaster Risk Reduction, DOI 10.1007/978-4-431-55117-1_15

prioritized and invested by all the donors. Post-disaster housing reconstruction is extensively studied in which issues of people's participation, ownership, and accountability are discussed. (Jigyasu 2001; Olshansky 2005; Duyne Barenstein 2013).

On 26th December 2004, a strong earthquake of magnitude 9.0 on the richer scale hit Northwest of Sumatra Island, Indonesia caused the Indian Ocean Tsunami. Aceh is the city closest to the epicentre where nearly 130,000 deaths, 500,000 homeless and extensive damage to life, property and infrastructures occurred (BRR 2005). According to the assessment carried out by the ministry of public works in May 2005, 116,900 houses were destroyed and 14,200 houses were heavily damaged in Aceh province (ACARP 2007; BRR 2005). The government of Indonesia (GOI) estimated that 85,000 new houses needed to be built and 115,000 houses needed to be repaired. One year later, the report from BRR estimated that 130,000 new houses needed to be built and 85,000 homes needed repair (UNDP 2006; World Bank 2007).

The people's participation in housing reconstruction has been a critical issue and the key factor for successful adaptation to their new houses. However, to ensure their participation, many issues like speed, quality, and management are critical to the post disaster reconstruction. More importantly, how people adopted to their newly built environment is crucial for the overall sustainability of their life and community.

In this chapter the participatory housing reconstruction efforts in Aceh and the changes that have been made over past several years are discussed.

15.2 Methodology

The following sections are based on field surveys that have been carried out from 2005 to 2008 and a re-visit of the two sites between 2009 and 2011. The early stages of the study focused on participatory housing reconstruction effort by the World Bank (WB). Later studies were made as a follow up to the WB and the government housing efforts after 5–8 years. The summary of issues and difficulties related to the housing reconstruction in Aceh during the early stage of rebuilding are discussed. And also, the latest conditions of the housing and of the people's adaptation to their new housing will be reported based on site observations and interviews.

In the case of Aceh, reconstruction, relocation and resettlement were important issues, however, this chapter deals with in-situ reconstruction.

15.3 Why Participatory Method?

Reconstruction efforts after a disaster have been criticized as a failure of conventional top-down approaches that concentrated their attention on speed, standardisation and technological oriented solutions (El-Masri and Kellett 2001). On the other hand, bottom-up approaches, also called participatory approach or owner-driven approaches have been advocated in the last two decades (Davis 1981; Cury 1981; Cockburn and Barakat 1991; IDNDR 1996). The bottom-up approaches puts importance on the involvement of the user, as well as educating groups to come together to participate in individual and community concerns and building outcomes such as the use of appropriate technologies and decisions concerning many other aspects of their lives (El-Masri and Kellett 2001). Carola Mantel refers to the notes criticizing that top-down approach does not fulfil the needs of the people but has more of a political interest. When the owner is given more control over the housing reconstruction, it is possible to have a higher level of satisfaction as well as a higher chance to restore a sense of pride and to integrate their indigenous ways of building (Mantel 2013). Comerio stated that self-help housing is now the current model for rebuilding after disasters in developing countries (Comerio 1998).

Arnstein (1969) proposed the idea of a "ladder of community participation" and it was later modified by Choguill (1996). Then Davidson et al. (2007) combined these two theories and revised them to fit in the context of post-disaster housing reconstruction. The ladder consists of five levels; manipulating, informing, consulting, collaborating and empowering. This means when post-disaster housing reconstruction is conducted in an appropriate way, it gives a chance to empower people and community.

In practice, involving people in the reconstruction process has not been easy. Because it increases the possibility of unexpected outcomes. However, still many scholars are convinced that community participation into the recovery processes is important (Rubin et al. 1985; Oliver-Smith 1991) because it would lead to building a sustainable community in long run (Berke et al. 1993).

Difficulties of involving the community are usually found in (a) quality control of the housing projects, (b) ensuring that the decision making processes are just and fair, and (c) the time constraints often disrupt scheduling and the proper allocation of money. Rubin et al. noted that the process of recovery and how to improve both its speed and quality needed to be evaluated (Rubin et al. 1985). For the participatory reconstruction to work properly requires careful planning and a collaboration of the supporting organizations to facilitate a successful reconstruction process. Also, it requires identification of issues, prioritizing the projects, and finding solutions to problems (Oliver 1987; Cuny 1981; Oliver-Smith 1991; El-Masri 1992).

15.4 Housing Reconstruction in Aceh

15.4.1 Outline of Disaster and Housing Damage

As mentioned above, the ministry of public works in May 2005 reported that 116,900 houses were destroyed and 14,200 houses were heavily damaged in Aceh province (ACARP 2007; BRR 2005). Many organizations such as national governments, international organizations, United Nations, bilateral organizations, NGOs supported housing reconstruction.

The Aceh and Nias rehabilitation and reconstruction agency (BRR) initially planned ambitious goals to complete the reconstruction and rehabilitation of the human settlements by December 2006 at the latest. However, it turned out that it took until the end of 2008. After 4 years of efforts by many people, organizations and contractors, over 140,300 houses have been built in Aceh alone (BRR 2008).

15.4.2 Issues on Housing Reconstruction in Aceh

Post disaster housing reconstruction always faces several issues such as land titling and tenure, identification of victims, duplication of built houses, community involvement processes, quality control of buildings and so on. Since the magnitude of the disaster was extremely vast, it added many issues and difficulties to the housing reconstruction process in Aceh. Major ones will be discussed briefly below.

(1) Land tenure and land loss
Areas affected by the earthquake and tsunami stretched along the coast of Sumatra Island limited to within the 500 m to 5 km from the seashore. Many affected areas were permanently inundated and required people to be relocated. About 12,000 families could not rebuild their houses on their land (Steinberg 2007). Large scale relocation and resettlement was needed.

Also, since most of the local offices were destroyed, official documents were not available. Therefore, identifying the land tenure and titling took longer than it was expected. Steinberg mentioned that land disputes among community members and previous absent family members complicated the situation. The fact that many villagers have never held formal land titles but only customary tenure added to the complications (Steinberg 2007).

(2) Identification of victims and beneficiaries
Most agencies and BRR verified entitled beneficiaries but it was not an easy process. The government announced that they will provide one housing unit per household. There have been many claimants who misrepresented themselves by submitting multiple claims. This led to the issue of double counting or even triple counting of beneficiaries and an excess of houses were built.

(3) Participation in the decision making process
People evacuated and settled in the temporary shelters. Many communities were divided into camps, temporary shelters, or moved to their relative's houses in the city or outside of the city. This situation made it difficult for people to meet and discuss community planning. In addition, in many places, village heads or religious leaders had died and it took some time to elect new persons who were responsible and respected, to represent their village.

(4) Recovery of infrastructure and city plan
The recovery and reconstruction of the basic infrastructure could not keep up with the pace of housing, thus caused many problems with water supplies, drainage, electricity and road accesses.

(5) Cost escalation of materials and labor

As great numbers of housing reconstruction activities were going on, the price of materials such as wood, steel, cement and paint increased as well as energy costs. The rise in the shortage of experienced labor added to the problem. Some of the work was not finished because groups of laborers tended to shift from one site to another when payments were delayed or where better deals were offered.

(6) Disorganization of aid agencies and their lack of experience in housing
According to the BRR report, about 100 agencies contributed to the housing projects but many of them were not experienced in housing reconstruction and wanted only to contribute in good will (BRR 2005). Some of housing was constructed very close to the seashore and later destroyed by high tides. Some projects ended up not completed because they did not succeed in complying with new building standards and the community's basic needs.

In addition to the issues listed above, there were land issues, technical issues, governance issues and, issues about traditional architecture vs. the new construction, community planning and landscaping (Okazaki and Narafu 2007; Steinberg 2007).

15.4.3 Building Standard and Perception of Beneficiaries

The BRR recommended newly reconstructed houses to have a main room, two bedrooms as well as a bathroom and toilet at the back of the house with at least 36 m^2 area. Most organizations adopted this basic plan accordingly. But there were still a great variety of size and types of houses built ranging from 27 to 70 m^2 (The Asian Foundation).

The quality of the construction varied from organization to organization or program to program. Some of the houses were well received by the beneficiaries and some were not. The perception of beneficiaries varied based on their original location, social class, income, family size and so forth. Some people in rural areas found the newly constructed house as a real improvement compared to their original wooden house without sanitation. On the other hand, in the urban areas where they used to have larger houses or traditional long houses, they felt the houses were temporary and unsatisfactory.

15.4.4 World Bank's Housing Reconstruction Project

Although over 100 agencies were involved in the housing reconstruction project, the 15 largest agencies account for over 80 % of the houses rebuilt (BRR 2005). Among those, World Bank (WB) as the main coordinated body of Multi-Donor-Fund (MDF) played a major role in establishing the implementation structure of community-based methodology for the housing reconstruction. This implementation structure was later adopted by the Government of Indonesia (GOI) and applied in other post disaster reconstruction cases in Indonesia.

Based on previous experiences of poverty alleviation projects in Indonesia, the WB established a coordinating body linking the government, the international donors and local communities. Housing facilitators (HF) who had civil engineering or architectural background were hired to coordinate the community committee with the smaller housing groups (10–15 household) that were responsible to make decisions on the housing projects. The housing group was the core group to implement all the project processes working together with HF by opening the group bank account to manage the fund, deciding the design and preparing the housing permit and so forth. The tasks of HF varied from explaining the concept of community-based housing reconstruction, helping the community to organize housing groups, making house designs and implementation plan, preparing the proposal for the local government and checking the quality of the housing. The housing group was the core group to implement all project processes. Working together with HF, residents collaborate in design decisions, pulling building permits and opening a group bank account to manage funds for materials and labor. HF played an intermediate role linking community to local government and donor agencies who funded the project.

In this scheme, beneficiaries in the community were asked to conduct several tasks such as self-survey to identify the damage and losses of their houses, discuss community plan and housing design to match their needs, and checking the quality of construction. All these processes involve a great deal of community participation. Their own decision making processes helped to reflect their needs, increased their responsibility and sense of reconnecting to their communities. Figure 15.1 shows

Fig. 15.1 A house built by World Bank's housing reconstruction project in Aceh

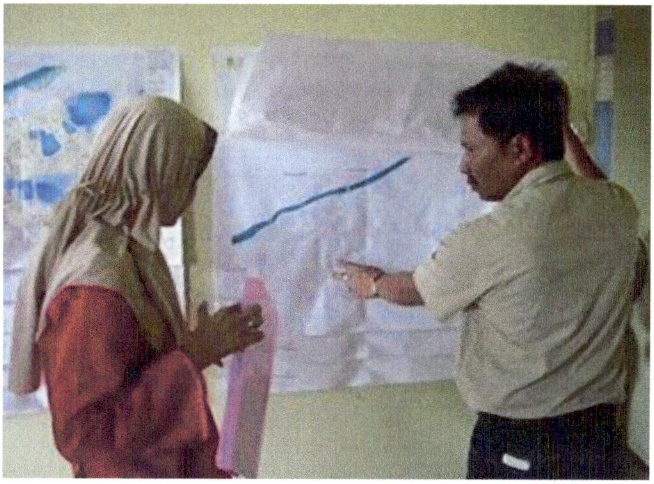

Fig. 15.2 A Housing Facilitator explaining the community plan

the newly constructed house by the project. Figure 15.2 shows community plan being explained by the housing facilitator.

Based on the questionnaire survey conducted in 2006–2007 targeting the beneficiaries of WB constructed houses and HF reveals that many people participated in the housing reconstruction processes fully or in parts of the processes. The level of participation varied from community to community and from person to person. These differences were caused by the varying characteristics of villages, leaders and leadership, gender, and location of the beneficiaries. It also depended on the communication skills of the HF sent to the site. In short, the participatory housing reconstruction methodology and implementation structure were evaluated as effective in some contexts but not versatile enough for all.

In a mass scale post housing reconstruction, certain levels of people's participation, an understanding of local and social requirements and their technical capacity are essential for a successful outcome (For more details, please refer to WB report World Bank (2012) and Ochiai and Shaw, Chap. 12, 2009).

15.4.5 Evaluation of Participatory Housing Reconstruction

In the early stages of the reconstruction process, the Indonesian government acknowledged the importance of the participation of people in the housing reconstruction processes and overall redevelopment of the region (BAPPENAS and The International Donor Community 2005). There have been several reports and papers discussing the housing reconstruction issues. Many other studies have been written about Aceh reconstruction project and have pointed out the many problems they have faced such as the delays in meeting due dates, poor construction methods and materials, low satisfaction and low accountability, and not enough community participation.

(1) Pros on participatory housing
According to the study by Dercon and Kusumawijaya (2007) evaluating the survey conducted by UNSYIAH (UNSYIAH and UNHABITAT 2006), they concluded that the quality and satisfaction of houses and accountability of community participation got a higher score in comparison with the contractor based approach. It also stated that the community-based approach had proven to be faster in Aceh. From several case studies of the participatory housing projects, they noted that community based housing reconstruction gets higher achievement because it responds more quickly to the urgent needs of the people and mobilizes solidarity among the members of a community and created social capital (Dercon and Kusumawijaya 2007).

Another positive point in the community based approach was fund management. Fund management gives opportunities for people in the community to get together to plan and work so it increases their sense of belonging (Ophiyandri et al. 2011, 2013).

Multi Donor Fund (MDF) (2008) reported that the community driven approach has proven an efficient means not only to build houses but also to create a sense of ownership and pride among beneficiaries.

(2) Cons on participatory housing
Despite the good intentions of the community-based approach, it was difficult to implement it optimally (Steinberg 2007). He mentioned that some donors kept insisting on a full cycle of the community consultation starting from the project identification to detailed action plans, which is normally used but does not meet their needs in an emergency circumstance. There have been cases where communities demanded organization to provide immediate housing rather than going through time consuming consultations. Steinberg stated that Aceh and Nias are far from ideal in the implementation of community based processes. Overstretching of the paradigm of participation caused more damage than good.

In addition, many studies have pointed out the discrepancies in the quality of construction techniques and lack of consistency. Mantel (Mantel 2013) draw a case study from one of international NGO (INGO) in one village and said participatory approach did not work the way they expected. INGO expected people to participate long hours of meetings to discuss their needs, wishes and opinions but few were never implemented. Also, people were asked to participate in construction activities who had no experience and the training was poorly organized. As Aceh is a merchant community, the people did not totally understand the concept of sharing the house rebuilding processes and were confused when asked to work on someone else's house. This caused mistrust and resentment between INGO and beneficiaries.

(3) Difficulties of participatory housing reconstruction
There are many positive sides to the participatory approach adopted in the housing reconstruction processes. It extends from the satisfaction of building one's

own house to the confidence gained in participating in a community. On the other hand, time consuming processes frustrated many since they were in an emergency situation and needed a safe and comfortable place to live as soon as possible. The majority of the people spent over 2 years in temporary shelters. They were frustrated with unfamiliar living environment and were anxious about not knowing when their houses would be completed.

Based on a questionnaire in two villages in Aceh, Mantel concluded that satisfaction with the new houses was influenced by the approaches pursued by the organization, the leadership of the organization and the timing of people's return to the village. Experienced NGO were able to offer more choices of construction methods and give people full control over the rebuilding of their houses (Duyne Barenstein 2013).

As a result, Strinberg stated that some communities have achieved extraordinary results while others have felt overloaded with too many responsibilities. Therefore, it may well be stated that community participation cannot be instrumented as if it were a panacea for all the shortcomings of actions by the government or contractors. As long as survival, livelihood, and well-being are so much at stake, the scope and potential role of community participation cannot, and should not, be overrated (Steinberg 2007).

15.5 Adaptation and Current Situations on Houses and Environment

The main fieldwork and survey was conducted from 2005 to 2008 and the summary of the findings were documented above. Since then, the housing reconstruction and the people's adaptation to their newly built houses and community have been monitored regularly. In 2009, 2010 and 2011 Aceh was revisited to observe and conduct interviews to update the most recent housing situation.

Observations and interview results showed that many people were returning to their villages. This section discusses the changes in housing and community after several years of Tsunami disaster in Aceh.

(1) Basic infrastructure
Most housing reconstruction efforts and basic infrastructures like road, water supply, mosks and tsunami evacuation tower throughout the city have been completed. Paved and wider roads are completed. Schools have been rebuilt and students are again walking through the village. It seems like the regular routine of life is returning.

Figure 15.3 shows new houses in the village. Figure 15.4 shows new tsunami evacuation tower and paved road. Figure 15.5 shows newly constructed school in the village and students walking through the village. On the other hand, many of inundated areas did not see any development nor improvement.

(2) Vacant and abandoned houses
Major issues regarding housing have been the large number of vacancies. Top-down or donor-driven housing solution in which houses were built and handed

Fig. 15.3 Newly constructed houses in one community

Fig. 15.4 Tsunami evacuation tower and widend-paved road

over to the beneficiaries and community were often criticized. It did not include careful consideration of their livelihood, family size, type of architecture, infrastructure and essential services like market, schools and road accesses. In addition some houses were poorly built sometimes using substandard materials which were then rejected by the people living there.

According to observations, some of the newly constructed houses were vacant with over grown grass and doors locked. The reason is often discussed as a double or triple counting of beneficiaries. However, there are also other

Fig. 15.5 Newly constructed school in the village and students walking by

reasons that interviews with residents revealed. Acehnese people are known as a mobile society (Maki 2011). At the time of the disaster, many families were living in different cities and towns for better business opportunity and better schools. Post disaster field surveys mentioned that they worked and lived in nearby cities like Medan, and cannot participate in the discussion. Also, many adults were killed by Tsunami and their offspring are not able to move back and often live with neighbours or relatives. These multiple reasons are contributing to the vacancy of houses.

(3) Modification of houses to personalize their space

In order to meet their needs and lifestyles, some houses were modified and upgraded, for example repainting with their preferred colors, using improved materials for doors and windows, connecting two houses together, making a front porch to create more space to relax, and extending roofs to give more shade in front of their house (Fig. 15.6 shows the modification of houses). Interviewees answered that these modifications and upgrades were necessary to meet their life style. They use their own money to purchase materials and hire labor or work on their own.

Also improvements in landscaping were noted. Trees, flowers and gardens give the outside spaces a more lived in and relaxed feeling.

15.6 Discussion and Conclusion

The large scale of an earthquake and tsunami cause huge losses in housing and infrastructure. Many efforts have been made to rebuild not only houses but also the community.

Fig. 15.6 The modification made to the built house

In the case of Aceh, the adaptation of the participatory method into their reconstruction efforts was suggested by the government and donor agencies from the start. Many people benefitted from the participatory method by increasing their ties to the communities and involving them in the rebuilding of their own homes. On the other hand, according to the studies, problems of quality control and lengthy discussion processes are viewed as failure of the participatory methods not appropriate for post disaster situation. These points suggest that the participatory method is not always the best way to solve the overall housing problems. We need well planned support and monitoring systems in place to get better results.

Overall, it is obvious that many people participated in the process of housing reconstruction and their knowledge and preferences were adopted into their newly constructed houses. To meet people's demand to rebuild their houses as quickly as possible sometimes contradicted the participatory process. However, managing these processes have a long term benefit of sustainability. Figure 15.7 shows the conceptual image to achieve a sustainable housing and community in housing reconstruction. Facilitation between national government and local communities was the key to managing the reconstruction process.

This coming year commemorates the tenth anniversary of the Indian Ocean Tsunami. The residents are slowly but surely regaining a sense of normalcy. Self-funded improvement projects are self-evident in individual house and landscaping details. Yet, still, many houses remain vacant. The development and improvement of the housings and the community will not end here but continues into the future. The people of Aceh, their resilience and will to survive will forever teach us that the power and spirit of a people can overcome many difficulties.

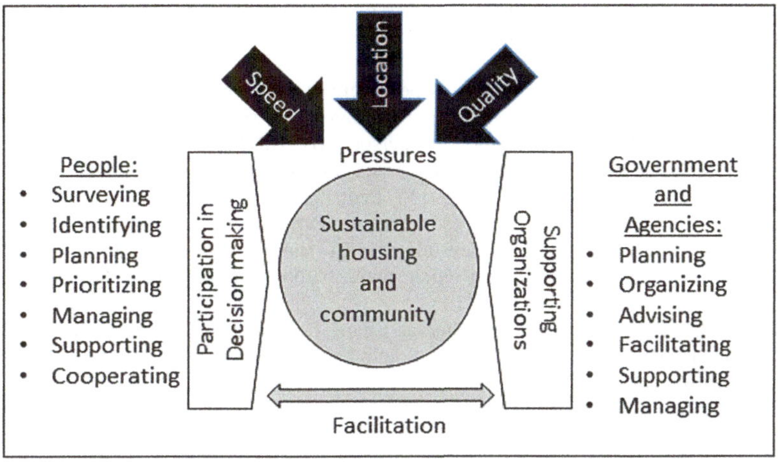

Fig. 15.7 The conceptual image of roles of people and government to achieve a sastanable housing and community

References

ACARP (2007) The Achenese Gampong three years on. Assessing local capcity and reconstruction assistance in post tsunami Aceh. Report of the Aceh Community Assistance Research Project. December 2007

Arnstein SR (1969) A ladder of citizen participation. JAIP 35(4):216–224

The Asian Foundation, The Aceh Rehabilitation and Reconstruction Appraisal (ARRA), Complete Findings Report http://asianfoundation.org/resources/pdfs/ARRAreport.pdf. Accessed August 2013

The Aceh Rehabilitation and Reconstruction Appraisal (ARRA), Complete Findings Report http://asian foundation.org/resources/pdfs/AEEAreport.pdf. Accessed August 2013

BAPPENAS, The International Donor Community (eds) (2005) Notes on reconstruction, The December 26, 2004 natural disaster. Jakarta, 19 January. BAPPENAS, Blue print for the reconstruction of Aceh and Nias, March 2005

Berke PR, Kartez J, Wenger D (1993) Recovery after the disaster: achieving sustainable development, mitigation and equity. Disasters 17(2):93–109

BRR (2005) Ache and Nias one year after the tsunami: the recovery effort and way forward. BRR and International Partners

BRR (2008) Strengthening the narrative of community life: two year report executing agency for the rehabilitation and reconstruction of Aceh-Nias. BRR report

Choguill MBG (1996) A ladder of community participation for underdeveloped countries. Habitat Int 20(3):431–444

Cockburn C, Barakat S (1991) Community prosperity through reconstruction management, architecture and design (January–February). pp 60–65

Comerio M (1998) Disasters hit homes: new policy for urban housing recovery. University of California Press, Berkeley

Cuny F (1981) Disasters and the small dwelling: the state of art. In: Davis I (ed) Disasters and the small dwelling. Pergamon, Oxford

Davidson CH, Johnson C, Lizarralde G, Dikmen N, Sliwinski A (2007) Truths and myths about community participation on post-disaster housing projects. Habitat Int 31:100–115

Davis I (1981) Disasters and the small dwelling. Pergamon, Oxford

Dercon B, Kusumawijaya M (2007) Two years of settlement recovery in Aceh and Nia: What should the planners have earned? 43rd ISOCARP congress. Antwerp, pp 19–23, September 2007, Congress papers
Duyne Barenstein J (2013) Community's perspectives on housing reconstruction in Gujarat following the earthquake of 2001. In: Duyne Barenstein JE, Leemann E (eds) Post-disaster reconstruction and change—community's perspectives. CRC
El-Masri S (1992) Reconstruction after disaster; a study of war-damaged villages in Lebanon: the case of al-Burjain. Habitat Int 25(2001):535–557, Pergamon
El-Masri S, Kellett P (2001) Post-war reconstruction. Participatory approaches to rebuilding the damaged villages of Lebanon: a case study of al-Burjain. Habitat Int 25:535–557
IDNDR (1996) Cities at risk: making cities safer…before disaster strikes, Geneva Indonesia-rebuilding lives. Habitat Int 31:150–166
Jigyasu R (2001) From Marathwada to Gujarat, Emerging challenges in post-earthquake rehabilitation for sustainable eco-development in South-Asia. Presented on international conference on post-disaster reconstruction, Florence
Mantel C (2013) Chapter 3 Ownership, control, and accountability in post-tsunami housing reconstruction processes in Aceh, Indonesia. In: Duyne Barenstein JE, Esther (eds) Post disaster reconstruction and change; Communities' perspectives
Multi Donor Fund (MDF) (2008) Investing in institutions: sustaining reconstruction and economic recovery, four years after the tsunami
Maki N (2011) Housing after the natural disaster; relocation and recovery. Kajima Institute Publishing
Ochiai C, Shaw R (2009) Chapter 12 Post-disaster reconstruction in urban areas in Aceh, community, environment and disaster risk management. Emerald Group Publishing Limited
Okazaki K, Narafu T (2007) Housing reconstruction practices in Aceh, Indonesia. J Architect Build Sci (619):257–262
Oliver P (1987) Dwellings: the house across the world. Phaidon, Oxford
Oliver-Smith A (1991) Successes and failures in post-disaster resettlement. Disasters 15(1):12–23
Olshansky RB (2005) How do communities recover from disaster? A review of current knowledge and an agenda for future research. 46th annual conference of the association of collegiate schools of planning, Kansas City, October 27 2005
Ophiyandri T, Amaratunga D, Pathirage C (2011) Community based post disaster housing reconstruction: Indonesian perspective. http://www.umb.edu/editor_uploads/images/centers_institutes/center_rebuild_communities/Registration_Form.pdf. Accessed August 2013
Ophiyandri T, Amaratunga D, Pathirage C (2013) Project risk management for community-based post-disaster housing reconstruction project. International conference on building resilience 2013: individual, institutional and societal coping strategies to address the challenges associated with disaster risk, Ahungalla, 17–19 September 2013
Rubin CB, Saperstein MD, Barbee, DG (1985) Community recovery from a major natural disaster, Monograph No41, Program on environment and behavior. Institute of Behavioral Science, University of Colorado, Boulder
Steinberg F (2007) Housing reconstruction and rehabilitation in Aceh and Nias. The Asia Foundation, The Aceh Rehabilitation and Reconstruction Appraisal (ARRA), Complete findings report
UNDP (2006) Accelerating Recovery; Technical Support for BRR, First progress report to the multi Donor Fund.
UNSYIAH, UN-HABITAT (2006) Post tsunami settlement recovery monitoring in Aceh
World Bank (2007) Brief: housing financing and progress report
World Bank (2012) REKOMPAK, Rebuilding Indonesia's communities after disasters

Chapter 16
Vulnerability Assessment and Retrofitting of Existing Buildings in Aceh and Transfer of Knowledge to the Community

Hari Darshan Shrestha

Abstract As some of the buildings in Aceh, Indonesia constructed after the 2004 Tsunami disaster were found vulnerable they were retrofitted to make them safe against earthquake. The retrofitting was carried out by many agencies; one among them being Save the Children. The retrofitting work was carried out to raise the awareness of the people and also to sensitize them about earthquake safe buildings construction techniques and also to contribute to earthquake disaster risk reduction. Against the widely held perception that it is expensive and complicated process the retrofitting work was found cost effective in comparison to the cost of demolition and reconstructing new building. One of the key findings was retrofitting can be achieved through the use of simple methods, tools and equipment and local human resource. The retrofitting process and onsite training can also be an effective medium for dissemination of best construction practices in the community. In order to assess the impact of retrofitting work, a survey was carried out in 2009 in the area where they were carried out during 2005–2008. A total of 104 people who were involved in retrofitting/rehabilitation process were interviewed. The survey results show that the respondents felt that the impact of retrofitting work in transferring knowledge in the community is not significant. However, the respondents felt that the retrofitting work had definite impact in raising awareness of disaster risks and measures for disaster risk reduction. The findings can be replicated in the areas having similar type of buildings. It can be stated after its findings that retrofitting of school building does not only make the schools safer and reduces the risk of children at risk but also is cost effective compared to new construction.

Keywords Awareness raising • Disaster risk reduction • Impact assessment in communities • Knowledge transfer • Retrofitting • Vulnerability assessment

---

H.D. Shrestha (✉)
Institute of Engineering, Pulchowk Campus, Kathmandu, Nepal
e-mail: harisunita@gmail.com

R. Shaw (ed.), *Recovery from the Indian Ocean Tsunami: A Ten-Year Journey*,
Disaster Risk Reduction, DOI 10.1007/978-4-431-55117-1_16

16.1 Introduction

The December 2004 Indian Ocean Tsunami that struck the Aceh Province of Sumatra Island, Indonesia was catastrophic with unprecedented human losses as well as destruction causing colossal economic losses. The death toll of human lives crossed 200,000 with many more people injured. The extent of damage was so severe that some of the villages were completely washed out leaving very few survivors. In February 2005 the International Organization for Migration (IOM) led a multi-agency assessment to determine the shelter and livelihood needs and aspirations of those displaced by the tsunami. The scale of damage caused by the tsunami is illustrated by these statistics from the damage assessment in April 2005 (IOM 2005);

- Damaged housing and settlements—Settlement areas: 173,673 ha (34.8 % destroyed)—Houses: 116,880 units (57 % destroyed)
- Damaged public buildings—Health facilities: 693 units (66 % destroyed)—School buildings: 1,662 units (46 % destroyed)—Government buildings: 1,412 units (70.6 % destroyed)—Markets/kiosks: 1,416 units (75 % destroyed)
- Damaged infrastructure—Arterial roads: 654 km (27.5 % destroyed)—Provincial highways: 603 km (38 % destroyed)—Bridges: 2,267 units (66.5 % destroyed)

The disaster brought various international aid agencies in the region to support the government in the reconstruction and rehabilitation of the tsunami affected areas. The aid agencies in limited time with scarce human resource and construction materials constructed many shelters, schools, and health facilities under close coordination, supervision and monitoring from the "Agency for Rehabilitation and Reconstruction of Aceh and Nias" (BRR).

With huge investment required to cope with the enormous works and failure to formulate a strong sustainable work plan for the reconstruction and rehabilitation by the local government, the central Indonesian government instituted "Agency for Rehabilitation and Reconstruction of Aceh and Nias" (BRR) with policies and guidelines to standardize the construction practices undertaken by different agencies. The donors and their respective implementing agencies were under tremendous pressure to construct schools, health facilities and shelters with their limited expertise in the sector (Kennedy et al. 2008). It became clear that "humanitarian agencies lacked the know-how and expertise to lead a mass building exercise" (Oxfam International 2005). The outcomes of these constructions were later found to be of poor quality and vulnerable to future hazards such as earthquakes and floods requiring vulnerability assessments of the structures.

Some of the implementing agencies were of the view that the vulnerable buildings should be retrofitted and strengthen without demolishing to construct new buildings instead. The retrofitting was a relatively new technology for the concerned agencies as well as to the BRR, and the major concern was focused on the cost effectiveness

Save the Children (SC) conducted a case study of retrofitted buildings to measure the effectiveness of the technology (Fig. 16.1). The case study report revealed that retrofitting work is cost effective if carried out with proper measures. SC opined that

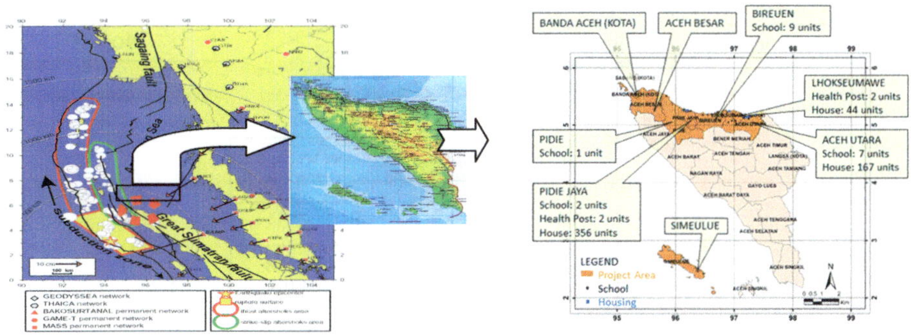

Fig. 16.1 Location map of the retrofitting project sites of SC which includes schools, houses and health posts

the retrofitting work undertaken will bring more participation of the community in the process and will have impact on the knowledge transfer of safer construction and raise the awareness on disaster risk. The field study covers retrofitting of schools as part of the projects that included many public and private facilities and shelters in Aceh.

16.2 Retrofitting of Existing Vulnerable Buildings

Buildings are designed to perform at required performance level throughout its life. The material degradation due to aging and alterations carried out during use over time necessitates the operations like Repair, Restoration and Retrofit. The decay of building occurs due to original structural inadequacies, weather, load effects, earthquake, etc.

The objective of the retrofitting is to reduce the risk of loss of life and injury to the people. This is accomplished by limiting the likelihood of damage and controlling the extent of damage to the buildings. When the existing building is incapable of withstanding the earthquake forces, it requires to be re-strengthened for safety. The complete replacement of such buildings in a given area may not be possible due to the historical importance or due to financial problems. Therefore, seismic strengthening of existing undamaged and/or damaged buildings is a definite requirement. The strengthening works must also be fully justified from the perspective of effectiveness.

Retrofitting is undertaken to enhance the original strength of building so that the desired safety is achieved and protection of lives can be guaranteed as per the current codes of practice against possible future earthquakes. Retrofitting of a building will involve either component strength enhancement or structural system modification or both. It is expected to improve the overall strength of the building.

The vulnerability assessment and retrofitting consists of threefolds of major activities—rapid and details structural condition evaluation and design of retrofitting and re-strengthening measures (FEMA 154, 155, 178, 274 & ATC 21, 40).

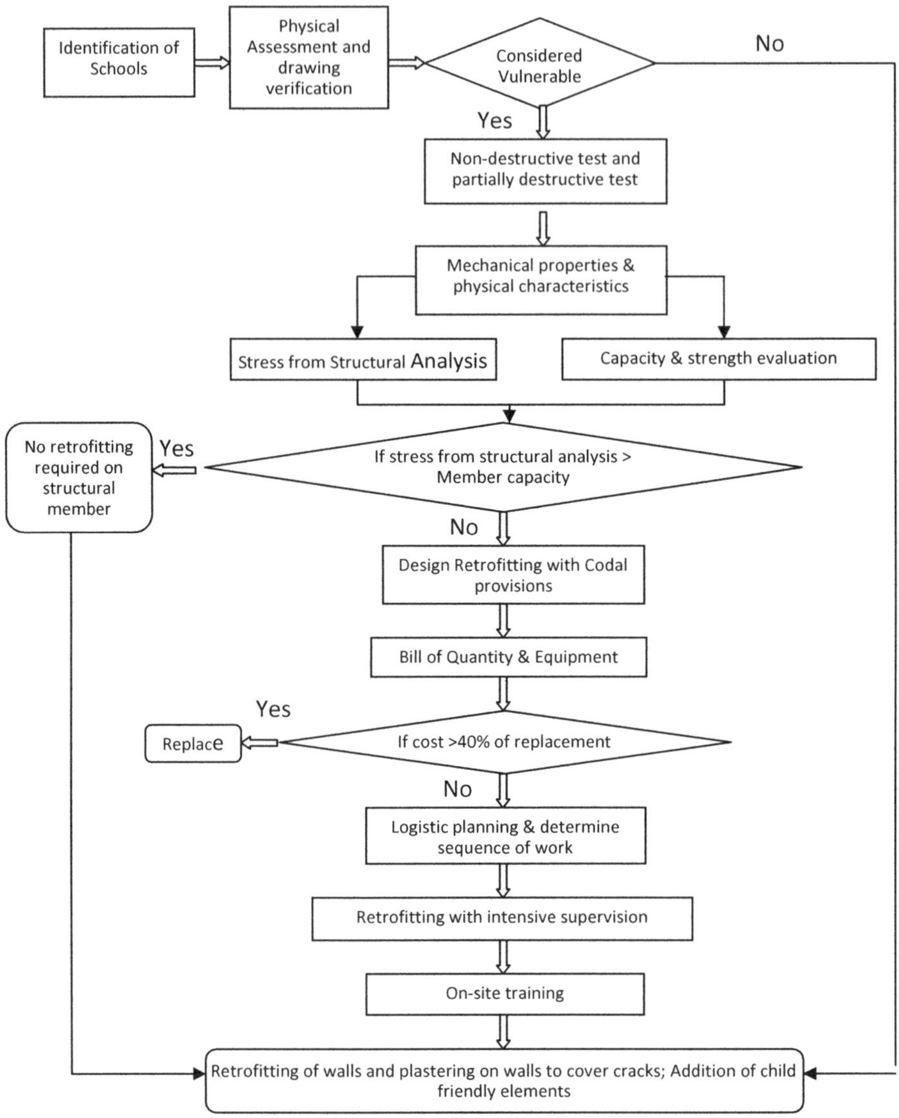

Fig. 16.2 Flowchart of methodology adopted for retrofitting. *Source*: Shrestha et al. (2012)

The integrated approach is shown schematically as flowchart in Fig. 16.2 (Shrestha et al. 2012).

The building damages by earthquake can be retrofitted during early recovery and recovery phase after vulnerability assessment and structural evaluation. This will reduce the rehabilitation cost significantly and will minimize the pressure on need of emergency and temporary shelter as the existing building under the category

Inspected and Limited entry identified by structural evaluation team after carried out of Rapid Evaluation Safety Assessment (ATC 20) could be allowed to occupy with minimum repair and retrofitting works. The approach has been practiced during post disaster phase in developed country but still lacking in developing countries. Hence, vulnerability assessment and retrofitting technology could be applied widely during post disaster management phase and will help to bring back situation to normal and to achieve built back better principle.

16.3 Retrofitting Work in Aceh

The vulnerability assessment team found that the construction quality is of low standard and design does not satisfy the new codal provision of Indonesian Standard (SNI). The team did the detail technical assessment and structural analysis to obtain mechanical properties of material and structural capacity of the members. After this, the team recommended for re-strengthening the weak and vulnerable members with simple retrofitting techniques by adopting local technology, local resources, and local artisans. In Total 58 School buildings of 19 schools, 4 existing health facilities buildings and 523 permanent shelters were retrofitted.

The location of retrofitting projects carried out is shown in Fig. 16.1. The area includes Pidie, Pidie Jaya, Bireuen, and Aceh Utara Districts of Aceh Province, Indonesia. In the study only the sites with retroåtting of school buildings were considered.

The retrofitting work was carried out to reduce vulnerability of the structures and also with the assumptions by the decision makers that the process of retrofitting will raise interest from the communities and attract their participation. The participation was assumed to have impact to the communities on knowledge of safe construction practices and raise communities' awareness on disaster risks and measures for disaster.

Schools were selected for the study purpose as such work involved participation of different cross-section of people such as representatives of school management bodies, teachers, and local people from the communities.

16.3.1 Rationale on Need of Retrofitting of Existing Building

An assessment showed that many school, health buildings and shelter constructed by different organizations didn't meet the seismic safety standards and the quality of materials and quality of construction were compromised which resulted in very poor quality buildings Figs. 16.3, 16.4, 16.5, and 16.6. This makes buildings vulnerable to earthquake. There was a dilemma on whether to dismantle all those vulnerable buildings or to retrofit them to meet the seismic safety standard. Dismantling the buildings could have spread negative message for the whole

Fig. 16.3 Poor quality concrete work

Fig. 16.4 Size of member less than required

reconstruction process and initially the retrofitting option was considered economically and technologically not feasible. As there are not many studies on cost benefit aspect of retrofitting, the decision was made with rigorous vulnerability analysis, structural evaluation, design and economical consideration compared with replacement cost

Fig. 16.5 Poor quality reinforcement detail

Fig. 16.6 Verification of foundation

16.3.2 Vulnerability Assessment

A preliminary physical investigation team from Save the Children (SC) and team from Syah Kuala University carried out the rapid visual assessment of buildings and concluded that buildings constructed during emergency response and early recovery stage after the 2004 tsunami and existing health facilities building in project area were vulnerable. This raised questions considering the safety of the buildings;

- Do the buildings meet seismic safety standard required for Aceh region?
- If not, is strengthening of the building technologically and economically viable option?

In order to address those questions, detailed vulnerability assessment of existing school, health facilities buildings and permanent shelter- buildings built before and after Tsunami was conducted. A total of 58 buildings from 19 schools, 4 health facilities building and 523 permanent shelters were selected for intervention. The assessment, which was made over a period from September, 2006 to February, 2007, had following main objectives;

- Evaluate the design, specification and drawings of the school buildings
- Determine the state of structure by field assessment and structural analysis
- Design suitable and affordable means of strengthening the members and retrofitting technique
- Transfer of knowledge on DRR and safe construction techniques to the communities through the retrofitting

The assessment includes review of drawings, on-site physical verification, on-site test and capacity assessment using analytical model. The process started with reviewing the drawing and specification followed by on-site verification. The structural drawings were reviewed in order to check the structural frame and truss system required for analytical models.

16.3.3 Detail Assessment and Field Investigation

The process started with reviewing the drawing and specification followed by on-site verification. The review team checked the drawings and compared it to seismic code requirements and was verified with on-site physical condition of the buildings. The structural drawings were reviewed in order to check the structural frame and truss system required for analytical models.

For the existing buildings, Compressive strength of a concrete and reinforcement details in building determines its load carrying capacity which is a major determinant of strength of the building. The onsite tests included non-destructive tests such as Rebound Hammer Test and Profometer and partially destructive tests such as Core Drilling and Static Penetration Test (SPT). The onsite tests were carried with objectives to define physical and mechanical properties of material in the existing buildings. This information was used later for further analysis of the buildings which was used for design of retrofitting.

Field assessment team conducted a detailed assessment of buildings—physical verification on quality of works, size and shape, bearing capacity non/partial destructive test Schmidt hammer (Fig. 16.7), Profometer (Fig. 16.8), core drilling (Fig. 16.9) and SPT (Fig. 16.10) to determine the properties of materials.

Fig. 16.7 Schmidt hammer Test

Fig. 16.8 Profometer Test

Compressive strength of concrete was obtained by non-destructive tests using Schmidt hammer test and partially-destructive test by taking out concrete sample from the building elements. The concrete sample was thereafter tested on Universal Testing Machine to obtain compressive strength and the strength was compared with that obtained from Schmidt hammer. Reinforcement in structural member was carried out by concrete scanner—profometer and bearing capacity of soil by SPT.

From physical assessment and on-site visits, most of the school buildings were found symmetric, rectangular in plan and were single storey RC and masonry structure.

Fig. 16.9 Concrete core sample for test

Fig. 16.10 Static penetration test

16.3.4 Structural Analysis and Capacity Analysis

The analysis was done for determining the internal forces in the structure. The building was modeled in SAP 2000 to obtain level of bending, shear and compressive stresses in earthquake loading. The properties of concrete, steel and the soil bearing capacity were determined from field assessment and the values calculated are used in the analytical model. The modulus of concrete is calculated on the basis of compressive strength of concrete in situ. The relation between modulus of elasticity and compressive strength, loading combination was adopted as defined by Indonesian Standard for Design of Concrete Structures SNI 2847 .3 2002. The calculation for earthquake load on building is based on SNI 1726 .3 2003 (Indonesian Standard of Earthquake Resistant Building) with the assumption that the earthquake zone is 4 and the bearing capacity of soil is moderate. The outcome is compared with the respective capacities as obtained from direct compression tests, bending capacity (from Eq. (16.1)) and shear capacity (Eq. (16.2)).

The bending capacity and shear capacity is derived from compressive strength using following Eqs. (16.1) and (16.2), respectively, as given in SN 2847 .3 2002 Indonesian Standard for Design of Concrete Structures.

$$Mn = \Upsilon\left\{Cc\left(d - a/2\right) + Cs\left(d - d'\right)\right\} \tag{16.1}$$

where:
Mn = bending capacity,
Υ = reduction capacity factor for bending analysis (from SNI 2847 .3 2002 Indonesian Standard for Design of Concrete Structures)
d = effective height of member section
Cc = Concrete compressive force, and Cs = Steel compressive force
a = equivalent block stress deep
d′ = distance between the center of compressive rebar and extreme concrete compressive fiber

The shear capacity contributed by concrete and stirrup is calculated based on formula recommended by SNI 2847 .3 2002 (Indonesian Standard for Design of Concrete Structures).

$$Vn = \Upsilon\left(Vc + Vs\right) \tag{16.2}$$

where:
Vn = shear capacity,
Υ = reduction capacity factor for shear analysis (from SNI 2847 .3 2002, Indonesian Standard for Design of Concrete Structures)
Vc = shear capacity contributed by concrete, obtained from SNI 2847 .3 2002,
Vs = shear capacity contributed by steel (stirrup), obtained from SNI 2847.3 2002.

If the capacity obtained for individual member (obtained from direct tests, Eqs. (16.1) and (16.2)) is less than the stresses obtained from analysis, it requires strengthening and the building is recommended for retrofitting (Ref. Fig. 16.2 for the flow chart). Based on the site assessment on the quality of the materials used in the buildings are the comparison of capacity with the analytical results, it has been found that majority of the buildings were vulnerable and need to be strengthened (Table 16.1).

Table 16.1 Vulnerability assessment summary from physical inspection, non-destructive test, partially destructive test and structural analysis

Findings	No. of school	No. of building	Intervention required
Compressive strength <20 MPA on columns and beams	5	17	Major Intervention required
Compressive strength <20 MPA on columns only	4	14	Major Intervention required
Compressive strength <20 MPA on beams only	3	8	Major Intervention required
Size of structural members columns, beam and foundation less than required	2	6	Major Intervention required
Size of Column and foundation less than required	3	10	Major Intervention required
Size of beam only less than required	2	5	Major Intervention required
Size of foundation only less than required	1	3	Major Intervention required
Settlement of foundation	2	6	Major Intervention required
Cracks on gable wall, walls only	6	17	Minor Intervention required
Compressive strength >20 MPA			

16.3.5 *Design of Retrofitting*

The retrofitting or strengthening strategy includes enlarging the column size, providing additional ring bars and connecting the walls with the columns (Fig. 16.11a,). In few cases horizontal beams were inserted in order to limit unrestrained height of walls and beams were strengthened in case they didn't meet the required capacity (Fig 16.11b), Furthermore, physical investigation of the buildings revealed that many buildings had weak foundations and intervention was made to strengthen the foundation (Fig. 16.11c).

The intervention was made in 58 buildings of 19 schools as listed in Table 16.2. The scale of intervention varied from building to building, and the detailed drawing, process, materials, techniques and cost estimates were prepared for each school buildings before starting the intervention. The general process and approach of retrofitting for safer and child friendly schools adopted is documented in a manual after the retrofitting work.

The retrofitting intervention is divided into two categories; viz major and minor, depending on whether the intervention required the strengthening of major structural elements or not. The summary of retrofitting intervention is given in Table 16.2. Major intervention includes extension of foundation, jacketing in column and beams, connection between column and walls and minor intervention include fresh plastering with wire mesh and removing vulnerable elements. Local materials, light and simple equipment and local technology were used according to the available technical know- how of local masons. Detail implementation plan for retrofitting process was prepared, which included material breakdown, equipment list, procurement and logistic plan, flow diagram on sequence of work, onsite training and requirement of human resources. Special attention was given on sequence of work

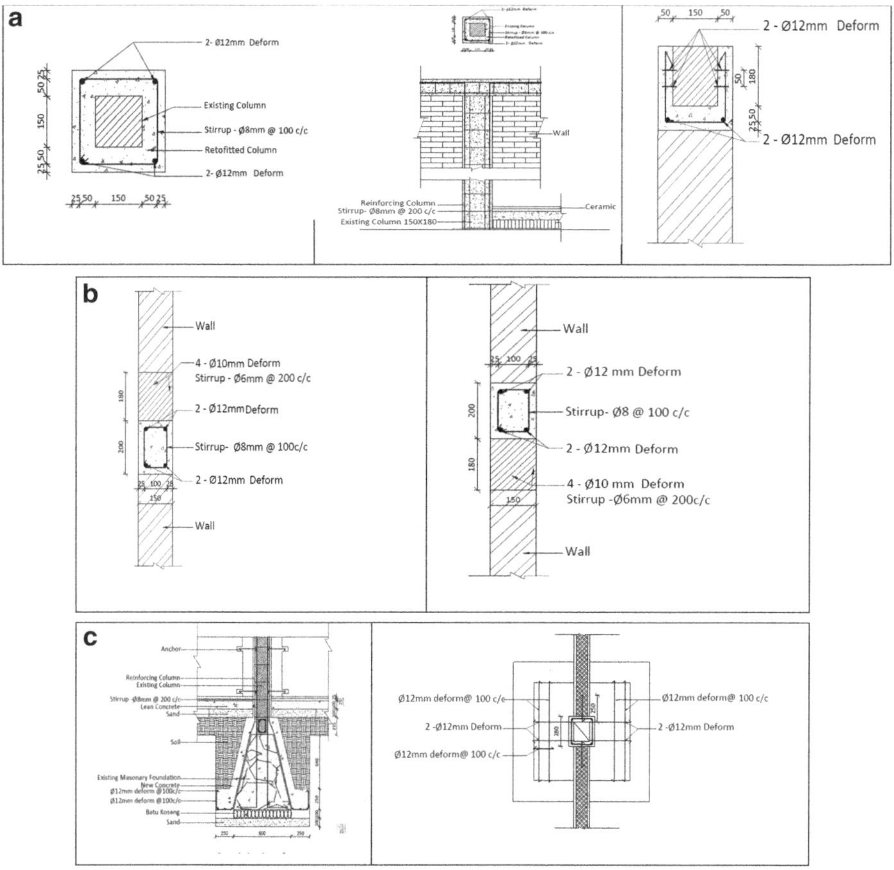

Fig. 16.11 Retrofitting detail. (**a**) Column and column wall retrofitting. (**b**) Beam retrofitting. (**c**) Foundation retrofitting. *Source*: Shrestha et al. (2012)

mainly on dismantling and strengthening, in order to avoid untoward incidents causing any casualties to the workers. A school building during retrofitting phase and retrofitted is shown in Fig. 16.11.

16.3.6 Cost Analysis and Recommendation for Retrofitting Work

After design and drawings of retrofitting works of each building, cost estimate of retrofitting works, replacement cost has been carried out and collected the initial construction cost. The decision has been made to retrofit only those buildings with

Table 16.2 Retrofitting Intervention in Schools

Level of intervention	Intervention	No of building	Name of schools
Major	Retrofitting of columns, beams and foundations	9	SDN 2 Peudada, SDN 3 Tanah Pasir, SDN 13 Syamtalira Arun
	Retrofitting of columns, and foundations	6	SDN Muhammadiyah, SDN Teupin Pukat
	Retrofitting of columns and beams	11	SDN 3 Pandrah, SDN 2 Kuala, SDN 7 Peusangan,
	Retrofitting of columns only	7	MIN Keudee Asan, SDN 1 Kuala
	Retrofitting of beams only	8	SDN 7 Tanah Pasir, SDN 12 Alue Dua, SDN 14 Kuta Makmur
Minor	Retrofitting of walls, gable walls, fixing the cracks	17	SDN 3 Peulimbang, SDN 8 Jeumpa, SDN 11 Syamtalira Arun, SDN 11 Sawang, SDN Suka Jaya, MIN Awe Geutah
Major	Retaining structure to protect foundation	3[a]	SDN 2 Peudada, SDN 12 Alue Dua

Fixing of plastering, removing of vulnerable elements and addition of child friendly elements done for almost all schools

[a]Schools same on above major intervention

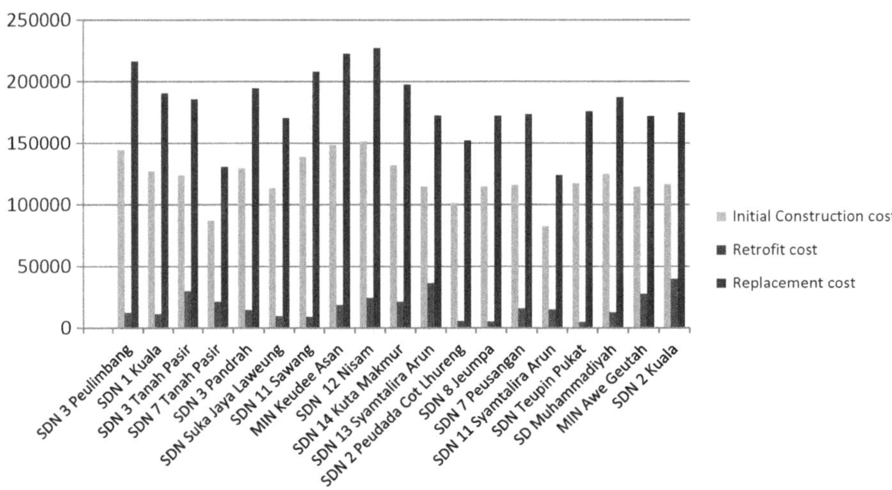

Fig. 16.12 Cost for retrofitting compared with initial and replacement cost

cost of retrofitting works less than 40 % of initial construction cost and 30 % of replacement cost whichever is less. Retrofitting design done on some testing and assumptions, which may differ during retrofitting works and estimated retrofitting cost may increase. Estimated cost of the retrofitting works is compared with replacement cost of construction and initial construction cost (Fig. 16.12) and retrofitted only those buildings which retrofitting cost is less than 40 % of initial cost and 30 %

Fig. 16.13 Child friendly features—Hand wash sink and ramp for disable

of replacement cost… In total 58 buildings were retrofitted; most of the retrofitting work required maximum of 15 % of the initial construction cost and only 10 % of the total cost of rebuilding. Retrofitting/repair cost for each building is different, and ranges from 6 to 23 % of replacement cost as shown in Fig. 16.12 (Shrestha et al. 2012)

16.3.7 Implementation Strategy

It is not an easy task to start the retrofitting of existing building, as community are not aware about the risk due to vulnerable buildings. At the same time, this challenges covert to opportunity as the retrofitting process can be the tools to demonstrate the risk due to poor design and construction and also mitigation approach to make the buildings safe against earthquake.

Project has been design with clear role and responsibilities and the clear steps so that community will see the cause and effect and need. Ultimately the goal was to raise awareness about the disaster and build the capacity of local community to make safe and child friendly school buildings. Pearce (2003) suggests that for successful implementation of imitative strategies, disaster management process must incorporate community participation from the local decision making level

The structures were not made only safe but also made child friendly (Shrestha et al. 2009). Some of the child friendly features included are ramp for disabled children, rounded corners of external walls, and hand washing sink with reduced height as shown in Fig. 16.13.

Communities were involved as observer in overall retrofitting process from vulnerability assessment to design and construction. Why the buildings are not safe and need of intervention and how it can be fixed are discussed with community, student and teachers from the beginning. Community, teacher and student were encouraged to observe the retrofitting process and engineer on site describe in detail why and how.

Retrofitting is not as simple as new construction; has to work on existing structure without affecting other elements. Keeping this in mind and to include the communities, teacher and student in the process it is decided not to hire contractor and do the work with the crew. Five engineers and thirty tradesmen (skilled labor) from the area were recruited and decided to use simple methods and tools familiar in area such as concrete cutter, electrical drilling machine, jack hammer, chisel, hammer and gun for adhesive etc. The retrofitting processes were started in two schools as a model and provide onsite training. After completion of the two building as a model and training, each engineers with six tradesmen were given the responsibility on five other buildings and more tradesman were trained. This process was continued till retrofitting works of all buildings were completed as shown in Fig. 16.14a–f.

With this process seven Engineer and about 200 tradesmen are trained on retrofitting technique. They have now local capacity on retrofitting technique and at the same time communities are now aware about the quality construction, cause and effect of substandard work, poor material and need of professional input and supervision to ensure the construction of safe and child friendly school building.

58 school buildings, 4 health facilities and 523 permanent shelters are made safe and reduced the number of children at risk, at the same time by involving in the retrofitting process; the level of awareness on disaster risk reduction of the communities, teachers and students has been raised.

16.4 Impact of Retrofitting Work

The retrofitting work was designed with the assumptions that the process of retrofitting will raise interest from the communities and attract their participation. The participation was assumed to have impact to the communities on knowledge of safer construction and also to raise their awareness about disaster risk. One of the five priorities of the Hyogo Framework for Action 2005–2015 (HFA), to build resilience of nations and communities by 2015, is use of knowledge, innovation, and education to reduce underlying risks (UNISDR 2005). HFA underscores community participation and technology transfer as two cross cutting issues for disaster risk reduction. Successful implementation of mitigative strategies, disaster management process can be achieve by engaging local community from the local decision making level, Pearce (2003)

Subedi (2010) suggested that participation in community based disaster preparedness may not be effective in raising awareness to the extent of reducing vulnerability, as receiving information only does not ensure that the information is used for making decisions that effectively reduce the vulnerability. There are also many factors that affect the effectiveness of community participation programs. Karanci et al. (2005) argue that male community members with higher level of education and smaller household size who are concerned about future disasters and participated in a disaster awareness program contributed to disaster preparedness behaviors.

Fig. 16.14 A building during retrofitting and after completion. (**a**) Retrofitting of column. (**b**) Retrofitting of foundation and form work for column. (**c**) Retrofitting of beam. (**d**) Joint between column and beam. (**e**) Building on retrofitting process. (**f**) Building after retrofitting

A survey was conducted to evaluate the impact of retrofitting work in raising awareness and transferring know-how of safer construction works in the communities after the shelters, schools and health facilities were retrofitted and occupied in the five districts—Pidie, Pidie Jaya, Bireuen, Aceh Utara and Lhokseumawe (Fig. 16.14). Data were collected by a team of community mobilizers and technical

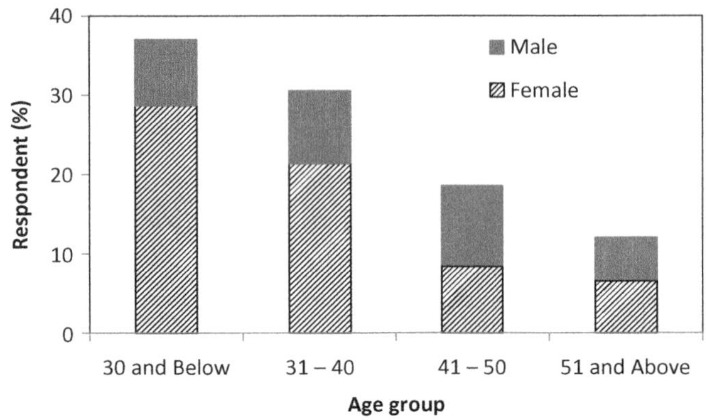

Fig. 16.15 Age distribution of the respondents

staffs that were trained and oriented to conduct the questioning to the interviewees, through interviews with individual household members, teachers, and community leaders involved in the retrofitting process. Mason, labour and staff of implementing agency who were directly involve in the construction/retrofitting works were excluded in survey. The respondents were among the community members, teachers, and school management persons who were encouraged to observe the retrofitting process and to whom the details were explained regularly. Most of survey respondents were female teacher and female beneficiaries as they were regular observer of retrofitting works. Although their degree of participation was not recorded during the survey, the respondents participated in the events during the retrofitting works.

This study aimed to review/analyze the impact of retrofitting works on participating communities. People from different backgrounds participated in the retrofitting process; from planning to execution stages. Among the participants, a total number of 104 people were interviewed. The survey results were tabulated and analyzed to investigate impact of their involvement in transferring know-how of the retrofitting and also in raising awareness of disaster risks and measures for disaster risk reduction.

Out of surveyed people 64.8 % are female, 37 % are aged 30 and below and only 12 % are aged 51 or above shown in Fig. 16.15. Among the respondents a majority of them have school education or above (83 %) and only 5 % are illiterate and 44 % of respondents are employed, 30 % self employed, 10 % unemployed and 16 % have not disclosed their occupation (Shrestha et al. 2012).

16.4.1 Transferring the Knowhow to the Community

One of the questions asked to the respondents was about the effectiveness of the retrofitting work as a means for transferring the know-how to the communities. Among all respondents, 28 % felt that their involvement in retrofitting was effective in transferring the know-how to the communities and another 14 % felt that it had positive but marginal impact. About 29 % of the respondents felt that involvement in the retrofitting work had no contribution to the transfer of knowledge to the communities and an equal percentage of people were reluctant to divulge their opinion about effectiveness of retrofitting work to the transfer the know-how to the communities

16.4.2 Transferring the Knowledge to the Community

Of the 39 % of the respondents whose houses were severely damaged, 13 %—out of the total 104 respondents—agree that the retrofitting work definitely transfers the knowledge to the community, 8 % agree that the retrofitting work has marginal impact in knowledge transfer, and 11 % feel that it has no impact at all.

The results show that involving people in the communities in retrofitting work to transfer the know-how to the communities was not very effective. This may be because of the fact that even after being completely involved in the process, the respondents did not feel confident about their knowledge on retrofitting to execute it later independently. The retrofitting work is a technically challenging process and in the lack of traditional knowledge in it, it is understandable that the communities were not confident about carrying out the work by themselves independently.

16.4.3 Raising the Community Awareness

The respondents felt that the retrofitting work is more effective in raising the community awareness of disaster risks and risk reduction measures. Among all respondents, 55 % felt that the retrofitting work was effective in raising awareness with 42 % saying the work had definite impact and only 20 % said that it had no impact at all, while 25 % of the respondents offered no opinion on this question. The results indicate that although communities' participation is not panacea to all disaster management problems as pointed by Allen (2006), their participation helps to raise awareness. Awareness is one of the key factors that contribute in making decisions to implement disaster risk reduction measures (Subedi 2010).

16.5 Conclusion

On-site inspection of the building was carried out to assess the strength parameters and after structural analysis suitable retrofitting measures were recommended. Out of the 19 schools (58 numbers of buildings) surveyed 13 schools (41 buildings), were found vulnerable and required immediate intervention. Although 6 schools (17 buildings) did not require major structural intervention, minor retrofitting and repair works were done and additional measures to make child friendly were installed. This exercise carried out in Ache revealed that retrofitting could be better option for making vulnerable buildings like schools and hospitals safer from future seismic events. It consumes less resource in terms of time, technology and money. On the basis of cost comparison, it has been found that retrofitting is a cost-effective measure, to reduce earthquake vulnerability of the poor quality buildings in Ache.

Methods, process, materials and technology adopted for retrofitting is another important aspect in the exercise. The involvement of local community in initial screening and retrofitting implementation process increased the interest of the people in retrofitting and brought the awareness among the communities to the next level. This exercise also provided confidence to the implementing agencies to carry out similar exercises in other places. The post construction e survey revealed that there is a positive impact on people who participated in the retrofitting work in raising their awareness of disaster risks and measures for disaster risk reduction. A large number of respondents felt that the retrofitting work has had definite impact on raising their awareness. However, the survey also showed that participation of communities did not help to transfer technology of retrofitting to the communities. This may be because of the fact that the retrofitting is new concept and it also requires experience and skill. This shows that capacity development efforts have to be continuous to establish the retrofitting work in the local culture of construction technology and methods even though construction is a regular activity in the community and they have traditional knowledge in it. The results show that engaging communities in disaster risk reduction activities, such as retrofitting, is effective in raising awareness, however Technology transfer for retrofitting that are relatively new to the communities, cannot be achieved as it requires certain degree of analytical and construction skill which can be acquired only after understanding the concept and approach of retrofitting and repetitive involvement in similar exercise. Effectiveness in raising awareness also varies among people from different age groups and education background. Therefore, packaging participation in mitigation projects along with other approaches such as training and regular drills can be effective ways to increase awareness among all cross-sections of people.

The methods and processes used for retrofitting works in Ache could be one of the best practices in community based retrofitting of poorly engineered or non engineered buildings. The method becomes more relevant in context of developing nations which are prone to seismic events. It can reduce the risks and ensure the functioning of essential services in post disaster scenario. It is a low cost retrofitting mechanism with intention to instill the skill and technology in local construction technology process.

References

SNI 2847 .3 (2002) Indonesian Standard for Design of Concrete Structures
SNI 1726 .3 (2003) Indonesian Standard of Design of Senior High School Building

Allen KM (2006) Community-based disaster preparedness and climate adaptation: local capacity-building in the Philippines. Disasters 30(1):81–101

Applied Technology Council (ATC) (1988) Rapid visual screening of buildings for potential seismic hazard: a handbook (Report ATC -21). Applied Technology Council, Redwood City

Applied Technology Council (ATC) (1996) Seismic evaluation and retrofit of concrete buildings. In: ATC-40, vols 1 and 2. Seismic Safety Commission, State of California

Federal Emergency Management Agency (1988a) NEHRP commentary on the guidelines for the seismic rehabilitation of buildings. Building Seismic Safety Council, Washington

Federal Emergency Management Agency (1988b) Rapid visual screening of buildings for potential seismic hazard: supporting document. Building Seismic Safety Council, Washington

Federal Emergency Management Agency (1988) Rapid visual screening of buildings for potential seismic hazard: a handbook, Building Seismic Safety Council. In: FEMA 154. Federal Emergency Management Agency, Washington

Federal Emergency Management Agency (1992) Handbook for the seismic evaluation of existing buildings. Building Seismic Safety Council, Washington

IOM (International Organization for Migration) (2005) Post disaster damage assessment in Nanggroe, Aceh

Karanci AN, Aksit B, Dirik G (2005) Impact of a community disaster awareness training program in Turkey: does it influence hazard-related cognitions and preparedness behaviors? Soc Behav Pers 33(3):243–258

Kennedy J, Ashmore J, Babister E, Kelman I (2008) The meaning of "Build Back Better": evidence from post-tsunami Aceh and Sri Lanka. J Contingencies Crisis Manag 16:24–36

Oxfam International (2005) A place to stay, a place to live: challenges in providing shelter in India, Indonesia, and Sri Lanka after the tsunami. Oxfam Briefing Note, Oxford

Pearce L (2003) Disaster management and community planning, and public participation: how to achieve sustainable hazard mitigation. Nat Hazards 28(2–3):211–228

Shrestha HD, Pribadi KS, Kusumastuti D, Lim E (2009) Manual on retrofitting of existing vulnerable school buildings—assessment to retrofitting

Shrestha HD, Yatabe Y, Bhandary NP, Subedi J (2012) Vulnerability assessment and retrofitting of existing school buildings: a case study of Aceh. Int J Disast Resilience Built Environ 3(1): 52–65

Shrestha HD, Yatabe Y, Bhandary NP, Subedi J (2013) The impact of retrofitting work on awareness raising and knowledge transfer in Aceh Province, Indonesia. Int J Disaster Risk Sci 4(4): 182–189

Subedi J (2010) Disaster informatics: information management s a tool for effective disaster risk reduction. In: Advanced ICTs for disaster management and threat detection: collaborative and distributed frameworks. IGI Global, Hershey, pp 80–94

UNESCO U (2007) Manual for restoration and retrofitting of rural structures in Kashmir. University of Roorkee (1994), Earthquake problem: do and don'ts for protection

UNISDR (United Nations International Strategy for Disater Reduction) (2005) Hyogo framework for action 2005–nations and communities to disasters

UNRC/BRR (2009) Tsunami recovery indicator package

Chapter 17
Learning on the Safety Issues of Reconstructed Houses from the 2004 Great Indian Ocean Earthquake and Tsunami in Aceh, Indonesia

Kenji Okazaki, Krishna S. Pribadi, and Dyah Kusumastuti

Abstract The 2004 Great Indian Ocean Earthquake and Tsunami has provided an opportunity to build a safer Aceh, Indonesia, through reconstruction of houses, buildings and infrastructures during the recovery process. A research conducted in 2006 to investigate and assess the house rebuilding practices of various actors in the reconstruction process produced the following findings. To improve the safety of the houses, planning and design should be improved so that the needs of the communities match with the reconstruction efforts, proper seismic designs should be ensured to avoid severe damage from future earthquakes, building materials should be improved to meet minimum requirements, quality of workmanship should be enhanced and good project management and supervision should be introduced to improve structural quality.

It was expected that the lessons learned from the study and the recommendations based on the findings were implemented to make Aceh houses safer from future earthquakes. However, almost a decade later, recent earthquakes in Aceh have shown that these lessons were hardly incorporated in the building practices in the area, and that more serious efforts are needed to really address the safety issues of houses. Stronger measures should be put into place to ensure continuously that appropriate building practices are observed in all over the province.

Keywords Aceh • Houses • Reconstruction • Structural safety

K. Okazaki (✉)
Graduate School of Global Environmental Studies, Kyoto University, Kyoto, Japan
e-mail: okazaki@archi.kyoto-u.ac.jp

K.S. Pribadi • D. Kusumastuti
Institute of Technology Bandung, Bandung, Indonesia

R. Shaw (ed.), *Recovery from the Indian Ocean Tsunami: A Ten-Year Journey*, Disaster Risk Reduction, DOI 10.1007/978-4-431-55117-1_17

17.1 Introduction

During 4 years (from 2005 to 2008), the Executing Agency for Rehabilitation and Reconstruction (BRR) of Aceh–Nias accomplished the reconstruction and rehabilitation of around 140,000 houses, exceeding the need to recover from the Great Indian Ocean Earthquake and Tsunami in December 2004 and the Nias Earthquake in March 2005, which had destroyed 139,000 housing units (BRR 2009).

One year after the disaster, progress was achieved in the reconstruction projects, albeit slower than expected, where 16,200 new houses had been completed and other 13,200 were underway construction (BRR 2005). School buildings were also reconstructed at a rate almost identical to that of houses. However, in the effort to achieve reconstruction at the expected pace and produce the targeted number of buildings, it seems that the quality control for the newly reconstructed houses and buildings was being overlooked. In the ongoing reconstruction process, the quality of the structures should have been taken as a serious issue, in addition to the issue of the large quantity of structures to be built in the region. Recognizing the geological and seismological conditions of Aceh Province as one of Indonesia's most earthquake-prone areas, "rebuilding a safer Aceh" should have been pursued as the ultimate goal of the post-disaster reconstruction process. To improve the safety of Aceh's communities, it was very important to ensure that building regulations should be implemented to minimize the earthquake vulnerability of the newly reconstructed buildings.

A look back reflection on whether the lessons from the reconstruction process after the 2004 Great Indian Ocean Tsunami and Earthquake were learned effectively or otherwise will be presented. In 2006 a research with the objectives of investigating and assessing the capacity and building practices of various actors during the ongoing reconstruction process, focusing on the vulnerability of houses, was conducted, and recommendations for technological interventions to prevent unsafe practices in future building construction in the province were proposed (Okazaki et al. 2008, 2011). Yet almost a decade later after the catastrophe, recent earthquakes in the province showed however that there were still more efforts are needed to realize and implement the recommended measures in order to make Aceh safer against earthquakes.

17.2 Survey in 2006

17.2.1 Outline of the Survey

The field survey was conducted among various reconstruction projects in Banda Aceh and Aceh Besar from January to February of 2006. Due to the limited time frame and the vastness of reconstruction activities in the area, the survey was limited to several selected samples of housings, school buildings, and small scale public buildings.

The buildings were randomly selected, and included those constructed by the communities, local government, and donors (NGOs). In total, 53 buildings were visited for the field survey (34 houses, 12 schools, 2 mosques, 5 other types).

The survey intended to obtain general pictures of ongoing reconstruction projects and their actual building practices. Therefore, field testing for building materials and interviews with various parties involved in the reconstruction process were conducted. Documentations including structural drawings, specifications, pictures, and notes were also obtained in the survey. A few samples of building materials were collected to be analyzed in the laboratory. Local actors—consisting of workers, contractors, related local government agencies, NGOs, and local community groups—were interviewed to obtain information concerning building technology capacity as well as the building procurement process and mechanisms. Input from the owners or future occupants of the buildings as well as other interested parties were also collected.

17.2.2 Assessment of Quality of Structures

The purpose of the assessment was to examine several aspects of the building practices: structural design, project management, materials, workmanship, infrastructure and supporting facilities, and policies, in order to understand whether the buildings were reconstructed with adequate safety or otherwise during the reconstruction process, and to find out the difference between existing and post disaster construction. The quality of structural design in terms of earthquake resistant design was analyzed based on the buildings' documentation. The structural drawings and specifications were reviewed to ensure that the original documents endorse seismic building design. The project management aspect, which includes construction technology, procurement process, and tools used, was reviewed to assess the effectiveness of the project. These assessments were followed by the analysis of the quality of the materials used in the structures. In addition to the results from the field testing, several material samples were tested in the laboratory to determine their properties and structural qualities. The quality of workers was analyzed to observe the adequacy of workmanship.

In addition to the buildings, the environment required for conducting a reconstruction project was also reviewed. Infrastructure and other supporting facilities such as roads, electricity, and drainage, necessary to ensure the smoothness of the reconstruction projects and to support the communities after the completion of the projects were investigated. Local/national codes and other policies used for the reconstruction projects were also studied to determine the adequacy of regulations for these projects.

Based on the results from the field survey and laboratory tests of materials, an analytical study was conducted to examine the structural performance of the buildings. A few buildings from the reconstruction projects were selected to examine their weaknesses and to find ways to improve the quality of the structures.

Structural analysis was conducted and the building performances were analyzed based on the calculated force/stress and displacement responses.

From the field observation and structural analysis, an appropriate technological intervention was developed to reduce the vulnerability of the buildings. This intervention is based on the available capacity, culture, local materials and tools, and also the supporting environment. A workshop was then conducted in Banda Aceh on 22 February 2006 to share the findings from the survey and analytical results.

17.3 Findings from the 2006 Survey

17.3.1 Construction Practices in Aceh

During the 2006 survey, it was found that, from an engineering perspective, about two thirds of the buildings in Aceh were non-engineered buildings, meaning that they were built traditionally with little or no assistance from qualified engineers (IAEE Committee on Non-Engineered Construction 1986). Most masonry houses fall into this category. Engineered buildings are characterized to be properly designed and constructed for satisfactory performance under applicable loading conditions. Most public facilities fall into this category, including school buildings, government offices, and mosques.

Most buildings in Aceh in the time of the survey were found to be masonry structures with one or two stories. This type of structure gained popularity during the 1970s and continues to be the first choice in Indonesia. The existing structures show that they can survive earthquakes with little or no damage, provided that they were built properly using materials and workmanship of good quality. The survey, consisting of field surveys on damaged houses and a series of interviews of researchers, workers, and government officials, reveals that masonry structures built in the 1970s or 1980s performed better than those built in the 1990s. Almost all damaged structures from the 1990s were not designed properly according to the seismic codes.

17.3.2 Planning and Design

The planning and design aspect relates to the needs and demands of the reconstruction projects. The survey showed that some problems arose due to limited study of feasibility prior to implementing reconstruction projects. Examples of planning problems include temporary and permanent houses being built at the same time as well as construction plans/schedules not being well coordinated with those of infrastructure for housing. There were several completed projects that were uninhabitable because of lack of infrastructure such as electricity and drainage, or because the community refused to live in the project area. The interviews in communities revealed several serious problems. First, some communities were relocated apart

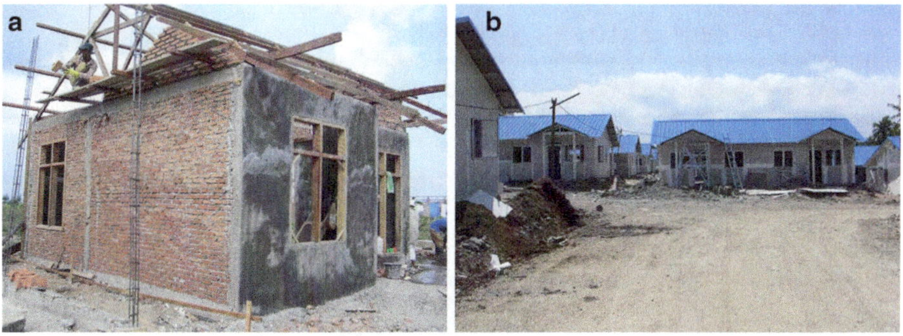

Fig. 17.1 (**a**) Type 36 houses under construction by BRR. (**b**) Houses with imported materials

from their former villages and were thus uprooted from their previous background. A typical example is the relocation of a fishing village to a location about 2 km from the coastline. The fishermen refused to be moved because easy access to their boats and to the sea is part of their lifestyle. Second, as mentioned above, many houses were found to be uninhabitable because of lack of electricity, drainage, or water supply. Third, the lack of planning also leads to houses being built very far from public facilities. Finally, some communities refused to move due to inadequate delivery of correct information and non-transparency in official procedures.

The design of structures for housing and schools generally follow typical design structures. Most school buildings surveyed showed that one- or two-story buildings were commonly used for school buildings, and the two-story buildings were often engineered buildings while most of the one-story buildings were categorized as non-engineered buildings. The houses were mostly what is called type 36 houses, meaning total floor area is 36 m^2, most of them can be categorized as non-engineered, one story buildings. Type 36 houses were endorsed by the government as replacement houses for the victims. Most buildings were made of burnt brick walls with reinforced concrete columns and beams. Roofs were constructed with wood trusses and galvanized iron sheets or roof tiles. Because these materials are commonly used for buildings in Indonesia, these buildings conformed with the local culture. Figure 17.1a shows a typical example of this type constructed by BRR. However, some projects used imported materials and technology, which resulted in questionable maintenance and sustainability (Fig. 17.1b).

17.3.3 Contracting

The construction industry in Aceh was not as powerful nor active as that in Java Island or other more developed regions. The political conflict in Aceh for more than three decades limited the growth of private sector investment, resulting in a relatively smaller private sector construction market. The main construction market

Fig. 17.2 Local builders in housing reconstruction

consisted of government infrastructure investment, while large international investment (oil and gas processing and related industry) was limited to large scale contractors, usually coming from outside the province. Many local contractors depended mainly on government works, resulting in tight and unhealthy competition among the local contractors. As a result, the local contractors did not invest in developing their human resources, construction plants, and technology. The tsunami disaster worsened the situation, as it caused the loss of contractors and family members, construction facilities, skilled human resources, and financial networks, reducing capacity and competitiveness.

Local small scale contractors (SSCs) were generally involved in small/medium sized infrastructure works such as road and small bridge construction and their maintenance, small scale irrigation schemes and their maintenance, residential and non-engineered public facilities (school buildings, government offices, tertiary health centers, etc.), and building maintenance and rehabilitation (Fig. 17.2). A typical SSC employed only a few (generally less than ten) permanent staff members (managerial and technical); most of them were not skilled technical personnel. This situation did not allow SSCs to develop their capacity and accumulate experiences and knowledge. Another problem for SSCs was the complicated procedures (proposals, supporting letters, collateral, etc.) that made it difficult for them to obtain financing from formal financing institutions.

Community contracting by implementing agencies was also used in reconstruction of Aceh, particularly in donor-driven projects. This process was mostly utilized together with a community participatory and bottom-up planning process, where the implementing agency provided a block grant and the community decided what it wanted at the local level. These programs became popular in recent years and the community perceived that they were efficient and cost effective. Community contracting were implemented in the following three modalities: (i) Community procures the services of (local) contractors to implement works using their own funds provided by donors; (ii) Community is contracted by government or donor agencies to implement infrastructure construction such as drainage and sewage; (iii) Community is contracted by the contractors who are requested to sublet part of the work to local communities.

17.3.4 Construction Aspects

(1) Materials and workmanship
The procurement of materials for the projects was crucial but the supply of materials was scarce. The survey found some projects were halted or cancelled due to the unavailability of materials. Materials found in the area could be divided into two categories: local and imported. The local materials supplied from Aceh included sand, gravel, bricks, and wood. Imported materials from outside Aceh came mostly from North Sumatra, but some were from Java, or even as far away as Australia, South Africa, or Europe. Usually, these imported materials were chosen by the donors. Examples included bricks from North Sumatra and Jakarta, cement from West Sumatra, roof tiles from East Java, and metal frames from Australia. As long as the materials were widely used in Aceh Province, maintenance and sustainability of the imported materials could be ensured. However, if the materials used were new for the locals, maintenance and sustainability would be a problem in the future. Some projects were indeed found to be using imported materials that were new and rare in Aceh, such as prefabricated steel frames and lightweight concrete panels.

In terms of materials and structure, most of the houses and school buildings conformably used typical local materials. The houses and school buildings were constructed in burnt brick confined masonry structure with sand and portland cement mortar, with RC columns and beams confinements. A typical flat footing was usually used for the column foundation in two-story buildings, while a stone masonry footing was commonly used for one-story buildings. Metal roof decks were the most popular choice for roofs due to their cost, workability, and durability.

The design of houses was similar to the designs used prior to the disaster. Post-earthquake design was not much improved, including the seismic aspect. Moreover, in most drawings, structural details that are important for seismic design were still left unclear, such as the beam-column connection, spacing for hoops, seismic stirrups, lap splices, and anchorage, so workers were left to "improvise" on these details. On the other hand, most designs for two-story school buildings showed a conservative approach regarding seismic loading,

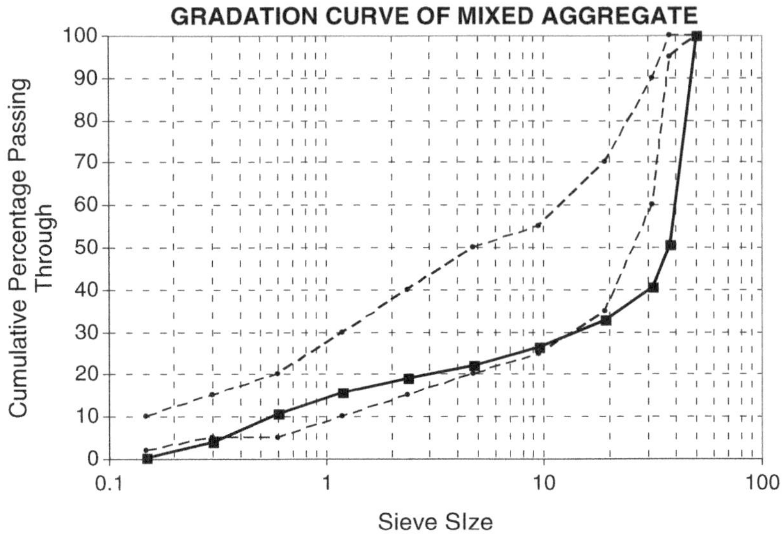

Fig. 17.3 Gradation curve of mixed aggregate of three samples (The two *dashed lines* show the range of ASTM C33-90)

emphasized by the large number of columns and the substantial dimension of structural elements. It should be noted that there was another big problem of the quality of construction. The field observations found that not a few of the materials used were different from the specifications and requirements of the drawings.

The concrete mixture was a serious problem for reconstruction of houses. The materials used for the mixture were adequate in most cases. However, almost no projects used the proper composition for the concrete mixture (volume ratio of 1 cement: 2 sand: 3 gravel). The most common mixture is cement and sand-gravel (sand with some gravel) with composition of 1 bag of cement for 3 carts of sand-gravel (about 1:5.25 in volume ratio). No sieving process was done to comply with the standard material gradation used for concrete mixture and no effort was made to avoid large aggregates getting into the mixture, which lead to a mixture having a very poor gradation with a high percentage of coarse aggregates. The laboratory tests conducted clearly illustrated this fact, as seen in Fig. 17.3.

Another critical issue concerning concrete was cement/water ratio. Excessive water was found in most of the cases. Proper curing for concrete was not common in the area. Considering all these facts, it was concluded that the quality of concrete was one of the most serious problems in the reconstruction projects.

The usual reinforcements used in the buildings were plain rebars of diameter 10 mm for longitudinal, and plain rebars of diameter 6 mm for hoops. In some construction sites, even weaker reinforcements were found: diameter 8 mm for

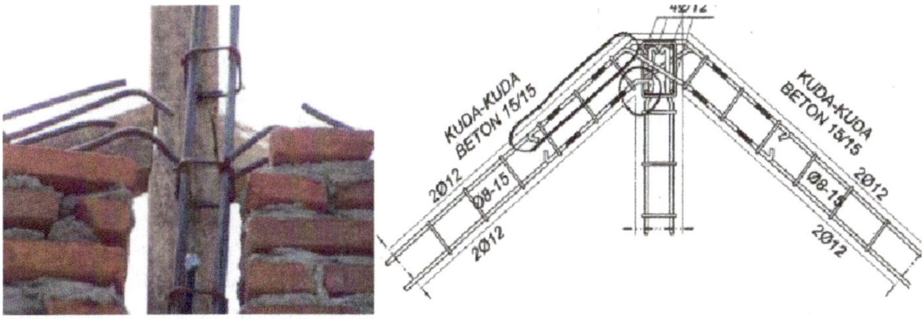

Fig. 17.4 Construction does not follow the drawing

Fig. 17.5 (**a**) Wide spaced hoops. (**b**) Melting bricks

longitudinal plain rebars and diameter 4 mm for hoops. The Indonesian seismic code (Badan Standardisasi Nasional 2004a) specifies the following minimum diameters for materials to be used in earthquake resistant buildings: 12 mm deformed bars for column longitudinal reinforcements, 10 mm deformed bars for beam longitudinal reinforcements, and 8 mm plain bars for hoops. Most of the reinforcements found in the field survey violated the building codes (Badan Standardisasi Nasional 2004b). Even worse, detailing of reinforcement such as bending work did not satisfy seismic demands for safety. Figure 17.4 shows the contrast between the drawing (the intended design) and the actual situation, a difference that was quite common in the area. The spacing of the hoops also posed a problem, with spaces averaging 200–250 mm, in contrast to the code's requirement of minimum spacing of 150 mm or less (see Fig. 17.5a). The quality of bricks used in the area was very low. The lowest class of bricks in the Indonesian code is Class III, which requires strength of minimum 25 kg/cm2. Most bricks fell below the minimum strength and even "melt" when soaked with water (Fig. 17.5b).

Table 17.1 Tensile tests of steel reinforcements

Type of reinforcement	Average yield strength (kg/mm^2)	Standard yield strength (kg/mm^2)
Bj. TP 24	29.56	39
Bj. TP 30	43.95	49

The lab tests showed that the strength of the steel bar was also below standard (see Table 17.1). All these facts indicated that the strength capacity of the structural elements for bending and shear in the buildings being constructed could be much lower than what was expected.

The mortar quality was also questionable because of poor composition: too little cement and too much water. The bricks observed at construction sites were not installed in straight lines and mortar was often excessive (spacing of more than 15 mm). The structures were also lacking in structural integrity, due to insufficient connections between structural components. For example, connection did not comply with the requirements in the codes/drawings, and no anchorage was provided from the walls to the columns.

(2) Workers

While the low quality of materials caused low structural quality, the low quality of workmanship amplified these problems. With the huge amount of work arising from reconstruction activities, trained and qualified workers were scarce in the area. The results of most projects showed that the workers were not adequately skilled for the job. They tended to ignore the important issues for producing safe structures, such as the sieving process for aggregates, curing of concrete, soaking of bricks before placement, water ratio for concrete mixture, and bending details. All these problems arose due to the workers' lack of the concept of quality of structures. The workers simply did what they thought of as the easiest way to construct the buildings—without concern for the quality—because they were not equipped with knowledge of proper construction methods nor basic concepts of quality of structures.

(3) Project management and supervision

The investigation showed that there were two types of reconstruction projects. In the first type, NGOs or other donors assigned the construction project to a contractor. This type of reconstruction project is usually called the "project type". In second type, the NGOs or other donors provided full or partial funding to owners or groups of owners while the construction project is carried out. This type of project is called the "community-driven type".

Each method shows advantages and disadvantages. The project type is prone to sub-contracting, which is done an average of 2–3 times, and in some cases up to 4–5 times. Every time a project is sub-contracted, each involved party gains a profit of a certain percentage of the original budget. This causes serious friction between parties, which often caused political upsets and scandals.

In contrast, the community-driven type is affected by the shortage of qualified construction workers and supervisors.

The survey also showed that quality controls were not implemented due to the shortage of trained and qualified supervisors. The shortage of supervisors made daily inspections difficult and resulted in less frequent inspections; cases of weekly, bi-weekly, or even monthly inspections were found. Unfortunately, the building permit system, which can be one of the mechanisms to ensure the quality of the structure, was suspended due to the massive number of construction projects and the limited capacity of the authorities. Thus, houses and other facilities damaged by the earthquake and tsunami were allowed to be repaired or reconstructed without building permits.

(4) Problems in official procedures

The victims of the earthquake and tsunami disasters were eligible to receive assistance for rebuilding houses based on their own reporting. The victims were requested to report the damage to their houses to the head of the village (known as Keuchik) or other village officials. However, this was not an easy procedure for the victims: because most of their official documents were lost or damaged in the disasters, it was difficult for them to provide proof of their ownership of land and houses. Many people had difficulty in securing assistance or had to wait a very long time for their assistance to be provided. On the other hand, governmental agencies had difficulties in confirming that each victim received only a single unit of housing for each family, causing some people to end up with no houses while others received two or more units.

17.3.5 Structural Analysis

A case study of structural analysis on a house funded by a donor was conducted. The type was selected mainly because of the availability of complete documentation and technical information. The house had a floor area of 45 m^2 and was a single story confined masonry structure with burnt brick walls. All of its structural elements were common in Aceh. Modal analysis in an elastic range was employed for the study. For structural analysis, the house was modeled as a confined masonry structure in which both the frame and walls supported earthquake loads. All structural components including columns, beams, roof, and walls were modeled as frame and shell elements. The properties for these materials were taken from the results of the field survey and material testing. The analysis found that the house for the case study had sufficient structural capacity in terms of the Building Codes and Standards of Indonesia when it was constructed according to the structural drawing and specifications with adequate quality of materials and construction. This meant that safe structures could be realized under the severe conditions in Aceh and many other places in Indonesia if appropriate designs, materials, and construction work were made available.

Fig. 17.6 Workshop on dissemination of the survey results

17.3.6 Stakeholders' Feedback

A one-day workshop was conducted on 22 February 2006, with the purpose of disseminating facts obtained from the survey and preliminary analyses, to obtain input from all related parties in the reconstruction process, and to find solutions for problems in the reconstruction process (Fig. 17.6). During the workshop, findings from the field survey, material tests, and structural analyses were presented to the audience.

The following recommendations after incorporating the input from the workshop were produced:

- Community awareness about building safety should be raised so that the community becomes involved in improving the quality of the structures.
- Good communication and coordination between donors, community, and government agencies should be built so that the society can participate.
- Building codes and standards should be enforced.
- Internal mechanisms for quality control should be set up by each related party, such as owner, consultant, contractor, donor, and government agency. If necessary, each party should have its own supervisor for each project to ensure the quality of the structure.
- Appropriate building construction methods should be disseminated.
- Feasibility studies should be done prior to conducting reconstruction processes. These should include the participation of the community so that the reconstruction efforts match the needs and demands of the community.
- Training for workers and supervisors should be carried out to ensure proper construction methods, which will lead to better construction.
- A pocket guide for building construction should be developed to facilitate dissemination of proper construction methods.

17.4 Experiences from Recent Earthquakes in Aceh

The following question arose when there were several recent earthquakes in Aceh Province: Did Aceh learn from the 2004 Great Indian Ocean Earthquake? A typical answer to the question can be concluded from the following surveys in two earthquakes both of which occurred in the province in 2013.

17.4.1 Earthquakes and Impacts on Houses

Two earthquakes which occurred respectively in Central Aceh and Bener Meriah Districts (M 6.1 on 3rd July, 2013 at 10 km depth) and Pidie District (M 5.3 on 22nd October, 2013 at 48 km depth) are presented to illustrate the earthquake vulnerability situation in typical earthquake prone areas in Indonesia. The distance between the two epicenters is about 80 km (Fig. 17.7). The first earthquake affected 370 out of 553 villages in the two districts, killing 49 people and injured more than 2,000. It also caused the damage (in various levels) of more than 18,000 houses and more than 1,000 public facilities (schools, hospitals, mosques and churches etc.). The second earthquake, with much lower energy, affected 19 villages in the district. It has nevertheless caused the damage of 616 houses and 47 public facilities, at mostly light and medium damage level.

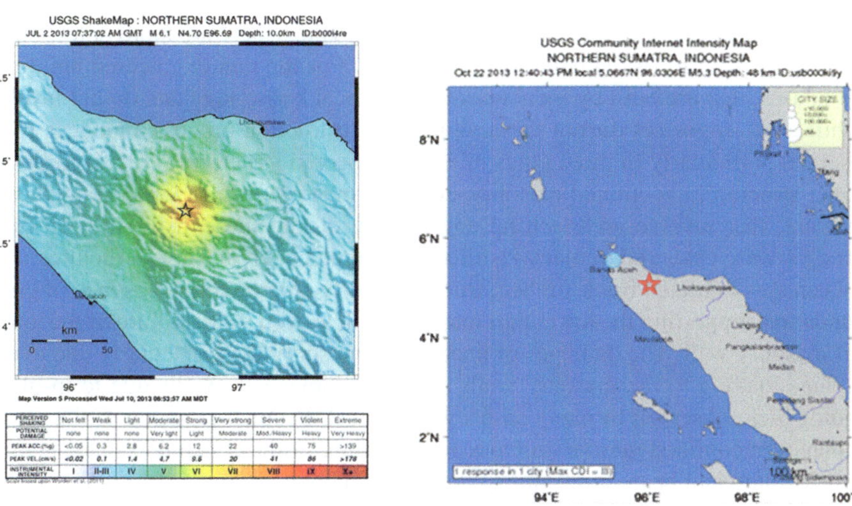

Fig. 17.7 Two earthquakes in Aceh Province (*left*: 3rd July 2013, *right*: 22nd October 2013). *Source*: www.usgs.gov

Fig. 17.8 Typical damaged houses (mixed masonry and timber house, cement brick material, diagonal cracks, collapse of adjoining walls)

17.4.2 Field Surveys in 2013

Surveys to the affected sites in 2013 revealed the characteristics of the damages. Despite the difference in the earthquakes characteristics, there are more or less similar damage patterns of the buildings and houses. For the non-engineered buildings and houses, that are built by local masons with local knowledge and do not involve architects or engineers during the design and construction process, damages were observed particularly in three types of non-engineered structures, i.e. mixed brick-timber structure, unconfined masonry, confined masonry, while for the engineered structure, that involve architectural engineering during the construction process, damages were observed in many in-filled reinforced concrete frame structures.

Damages in mixed brick-timber structure mostly are due to the absence of good foundation supporting the lower part masonry wall as well as the failure of the lower part masonry wall. The damages of unconfined masonry are particularly due to out of plane loading to the masonry walls although damages to walls due to in-plane loading were observed in many cases, with typical diagonal cracking of the walls. Damages in confined masonry are due to the failure of the concrete frames, mainly in the column and beam joints due to the bad connection of the main reinforcement bars as well as shear failure of the column due to lack of stirrups. The failure of the gable walls due to the out of plane loading are often observed in many unconfined and confined structures, due to inadequate framing (Fig. 17.8). Damages in in-filled

Fig. 17.9 Soft story failure of two storied house

Fig. 17.10 Damaged school buildings

RC buildings (mostly two stories) are due to the failure of the RC frames, as well as the weakness of the lower story (soft story effect), where the weak first story absorbs most of the lateral forces transferred by the more rigid upper story (Fig. 17.9).

Damages due to short column effect are also observed in many buildings. Some building failures due to ground failure (landslide) are particularly observed in the first earthquake site, as the site is particularly hilly and many houses were built on steep slope. Typical school building damages are also observed (Fig. 17.10).

Further field investigations show the following facts:

- Good earthquake resistant building practices as well as the prevailing building codes in Indonesia were not observed in many of the damaged buildings and houses. The size and quality of reinforcement bars, proper dimension and spacing of hoops, quality of construction materials, in particular concrete materials, do not comply with the minimum key requirements such as those found in government guidelines.
- Many damaged houses were found to be using heavy concrete canopy in front of the house, tied to the small RC tie beams that connect the walls to the roofs.
- Most of the local builders, masons, carpenters, concreters, and steel bar benders have very limited knowledge on earthquake resistance technology.

- Most of the house owners either build their house by themselves or assign builders to build their houses without awareness of the earthquake risk in the area. They just trust the local builders to design the structural features of the houses, without the capacity of ensuring whether the masons understand or not earthquake resistant technology.

The surveys found out that many of the damaged houses and buildings were built post-2004 following the recent economic growth brought by the post-disaster (and post-conflict) recovery programs. Some of the schools were even built with the government reconstruction aid, in the areas which were not affected by the 2004 Earthquake. This finding eventually shows that there is little learning from the past experiences, as mistakes in building construction similar to those occurring in the post-2004 earthquake reconstruction were found in recently constructed buildings.

17.5 Conclusions

Several important findings were obtained from this reflection. The 2006 survey on planning and design aspects showed that improvement was needed so that the reconstruction efforts would match the needs and demands of the communities. The study of design aspects found that proper seismic design should be ensured to avoid severe damage from future earthquakes. In this context reliable technical information should be delivered to engineers through seminars, publications, or other means. The construction aspect showed that materials used in the reconstruction projects should be improved so that they would meet the minimum qualifications for building materials as specified by the codes/standards. The quality of workmanship should be improved to reduce the vulnerability of the structures. For this purpose practical training programs should be implemented. Good project management and supervision should be conducted to ensure that the components in a building can work properly and together form an integrated structure. Studies by Pribadi et al. (2011) and Okazaki et al. (2012) showed that similar situations regarding non-engineered buildings were also observed in many other developing countries and made generic recommendations to address the issues of non-engineered building vulnerabilities in a similar approach.

Simple technical guidelines and manuals on earthquake resistant non-engineered houses are already available for dissemination to local builders and house owners (Boen 2010; BRR 2007; Ministry of Public Works 2006; Ministry of Public Works and JICA 2009). However, two earthquakes in Aceh Province in 2013 showed that the lessons learned from the 2004 Great Indian Ocean Earthquake and Tsunami and following reconstruction process were not adequately pursued in the other parts of the province which are prone to earthquakes, where building practices still produce unsafe houses. It is then again strongly recommended to look back at what lessons have been learned in the past and seriously take actions to really pursue the goal of a safer Aceh, through a series of systematic actions aimed at building the aware-

ness of the relevant stakeholders to safer building practices in a more massive and sustained way. Simple guidelines and manuals need to be seriously communicated to the local communities in order that safer building practices can be internalized by the communities as part of their way of life.

References

Boen T (2010) Method for repairing non-engineered building damaged by earthquake – in Bahasa, Indonesia. AIFDR-BNPB-IDEP, Jakarta

BRR (2005) Aceh and Nias one year after the tsunami, the recovery effort and way forward, a joint report of the BRR and international partners, December 2005. http://siteresources.worldbank.org/INTEASTASIAPACIFIC/Resources/1YR_tsunami_advance_release.pdf. Accessed 31 January 2014

BRR (2007) Handbook on good building design and construction Aceh and Nias islands. UNDP Regional Centre, Bangkok. http://www.preventionweb.net/files/1525_15255681757.pdf. Accessed 10 December 2013

BRR (2009) 10 Management lessons for host governments coordinating post-disaster reconstruction. Executing Agency for Rehabilitation and Reconstruction (BRR) of Aceh–Nias 2005–2009

IAEE Committee on Non-Engineered Construction (1986) Guidelines for earthquake resistant non-engineered construction. International Association for Earthquake Engineering (IAEE)

Badan Standardisasi Nasional (2004a) Design requirements for earthquake resistant buildings SNI03-1726-2002 – in Bahasa, Indonesia. National Standardization Agency of Indonesia, Jakarta

Badan Standardisasi Nasional (2004b) Design requirements for reinforced concrete buildings SNI03-2847-2002 – in Bahasa, Indonesia. National Standardization Agency of Indonesia, Jakarta

Ministry of Public Works (2006) Technical guideline on earthquake resistant building and houses construction and repair and rehabilitation method – in Bahasa, Indonesia, Ministry of Public Works, Jakarta. http://luk.staff.ugm.ac.id/gapai/PU2006TahanGempa.pdf. Accessed 15 July 2014

Ministry of Public Works, JICA (2009) Key requirements on safer non-engineered construction-in Bahasa Indonesia, Ministry of Public Works and JICA, Jakarta

Okazaki K, Narafu T, Kusumastuti D, Pribadi K (2008) Reconstruction practices in Aceh after 2004 tsunami disaster. In: 14th World conference on earthquake engineering, Beijing

Okazaki K, Pribadi K, Kusumastuti D, Boen T, Ando S (2011) Safety issues of the reconstructed houses in Aceh after the 2004 tsunami, surveys and activities on post-earthquake disaster. UNESCO, Paris

Okazaki K, Pribadi K, Kusumastuti D, Saito T (2012) Comparison of current construction practices of non-engineered buildings in developing countries. In: Fifthteenth world conference on earthquake engineering, Lisbon

Pribadi K, Okazaki K, Kusumastuti D, Saito T, Lim E, Utami A (2011) Research findings of the survey on non-engineered houses in developing countries, surveys and activities on post-earthquake disaster. UNESCO, Paris

Chapter 18
The Role of Islamic Teachings in Encouraging People to Take Tsunami Preparedness in Aceh and Yogyakarta Indonesia

Wignyo Adiyoso and Hidehiko Kanegae

Abstract The role of religion in post-disaster such as reducing psychological its impact, faith-based organizations (FBO) relief and assistance, have been widely acknowledged, yet studies on positive aspect of religion on pre-disaster have been limited although there is an indication that religious teachings have a positive aspect in preparedness. Given the fact that Indonesia is a tsunami prone country and its people is holding religions, study on the relationship between the role of religious teachings and natural disaster is urgently important. This study aims at comparing the effect of Islamic teaching belief on tsunami preparedness action in Aceh and Yogyakarta, Indonesia. The concept of Tsunami Resilient Preparedness (TRP) consisting of Tsunami Early Warning System (TEWS), Emergency Plan and Capacity based on social level/interaction such as individual, family, community and society is introduced. This study has involved 173 residents in Yogyakarta and 305 in Aceh Indonesia as a subject. Findings show that optimistic view significantly predicted most of TRP meaning that both communities have similar views that taking TRP was in accordance with the Islamic teachings and not against God's will. Variables of believe in religious leader significantly predicted most TRP except for Capacity-society and TEWS-individual in Yogyakarta explaining that there is difference impact of residents' interpretation of Islamic teachings. In Aceh, the role of neighbour is very important to influence people to take TRP mostly in TEWS and Emergency Plan but limited in Capacity. This study is very important to challenge the dominant view that natural disaster is caused by god's punishment. Finally, Islamic teachings should be considered as one of the essential aspects to be included in the disaster management policy, especially in Indonesia where majority of the community is holding Islamic religion.

Keyword Aceh • Islamic teaching • Preparedness • Religion • Tsunami

---

W. Adiyoso (✉)
National Development Planning Agency (BAPPENAS), Jakarta, Indonesia
e-mail: wignyoa@yahoo.com

H. Kanegae
Ritsumeikan University, Kyoto, Japan

R. Shaw (ed.), *Recovery from the Indian Ocean Tsunami: A Ten-Year Journey*,
Disaster Risk Reduction, DOI 10.1007/978-4-431-55117-1_18

18.1 Introduction

Disaster preparedness has become a central issue on disaster management since the increase in unpredictable natural disasters events for a decade over the world. It can reduce the risk of loss lives and injuries as well as preventing loss of properties. However, evaluation of disaster preparedness indicates that people living in disaster prone areas shown the low of level preparedness. Coppola and Maloney (2009) indicates that the failure of people in taking preparedness for natural disaster was caused the ineffectiveness of public education programs for lacking in understanding community characteristic, culture and view of disasters. The predominant view of hazard relied on geophysical perspective "tend to explain risk and disaster terms of external agents and their impacts" (Smith 2003). Consequently, preparedness that tends to top down and lack in resilient perspectives leading to people behave in reactive.

Although more than 30 years, dominant view of natural hazard relying on geophysical and geotechnical has been shifting to social perspectives (Phillips and Fodham 2010), less study paid attention to cultural and psychological motivations of people in viewing and responding natural disaster (Harris 2012). As natural hazard is a social construct, it requires understanding people in perceiving, viewing, and responding to natural disaster (Paton 2003). As Kulatunga (2010) stated that human view the hazard is influenced by personal and cultural factors including religion.

The role of religion in post-disaster such as reducing psychological its impact, faith-based organizations (FBO) relief and assistance, have been widely acknowledged (Adiyoso and Kanegae 2012; Kemkens 2013). In contrast religious teachings have often been perceived as a constraint factors for individuals in protecting their life from natural disaster, called fatalistic view. However, although studies on positive aspect of religion have been limited, there is an indication that religious teachings have a positive aspect in pre-natural disaster.

Given the fact that million religious people in Indonesia are living in the disaster prone areas, study to investigate the relationship between disaster preparedness and religious teachings is important. This study aims at comparing the effect of Islamic teachings on tsunami preparedness action in Aceh and Yogyakarta, Indonesia. Tsunami Resilient Preparedness (TRP) as a list of knowledge, attitude and behavior should be equipped by individual, family, community and society to prevent form tsunami disaster is introduced.

Aceh and Yogyakarta shared the geographic that prone to natural disaster including earthquake and tsunami. The 2004 December tsunami triggered by earthquake measuring 9.0 on Richter scale in northern Sumatra Island had severely impacted on social, economic and people's livelihood of Aceh community. Among countries situated along Indian Ocean affected by tsunami, Aceh was the worst areas leaving 123,000 people killed, 113.000 people missing and 406,000 people displaced (Rachmalia et al. 2011). After 2004, Aceh has experienced big earthquake triggered small tsunami. Similar to Aceh, Bantul Yogyakarta experienced small tsunami in 2006, 3 months after 2006-Yogayakarta earthquake leaving 5,778 people killed, 1,649,420 people displaced and more than 50,000 people injured (UNDP-Indonesia 2008). In terms of categoriza-

tion of disaster zone, as stated by Hamzah (2006), Bantul Regency Yogykarta is a part of Zone B or East Sunda arc covering the regions from eastern part of the Sunda Strait to Sumba strait.

However, Aceh and Yogyakarta communities are different characteristic in terms of social, culture and religion. After achieving agreement between central government and Aceh Free Movement in 2006, Aceh province became the only region of Indonesia where Islamic shari'a law was officially in effect. Historically, Aceh was the region as the "first gate" of spreading Islamic in Indonesia about 16 century ago. Islamic kingdom has ruled before Dutch colonial occupied Indonesia. Therefore this province has long been called as "Veranda to Mecca". Islamic environment dominated Acehnese life in all aspects. The roles of religious leaders are also very important. They control the Mosques or small Mosque (*Meunasah*) where all of community activities have been placed (Sulasman 2012). Meanwhile, majority of community in Yogyakarta is Javanese. Javanese cultures have been identified as civilization influenced by more than a millennium of interactions between the native animism and the Hindu and Buddhism culture in Java Island. In terms of religion, although they are formally holding Islam, in practice, they combine animistic beliefs in spirits and various superstitions.

18.2 Religion Is More Contribution in Post-disaster But Less in Preparedness

Religious view of disaster still exists in all major religious across the world (Chester et al. 2008). In Buddhism, Karma belief sees that current actions are the seed for future happiness. Hindu community believes that gods are powerful, so they accepted disaster as part of god creation (Duncan et al. 2012). Chester et al. (2008) stated that Shinto, the way of 'kami", which is the way of god spirits, a mystical figure and force of the natural world. Islamic religion, which is similar to Christianity and Judaism, has same theodicy about viewing natural disaster.

There is increasing research investigating the relationship between religious and traditional belief on natural disaster. Duncan et al. (2012) for example, compiled an archive about religious reaction to a major earthquake and volcanic eruption that occurred between 1900 and 2008 in countries predominantly Christian. They found that of 61 discrete events, seventy percent show evidence that people responding using religious frame or reference (Duncan et al. 2012). Levy and Gopalakrishnan (2005) conducted study focusing on effect religious belief toward recovery process among victim of tsunami 2004 in Sri Lanka after 2004 tsunami. They examined whether certain causal beliefs, attributional style and karma, a Buddhist's concept, associated with tsunami survivor experiencing PTSD and poor health. As they expected, belief in karma predicted worse health and a pessimistic explanatory style both poor health and PTSD. Another finding showing that the religion plays a significant role in disaster study conducted by Gillard and Paton (1999) in Fiji Island. Religious denomination was perceived important in assisting people to recover from hurricane. Of 89 % Fijians Christians, 76 % of Hindus, and 63 % of Moslem reported that religious belief were helpful during crisis.

There has been growing a number of Faith-based Organization (FBO) in the context of disaster recovery and response. After 2004 tsunami, the number of international and local FBO deployed to Aceh for helping victims. The role of religion in post-disaster rooted from the religious philosophy of inspiration in helping each other (Kemkens 2013). According to Timmerman (2014) "FBOs are motivated by spiritual compulsion seeking to meet not only the physical and material needs of disaster victims but also the spiritual and emotional needs". The advantages of FBO in the context of disaster recovery are that it offers psychosocial-development, recovery, less dependence on donor funding and close to the local community context (Kemkens 2013. p. 94). In this context, FBOs are often responsible for providing spiritual guidance, fulfilling the community's needs and working with the local government and local community organization.

On the other hand, a view that natural disaster as God's punishment still exists although people has a massive public education. Given that public disaster education widespread after 2004 tsunami in Indonesia, Act of God's view of natural disaster is still dominant among public and students especially in Aceh. Study comparing two different schools implementing disaster topic curriculum into elementary schools in Aceh found that the belief that the tsunami was because of God's wrath remain equally high in both school adopting and non-adopting disaster into curriculum with 31 and 40 % (Adiyoso and Kanegae 2012).

Similar finding confirms that disaster public education failed to change the belief of fatalistic view among students either in school adopting and non-adopting public disaster education. Study conducted by Adiyoso and Kanegae (2013) in two junior high schools implementing school disaster education projects in Yogyakarta revealed that students in adopting disaster curriculum is still holding fatalistic view (22 %) comparing to 43 % in school non adopting disaster education.

Kemkens (2013) identifies how people reacted to actual disaster as God's action and punishment. For example, the impact of 1775 Great Lisbon Earthquake Portugal, people believed was a God punishment. Other example was the mount eruption in Etna southern Italy. Roman Catholic communities living in vicinity Etna believed that the eruption of Etna is associated with the roles of saints. Elsewhere, when disaster struck some people always raise question why God punished them (Kemkens 2013).

In addition, religious teachings, especially Judaism, Christianity and Islam view natural disaster as god punishment as told in the holy book. Islamic leaders often argue that disaster as God's punishment because of human sin (Kosim 2012). They refer to the holy Qur'an containing the story where non-believer has punished by God through disaster. According to Kosim (2012), there are mentioned in Holy Qur'an about different disaster such as earthquake (al-A'raf/7: 78, 91, 155, al-Ankabut/29: 37), floods (al-Ankabut/29: 14; danSaba'/34: 16), stone rain (al-A'raf/7: 84; an-Naml/27: 58), typhoon (al-Haaqqah/69:6), drought and famine (Qs. al-A'raf/7: 130). The all of stories, according to Kosim, God gave punishment because humankind against God. He also refers another message in the Hadith

(At-Thabrani, Al-hakim and Al-Baihaqi) explaining that immoral deeds will result in God's wrath (Kosim 2012).

On the other hand, there are believers that Holly Koran acknowledged the importance of science and to encourage humankind to study about scientific phenomena (Nahadi et al. 2011: 3). For example, in Qura'n (Al-Jatsiyah) 45: 3–5 it is mentioned "....nd [in] the alternation of night and day and [in] what Allah sends down from the sky of provision and gives life thereby to the earth after its lifelessness and [in His] directing of the winds are signs for a people who reason".

As argued by Nasution (2011), in Holy Koran, there are many verses that encourage people to take preparation for disaster. For example, Koran (Ali 'Imran) 3: 200 mentioned "O you who have believed, persevere and endure and remain stationed and fear Allah that you may be successful". Another the need for preparation is mentioned in Qura'n (An'aam) 6:131 and (Al-Hasyr) 59: 18. These verses can be interpreted that it is necessary for people living in the disaster-prone areas should be prepared from disaster. Nasution (2011) explained that the early warning in disaster can also be found in Holy Qura'n (AsySyu'ara') 42: 26

Similar views have been supported by Ghafory-Ashtiany (2009) where Islamic teaching can be guided principally to reduce the impact of earthquake by understanding Islamic views on disaster, nature, God will, God's kindness and wrath, and human life. This study concluded that God who is kind rather than wrathful encourage humankind to develop and use knowledge to improve humankind life.

Building from the previous discussion, this study proposes three positive aspects of Islamic teachings in supporting tsunami disaster risk reduction. *Firstly*, given the dominance of fatalistic views on natural disaster, however, there is increase in changes of Moslem views that although natural disasters is believed as God's will creator, promoting positive aspect of Islamic teachings will motivate people to take disaster preparedness. In this study, such concept, I called "optimistic view" is used as a view or belief that Islamic teachings encourage people to protect their life and taking such preparedness efforts are not against God's will. *Secondly*, religious follower still largely depended on religious leader as a reference to the issues related to religion and other problems. Using concept of social pressure in the Theory of Planned Behaviour from the Ajzen and Fishbein (2010), therefore, believe in religious leader that he/she believe that Islamic teachings encourage people to protect their life and taking such preparedness efforts are not against God's will and in turn motivate resident to take disaster preparedness. This concept is in accordance with subjective norm concept proposed by Ajzen and Fishbein (2010) stating that human behavior is as a result of perceived social pressure to perform (not to perform) behavior. *Lastly*, similar to believe in religious leader, believe in neighbor that he/she believe that Islamic teachings encourage people to protect their life and taking such preparedness efforts are not against God's will and in turn motivate resident to take disaster preparedness.

In conclusion, although the previous studies have been limited focusing on the understanding of Qur'an verses-disaster management relationship, theoretical perspectives on how Islamic teachings view on science and natural disaster give us a framework to this study to investigate the Islamic teachings used in the disaster risk reduction focused on tsunami disaster preparedness.

18.3 Three Dimensions of Tsunami Preparedness

Disaster mitigation and preparedness are central issues in disaster risk reduction in recent decades. While the mitigation measures undertaken to reduce the impact of hazard before a disaster occurs, the preparedness is any activities related to anticipation of the hazard events. However, it is often difficult to distinguish the differences of mitigation and preparedness in practical.

Concept of preparedness refers to measures which taken before a disaster occurs to ensure a proper response to the impacts of hazards. It can be done through "effective precautionary measures that ensure a timely, appropriate, and efficient organization and delivery of response and relief action" (Coppola 2007). Preparedness is a mediate action to prevent such a potential hazard. Preparedness action such as developing hazard maps, preparing disaster kits in the home, participating in the emergency drills, and involving in the early warning activities will reduce the loss of lives and injuries. This value of preparation is apparent especially when community encountering unpredictable disaster such as earthquake (Shaw et al. 2004).

However the concept of preparedness that spreading used in disaster management has been criticized by many experts. *Firstly,* much disaster preparedness planning is based on military lines (Smith 2003). Preparedness is viewed as running battle and in stress condition that associated with communication, logistic and security. Although such approach is important, the command and control model that tend to top down and hierarchical control is not always appropriate. Therefore, preparedness program should be based more locally cultural approach. *Secondly*, it is often said that preparedness concept is associated with the resilience concept however resilience does not cover all phases of emergency management such as preparedness and response (McEntire et al. 2010). Looking back on the definition of resilience containing "bounce back", that there has been tendency only covering post-disaster, resilience neglect community that does not have experience with disaster. Resilience has been generally defined in two broad ways: as a desired outcome(s) or as a process leading to a desired outcome(s) (McEntire et al. 2010). However, as argued by Mayunga (2007), "use of the terms cope, bounce back, withstand or absorb negative impact, return to normal within the shortest possible time, tend to emphasize a reactive stance". This problem is especially applying in tsunami disaster-resilient community, due to infrequent events, some certain community has never been experienced disaster. Therefore, it is necessary that a concept of tsunami should be expanded more than just resilience, where also apply for people who have not experienced natural disaster. It is argued that redefine a "resilient preparedness" in the tsunami case is needed.

This study, therefore, argues that the concept of preparedness should not merely be focused on capacity to respond during emergency (Hemond and Robert 2012) and recovery after disaster but it should also cover with the "resilient preparedness" perspectives. This study proposes the concept of tsunami preparedness that accommodates a concept of "disaster culture" in which it is similar to the resilience concept, but it focuses on preparedness.

An important point of disaster culture has been raised by Murata et al. (2010) in the context of tsunami. Disaster culture refers to having accurate knowledge how to react to tsunami that integrated in daily habits or "lifestyle culture". Murata et al. (2010) emphasize how people should learn and acquire correct knowledge about tsunami behavior and how to respond it. In general it can be recognized that culture of disaster should have two important things: core capacities (eg. knowledge, skill) and change of stage. While core capacities have been discussed in the variety of natural disaster, different capacity also more suggested than proposed by Murata et al. (2010).

Literature references and best practices provided abundant prescriptive disaster preparedness, however it is limited for tsunami disaster. Therefore three dimensions of tsunami preparedness including Tsunami Early Warning System (TEWS), Emergency Plan, and Capacity are introduced.

18.3.1 Tsunami Early Warning System (TEWS)

In terms of tsunami hazard, early warning system is the foremost important factor in reducing the possibility of fatalities event such as injury and loss of life (ISDR 2005). The ISDR underlined the important of TEWS apply four interrelated-elements of the model: (i) understanding and detecting hazard (risk knowledge), (ii) checking and developing warning tools; (iii) the communicating risk and how respond (dissemination of communication), and (iv) ability to respond.

As the main objective of TEWS based on people-centered is to give information to people at risk, so it should detect the risk and make decisions to take action. According to Sorensen and Rogers (1988) "this definition is simple, but warning systems are complex because they link much specialties and organizations—science (government and private), engineering, technology, government, news media, and the public".

Within this concept, it should be emphasized not only community understand the meaning of TEWS and how to respond it, but it should also change their belief about the useful of TEWS and increase awareness. Effectiveness of TEWS should be also supported by practicing disaster drill organized both by government and community. The use of multiple communication channels is also a key of tsunami preparedness in the context of TEWS as during disaster TEWS facilities may damage.

18.3.2 Emergency Plan

Emergency plan is another critical element in tsunami preparedness. As stated by Sutton and Kathleen (2006) that as tsunamis can occur at any time, all family members should understand warning system and how to respond it. In Japan, for example, the word "Tendenko" is a measure suggested that people should evacuate by themselves without waiting other family member wherever they located when disaster occurs (Fraser et al. 2012). Family emergency plan should also ensure that family members know the safer place in case of separated and know emergency and other family members phone number to be contacted. However, this definition of Emergency Plan is not limited on family, but it is also apply for wider community members.

Most of conventional tsunami preparedness measures recommended only focus on TEWS and evacuation plan. In this concept, it is indispensable to prepare such disaster kits as recommended also in earthquake measure. However, it is important to note that disaster kits in tsunami preparedness should be only limited the most important goods and easily to carry when tsunami struck.

18.3.3 Capacity

The meaning of capacity in this context is related to ability of community in maintaining awareness, attitude and behavior before tsunami occurs, ability to respond during crisis and ability to recover after disaster. The key issue in tsunami preparedness capacity dimension is that tsunami preparedness integrated in the community activities. Therefore, community involvement is very important. Community involvement is prerequisite for building disaster resilient-community. Referring to the concept of community development, the main reason why community involvement is important is that collective action can generate better solution (Ife 2002). Therefore, community involvement ultimately addresses the problem of the community to look for alternatives ways of providing a more sustainable basis for the meeting of human needs based on ecological perspectives as a foundation of community resilience (Adiyoso 2010).

GeoHazard International (2007) recommends that making tsunami hazard and evacuation maps are the best way to begin preparedness efforts for short and long-terms. Evacuation map basically is a route for people to use in evacuations. This route should be familiar with the community members and can guide people to the safest and fastest routes when tsunami coming. The best way to develop evacuation routes is to involve communities started from identifying the safest places to the creating a map, conducting community workshop, disseminating maps to other community members and exercising or practicing evacuation routes. However due to the long frequency of tsunami events, updating the map and regularly practice the evacuation route is recommended (GeoHazard International 2007).

Previous research on disaster has acknowledged that the role of family and community-level is important (Phillips and Fodham 2010). The role of individual, family and community is in accordance with the concept of self-help, mutual-help and public-help on how people should respond the disaster. Based on three tsunami preparedness dimension and the role of individual, family, community and society, Tsunami Resilient Preparedness (TRP) concept is proposed. As discussed previously, that basically, the effectiveness of tsunami preparedness that consist of TEWS, Emergency Plan and Capacity can be done by individual, family, community and society focused on activities range from the simple to complex preparedness.

18.4 Method

This study has involved 173 residents in Yogyakarta and 305 in Aceh conducted from December 2012—January 2013 or almost 6 years after 2006 Yogyakarta Earthquake and 8 years after 2004 tsunami Sumatera island. Although the risk perception is not constant, as Aceh and Yogyakarta have been experiencing big earthquakes many times, people awareness on tsunami is expected as a moderate level. There are some factors contribute to the change of risk perception including memory, experience, knowledge, mood, exposure and social context (Krallis and Csontos 2014).

A set of questionnaires consisted of respondent's characteristic, tsunami experiences, independent and dependent variables were used to gather information from the residents living in the tsunami prone areas. In terms of age, majority of respondents in Yogyakarta has age more than 36 years accounted for 79 %, in contrast, more than a half of Acehnese respondents is dominated by less than 37 ages. The proportion of sex and religion has no differences between communities, in Yogyakarta male at 60 % and female 40, while in Aceh male 54 % and female 46 %. Another difference is living with children under 12 years old which is categorized as vulnerable groups (Birkmann 2006). The majority of Aceh respondents have children under 12 year old (70 %) while in Yogyakarta only 30 %.

In terms of living length, majority of respondents (70 %) in Yogyakartahas been living in their areas more than 15 years, while in Aceh, they (61 %) have lived in areas less than 10 years. On the other hand, due to land topography, all of respondents' houses (100 %) in Yogyakarta lies within one kilometer from the coast line, while in Aceh is about 65 % of respondents' house is located within 1 km from coastal line. Educational background of respondents in two communities is not differences, more than half of them graduated from the secondary and senior high schools. In terms of monthly income, both communities are similar. Residents' income is ranging from one to two million IDR monthly (equal to 222 USD with exchange rate in January 2013 1 USD = 9,000 IDR).

As discussed in the previous part that there are many factors that can predict people in taking TRP from demographic, economic, social, psychological and cultural factors. Although this study focused on religious factors it also includes personal and involved in disaster group factors (social). Personal factors such as age, gender, education were included because these variables were often as confounded factors (Mishra 1999).

To develop model there are seven combinations of predictors or independent variables consists of: (1) Age (X_1). It was measured using real data reported by respondent, (2) Gender (X_2). It was dummied into two categories, 0 = female and 1 = male, (3) Education (X_3). Education was dummied into two categories, 0 = lower education (elementary school and no school) and 1 = higher education (secondary school and upper), (4) Optimistic view (X_4). It was evaluated by two items consisted of (a) "Preparing tsunami (protecting my life) is not against God's will" and (b) "Preparing tsunami (protecting my life) is in accordance with Islamic teachings".

Response range from "not strongly agree" with 1 point to "strongly agree with 5 points, (5) Believe in religious leader (X_5). It was assessed by one items that was "I believe in religious leader that he/she think that preparing tsunami (protecting my life) is not against God's will and in accordance with the Islamic teachings and not against God's will". Response range from "not strongly agree" with 1 point to "strongly agree with 5 points, (6) Believe in neighbour (X_6) was measured by one item "I believe in neighbor that he/she think that preparing tsunami (protecting my life) is not against God's will and in accordance with the Islamic teachings and not against God's will". Response range from "not strongly agree" with 1 point to "strongly agree with 5 points, (7) Involved in disaster group (X_7). It was dummied into two categories, 0 = not involved in disaster group, 1 = involved in disaster group.

The dependent variables (Y) are tsunami resilient preparedness (TRP) that comprised 12 variables derived from 30 indicators that divided into three dimensions (TEWS, Emergency Plan and Capacity) consists of individual, family, community and society. Basically, each indicator of TRP is as follows: (1) TEWS-individual comprises 2 indicators: knowing natural sign of tsunami and knowing communication mean for TEWS, (2) TEWS-family comprises 2 indicators Sharing/discussing information of TEWS with family members and sharing/discussing information about evacuation route/shelter with family members, (3) TEWS-community comprises 2 indicators: understanding the meaning of TEWS developed by community and participated in tsunami drill organized by community or visited/practiced evacuation route/shelters in their areas, (4) TEWS-society comprises 2 indicators: understanding the meaning of TEWS developed by government and participated in tsunami drill organized by government, (5) Emergency Plan-individual comprises 2 indicators: knowing tsunami signboards, evacuation route/shelters in their areas and knowing and keeping emergency phone number, (6) Emergency Plan-family comprises 8 indicators: preparing disaster kits such as flashlight, transistor radio, food, water, first aids, multi purposes tools, communication means and documents, (7) Emergency Plan-community comprises 2 indicators: discussing issues related tsunami and how to prepare for tsunami with neighbor/community and understanding safe places or meeting point in their areas, (8) Emergency Plan-society comprises 2 indicators: understanding how to contact local government and to find out information before and during tsunami disaster and sharing phone number with relative from outside community, (9) Capacity-individual comprises 2 indicators: Knowing about causes of tsunami and knowing tsunami hazard (map) in their areas, (10) Capacity-family comprises 2 indicators: Sharing/discussing tsunami hazard (map) in their areas with family members and discussing/sharing past tsunami with family members, (11) Capacity-community comprises 2 indicators: participating/attending community meeting organized by local community and visiting/finding information in tsunami facilities (poles, escape building, museum) in community areas, (12) Capacity-society comprises 2 indicators: attending/participating meeting (eg. dissemination, workshop, and training) organized by non-community at least one time a year and regularly searching/updating tsunami information from different source information/media.

All indicators were measured using three answers “Yes”, “No” and “I am not sure”. The answer “Yes” meant that residents have known/understood or done the TRP asked, while “No” answer meant that residents have not known/understood or done the TRP asked. “I am not sure” answer meant that residents have known/ understood partially or done some activities the TRP asked. However, in analyses score was given only the answer “Yes” with 1 and “No” and “I am not sure” answer” was score 0. The more residents answer “Yes” the more they are prepared. Multiple regressions are used to analyze the influence of independent variables on dependent variables. Basically, multi regressions are statistical analysis to predict dependent variable from the independent variables.

18.5 Finding and Discussions

The main hypothesis is that model proposed is fit the data (Ghozali 2011), or in other words, independent variables such as age, gender, education, optimistic view, believe religious leader, believe in neighbor and involved in disaster groups will predict people in adopting TRP TEWS, Emergency Plan and Capacity in the individual, family, community and society level.

18.5.1 Comparison Predictor Variables on TRP TEWS in Aceh and Yogyakarta

As presented in Table 18.1 multiple regression was conducted separately for data from Yogyakarta and Aceh. In general all of models were significantly fit with the data both in Aceh and Yogyakarta. In Yogyakarta model accounted for 22 % in explaining the variance in TEWS-individual ($F=7.646$, $df=169$, $Sig=0.000$), 26 % in TEWS-family ($F=9.661$, $df=169$, $Sig=0.000$), 22 % in TEWS-community ($F=7.899$, $df=169$, $Sig=0.000$), and 18 % TEWS-society ($F=6.466$, $df=169$, $Sig=0.000$).

On the other hand, in Aceh, model was fit with the data, with only 3 % accounted in explaining in TEWS-individual variance ($F=2.456$, $df=292$, $Sig=0.018$), 6 % in TEWS-family ($F=3.781$, $df=292$, $Sig=0.011$), 10 % in TEWS-community ($F=5.451$, $df=292$, $Sig=0.000$), and 6 % in TEWS-society ($F=3.480$, $df=169$, $Sig=0.001$). The equation model of TEWS-individual can be seen in appendix together with all of other equation models related to split data in Aceh and Yogyakarta. Further analysis in each predictor variable, in Yogyakarta, optimistic view significantly predicted all of TRP TEWS including TEWS-individual ($\beta=0.208$, $p<0.050$), TEWS-family ($\beta=0.262$, $p<0.050$), TEWS-community ($\beta=0.409$, $p<0.050$), and TEWS-society ($\beta=0.399$, $p<0.050$). In Aceh optimistic view significantly contributed TEWS-individual ($\beta=0.117$, $p<0.050$), TEWS-family ($\beta=0.140$, $p<0.050$), and TEWS-community ($\beta=0.152$, $p<0.050$).

Table 18.1 Multiple regressions of personal, religious, involved in disaster group variables on TRP TEWS in Yogyakarta and Aceh

Dependent variables (Y)		TEWS-individual		TEWS-family		TEWS-community		TEWS-society	
Yogyakarta		df = 169, F = 7.646		df = 169, F = 9.661		df = 169, F = 7.899		df = 169, F = 6.466	
		Sig = .000, α = −0.391		Sig = 0.000, α = −0.791		Sig = 0.000, α = −0.173		Sig = 0.000, α = 0.193	
		Adjusted $R^2 = 0.216$		Adjusted $R^2 = 0.254$		Adjusted $R^2 = 0.222$		Adjusted $R^2 = 0.185$	
Aceh		df = 292, F = 2.456		df = 292, F = 3.781		df = 292, F = 5.451		df = 292, F = 3.480	
		Sig = 0.018, α = 0.150		Sig = 0.001, α = 0.096		Sig = 0.000, α = 0.018		Sig = 0.001, α = 0.201	
		Adjusted $R^2 = 0.034$		Adjusted $R^2 = 0.063$		Adjusted $R^2 = 0.096$		Adjusted $R^2 = .056$	
Independent variable		SE	β	SE	β	SE	β	SE	β
Age (X_1)	Yogyakarta	0.002	−0.081	0.002	0.000	0.002	−0.111	0.002	−0.110
	Aceh	0.002	−0.034	0.002	−0.156*	0.002	−0.047	0.002	−0.017
Gender (X_2)	Yogyakarta	0.044	−0.003	0.052	0.018	0.045	0.108	0.042	0.030
	Aceh	0.040	0.032	0.048	0.011	0.032	−0.073	0.036	0.031
Education (X_3)	Yogyakarta	0.057	0.083	0.067	0.098	0.057	−0.043	0.054	−0.069
	Aceh	0.091	0.070	0.109	0.005	0.071	0.006	0.081	0.014
Optimistic view (X_4)	Yogyakarta	0.016	**0.208***	0.019	**0.262***	0.016	**0.409***	0.015	**0.399***
	Aceh	0.015	**0.117***	0.018	**0.140***	0.012	**0.152***	0.013	0.011
Believe religious leader (X_5)	Yogyakarta	0.029	**0.348***	0.034	**0.363***	0.029	0.097	0.027	0.000
	Aceh	0.020	−0.007	0.024	0.079	0.016	0.004	0.018	0.008
Believe neighbors (X_6)	Yogyakarta	0.018	0.038	0.021	0.015	0.018	0.058	0.017	0.055
	Aceh	0.015	−0.002	0.018	0.070	0.011	**0.138***	0.013	**0.170***
Involved in disaster group (X_7)	Yogyakarta	−0.009	−0.009	0.107	0.046	0.091	0.037	0.086	0.124
	Aceh	0.075	0.188*	0.090	0.129*	0.059	0.257*	0.067	0.208*

Source: Analysis (Survey date, December 2012–January 2013)
Note: *$p < 0.05$

Interestingly, in Aceh, involved in disaster group significantly predicted all of TRP TEWS such as TEWS-individual ($\beta=0.188$, $p<0.050$), TEWS-family ($\beta=0.129$, $p<0.050$), TEWS-community ($\beta=0.257$, $p<0.050$), and TEWS-society ($\beta=0.208$, $p<0.050$).

While in Yogyakarta believe in religious leader significantly influenced TEWS-individual ($\beta=0.348$, $p<0.050$) and TEWS-family ($\beta=0.363$, $p<0.050$), in Yogyakarta believe in neighbor significantly predicted TEWS-community ($\beta=0.138$, $p<0.050$), and TEWS-society ($\beta=0.170$, $p<0.050$). All of models significantly predicted TRP TEWS. This finding shows different feature of role religious factors in two communities. The role of optimistic view and involved in disaster groups were the most predictor variables in Aceh while in Yogyakarta optimistic view and believe in religious leader were predictor factors in Yogyakarta.

18.5.2 Comparison Predictor Variables on TRP Emergency Plan in Aceh and Yogyakarta

As presented Table 18.2, in general all of models were significantly fit with the data both in Aceh and Yogyakarta. In Yogyakarta, model accounted for 10 % in explaining the variance in Emergency Plan-individual ($F=3.869$, $df=169$, $Sig=0.001$), 24 % in Emergency Plan-family ($F=8.716$, $df=169$, $Sig=0.001$), 23 % in Emergency Plan-community ($F=8.257$, $df=169$, $Sig=0.000$), and 7 % Emergency Plan-society ($F=2.722$, $df=169$, $Sig=0.011$).

Moreover, in Aceh, model was fit with the data, with 4 % accounted in explaining in Emergency Plan-individual variance ($F=2.696$, $df=292$, $Sig=0.010$), 6 % in Emergency Plan-family ($F=3.442$, $df=292$, $Sig=0.001$), 3 % in Emergency Plan-community ($F=2.435$, $df=292$, $Sig=0.019$), and 8 % in Emergency Plan-society ($F=4.806$, $df=169$, $Sig=0.000$). The contribution of each predictor variable in Yogyakarta shows that the most predictor variables significantly predicted TRP was optimistic view that influencing in Emergency Plan-individual ($\beta=0.311$, $p<0.050$), and Emergency Plan-community ($\beta=0.240$, $p<0.050$). Believe in religious leader significantly influenced Emergency Plan-family ($\beta=0.335$, $p<0.050$), Emergency Plan-community ($\beta=0.323$, $p<0.050$), and Emergency Plan-society ($\beta=0.252$, $p<0.050$). Believe in neighbor significantly influenced Emergency Plan-family ($\beta=0.120$, $p=0.000$).

In Aceh, the predictor that significantly contributed Emergency Plan was believe in neighbor that determined Emergency Plan-family ($\beta=0.163$, $p<0.050$), and Emergency Plan-society ($\beta=0.194$, $p<0.050$). Second variable was involved in disaster group that significantly predicted Emergency Plan such as Emergency Plan-individual ($\beta=0.205$, $p<0.050$) and Emergency Plan-society ($\beta=0.161$, $p<0.050$). Optimistic view influenced in Emergency Plan-community ($\beta=0.156$, $p<0.050$). In summary, all of models significantly predicted Emergency Plan. Again, the finding in Yogyakarta is similar to TRP TEWS, optimistic view and believe in religious leader significantly predicted some of TRP Plan, while in Aceh, believe in neighbor and involved in disaster predictor variables that influenced in TRP Plan.

Table 18.2 Multiple regressions of personal, religious, involved in disaster group variables on TRP emergency plan in Yogyakarta and Aceh

Dependent variables (Y)		Emergency Plan-individual		Emergency Plan-family		Emergency Plan-community		Emergency Plan-society	
Yogyakarta		df = 169, F = 3.869		df = 169, F = 8.716		df = 169, F = 8.257		df = 169, F = 2.722	
		Sig = .001, α = 0.184		Sig = 0.001, α = −0.241		Sig = 0.000, α = −0.506		Sig = 0.011, α = −0.132	
		Adjusted R^2 = 0.106		Adjusted R^2 = 0.242		Adjusted R^2 = 0.231		Adjusted R^2 = 0.067	
Aceh		df = 292, F = 2.696		df = 292, F = 3.442		df = 292, F = 2.435		df = 292, F = 4.806	
		Sig = 0.010, α = 0.312		Sig = 0.001, α = 0.317		Sig = 0.019, α = 0.125		Sig = 0.000, α = 0.305	
		Adjusted R^2 = 0.039		Adjusted R^2 = 0.055		Adjusted R^2 = 0.033		Adjusted R^2 = 0.084	
Independent variables		SE	β	SE	B	SE	β	SE	β
Age (X_1)	Yogyakarta	0.002	−0.099	0.002	−0.065	0.002	−0.056	0.002	−0.017
	Aceh	0.001	−0.074	0.002	−0.127*	0.002	−0.112	0.002	−0.104
Gender (X_2)	Yogyakarta	0.040	0.096	0.037	0.087	0.039	0.077	0.051	0.000
	Aceh	0.029	0.066	0.035	−0.057	0.039	−0.008	0.035	−0.060
Education (X_3)	Yogyakarta	0.052	−0.080	0.047	−0.028	0.050	0.043	0.066	−0.060
	Aceh	0.064	0.018	0.079	−0.071	0.088	−0.018	0.079	−0.052
Optimistic view (X_4)	Yogyakarta	0.014	**0.311***	0.013	0.216	0.014	**0.240***	0.018	0.144
	Aceh	0.011	0.072	0.013	0.062	0.014	**0.156***	−0.007	−0.032
Believe religious leader (X_5)	Yogyakarta	0.026	0.041	0.024	**0.335***	0.025	**0.323***	0.033	**0.252***
	Aceh	0.014	0.022	0.018	0.080	0.020	0.079	0.030	0.102
Believe neighbors (X_6)	Yogyakarta	0.016	0.072	0.015	**0.120***	0.016	0.097	0.021	0.013
	Aceh	0.010	0.037	0.013	**0.163***	0.014	0.039	0.043	**0.194***
Involved in disaster group (X_7)	Yogyakarta	0.082	0.017	0.075	0.062	0.079	−0.006	0.104	−0.103
	Aceh	0.053	0.205*	0.066	0.007	0.073	0.020	0.186	0.161*

Source: Analysis (Survey date, December 2012–January 2013)
Note: *$p < 0.05$

18.5.3 Comparison Predictor Variables on TRP Capacity in Aceh and Yogyakarta

Table 18.3 displays a result of multiple regression that conducted separately for data from Yogyakarta and Aceh. In general, all of models were significantly fit with the data both in Yogyakarta, but only two models were fit in Aceh. In Yogyakarta, model accounted for 10 % in explaining the variance in Capacity-individual ($F=3.593$, $df=169$, $Sig=0.000$), 19 % in Capacity-family ($F=6.739$, $df=169$, $Sig=0.001$), 14 % in Capacity-community ($F=4.939$, $df=169$, $Sig=0.000$), and 9 % Capacity-society ($F=3.467$, $df=169$, $Sig=0.002$). In Aceh, model was not fit with the data on Capacity-individual and Capacity-community. Data shows that 4 % accounted in explaining in Capacity-family ($F=2.640$, $df=292$, $Sig=0.001$), and 4 % in Capacity-society ($F=2.548$, $df=292$, $Sig=0.015$).

In addition, analysis in each predictor variable shows that, in Yogyakarta believe in religious leader significantly influenced TEWS including Capacity-individual ($\beta=0.288$, $p<0.050$), Capacity-family ($\beta=0.280$, $p<0.050$), Capacity-community ($\beta=0.172$, $p<0.050$), and Capacity-society ($\beta=0.264$, $p<0.050$). The second predictor that influenced TRP Capacity in Yogyakarta was optimistic view that significantly predicted TRP Capacity including Capacity-family ($\beta=0.199$, $p<0.050$), Capacity-community ($\beta=0.280$, $p<0.050$), and Capacity-society ($\beta=0.155$, $p<0.050$).

In Aceh, only limited predictor variables significantly determined TRP. Optimistic view significantly contributed Capacity-family ($\beta=0.139$, $p<0.050$), believe in religious leader influenced Capacity-family ($\beta=0.126$, $p<0.050$) and involved in disaster group affected Capacity-society ($\beta=0.151$, $p<0.050$). In conclusion, all TRP Capacity in Yogyakarta and two TRP Capacities in Aceh were rejected. In contrast, two TRP Capacities in Aceh failed to reject. In case of Yogyakarta, this finding is consistent with TRP TEWS and Emergency Plan where the role of optimistic view and believe in religious leader were very influence factors in most TRP. However, in Aceh, only TRP Capacity family was influenced by belief in religious leader and TRP Capacity-society was predicted by involved in disaster group.

Comparable data analysis between Aceh and Yogyakarta shows that optimistic view significantly predicted most of TRP. This meant that both communities have similar views that taking TRP was in accordance with the Islamic teachings and not against God's will. Although both communities agreed that natural disaster are viewed as a God's will but they also believe that preparing for disaster does not against God's will. The two ways of thinking among residents are consistent with the use of parallel interpretation and practice stated by Kemkens (2013).On the other hand, variable of believe in religious leader significantly predicted most TRP except for Capacity-society and TEWS-individual in Yogyakarta. This finding can be explained that because the Islamic rules in Yogyakarta less strong compare to Aceh, residents' interpretation of Islamic teaching is less strict including religious leader, therefore because between resident and religious leader have no different belief with religious leader, so residents trust to religious leader in dealing with taking TRP.

Table 18.3 Multiple regressions of personal, religious, involved in disaster group variables on TRP capacity in Yogyakarta and Aceh

Dependent variables (Y)		Capacity-individual		Capacity-family		Capacity-community		Capacity-society	
Yogyakarta		df = 169, F = 3.593		df = 169, F = 6.739		df = 169, F = 4.939		df = 169, F = 3.467	
		Sig = 0.001, α = −0.302		Sig = 0.000, α = −0.297		Sig = 0.000, α = −0.281		Sig = 0.002, α = −0.282	
		Adjusted R^2 = 0.097		Adjusted R^2 = 0.192		Adjusted R^2 = 0.140		Adjusted R^2 = 0.093	
Aceh		df = 292, F = 1.358		df = 292, F = 2.640		df = 292, F = 0.667		df = 292, F = 2.548	
		Sig = 0.223, α = 0.292		Sig = 0.012, α = 0.010		Sig = 0.700, α = 0.416		Sig = 0.015, α = −0.136	
		Adjusted R^2 = 0.009		Adjusted R^2 = 0.038		Adjusted R^2 = −0.008		Adjusted R^2 = 0.036	
Independent variables		SE	β	SE	β	SE	β	SE	β
Age (X_1)	Yogyakarta	0.002	−0.034	0.001	−0.069	0.002	−0.045	0.002	0.012
	Aceh	0.002	−0.054	0.002	−0.082	0.002	−0.094	0.001	0.074
Gender (X_2)	Yogyakarta	0.036	−0.039	0.032	0.132	0.042	0.097	0.034	0.077
	Aceh	0.045	0.081	0.038	−0.002	0.036	0.025	0.027	0.089
Education (X_3)	Yogyakarta	0.046	0.027	0.041	−0.029	0.054	0.073	0.044	−0.014
	Aceh	0.102	0.098	0.085	−0.034	0.082	−0.040	0.061	0.115
Optimistic view (X_4)	Yogyakarta	0.013	0.051	0.011	**0.199***	0.015	**0.280***	0.012	**0.155***
	Aceh	0.017	0.079	0.014	**0.139***	0.013	−0.013	0.010	0.078
Believe religious leader (X_5)	Yogyakarta	0.023	**0.288***	0.021	**0.280***	0.027	**0.172***	0.022	**0.264***
	Aceh	0.023	−0.020	0.019	**0.126***	0.018	−0.059	0.014	−0.020
Believe neighbors (X_6)	Yogyakarta	0.015	0.116	0.013	0.008	0.017	0.009	0.014	−0.023
	Aceh	0.016	0.012	0.014	0.046	0.013	0.065	0.010	−0.006
Involved in disaster group (X_7)	Yogyakarta	0.074	0.034	0.066	0.139*	0.086	0.021	0.070	0.049
	Aceh	0.084	0.086	0.039	0.032	0.068	0.034	0.051	0.151*

Source: Analysis (Survey date, December 2012–January 2013)
Note: *$p < 0.05$

Interestingly, in Aceh, involved in disaster group and believe in neighbour variables significantly predicted TRP mostly in TEWS and Emergency Plan but limited in TRP capacity. It seems that residents in Aceh shared the same opinion of optimistic views with neighbor so when they decide to take or not to take TRP were influenced by neighbour. These findings imply that in community holding less Islamic rules like in Yogyakarta, promoting to take TRP through religious leader will be more effective than community hold strong Islamic rules like in Aceh. In contrast, promoting TRP in Aceh will be effective if involved neighbour and joining disaster group.

Considering that FBOs play important roles in disaster recovery and construction process, it is recommended that FBOs' roles in disaster management cycle should be emphasized on the preparedness step. Finding supports that FBOs have an opportunity in spreading positive interpretations of religious teaching on the disaster preparedness (Adiyoso and Kanegae 2012). Disaster manager should cooperate with FBOs to develop risk communication tools using religious teachings to influence community member in preparing tsunami disaster.

This finding is very important and challenges for the new perspective of disaster theory, as the dominant view in the disaster research that most findings revealed people who believe that natural disaster as a god's punishment (fatalistic view) will decrease in their level of preparedness (Duncan et al. 2012). In contrast, it is evidence that if residents hold an optimistic view they would take action for disaster anticipation. Disaster management managers in region with the majority of Moslem should think that including Islamic teachings and involving religious leader in promoting disaster risk reduction is necessary. Although there is still progressing debates among Islamic leader related to Islamic view about natural disaster, the fact that, in general, Islamic teachings are perceived by residents in Yogyakarta and Aceh in agreement with an idea to reduce the impact of natural disaster (Ghafory-Ashtiany 2009). Consequently, Islamic teaching should be considered as one of an essential aspect to be included in the disaster management policy, especially in Indonesia where most of community is holding Islamic religion.

18.6 Conclusions

This study examines how three religious factors influence people to take action for tsunami resilient preparedness (TRP). Religious factors are focused on the concept of positive interpretation of Islamic teachings on natural disaster instead of negative view, or called as fatalistic view. The three aspects of religious factors are optimistic view as a belief that Islamic teachings encourage human to prepare for disaster, believe in religious leader and neighbor that they also hold a view that there is no contradiction in religious teaching and preparing for disasters. Comparable data analysis between Aceh and Yogyakarta shows that optimistic view significantly predicted most of TRP meaning that both communities have similar views that taking

TRP was in accordance with the Islamic teachings and not against God's will. Variable of believe in religious leader significantly predicted most TRP except for Capacity-society and TEWS-individual in Yogyakarta that explaining that there is differences impact of residents' interpretation of Islamic teaching. In Aceh, the role of neighbour is very important to influence people to take TRP mostly in TEWS and Emergency Plan but limited in TRP capacity. This study is very important to challenge the dominant view that natural disaster is caused by god's punishment (fatalistic view). Finally, Islamic teaching should be considered as one of an essential aspect to be included in the disaster management policy, especially in Indonesia where most of community is holding Islamic religion.

References

Adiyoso W (2010) Participative planning in a community development project in Indonesia. Selected paper the 1st international conference on sustainable future for human security 2010, Kyoto. December 11–12, 2010. ISSN: 1884–8850

Adiyoso W, Kanegae H (2012) The effect of different disaster education programs on tsunami preparedness among schoolchildren in Aceh, Indonesia. Disast Mitigation Cult Herit Historic Cities 6(7):25–35

Adiyoso W, Kanegae H (2013) The preliminary study of the role of Islamic teaching in the disaster risk reduction (A qualitative case study of Banda Aceh, Indonesia). Procedia Environ Sci 17(2013):918–927, Elsevier

Ajzen I, Fishbein M (2010) Predicting and changing behavior: the reasoned action approach. Psychology, New York

Birkmann J (2006) Measuring vulnerability to promote disaster-resilient societies: conceptual frameworks and definitions. In: Birkmann J (ed) Measuring vulnerability to natural hazards: towards disaster resilient societies. United Nation University Press, Tokyo

Chester DK, Duncan AM, Dibben CJL (2008) The importance of religion in shaping volcanic risk perception in Italy, with special reference to Vesuvius and Etna. J Volcanol Geotherm Res 172(2008):216–228

Coppola D (2007) Introduction to international disaster management. Butterworth-Henemann, Amsterdam

Coppola D, Maloney EK (2009) Communicating emergency preparedness: strategies for creating disaster resilient public. Auerbach Publications, Taylor & Francis, London

Duncan AM, Chester D, Sangster H (2012) Religious interpretations of disaster. In: Wisner B, Gaillard JC, Kelman I (eds) The Routledge handbook of hazards and disaster risk reduction. Taylor & Francis, New York

Fraser S, Mtsuo I, Leonard GS, Murakami H (2012) Tsunami evacuation: lessons from the great east Japan earthquake and tsunami of March 11th 2011. GNS Science Report 2012/17 April 2012

GeoHazard International (2007) Preparing your community from tsunamis. Retrieved from http://www.geohaz.org/news/images/publications/Preparing%20Your%20Community%20for%20Tsunamis%20version%202%20low%20res.pdf. Accessed 1 August 2011

Ghafory-Ashtiany M (2009) View of Islam on earthquakes, human vitality and disaster. Disast Prev Manag 18(3):218–232

Ghozali I (2011) Aplikasi analisis multivariate dengan program IBM SPSS 19. Badan Penerbit Universitas Diponegoro, Semarang

Gillard M, Paton D (1999) Disaster stress following a hurricane: the role of religious differences in the Fijian Islands. Aust J Disaster Trauma Stud 1999-2. http://www.massey.ac.nz/~trauma/issues/1999-2/gillard.htm

Hamzah M (2006) The human dimension of early warning: why technology alone is not the answer. Stockholm Environment Institute, February 2006

Harris D (2012) The impact of cultural and religious influences during natural disasters (volcano eruptions). Scientific Paper. University College London, London

Hemond Y, Robert B (2012) Preparedness: the state of the art and future prospect. Disast Prev Manag 21(4):404. doi:10.1108/09653561211256125

Ife J (2002) Community development. Pearson Education, Frenches Forest NSW

ISDR (2005) HFA 2005–2015: Building the resilience of nations and communities to disasters. New York: UN-ISDR

Kemkens L (2013) On the connections between religion and disaster: a literature review. Center for Religious and Cross-Cultural Studies/CRCS Graduate School, GadjahMada University of Indonesia, Yogyakarta

Kosim M 2012) Bencana Moral Picu Bencana Alam, Berita Haluan. Retrieved from: http://www.harianhaluan.com/index.php?option=com_content&view=article&id=15062:bencana-moral-picu-bencana-alam-&catid=11:opini&Itemid=83

Krallis D, Csontos A (2014) From risk perception to safe behavior. Retrieved from: https://www.sia.org.au/downloads/SIGs/Resources/From_Risk_Perception_to_Safe_Behaviour.pdf

Kulatunga U (2010) Impact of culture towards disaster risk reduction. Int J Strat Property Manag 14(4):304–313

Levy JK, Gopalakrishnan C (2005) Promoting disaster-resilient communities: the Great Sumatra–Andaman earthquake of 26 December 2004 and the resulting Indian Ocean tsunami. Water Resour Dev 21(4):543–559, Routledge

Mayunga JS (2007) Understanding and applying the concept of community disaster resilience: a capital-based approach. A draft working paper for the summer academy for social vulnerability and resilience building, Munich

McEntire D, Crockers CG, Peters E (2010) Addressing vulnerability through an integrated approach. Int J Disast Resilience Built Environ 1(1):50–64. doi:10.1108/17595901011026472

Mishra R (1999) Globalization and the welfare state. Edward Elgar, Cheltenham

Murata S, Imamura F, Katoh K, Kawata Y, Takahashi S, Takayama (2010) Tsunami: to survive from tsunami. World Scientific, Singapore

Nahadi M, Sarimaya F, Rosdianti SR (2011) Hubungan Islam dengan Ilmu Pengetahuan Alam dalam Perspektif Sejarah. ATIKAN 1(1):2011

Nasution AI (2011) Siaga bencana dalam Islam, Serambi Indonesia. Retrieved from: http://aceh.tribunnews.com/2011/12/23/siaga-bencana-dalam-islam

Paton D (2003) Disaster preparedness: a social-cognitive perspective. Disast Prev Manag 12(3):210

Phillips B, Fodham M (2010) Introduction. In: Phillips BD, Thomas DSK, Fothergill A, Blinn-Pike L (eds) Social vulnerability to disaster. Taylor & Francis, Boca Raton

Rachmalia, Hatthakit U, Chaowalit A (2011). Tsunami preparedness of people living in affected and non-affected areas: a comparative study in coastal area in Aceh, Indonesia. Australas Emerg Nurs J 14:17–25

Shaw R, Shiwaku K, Kobayashi H, Kobayashi M (2004) Linking experience, education, perception and earthquake preparedness. Disast Prev Manag 13(1):39–49

Smith K (2003) Environmental hazards: assessing risk and reducing disaster, 3rd edn. Routledge, London

Sorensen JH, Rogers GO (1988) Local preparedness for chemical accidents: a survey of U.S. communities. Ind Crisis Q 2:89–108

Sulasman (2012) Kyai and Pesantren in the Islamic historiography of Indonesia. Int J Hist Stud 4(1):67–82

Sutton J, Kathleen T (2006) Disaster preparedness: concept, guidance, and research. Natural Hazard Center, Institute of Behavioral Science, University of Corolado, Boulder

Timmerman D (2014) The role of faith based organization in disaster response. Retrieved from http://www.google.com/url?sa=t&rct=j&q=&esrc=s&source=web&cd=4&ved=0CEUQFjAD&url=http%3A%2F%2Fwww.mnosynod.org%2FPower%2520Point%2FThe%2520Role%2520of%2520Faith%2520Based.ppt&ei=7Y8MU4OmJMfqkAXq9YHwCQ&usg=AFQjCNEFEsrFKkvJgd5VzTkXsI27syL7uQ&sig2=rw8DZGpaVW5Emq6zVWXivw
UNDP-Indonesia (2008) Project facts: early recovery assistance for Yogyakarta and Central Java (ERA). Retrieved from http://www.undp.or.id/factsheets/2008/CPR%20Early%20Recovery%20Assistance%20for%20Yogyakarta%20and%20Central%20Java.pdf

Part III
India

Chapter 19
Rhetoric and Ground Reality of Institutionalizing Disaster Risk Reduction

N. Vinod Chandra Menon

Abstract The Indian Ocean Tsunami of 26th December 2004 and the Hyogo Framework for Action (2005–2015) endorsed by more than 168 national governments at the World Conference on Disaster Reduction at Kobe in 2005 compelled the national leadership in many countries in South Asia to establish appropriate institutional mechanisms for strengthening disaster preparedness, institutionalizing disaster risk reduction and improving emergency preparedness in these countries. However, the pursuit of developmental priorities without factoring in the disaster risk and vulnerability profiles of these countries at the national, provincial and local levels is resulting in recurring disasters with enormous damage to property, assets and infrastructure as indicated by the recent disasters in the region. Lessons of past disasters and the Risk, Vulnerability and Hazard Profiles of disaster-prone areas are being ignored, which is resulting in repeated occurrence of sudden onset disasters affecting the lives and livelihoods of marginalized communities. The sterling examples of disaster risk reduction in the cyclone prone areas of Bangladesh and the success of early warning dissemination and large scale evacuation of disaster-prone communities from the way of harm before the Cyclone Phailin made its landfall in Odisha in India in October 2013, thereby reducing the loss of lives significantly, need to be replicated and scaled up in other developing countries by mobilizing community participation. This Chapter reviews the institutional mechanisms for disaster risk reduction in South Asian countries, analyses the post-tsunami policy pronouncements of countries in South Asia and evaluates the post-tsunami implementation of policy directives at the national and provincial levels. It is argued that the goals of disaster risk reduction and climate change adaptation will continue to elude us until the communities at risk empower themselves to internalize these concerns as a culture of preparedness, risk reduction and emergency response by strengthening community resilience through people-led, people-owned and people-managed social transformation initiatives.

Keywords Climate change adaptation • Community resilience • Disaster risk reduction • Economic and social impacts • Governance • Institutional mechanisms

---

N.V.C. Menon (✉)
Founder Member, National Disaster Management Authority (NDMA),
Government of India, New Delhi, India
e-mail: nvcmenon@gmail.com

R. Shaw (ed.), *Recovery from the Indian Ocean Tsunami: A Ten-Year Journey*,
Disaster Risk Reduction, DOI 10.1007/978-4-431-55117-1_19

19.1 Introduction

During the International Decade of Natural Disaster Reduction (IDNDR) which was observed by the United Nations during the 1990s (1990–2000), several initiatives were launched at the global level to address the need to reduce the enormous losses of lives, injury to people and damage and destruction of property, public infrastructure and assets caused by natural disasters. The Yokohama Strategy and Plan of Action for a Safer World adopted by national governments, international humanitarian organizations and the United Nations in May 1994 became the first ever global blueprint which outlined a strategy and plan of action for working towards strengthening disaster preparedness in the disaster prone countries. The IDNDR decade witnessed severe devastation caused by several disasters like the cyclone in Bangladesh in 1991, Hurricane Andrew in the Bahamas and the United States of America in 1992, Latur earthquake in Maharashtra, India in 1993, Northridge earthquake in Los Angeles, California in the United States of America in 1994, Kobe earthquake in Japan in 1995, Hurricane Mitch in Honduras and Nicaragua in Central America in 1998, Yangtse flood in China in 1998, Orissa super cyclone in Orissa, India in 1999, and the Marmara earthquake in Turkey in 1999. The adverse impact of these disasters on the lives and livelihoods of the poorer and weaker sections of the society in the affected countries provoked reflections among disaster management practitioners about the human induced causes which trigger natural disasters. The social dimensions of increasing disaster risk and vulnerability among the poorer and weaker sections of the society received the attention of the humanitarian assistance agencies because of the recognition that the weak coping capacities of the disaster affected communities tend to amplify and reinforce the disaster risk and vulnerability among such sections and make them repeatedly victims of disasters, especially where they are forced to stay in low lying flood-prone settlements, landslide-prone slopes or in highly fragile non-engineered houses in high seismic risk zones.

Several publications by multilateral and bilateral organizations and International Financial Institutions argued for the need for strengthening the resilience in disaster-prone countries through appropriate policy interventions by these new institutional mechanisms. Some of the most prominent of these include Asian Development Bank (2012), Asian Development Bank (2013), United Nations Development Program (2010), United Nations Economic and Social Commission for Asia and the Pacific and United Nations International Strategy for Disaster Reduction (2012), United Nations International Strategy for Disaster Reduction (2005a, b), United Nations International Strategy for Disaster Reduction (2013a, b), World Bank (2012a), World Bank (2012b), and World Bank (2012c). The Indian Ocean tsunami of 26th December 2004 and the Hyogo Framework for Action 2005–2015 endorsed by 168 national governments at the World Conference on Disaster Reduction at Kobe compelled national leaderships in several countries around the world to enact appropriate legislation for establishing institutional mechanisms at the national and provincial levels to strengthen disaster preparedness, institutionalize disaster risk reduction and improve emergency response capacities in these countries.

In spite of improvements in Early Warning Systems through the deployment of modern state-of-the-art scientific and technological solutions like Computer Aided Models, satellites, Doppler weather radars, etc. and the application of nowcasting techniques in weather monitoring, the increasing frequency and the enormous increase in disaster losses in the recent past makes it imperative to ensure that the institutions mandated to oversee disaster management at the global, regional, national and provincial levels strive hard to address the critical gaps in the effective management of disasters. The enormous loss of lives caused by earthquakes in Pakistan in 2005, China in 2008, Haiti in 2010 highlight the need for strict enforcement of earthquake-resistant building codes in countries vulnerable to earthquakes.

19.2 Megadisasters in the Recent Past

The International Strategy for Disaster Reduction (ISDR) which was announced to be launched by the United Nations in the new millennium continued to pursue disaster risk reduction efforts by engaging with national governments, international humanitarian assistance agencies, civil society partners, donor agencies, corporate sector and other stakeholder groups. The Gujarat earthquake in India in 2001, the Bam earthquake in Iran in 2003, Hurricane Ivan in 2004, the Indian Ocean tsunami in 2004, Hurricane Katrina in 2005, Muzaffarabad earthquake in Pakistan in 2005, Cyclone Sidr in Bangladesh in 2007, cyclone Nargis in Myanmar in 2008, Sichuan earthquake in China in 2008, Haiti earthquake in 2010, Pakistan floods in 2010, the triple disaster triggered by the earthquake, followed by a tsunami and nuclear emergency in Japan in 2011, Pakistan floods in 2011, Uttarakhand floods in 2013, cyclone Phailin in Odisha and Andhra Pradesh in India in 2013 and the typhoon Haiyan in Philippines in 2013 were a few of the most devastating disasters in the recent past. However, the Indian Ocean tsunami of 26th December 2004 has to be seen as a watershed event which triggered a series of actions by governments of the affected countries and emulated by a large number of national governments around the world due to the strong advocacy by the United Nations, donor agencies, international humanitarian assistance agencies, civil society partners, regional networks and other stakeholder groups.

19.3 Economic Consequences of Disasters

The Indian Ocean tsunami of 26th December 2004 triggered by a 9.3 Richter Scale magnitude earthquake reportedly killed 184,167 people, affected more than five million people and left 45,752 people missing in 14 countries in the Indian Ocean region. It is estimated that the 2004 tsunami had caused an economic damage of about US$ 10 billion. Quoting UNISDR and Munich Re, the web site www.dosomething.org reported that between 2000 and 2012, natural disasters caused economic damages estimated at US$ 1.7 trillion, affected 2.9 billion people and killed 1.2 million people. The UNISDR Infographic on Impact of Disasters since the 1992 Global Earth Summit at Rio de Janeiro mentions that natural disasters affected 4.4 billion people, caused economic damages estimated at US$ 2.0 trillion and killed 1.4 million people. It further stated that in 2011, worldwide there were 154 floods, 16 droughts and 15 cases of extreme temperature. According to the web site, 2012 marked the third consecutive year of worldwide natural disaster damage exceeding US$ 100 billion, with 2011 reaching a record high of US$ 371 billion in damages caused by natural disasters. Between 2000 and 2012, it is estimated that more than 1.1 million people were killed by natural disasters.

The Global Assessment Report 2013 published by UNISDR observes that direct disaster losses are at least 50 % higher than internationally reported figures. The total direct losses in 40 low and middle income countries was estimated to amount to US$ 305 billion over the last 30 years. However, of these more than 30 % were not internationally reported. (United Nations International Strategy for Disaster Reduction 2013a, b). According to the Global Assessment Report 2013, the total expected annual global loss from earthquakes and cyclone wind damage alone now amounts to US$ 180 billion per year. This figure does not include the significant cost of local disasters from floods, landslides, fires and storms or the cost of business interruption (United Nations International Strategy for Disaster Reduction 2013a, b).

According to UNISDR, "between 2002 and 2011, there were 4,130 disasters recorded, resulting from natural hazards around the world where 1,117,527 people perished and a minimum of US$ 1,195 billion was recorded in losses. In the year 2011 alone, 302 disasters claimed 29,782 lives; affected 206 million people and inflicted damages worth a minimum of estimated US$ 366 billion." UNISDR further observed that "the proportion of world population living in flood-prone river basins has increased by 114 %, while those living on cyclone-exposed coastlines have grown by 192 % over the past 30 years. Over half of the world's large cities, with populations ranging from 2 to 15 million, are currently located in areas highly vulnerable to seismic activity. Rapid urbanization will further increase exposure to disaster risk." (United Nations International Strategy for Disaster Reduction 2013a, b)

The World Disasters Report 2003 published by the International Federation of Red Cross and Red Crescent Societies observed that in the previous two decades alone, direct reported economic losses have multiplied fivefold in real terms to US$ 629 billion (International Federation of Red Cross and Red Crescent Societies 2003). According to a UNDP Report in 2004, the real economic losses have averaged US$ 75.5 billion in the 1960s, US$ 138.4 billion in the 1970s, US$ 213.9 billion in the 1980s and US$ 659.9 billion in the 1990s. In a recent World Bank Working Paper on the Cost Benefit Studies on Disaster Risk Reduction in Developing Countries, K C Shyam observed that "the losses from natural disasters to mankind are undoubtedly massive—on average, globally every year over 100,000 people were killed and some 246 million people affected by natural disasters during the period 2002–2011 and the estimated average economic loss was US$ 131 billion per year" quoting a CRED publication EMDAT CRED Crunch Newsletter No. 31, 2013 (Shyam 2013).

In spite of the explosive increase in the economic damages caused by natural disasters on property, infrastructure, assets and amenities in the past several decades, there has been very little recognition by national governments of the need to invest in disaster risk reduction and mitigation. Even in the case of donor grants given by humanitarian assistance agencies and international financial institutions to national governments affected by natural disasters, disaster risk reduction and mitigation interventions receive very little attention in post disaster rehabilitation and recovery frameworks.

It is estimated by the World Bank that the annual costs of damage caused by disasters vary from 2 to 15 % of the Gross Domestic Product (GDP) of the affected countries. According to the World Bank, during the period 1996–2000, India lost approximately 2.25 % of the Gross Domestic Product and 12.15 % of the revenue annually due to natural and man-made disasters. The economic losses caused by natural disasters are sometimes 20 times greater as a percentage of the GDP in developing countries compared to the developed countries. For instance, Hurricane Val cost 230 % of the Gross Domestic Product of Samoa in 1991. Similarly, the impact of Hurricane Ivan in 2004 cost the Caribbean island of Grenada 200 % of its GDP.

Recognising the multi-hazard risk and vulnerability profile of the South Asian countries, a 'Comprehensive Regional Framework on Disaster Management 2006–2015' was approved by the SAARC Council of Ministers in 2006. The publication "Mainstreaming Disaster Risk Reduction in Development" by the SAARC Disaster Management Centre (SDMC) released during the Workshop organized by SDMC and Sri Lanka's Disaster Management Centre on 14th to 15th November 2008 observed that "it is estimated that the countries of the region have been losing between 2 and 20 % of their GDP and 12–66 % of the revenues on account of disasters every year. These do not include losses in some of the informal sectors of the economy which generally go unaccounted or long term damage and loss of environment and ecology which can not be measured immediately. Some of the countries may not be spending as much on social sectors like public health or education as they may loosing due to disasters. On top of it, almost all the countries of the region are forced to divert scarce resources to disaster relief, rehabilitation and reconstruction, which create further set back to development. It is now quite evident that natural and manmade disasters in South Asia are one of the important barriers to the realization of the Millennium Development Goals of the United Nations." It has now been recognized that the goals of disaster risk reduction and climate change adaptation will continue to elude us until the communities at risk empower themselves to internalize these concerns as a culture of preparedness, risk reduction and emergency response by strengthening community resilience through people-led, people-owned and people-managed social transformation initiatives.

19.4 Emerging Concerns of Climate Change

While many national governments are yet to acknowledge the challenges posed by climate change, many countries are facing the adverse consequences of climate change and extreme weather events. The global risks advisory firm Maplecroft in its publication Climate Change Risk Atlas 2011 ranked 170 countries according to their Climate Change Vulnerability Index (CCVI), computed by evaluating 42 social, economic and environmental factors to assess national vulnerabilities across three core areas: exposure to climate-related natural disasters and sea-level rise; human sensitivity, in terms of population patterns, development, natural resources, agricultural dependency and conflicts; and future vulnerability by considering the adaptive capacity of a country's government and infrastructure to combat climate change. Based on the Climate Change Vulnerability Index, Maplecroft has identified 16 "extreme risk" countries: Bangladesh, India, Madagascar, Nepal, Mozambique, Philippines, Haiti, Afghanistan, Zimbabwe, Myanmar, Ethiopia, Cambodia, Vietnam, Thailand, Malawi and Pakistan ranked from 1 to 16 respectively. Maplecroft observes that the countries with the most risk are characterised by high levels of poverty, dense populations, exposure to climate-related events; and their reliance on flood and drought prone agricultural land.

According to Maplecroft, "over the next 30 years the vulnerability (of these countries) to climate change will rise due to increases in air temperature, precipitation and humidity. This means organizations with operations or assets in these countries will become more exposed to associated risks, such as climate-related natural disasters, resource security and conflict" (Maplecroft 2013).

The Climate Change Risk Atlas 2013 prepared by Maplecroft ranked 50 cities at risk from climate change and the extreme risk cities are Dhaka, Bangladesh; Manila, the Philippines; Bangkok, Thailand; Yangon, Myanmar; Jakarta, Indonesia; Ho Chi Minh City, Viet Nam; and Kolkata, India which were ranked 1 to 7 extreme risk cities respectively. The high disaster risk profile of these cities makes it important for the Urban Local Bodies in these cities to address their Urban Disaster Risk especially because of the very high population densities and the concentration of poor people in the urban slums.

19.5 Challenges at National, Provincial and Local Levels

At the national level, mainstreaming disaster risk reduction and climate change adaptation in the development planning and budgeting processes has to be negotiated with the Planning Commission and the nodal Ministries mandated with disaster management, finance and planning. Once the Central Planning agency accepts the rationale and initiates the process of mainstreaming disaster risk reduction and climate change adaptation in development planning, the administrative and executive orders can be issued by the relevant divisions in the Ministry of Finance. The Central

Planning agency must also issue appropriate instructions or guidance notes for the Provincial or State Planning Boards to replicate the same process at their levels. However, in a federal structure, it is likely that a few Provinces or States may be governed by political parties which may not necessarily belong to the ruling party running the Government at the Centre. It is therefore necessary that the concerns of disaster risk reduction and climate change adaptation receive adequate attention from all political parties and a consensus is created among all political parties to endorse these concerns at the national, provincial and local levels by the apex body for disaster management in the respective countries. The capacity building of elected representatives, the generation of public awareness among the non-officials and the efforts to make these opinion leaders champions and advocates of disaster risk reduction and climate change adaptation deserve the highest priority in the formative years of the new paradigm shift in disaster management.

Informed policy formulation in the fields of disaster risk reduction and climate change adaptation can only be guaranteed by ensuring that those mandated with the task of policy formulation are made aware of the rationale of investing in disaster preparedness, risk reduction and mitigation which will reduce the adverse impacts of disasters that within a few minutes can wipe out the developmental gains secured across years of relentless efforts. Investments in early warning systems, scenario analysis and modeling, remote sensing, dissemination of early warning alerts and last mile connectivity to the remote villages have proved to yield positive results by reducing the loss of lives and loss of property, assets, amenities and infrastructure in the event of a sudden disaster. Special attention needs to be paid to the dissemination of good practice case studies where traditional wisdom, indigenous knowledge and the applications of modern science and technology have saved the lives of people and strengthened community resilience in disaster-prone areas.

19.6 Institutional Reforms Triggered by the Indian Ocean Tsunami

The Indian Ocean Tsunami of December 2004 became a watershed for institutional reform in disaster management globally as it exposed the weaknesses in the development policies ignoring the multi-hazard risk and vulnerability profiles of locations where unchecked and unregulated development was taking place due to the compulsions posed by growing populations, increasing socio-economic deprivation among the poorer sections and unsustainable development initiatives. The national governments in most countries affected by the Tsunami in 2004 took the initiative in enacting disaster management legislations which mandated the establishment of appropriate institutional mechanisms at national, provincial and local levels to strengthen multi-hazard preparedness, disaster risk reduction and mitigation and improve the emergency response capacities of stakeholder groups. Several countries enacted disaster management legislations which sought to replace the hitherto reactive, post-disaster response strategies with a more pro-active paradigm of strengthening disaster preparedness, addressing disaster risk reduction and improving emergency response capacities through establishing dedicated first responder agencies, undertaking capacity building of stakeholder groups, and strengthening early warning systems, etc.

The UNISDR Publication titled "Implementation of the Hyogo Framework for Action—Summary of reports 2007–2013" observes that since 2005, 121 countries have enacted legislations to establish policy and legal frameworks for disaster risk reduction, 191 countries have established HFA focal points and 85 countries have set up National Platforms or national coordinating bodies for disaster risk reduction (United Nations International Strategy for Disaster Reduction 2013a). India established a new National Platform a few days before the Global Platform 2013, making the number of countries with national platforms for disaster risk reduction to go upto 86. While the intentions of the national governments to support the global initiative of disaster risk reduction need to be applauded, the effectiveness of these legislations and coordination mechanisms would be tested only when devastating disasters strike these countries. In the case of India, the unprecedented rainfall on 16th and 17th June 2013 in the mountain state of Uttarakhand caused floods and landslides which disrupted road networks, power, water supply and communication systems and trapped more than 100,000 pilgrims on the Char Dham pilgrimage routes. Tourists and villagers had to be evacuated using Indian Air Force helicopters and aircrafts over the next few weeks. This incident exposed the need to strengthen the monitoring, analysis and forecasting of weather events like cloudbursts during the monsoon season. The Table 19.1 above provides the list of institutional mechanisms established in South Asian countries after the Indian Ocean Tsunami of 2004.

In the South Asian countries, the post-tsunami reorganization of the disaster management mechanisms at the national level saw the emergence of the National Commission for Emergency and Disaster Management in Afghanistan, the National

Table 19.1 Relevant Legislations and institutional mechanisms for disaster management in South Asian countries

Country	Disaster management legislation	National institutional mechanism for disaster management
Afghanistan	Draft Law	Afghanistan National Disaster Management Authority
Bangladesh	Disaster Management Act 2008	National Disaster Management Council
Bhutan	National Disaster Management Act, 2008	National Committee for Disaster Management
India	Disaster Management Act 2005	National Disaster Management Authority
Maldives	Disaster Management Act 2006	National Disaster Management Council
Nepal	Disaster Management Act 2012	Central Natural Disaster Relief Committee
Pakistan	National Disaster Management Act 2010	National Disaster Management Commission
Sri Lanka	Disaster Management Act 2005	National Council for Disaster Management

Disaster Management Council in Bangladesh, the National Disaster Management Authority, the State Disaster Management Authorities and the District Disaster Management Authorities in India, the National Disaster Management Center (NDMC) in Maldives, the Central Natural Disaster Relief Committee (CNDRC) in Nepal, the National Disaster Management Authority (NDMA), the Provincial (Regional) Disaster Management Authorities, the Municipal Disaster Management Authorities, the District Disaster Management Authorities and the Town and Tehsil Authorities in Pakistan, the National Council for Disaster Management (NCDM) and the Disaster Management Centre (DMC) in Sri Lanka, the National Committee for Disaster Management (NCDM), the Dzongkhag Disaster Management Committee (DDMC), the Dungkhag Disaster Management Committee (DuDMC), the Thromde Disaster Management Committee (TDMC) and the Gewog Disaster Management Committee (GDMC) in Bhutan. While these are a few of the newly established institutional mechanisms, the effectiveness of these mechanisms to strengthen disaster risk reduction and mitigation remains to be seen. Most of these bodies at the national level and the provincial levels are headed by the Chief Executive, viz. the Prime Ministers, the Chief Ministers or the Presidents or Governors. In most cases, it has been observed that these bodies rarely receive the attention they deserve because the heads of Governments are unable to devote the time or attention to ensure the smooth functioning of these bodies. In most cases, these bodies continue to perform the role of coordinating the post-disaster response.

In the Indian context, the National Disaster Management Authority established by the Government of India recognized the need to engage with the Planning Commission to ensure that disaster management concerns are mainstreamed into development plans at the national, provincial, district and local levels if disaster risk reduction has to become successful. While the Eleventh Five Year Plan was getting formulated, a Working Group was constituted by the Planning Commission under the Chairmanship of one of the Members of the National Disaster Management Authority to work out the modalities of mainstreaming disaster risk reduction in development planning. This Working Group recommended that a few key mitigation projects be initiated to strengthen disaster preparedness, improve disaster risk reduction by addressing the critical gaps in the management of earthquakes, floods, cyclones, drought, landslides, tsunamis, etc. and strengthen emergency response capacities at the national, state, district, and local levels. The strategy of mainstreaming disaster risk reduction in the development planning and budgeting processes was envisaged as given in Fig. 19.1 below.

The Planning Commission, Government of India observed in the Eleventh Five Year Plan Report that "Mainstreaming disaster management into the development planning process essentially means looking critically at each activity that is being planned, not only from the perspective of reducing the disaster vulnerability of that activity, but also from the perspective of minimizing that activity's potential contribution to the hazard. Every development plan of a ministry/department should incorporate elements of impact assessment, risk reduction, and the 'do no harm' approach"; such as through urban planning and zoning, upgradation of building codes and their effective enforcement, adoption of disaster resilient housing designs

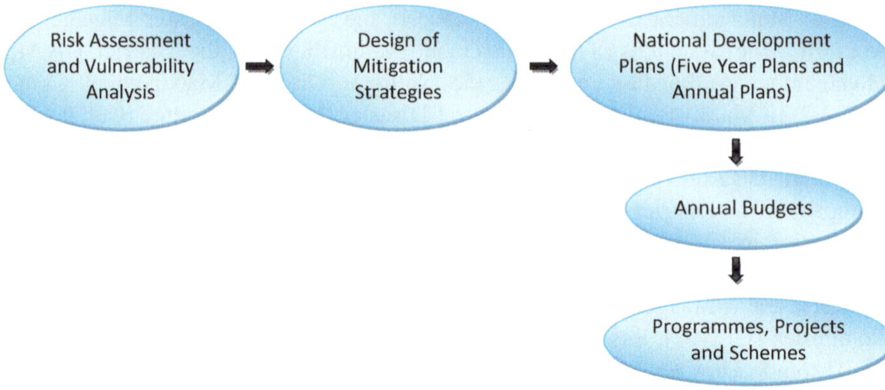

Fig. 19.1 Mainstreaming disaster risk reduction in planning and budgeting processes.

and construction of schools and hospitals, flood proofing, response preparedness planning, insurance, establishment of early warning systems for various types of disasters, generating community awareness, creating technical competence and promoting research among engineers, architects, health experts, and scientists. (Planning Commission, Government of India, 2007)

On 19 June 2009, the Finance Ministry sent a memo to all ministries and departments, saying that the Expenditure Finance Committee (EFC) and Detailed Project Report (DPR) formats had been modified to include disaster management concerns, and enclosing the questions and issues that needed to be addressed while framing the EFC memo and DPRs. The Ministries and departments were also asked to self-certify the correctness of the response to these issues. It was necessary for the ministries to ensure that if the project involved creation and/or modification of any engineering assets or existing land use plans, the cost of prevention and mitigation of possible disasters were included in the project cost. All possible risks were to be identified and their impact assessed, and preventive measures for these to be listed. Risk treatment options—both engineering and non-engineering—had to be identified, prioritized and included in the budget. Further, reference to and compliance with relevant NDMA guidelines, the National Building Code, the Bureau of Indian Standards codes and safety standards, land-use directives, and other guidelines of relevant technical bodies like the Indian Road Congress, Railway Board manual, etc. had to be confirmed. These were all identified and listed and appended to the memo. Finally, it had to be confirmed that the whole process of risk assessment was done based on the available information and that the mitigation measures were in conformance with the statutory and regulatory requirements and were the most viable ones available at the time.

A detailed check list was also attached to the memo, which included information about the project site, hazard risk–zoning information, proneness to various disasters based on historical evidence, presence of dams, etc., and nature of the project in terms of damage-risk, environs to be considered for impact assessment like railway lines, waterways, drainage lines, etc.

In July 2009, the Finance Ministry of the Government of India issued another memo which pointed out that the CNE memo format had been revised to incorporate NDMA's concerns about structural safety of new buildings, public assets and infrastructure proposed to be constructed in geographical areas with high disaster risk and vulnerability, and requiring the ministries to provide self-certification that the concerns had been met. The Revised Cost Estimate (RCE), EFC and the DPR formats were also modified to explicitly incorporate these concerns.

One of the very important steps adopted by the Government of India to ensure individual ownership and accountability in the planning process was to advise the officials submitting the proposals to self-certify that they have factored in the disaster risk and vulnerability of the geographical locations where the projects will be implemented. The damage caused to hundreds of thousands of houses built under the popular mass housing scheme Indira Awas Yojana for people below poverty line due to floods in several states in India every year and the damage and destruction of property, assets, infrastructure and amenities due to disasters like the Sikkim earthquake of 2011 and the Uttarakhand cloudburst, flashflood and landslides in 2013 show that the policy pronouncements and administrative instructions have remained only on paper and are not being implemented by officials at the cutting edge level.

The IFRC Publication entitled "Better Laws Safer Communities" observed that there is widespread agreement among policy makers that legal frameworks are a critical tool for governments to shape the choices for disaster risk reduction and mitigation—both for themselves and for others. The endorsement of the *Hyogo Framework for Action 2005–2015: Building the Resilience of Nations and Communities to Disasters* (HFA) by 168 national governments in 2005 echoed these sentiments. While HFA's first priority is to "ensure that disaster risk reduction is a national and a local priority with a strong institutional basis for implementation," notably through "policy, legislative and institutional frameworks for disaster risk reduction," the legislative route and the setting up of these institutions have been discounted by many disaster management practitioners. Several disaster management practitioners feel that the many new laws and policies that have been developed to address disaster risk reduction and mitigation have not succeeded in making the difference they promised, especially in mobilizing community participation and in reducing disaster risk and vulnerability at the local levels.

The IFRC publication concludes the following:

- While there have been many new laws and policies adopted by now, the resulting legal reforms seem to be less comprehensive than is generally assumed.
- The development of new legislations on disaster risk management can be a key enabler for awareness on disaster risk reduction through the public process of law-making, as well as by providing the content for clear institutional mandates and implementation of disaster risk reduction in these countries.
- While there have been many new safety legislations in sectors outside, disaster management laws hold the key to long-term reduction of underlying risks

(as identified in "Priority Four" of the HFA), but their implementation needs more support. To establish an integrated approach to disaster risk reduction, these sectoral laws also need to include disaster risk reduction criteria, and to be coordinated with disaster risk management systems.
- Partnerships between local government and communities are needed for effective implementation of disaster risk reduction safety regimes at the local levels. Challenges in implementation of safety regulations at the local levels, such as land use planning and building codes, emerge as the most common barrier to effectiveness.

19.7 "Manageable" Disasters: Role of Early Warning

The establishment of weather monitoring networks consisting of satellites, Doppler radars, etc. has significantly helped in facilitating the tracking of cyclones, hurricanes and typhoons in the recent past. The recent example of minimizing the loss of livelihoods during the landfall of Cyclone Phailin in Odisha and Andhra Pradesh in India by close monitoring of the cyclone's movements by the Indian Meteorological Department of the Government of India is clearly a success story which highlights that close monitoring, dissemination of early warning and alert messages to the disaster-prone communities, their evacuation to cyclone shelters, search and rescue, etc. can effectively reduce loss of lives.

However, in the case of earthquakes, the lack of early warning and forecasting systems will continue to pose serious challenges to policy makers. The devastating experience of the 7 magnitude earthquake in Haiti in 2010 which reportedly killed more than 250,000 people was followed a few days later by a 8.8 magnitude earthquake in Chile which killed only 497 people because of the latter's strong compliance and enforcement of earthquake-resistant building codes. These earthquakes highlight the need to ensure strict compliance of earthquake-resistant building codes in earthquake-prone areas. The increasing urbanization, unplanned and unregulated expansion of multi-storied buildings in urban, semi-urban and peri-urban areas and the construction of flyovers, bridges and rapid mass transport systems like metro-rail, etc. accentuate the problem of high seismic risk in megacities which are close to fault lines. Public awareness and capacity building of masons, contractors, builders and engineers deserve special attention in the context of increasing seismic risk in urban areas in developing countries with weak techno-legal regimes.

The coordinated efforts of 29 national governments around the Indian Ocean resulted in the establishment of a regional tsunami early warning system which has been operational since 2006 as part of a coordination plan by UNESCO-Intergovernmental Oceanographic Commission. The Government of India also took the decision to establish a Tsunami Early Warning System which is being coordinated by the Indian National Centre for Ocean Information Systems (INCOIS), Hyderabad which has deployed tidal gauges, buoys and bottom pressure recorders in several locations on the Indian Ocean and the Bay of Bengal for monitoring tsunamis.

19.8 Potential of Improved Preparedness

The increasing use of applications like Google Crisis Mapper, Person Finder, etc. and the use of social networking tools like Facebook, Twitter, Whatsapp, We Chat, etc. have enabled the effective use of crowd sourcing to disseminate ground level information about the disaster impact and the living conditions of affected communities, as we saw after Hurricane Sandy in New York, Haiti earthquake and more recently Typhoon Haiyan in Tacloban in the Philippines. The civil society organizations have made special efforts to centre-stage communities in strengthening disaster preparedness, disaster risk reduction, mitigation and emergency response in disaster-prone areas. We are also witnessing the post-disaster rehabilitation and recovery strategies in several countries where local stakeholder groups actively participate in evolving inclusive, gender sensitive and sustainable rehabilitation, reconstruction and recovery frameworks by involving affected communities.

One of the major challenges faced by national governments is to ensure that disaster risk reduction and climate change adaptation concerns are mainstreamed into the developmental planning and budgeting processes. According to the UNISDR Global Assessment Report 2009, the organizations mandated with disaster risk reduction often lack the political authority and the technical capacity to influence development sectors (United Nations International Strategy for Disaster Reduction 2009). The 2009 national report from Cambodia reporting on the progress of the implementation of the Hyogo Framework of Action 2005–2015 stated that "there is no common understanding of multi-sector integration approaches and lack of comprehensive understanding of disaster risk reduction and vulnerability reduction development agencies" (United Nations International Strategy for Disaster Reduction 2009).

The Independent Evaluation Group's review of the World Bank's assistance to disaster affected countries pointed out that even within disaster response projects funded by the World Bank, the World Bank did better in reconstructing damaged infrastructure and housing than it did in reducing disaster risk and vulnerabilities and in assessing their root causes. Till very recently, the World Bank did not insist on incorporating disaster risk reduction concerns while reconstructing damaged infrastructure in disaster affected countries. The Global Facility for Disaster Reduction and Recovery (GFDRR) played a key role in influencing this significant transition while drawing up post-disaster rehabilitation and recovery strategies funded with the World Bank assistance with national governments affected by natural disasters.

19.9 Lessons from Tsunami Recovery

In the publication entitled Key Propositions for Building Back Better, the United Nations Secretary General's Special Envoy for Tsunami Recovery outlined the following ten key propositions for recovery.

- Governments, donors, and aid agencies must recognize that families and communities drive their own recovery.
- Recovery must promote fairness and equity.
- Governments must enhance preparedness for future disasters.
- Local governments must be empowered to manage recovery efforts, and donors must devote greater resources to strengthening government recovery institutions, especially at the local level.
- Good recovery planning and effective coordination depend on good information.
- The UN, World Bank, and other multilateral agencies must clarify their roles and relationships, especially in addressing the early stage of a recovery process.
- The expanding role of NGOs and the Red Cross/Red Crescent Movement carries greater responsibilities for quality in recovery efforts.
- From the start of recovery operations, governments and aid agencies must create the conditions for entrepreneurs to flourish.
- Beneficiaries deserve the kind of agency partnerships that move beyond rivalry and unhealthy competition.
- Good recovery must leave communities safer by reducing risks and building resilience.

Groupe URD's 10 lessons Learnt from the Humanitarian Assistance Funded by the French State (Ministère des Affaires étrangères et européennes 2007) highlighted the following ten major lessons.

- 'Too much money detracts from common sense'
- Ensure solid coordination amongst donors at the outset
- Understanding constraints is just as important as analyzing needs
- Ensure that 'capacity analysis' is an integral part of the initial assessment
- Distinct, even conflicting objectives will need to be managed
- Post-crisis reconstruction is more complex than emergency relief
- SCAC teams (Cooperation and Cultural Action Sections within French Embassies) require support in specific resources for field monitoring
- Existing partnerships created some true added value
- It is important to confuse visibility with communication
- Humanitarian aid may have an impact on local politics and conflicts

Based on their experiences in several tsunami affected countries, a UNDP Report highlighted the following emerging principles which must be kept in mind while working with local governments affected by a disaster (United Nations Development Program 2006).

- The ability of local governments to play an effective part in the recovery effort will in large part depend on their capacity prior to a disaster.
- It is crucial to understand the legal and institutional framework for decentralization as this is likely to be an important factor in determining the effectiveness of local government in the post-disaster recovery phase.
- In order for local governments to be effective players in the recovery process they must have adequate financial resources in addition to capacity.

- Tax bases are typically reduced significantly due to the impact of the disaster on the local economy and reduction of user charges etc. from damaged local government infrastructure and compensatory mechanisms should be established as early as possible.
- Although there is a trade-off between a quick response to urgent needs and a carefully planned reconstruction process that allows for consultation, participation and capacity development, it is necessary to opt for the latter in order to ensure the effective use of resources, to ensure that recovery mirrors the actual demands of the affected communities and to develop capacity in the long term.
- Local government should be involved in the planning, design, implementation and monitoring of the recovery process. Involving local governments at all stages of the process is also an effective way to develop capacity.
- Local governments are likely to be the best placed to assess the needs of the affected communities.
- Local governments have a crucial role to play, through coordination and facilitation of participatory processes, in ensuring that the issue of land and property rights is dealt with in an equitable and timely manner. Whereas land management might not be a devolved responsibility of local governments, the elected representations can play an important role as a partner in solving problems related to e.g. ownership and use of land.
- It is crucial to involve communities in the recovery process at all stages. Communities are likely to be the most efficient in restoring some micro-infrastructure. Furthermore attention should be given to building community-based disaster preparedness.
- In planning for post-disaster recovery, attention should be paid to the opportunity to promote issues such as gender equality, conflict resolution and human rights.
- There is an urgent need to ensure UN and donor support for early recovery of local governance institutions as well as reconstruction of the infrastructure of local governments, which will allow them to re-establish delivery of economic and social services.

In a brief report (United Nations International Strategy for Disaster Reduction 2005b) published immediately after the tsunami, UNISDR also highlighted ten lessons learnt from the tsunami recovery as follows:

- We are all vulnerable to natural disasters.
- Careful coastal land use planning is essential to minimize risk.
- Public awareness and education are essential to protecting people and property.
- Early warning saves lives.
- Countries can work together ahead of time, as well as when disaster strikes.
- Reducing risk depends on close interaction between the scientific and technical community, public authorities and community-based organizations.
- Developing and respecting appropriate building codes can minimize exposure to risks.
- Humanitarian aid needs to invest more in disaster prevention in addition to immediate relief needs.

- Concrete action and good coordination is vital to ensure people's safety from disasters.
- Telecommunications and the media have a crucial role to play in disaster risk reduction.

In conclusion, the documentation of the Indian Ocean Tsunami recovery experiences by the Tsunami Recovery Impact Assessment and Monitoring System, the Tsunami Global Lessons Learned Project, the International Federation of Red Cross and Red Crescent Societies (International Federation of Red Cross and Red Crescent Societies 2013), Office of the UN Secretary General's Special Envoy for Tsunami Recovery (2006), Groupe URD's Report on the 10 Lessons Learnt from the Humanitarian Assistance Funded by the French State, UNISDR's 10 Lessons Learnt in Tsunami Recovery and UNDP's Local Governance in Tsunami Recovery: Lessons Learned and Emerging Principles, all provide us rich insights on the integral elements of an effective Recovery Framework. However, these publications also highlight the critical imperative for ensuring that international humanitarian assistance organizations recognize the need for creating greater stakeholder ownership for disaster risk reduction at the local level by empowering communities at the grassroots level. This is the only way in which we will succeed in saving lives and preventing natural hazards from becoming devastating disasters wiping out lives, livelihoods and destroying hard won gains of development.

References

Asian Development Bank (2012) Learning lessons: ADB's response to natural disasters and disaster risks. ADB, Manila

Asian Development Bank (2013) Investing in resilience: ensuring a disaster-resistant future. ADB, Manila

International Federation of Red Cross and Red Crescent Societies (2003) World disasters report 2003: from risk to resilience, helping communities cope with crisis. IFRC, Geneva

International Federation of Red Cross and Red Crescent Societies (2013) Stronger together: the Global Red Cross Red Crescent response to the 2004 Indian Ocean earthquake and tsunami. IFRC, Geneva

Maplecroft (2013) Climate change and environmental risk atlas. Bath, UK

Ministere des Affaires etrangeres et europeennes (2007) Indian Ocean tsunami 2004: 10 lessons learnt from the humanitarian response funded by the French State, Groupe URD: Francois Grunewald, Beatrice Boyer, Dr Hugues Maury and Peggy Pascal

Office of the UN Secretary General's Special Envoy for Tsunami Recovery (2006) Key propositions for building back better, lessons learned from tsunami recovery

Planning Commission, Government of India (2007) Eleventh Five Year Plan (2007-2012) volume I inclusive growth. Planning Commission, New Delhi

Shyam KC (2013) Cost benefit studies on disaster risk reduction in developing countries, EAP DRM Knowledge Notes. Working Paper No. 27, World Bank

United Nations Development Program (2006) Local governance in tsunami recovery, lessons learned and emerging principles. UNDP Regional Centre, Bangkok

United Nations Development Program (2010) Disaster risk reduction, governance and mainstreaming. UNDP, New York

United Nations Economic and Social Commission for Asia and the Pacific, United Nations International Strategy for Disaster Reduction (2012) Reducing vulnerability and exposure to disasters, the Asia Pacific disaster report 2012. UNESCAP and UNISDR, Bangkok

United Nations International Strategy for Disaster Reduction (2005a) Hyogo framework for action 2005-2015, building the resilience of communities and nations to disasters. UNISDR, Geneva

United Nations International Strategy for Disaster Reduction (2005b) 10 preliminary lessons learned from the Indian Ocean Tsunami of 26 December 2004. UNISDR, Geneva

United Nations International Strategy for Disaster Reduction (2009) Global assessment report 2009. UNISDR, Geneva

United Nations International Strategy for Disaster Reduction (2013a) Implementation of the Hyogo framework for action—summary of reports 2007-2013. UNISDR, Geneva

United Nations International Strategy for Disaster Reduction (2013b) Global assessment report 2013. UNISDR, Geneva

World Bank (2012a) Acting today for tomorrow: a policy and practice note for climate- and disaster-resilient development in the Pacific Islands region. World Bank, Washington

World Bank (2012b) Executive summary of knowledge note: learning from megadisasters. World Bank, Washington

World Bank (2012c) Sendai report: managing disaster for a resilient future. World Bank, Washington

Chapter 20
Role of GI Services in Emergency Response Management in India

Ajay Srivastava

Abstract Emergency response management is not a play of single agency or organization. Involvement of different agencies and organizations depend on the type and severity of the disaster. Disaster management organizations, agencies and emergency management personnel must be prepared in advance with all types of data and resources. GIS based systems and solutions allow the integration of data from various resources and provide a single window for viewing, querying and analyzing information. Applying models and recognition of patterns allows the detection of possible threats and provides an efficient tool for supporting decision makers in responding to any disaster. Analytical and modeling capabilities of GIS enable generation of new information, and permit the integration of data from multiple sources and thus improve the workflow in all phases of emergency management and support the requirements of the emergency management. At present, many initiatives have been taken by Government of India to develop trained and qualified responders supported by enhanced IT-GIS infrastructure. Development of a disaster management system with operational national operations center on its mission to carry out its daily operations is under development to support all the stakeholders during the time of disasters.

Keywords Crisis management • Disaster recovery • GIS • ICT • IT

20.1 Introduction

Disaster preparedness to a large extent is spatial in nature. An effective and real-time disaster management programme requires spatial data from various sources which should be collected, manipulated, analyzed, and displayed in a meaningful manner. A successful, practical and real-time disaster preparedness and management needs an up-to-date geographically tagged database. Geospatial analysis of frequency and magnitude of meteorological and geological extreme events enable us to determine the most frequent disaster prone areas. Detecting anomalies in sensor data based on pattern recognition using network-related information and analyzed

A. Srivastava, Ph.D. (✉)
GISP, Noida, G.B. Nagar 201301, U.P., India
e-mail: ajaysrivast@gmail.com

R. Shaw (ed.), *Recovery from the Indian Ocean Tsunami: A Ten-Year Journey*,
Disaster Risk Reduction, DOI 10.1007/978-4-431-55117-1_20

for any response planning is easier than before now. It is also possible to identify the areas more frequented by a disaster and also assess the combined vulnerability in a geospatial environment.

Time saved means lives saved, and IT-GIS systems enable us to quickly organize the massive volumes of mission-critical data created after a disaster and clearly present it to first responders and relief agencies. Agencies can then access the catalog and relevant map displays, and consume data and services in a variety of applications and mobile devices. Emergency Response relies on industry standard routing and addressing procedures. Most civilian agencies at local, regional, state level need to adopt these procedures and capabilities. Information & Communication Technology (ICT) adoptions have helped in creating a large network of seismographic centres, national warning centres, coastal and deep-ocean stations in place across the Indian Ocean to detect potential tsunamis and pass on warning messages to communities. Now, Indian Ocean Tsunami Warning and Mitigation System had also started functioning since March 2013 which made possible to issue timely warning of a possibility of a Tsunami and execution of rescue and evacuation plans.

20.2 Remote Sensing and GI Services

In the recent times, Geographic Information (GI) Services have been very effectively utilized for all phases of emergency management—planning, mitigation, preparedness, and recovery. Most of the emergency data requirements are of spatial nature hence Geographic Information System (GIS) can be utilized as an underpinning framework for integrating all kinds of relevant information for disaster preparedness. GIS is also an effective tool for emergency responders to access information in terms of crucial parameters for the disaster affected areas. The crucial parameters include location of the public facilities, communication links and transportation network at national, state and district levels. It is therefore desirable to have the right data in the right place at the right time to assess vulnerabilities. The data should be organized in a usable format for stakeholders to analyse, respond and take actions in case of an emergency. Geo-visualization, analysis and data dissemination capabilities make critical information readily available to all the stakeholders involved during the response phase. Advancements in the communication technologies, GPS, sensors, mobile-based GIS applications, social media and geo-portals have also greatly enhanced the coordination of response efforts (Shepherd 2006). GIS can provide the backbone during the entire disaster management cycle by providing data and information when and where it is needed using adequate and effective telecommunication infrastructure (Simons 2010).

- GIS technology is a powerful tool for:
  - Assessing potential emergencies and their probable location
  - Their potential impacts, damage and losses
  - Identifying critical infrastructure and populations at risk that require priority mitigation actions

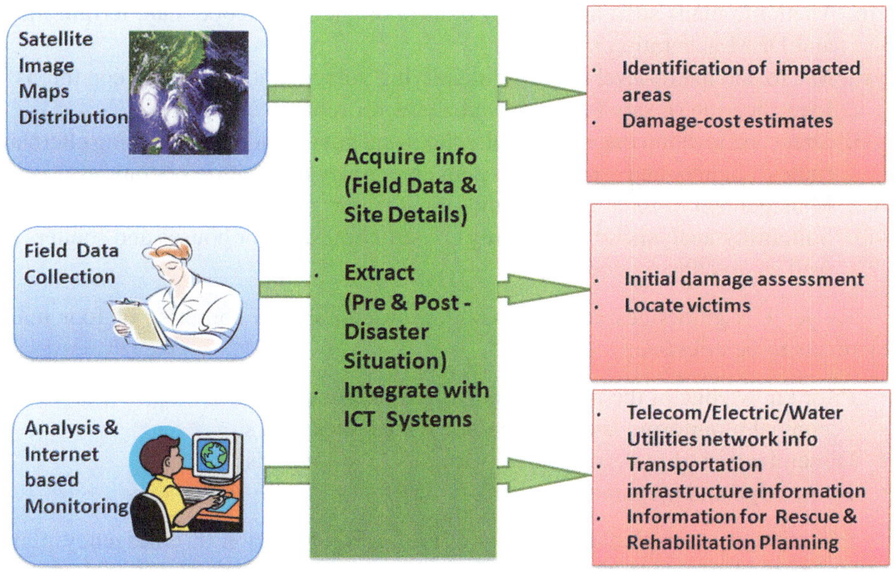

Fig. 20.1 Use of GIS and ICT in situational analysis and recovery planning

- GIS is essential for creating situational awareness for daily operations
  - Incident locations, tracking, sensors, video, traffic, hospital status, weather and other dynamic data integrated with GIS data (images, streets, critical infrastructure, etc.) to produce "Actionable Information"

20.2.1 GI Services as Disaster Response Tool

The true value of spatial information is realized when integrated with other relevant information for the purposes of gaining a better understanding of the combined dataset by relating it to a common location (Fig. 20.1). It enables those crucial questions to be asked:

- What lies at a certain height above mean sea level?
- What number of incidents happened at a certain location?
- What is the distribution of current clouds, wind and recent rainfall patterns?
- What structures, buildings are adjacent to a particular river, beach or road?
- Or what are the topographical characteristics of a particular area?

GIS database with various themes is helpful to disaster managers in decision making process when catastrophic event occur. GIS database include following information to assist in disaster management.

i. Use of different types of satellite images ex. Quickbird, Geoeye, Worldview, Cartosat etc. for GIS data creation
ii. Preparation of base maps using satellite images

iii. Thematic maps such as hydro-geomorphologic map, slope map, terrain map, and DEM generation in GIS
iv. Macro and micro-level maps for identifying vulnerability and threat conditions
v. Identification of safe locations and zones for rehabilitation
vi. Street and location maps (hospital, school and govt. building) for finding alternate routes, shelters and locations to plan emergency evacuation operations
vii. Management of rehabilitation and post-disaster reconstruction
viii. Suitability analysis for identifying construction sites for houses and shelters as well as no construction areas

As per UN guidelines an early warning system should comprise of four main elements to be effective:

- Prior knowledge of the risks faced by communities
- Technical monitoring and warning service for these risks
- Dissemination of understandable warnings to those at risk
- Knowledge and preparedness to act

Effective disaster planning relies on thorough integration of emergency plans at all levels of government. IT and GIS can provide valuable support backed with easily accessible near real-time data for planning and disaster response operations. IT-GIS systems help manage operations, increase efficiencies, and track scientific progress more accurately so that agencies can quickly respond when any area is threatened. The enterprise database delivers secure and unified data in a universal format for viewing and updating spatial data. This promotes better information sharing, coordination, and reuse, while preventing errors, and risks that arise when different agencies work in silos. Thus, geospatial solutions improve staff collaboration and communication during disaster planning operations.

An ideal disaster response support system should be based on GIS technology that manages and utilizes data generated during a disaster in the form of geospatial information. It also includes remote sensing technology for data gathering, and search and distribution technologies for communication. The system collects and collates information about the disasters and presents it on a map to provide visual representation of the situation. This includes information acquired from the disaster site or from associated agencies, and also meteorological observations, satellite images, aerial photographs, and other sensor data (Fig. 20.2).

GIS analysis functionalities provide rescue and recovery teams productive tools to proactively design evacuation and recovery actions. With precise spatial intelligence at their fingertips, officials can respond better by accurately predicting any event impact for a particular locality. Authorities can use enhanced imageries and accurate terrain data to update attributes in GIS. They can deliver and distribute that geospatial information via a geo-portal, and visualize Digital Surface Models (DSM) and create models for scenario-based analysis. Computerized numerical models combine that information with a complete view of the geospatial data and create precise 3D maps of the area of interest. Authorities can use GIS to simulate the impact of a given storm on city properties. The information contained within

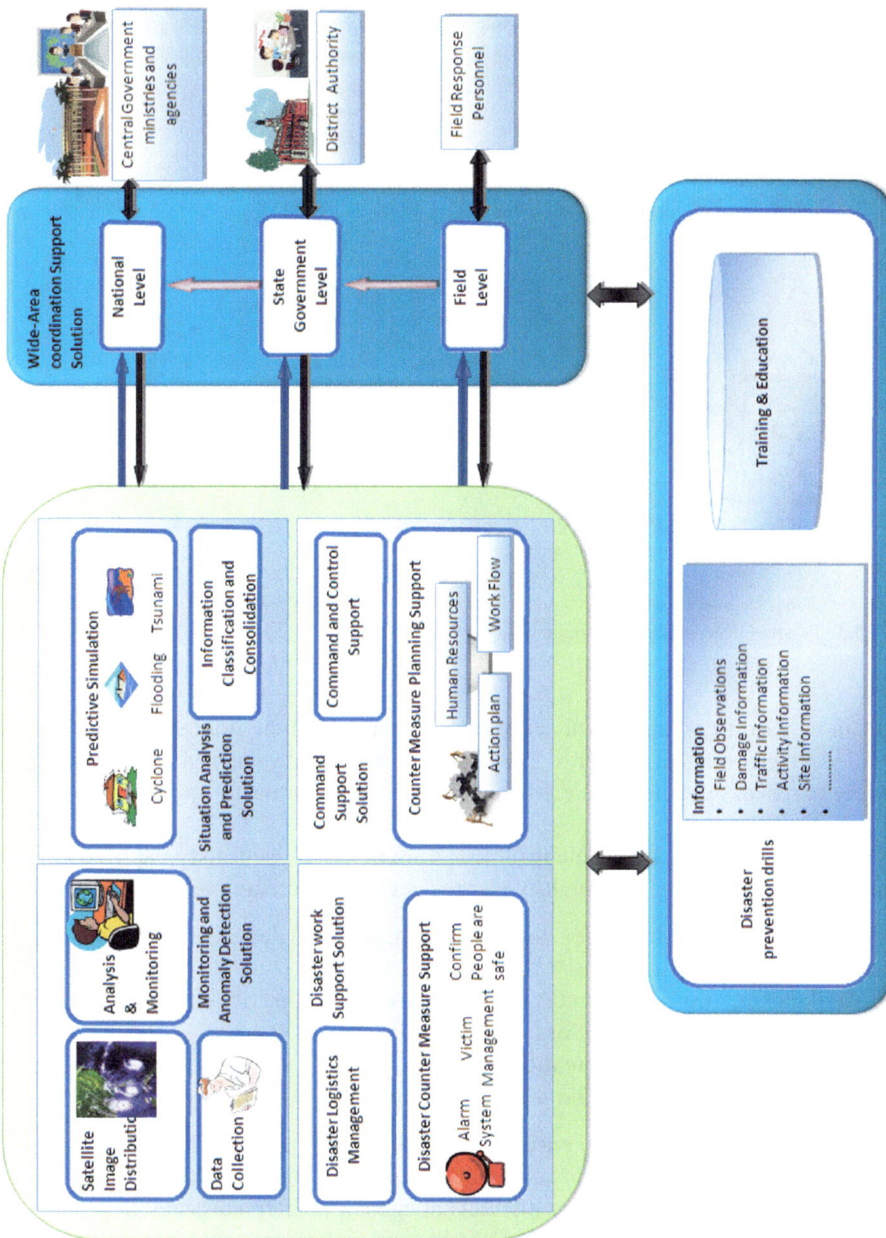

Fig. 20.2 A decision support framework for disaster management planning

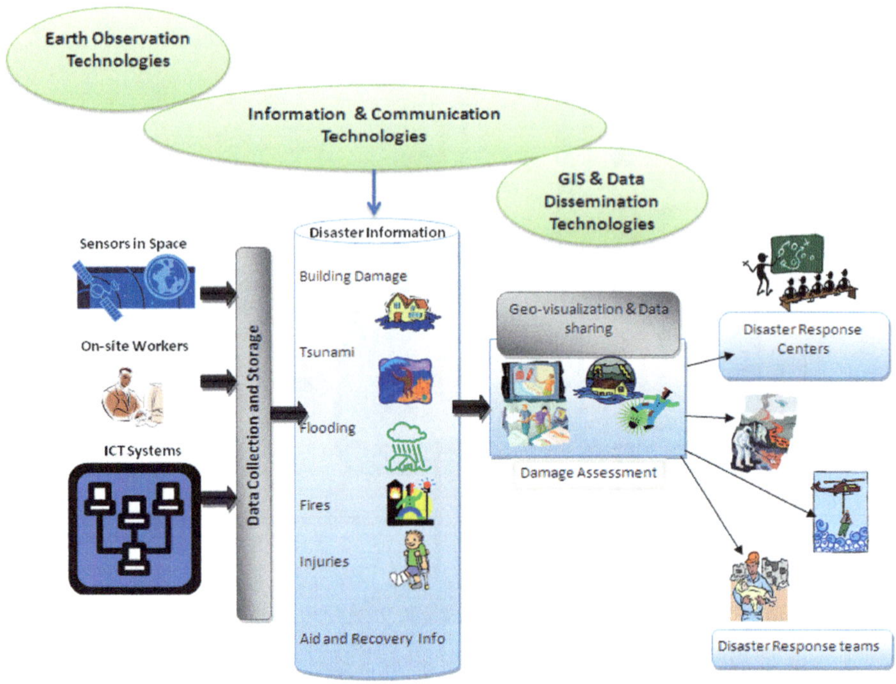

Fig. 20.3 IT and GIS in disaster management

GIS can be used to predict whether a particular parcel will be flooded based on its climatic and environmental records. The result—a geospatially enabled understanding of every parcel of city and privately owned land enhances proactive planning efforts and drives efficiency.

Disaster prevention management solutions are required that satisfy the following requirements for dealing with multiple large disasters occurring over a wide area.

i. Decision support solution for faster and more accurate decision making
ii. Command support solutions
iii. Wide-area coordination support solution for coordinating activities and sharing information across different organizations
iv. Disaster management information system
v. Training and education support solutions to improve disaster response capabilities

An essential requirement for ensuring a rapid reaction and an effective response when a disaster occurs is to make easily understandable information about the situation available and to share it with relevant agencies. After a disaster, imagery is traditionally collected from a host of platforms and sensors mounted on satellites or aircraft. Due to the large size of the image products, requested images are traditionally received and distributed via external hard disk. This traditional distribution system encourages unnecessary duplication of data, delays in the receipt of needed data, and the creation of non-standard GIS products (Fig. 20.3). Nowadays, crowd-sourcing data about city infrastructure is a new technique being used by forward-looking municipalities where incident information, can be made available to citizens as a free mobile apps that allows them to report issues. Municipalities can subscribe to receive the input by e-mail or through portals using the web services.

20.2.2 Decision Support Solution for Situational Analysis

GIS as well as Remote Sensing (RS) application cannot be developed from a single dataset. Hence, there is a move from information systems to information infrastructure (Georgiadou and Harvey 2007). Decision support solution collects information from sources such as sensors (seismometers, river level gauges, surveillance cameras, unmanned aerial vehicles, satellites, and so on.) and internet social networking services to enable functions such as status monitoring in which this collected data is integrated with geospatial information, and anomaly detection is performed using techniques such as difference extraction (Fig. 20.3). It uses information classification techniques to classify and collate data based on similarities and categories, making it easier to obtain the desired information. This can be used to generate intelligence such as warning that a particular area is at risk of a disaster in an hour's time.

20.2.3 Command Support Solution for Planning

Command support solution supports effective and efficient command and control for relief and recovery. For ex., it provides the disaster response headquarters with a map of the disaster situation that they can refer to as they assign people, organizations, and other resources in accordance with the evolving situation on the ground. This solution builds a database from which the data required for the tasks associated with the event and their execution can be accessed quickly based on an event model of the time when the disaster strikes, a disaster response model, a datamodel that specifies the relationship between data, and a disaster management model that links these other models together. This allows the "push" delivery of information based on user's circumstances and responsibilities.

Providing information with a reference map to associated agencies helps involved agencies share a Common Operational Picture (COP). Other important uses include identifying locations where major damage has occurred in order to provide the prompt assistance, and enabling different agencies to mount a well coordinated response.

20.2.4 Communication Support Solution

Such solutions use a GIS to supply processed information with stake holders for planning and support. The objective is to adopt and establish the industry standard which includes assigning street names, street ranges and addresses for all applicable facilities and structures. By sharing common information, the solution provides a consistent understanding of the situation while also presenting information in ways that suite the different response levels of each organizations. Information can be also provided in a range of forms and formats to facilitate its distribution, use and reuse. Geospatial technology can help authorities capture, process, deliver, and share data. This data can then, in turn, be used to plan and prepare for disasters. After disaster strikes, geospatial data can be used across divisions and departments to respond and rescue.

Key components of the system are the tabular and geospatial addressing data for the Computer- Aided Dispatch system. In addition, information sharing and collaboration for lasting partnerships is a key factor to the effectiveness of such systems. Such communication and coordination support system will allow affected citizens to communicate with emergency response personnel during time of an emergency by exchanging some of the following types of information:

From the citizen:

- Current location
- Current condition
- Need for assistance
- Request for status update
- Upload photo/video

From the EOC:

- Push specific messages to geo-targeted areas
- View individual citizen record
- Initiate contact directly with citizens
- Monitor groups based upon information provided

20.2.5 Disaster Management Information System

The Disaster Management Information System (DMIS) are developed in a web based environment to make them accessible from anywhere and more user friendly. GIS is an integrated and essential part of the system in order to display generic datasets. The overlay functionality of the GIS allows various databases and datasets to be overlaid in order to provide a valuable and powerful decision-making tool for the decision makers and other municipal stakeholders.

The system relies on data and different databases for disaster risk assessment. Spatial and attribute data is joined to provide a complete picture of disaster risk or what to expect whenever a disaster occurs. Vulnerability and hazard maps is a key output of the system as this can provide agencies with a better understanding of the dynamics of a specific ward. The DMIS is enabled to do hazard identification, vulnerability assessments, risk assessments, disaster risk reduction as well as planning for an incident. In order to conduct the above a vast amount of GIS information is required which ranges from baseline information (such as infrastructure, land use, topography and hydrology) to stakeholder contact details for contingency planning.

A functional element of the DMIS is the integration of various checklists and workflows which allow users to track tasks and responsibilities of all role-players. These checklists guide the user through a planning process which ensures that all information requirements are supplied by the various municipal departments and entities. Furthermore, all standard operation procedures, legislation, policies and by-laws documentations are incorporated in the system for ease of use and reference.

Any modern DMIS should have a SMS and E-mail facility to enable communication between all stakeholders and send bulk emergency alerts. Dash board type reporting to the planners and top management assist in keeping everyone on same page and ensure coordinated actions of all role-players. Recently, the Google Crisis Response Team has created 'Public Alerts' and 'Crisis Map' for some countries (Colombia, Australia, Canada, Japan, Taiwan, and the United States) to help people better prepare for these unfortunate situations. Such GIS based services provide people with access to useful information before, during, and after a natural disaster such as a tropical storm, hurricane, flood or landslide. Relevant information about extreme weather changes which threaten the safety appears on Google Public Alerts as well as emergency related information for impacted areas on Crisis Map. This information is also displayed in Google Search, Google Maps, Google Maps Mobile and Google Now.

20.3 Recovery Efforts and IT Implementations

To support the total cycle of disaster management for the country, in near real-time, the database creation is addressed through National Database for Emergency Management (NDEM). NDEM is envisaged to have core data, hazard-specific data, and dynamic data in spatial as well as aspatial form. Airborne Laser Terrain Modeling data acquisition is being carried out for the flood prone basins in the country. The Disaster Management Support (DMS) Programme of Govt. of India (GoI), addresses disasters such as tsunami, flood, cyclone, drought, forest fire, landslide and earthquakes. These include creation of digital database for hazard zonation, simulation and modeling, post-disaster damage assessment, monitoring of major natural disasters using satellite and aerial data; development of appropriate techniques and tools for decision support, establishing satellite based reliable communication network, deployment of emergency communication equipments and R&D towards early warning of disasters.

20.3.1 Recovery Planning Initiatives Using GIS

It has been accepted that GIS has the capability to create and efficiently integrate and manage strategic information such as topographic, bathymetric data, satellite images, vector maps, mesh data, transportation network data etc. It is a system to support the mission to provide and distribute operation maps from the collection, analysis to evaluation, and management of marine information processing. Satellite imagery has become an important tool for decision makers in getting alerts for disasters and in assessing the situation pre and post-disaster (Nanda Kumar 2012). GIS analysis and 3D-visualisation techniques can generate event-based scenarios to make full use of modeling and simulation technologies.

Post-disaster redevelopment plan can provide communities the opportunity to change previous development decisions that may no longer be desired, leap forward in implementing its vision for the future, and become more resilient to disasters by avoiding or mitigating development in hazardous locations. Implementing changes in land use after a disaster also must recognize private property rights and the financial burden that disaster survivors may face if required to rebuild to a higher standard.

20.3.2 Implementation of Early Warning Systems

Indian Meteorological Department (IMD) is mandated to monitor and give warnings regarding Tropical Cyclone (TC). Monitoring process has been revolutionized by the continuous availability of remote sensing satellite data. A cyclone intensity analysis and forecast scheme using satellite image interpretation techniques can facilitate forecasting of storm surges. Similarly, meteorological satellite has also made a tremendous impact on the analysis of cyclones. All developing cloud clusters are routinely observed through satellite cloud imagery & those showing signs of organisation are closely monitored for signs of intensification. IMD has installed Doppler radars and specially designed receivers within the vulnerable east coast areas for transmission of warning using broadcast capacity of INSAT satellites. There are 352 Cyclone Warning Dissemination System (CWDS) stations along the Indian coast, of which 100 digital CWDS are located along the Andhra Pradesh coast (CAG Report 2013). Similarly, satellite integrated automated weather stations have been installed on islands, oil rigs and all along the exposed coastal sites. Buoys for supplementing the surface data network in the tropical ocean have been deployed. The Government has also started a National Data Buoy Programme (NDBP) by deploying moored buoys in the northern Indian Ocean to provide meteorological and oceanographic data.

20.3.3 Indian Ocean Tsunami Warning and Mitigation System

As on March 2013, the Regional Tsunami Service Providers (RTSPs) of Australia, India and Indonesia are the only official source of tsunami advisories for the Indian Ocean. Indian Ocean tsunami early warning system was designed to combine data from underwater probes, orbiting global positioning system satellites, and floating buoys, to better detect a coming tidal wave. Within 2 min, we can now calculate the speed and know exactly when waves will reach the shore. Messages could be sent out immediately after the calculation. For emergency situations that become disasters, the scalability of GIS is invaluable in coordinating efforts of participating agencies. State government of Andhra Pradesh on 11 April 2013 launched a GPS-GIS based Central Emergency Response System in Hyderabad. The emergency response system is designed to enable police reach the area of complaint within minutes of receiving the complaint. The project will utilize the GPS-GIS based state of the art emergency response system that will operate on a single command and control centre for complete state.

20.4 Developments in Communication Systems

It has been observed that in any hazard, communication is the first casualty. It has been therefore decided to put in place multi-mode and multi-channel communication systems so that enough redundancy is available at the Emergency Operation Centers (EOCs). National Disaster Management Communication Plan to provide satellite based mobile voice/data/video communication between National/State/District EOCs/Mobile EOCs and remote emergency sites is under implementation and completion is expected soon.

20.4.1 India Disaster Resource Network

One major step towards strengthening of response system has been the launch of India Disaster Resource Network (IDRN). The network will ensure quick access to resources to minimise response time in emergencies. The system will give, at the touch of the button, location of specific resources as well as the controlling authority for that communications system resource so that it can be mobilized for response in the shortest possible time. The database will be available simultaneously at the district, state and national levels (CAG Report 2013).

Figure 20.4 is a graphical representation of the proposed information flow of mission-critical information between the different role-players in the case of an emergency. The flow of information is directly related to the physical set-up within the EOC. For purposes of coordination and different levels of decision-making the information flow is divided into two sections: information for tactical decision-making and information for strategic decision-making. The States are also being persuaded to set up emergency operations centres at the state and district levels for early warnings and hazard monitoring (Government of India, Ministry of Home Affairs Report on Disaster Management in India 2004). The control rooms, which will function round the clock, will be composite control rooms to look after law and order issues as well as disaster management.

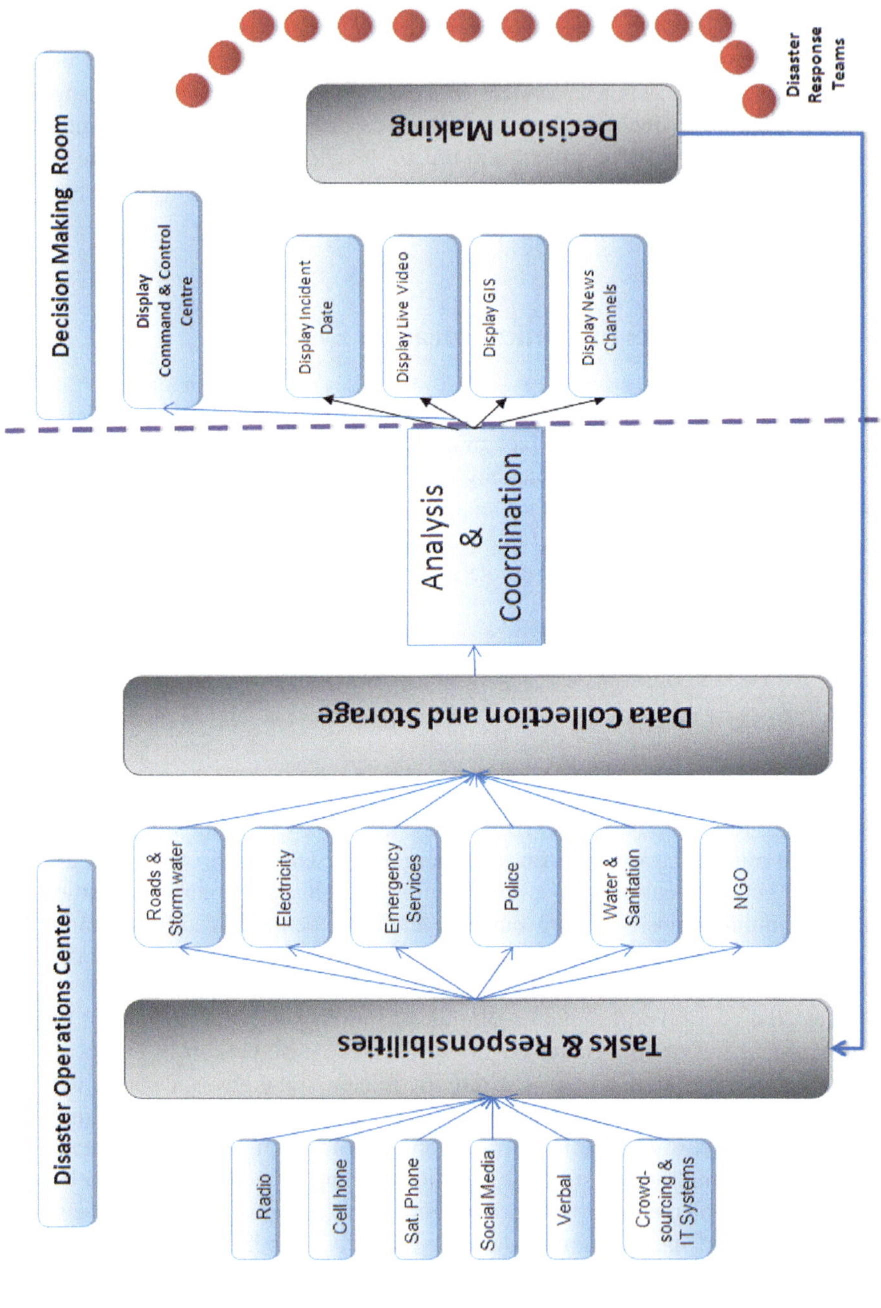

Fig. 20.4 Ideal communication flow in emergency operations centre

20.4.2 Incident Command System

Incident Command System provides for specialist incident command teams with an Incident Commander and officers trained in different aspects of incident management—logistics, operations, planning, safety, media management etc. The incident Command System has been finalized keeping in view the systems and procedures prevalent in India by dovetailing it in the existing governmental machinery already in position.

20.5 Status of the Major Projects

Disaster risk management has become the major challenge the entire world is faced with currently owing to the ever growing population, climate change, and the ever changing environment in general. Natural and man-made disasters result in economic as well as environmental losses which places a need for systematic approach to management of risks. Certain performance indicators (Table 20.1) have been identified to guide various organisations in establishing such systems.

Primary risk-reduction measure undertaken was the introduction of coastal buffer zones, which were areas of coastal land where construction was prohibited to provide protection from hazards. A new land-use zoning policy has been introduced

Table 20.1 Enablers for disaster management planning

S.N.	Action	Performance indicator
1	Communicate disaster risk management data and information to all relevant Divisions/departments/municipal entities.	Divisions/departments/municipal entities are communicating disaster risk management data and information on an on-going basis with written evidence.
2	Draft standard warning messages to be disseminated in the media for specific risks pertaining to the core function of the division/department/municipal entity.	Standard warning messages for specific primary, secondary or tertiary responsibility risks is drafted.
3	Diligently record all incidents, hazardous impacts and disasters in order to establish a historical record for forward planning.	Historical records of hazardous impacts and incidents pertaining to division/department/municipal entity responsibility are maintained.
4	Establish communication links.	Establish communication links with all relevant disaster risk management focal points.
5	Facilitate the development of response and recovery plans.	Contingency plans for prioritized risks have been developed by all divisions and municipal entities.
6	Maintain public awareness mechanisms.	Implement regular public awareness strategies in line with its indicative disaster risk profile.

Table 20.2 Status of various components of disaster management support programme and other communication programmes

Projects	Purpose	Start date	Status
National Database for Emergency Management (NDEM)	Sat. images and ground data	2006	Incomplete
Airborne Laser Terrain Mapping and Digital Camera System (ALTM-DC)	Mapping flood prone areas	2004	Incomplete
Disaster Management Synthetic Aperture Radar	Flood warning and tracking	2003	Incomplete
Satellite-based Communication Network for Disaster Management	From PMO to district level	2005	Incomplete
Doppler Weather Radars	Predicting cyclone and storms	2006	Incomplete
National Disaster Communication Network	Early warning	2007	Incomplete
National Disaster Management Informatics System (NDMIS)	GIS based DSS	2008	Incomplete

Source: CAG Report 2013

as a result of the lessons learnt from the tsunami experience. A project for this purpose is being drawn up with a view to utilize these technologies for the benefit of people in the coastal areas. The database will provide multi- layered maps on district-wise basis. These thematic maps along transportation network, hospital etc. will enable the district administration to carry out hazard zonation and vulnerability assessment, as well as coordinate response after a disaster. In the event of a disaster it is handy to know different routes through which affected people can be evacuated and relief material and medical aids can be dispatched. Identification of safer locations in case of evacuation of people from affected area is very important in the event of any disaster. Summary of important projects undertaken by the Government of India is given in Table 20.2.

Details about these projects, funding and implementing agencies can be found in the CAG Report 2013. Most of these projects were scheduled to be completed by the end of the Eleventh Five Year Plan ie. 2007–2012 but would be now continued during Twelfth Plan.

20.6 Disaster Management Capacity Building

Any IT system including early warning system are only as good as the nation's capacity to respond promptly to its messages. Therefore, we should look at the disaster management systems as a whole: preparedness, mitigation and prevention, response, and recovery. These elements inevitably cross into areas covered by government departments, civil societies and academia. No one is as yet satisfied with the manpower capacity and availability in the country. This may be because there is lack of institutions offering relevant courses to train manpower who understand latest trends in ICT and its application in domains like mathematical modeling, meteorology and Earth System science.

The goal of any Disaster Management Training Program should be to increase disaster management capacity among staff of government and non-government organizations, universities, private sector, other organizations and communities throughout the country. Increased capacity will contribute to the overall aim of safer and more resilient communities throughout the country.

Capacity building programmes should focus on the following objectives to improve preparedness for natural disasters:

i. Upgrade the capacity of people implementing disaster management activities
ii. Train a core group of disaster management master instructors/experts
iii. Conduct disaster management research
iv. Increase the public's awareness about disaster management

20.7 Conclusion

Warnings are useless unless hazard-resistant socio-economic infrastructure is in place. People need to know how to be prepared for disaster and how to react if a warning is given, including where to go to for safety. Presently the accessibility to spatial and non-spatial information lying under the control of many public sector organizations is a dilemma as plenty of information such as topography, demography, health, energy, climate, agriculture, biodiversity, water and weather exists in the organizations but it is not shared with other stakeholders when required. A common GIS-based framework makes data maintained by various agencies readily accessible so that timely information can be brought to bear on the developing situation. This framework also centralizes and organizes data for analysis and visually displays critical information during an emergency. Decisions that direct response efforts require answers to geography-based questions. Disaster creates opportunity for new development by destroying some existing development; however, the map is not blank since it comes with pre-existing property boundaries, competing interests. Local governments must also be kept in the data loop; seamless emergency response relies on data inside and outside the government agencies. As the authoritative source for addressing, local government entities may include the data, addressing office, planning department, or any GIS personnel. By working together, the multiple tiers of government agencies can accomplish the common goal of having the most timely and accurate life-saving data.

Based on the lessons from the 2004 Indian Ocean Tsunami, it is anticipated that even greater steps will be taken at a nationwide level to establish organizational capabilities and infrastructure to provide facilities and systems aimed at mitigating disasters in India. In particular, in the case of large disasters that affect wide area and in which the situation changes rapidly with time, it is vital that an effective response be mounted that include coordination with national and regional agencies, as well as the general public, in order to reduce the amount of damage and speed up the subsequent recovery and reconstruction. GI Services can be used to improve emergency response systems although it is no easy to standardize data and work together to make disparate data sources and standards compatible to each other.

References

CAG Report (2013) Performance audit of disaster preparedness in India. http://www.indiaenvironmentportal.org.in/files/file/cag%20report%20on%20disaster%20preparedness.pdf

Georgiadou P, Harvey F (2007) A bigger picture: information systems and spatial data infrastructure research perspectives. In: Wachowicz M, Bodum L (eds) Proceedings of the 10th AGILE international conference on geographic information science: the European Information Society: leading the way with geoinformation, Aalborg, 8–11 May 2007, AGILE, 2007, p 6. ISBN:978-87-819-3004-4

Government of India, Ministry of Home Affairs Report on Disaster Management in India (2004) National Disaster Management Division/MHA/GOI/28/06/2004. p 98

Nanda Kumar T (2012) Managing disasters in India, Yojana, vol 56. March 2012, p 5–10

Shepherd B (2006) The role of GIS services, data and portals in disaster management-planning, response and recovery. In: Seventeenth United Nations regional cartographic conference for Asia and the Pacific, Bangkok, 18–22 September 2006. http://unstats.un.org/unsd/geoinfo/RCC/docs/rccap17/ip/17th_UNRCCAP_econf.97_6_IP17.pdf. Accessed 27 Jan 2014

Simons W (2010) GIS for disaster preparedness and emergency response, ITU training workshop on disaster management. http://www.itu.int/ITU-D/emergencytelecoms/events/kigali/presentations/emergency_management_rwanda.pdf. Accessed 27 Jan 2014

Chapter 21
Healthy Ecosystems for Long Term Security and Sustainability of Natural Resource Management: Case of India

Irene Stephen Ravindran

Abstract Lessons from the devastating Indian Ocean tsunami of 2004 offered an opportunity for India to assess and monitor the resilience of its costal line and ecosystem. Which are rich in biodiversity, marine and land based resources relatively fragile to such extreme disasters. The framework discussed in this paper is associated with the ecosystem based adaptation approach that link ecosystems and coastal management for reducing risks from natural hazards. Broadly this framework integrates the management and use of ecosystem services and its conservation, as an important natural resource. This approach in India adopts the sustainable management, conservation, and restoration of the coastal ecosystem to provide services which help people adapt to both adverse environmental variability and its changing impacts.

Keywords Coastal management • Coastal zone regulation • Coastal zones • Conservation • Ecosystem • Ecosystem based adaptation • Ecosystem services • Natural resource

Abbreviations

CRZ	Coastal Regulation Zone
CZMP	Coastal Zone Management Plans
EIA	Environment Impact Assessment
ICRI	International Coral Reef Initiative
ICRZ	Island Coastal Regulation Zone
IPZ	Island Protection Zone
MFF	Mangroves for Future
NFF	National Fish workers Federation
SEZs	Special Economic Zones
SICOM	Society of Integrated Coastal Management

I.S. Ravindran (✉)
New Delhi, India
e-mail: irene.stephen@gmail.com

R. Shaw (ed.), *Recovery from the Indian Ocean Tsunami: A Ten-Year Journey*, Disaster Risk Reduction, DOI 10.1007/978-4-431-55117-1_21

21.1 Introduction

No one would argue that the seas have a "right" to violate their established boundaries and crash across the coastline. In fact, people agree that disaster occurs whenever the sea breaches the shoreline. On December 26 2004, an earthquake shook the whole earth. Many people didn't feel it, but the earthquake transformed the Indian Ocean into a catastrophic oscillation called "tsunami", felt across the South Asian region and parts of Africa. Approximately 2,260 km of the India's coastline along Tamil Nadu, Pondicherry and Andaman and Nicobar Islands was damaged. And more localized damage was along the coasts of Kerala and Andhra Pradesh. Environmental impacts of the tsunami varied according to factors—notably the bathymetry and geomorphology of the coastline, presence of natural protective features and differences in the topography of India's east and west coasts.

21.1.1 Coastal Ecosystem: Protective Shields

Nature has placed the sand as the boundary of the sea and provides a safety net in times of crisis. The presence of sand dunes and plantations at most of the tsunami affected locations played a vital role in protecting coastal villages in Tamil Nadu, India. Anecdotal reports of the buffering role of mangroves, reefs, sand dunes—the first line of defence from the tsunami, were sited during the post tsunami assessment. There is evidence, that coastal dunes have provided protection from tsunami. Examples of "South of Cuddalore, areas adjoined to the river mouths of Vellar, Chinna Vaikal (Pichavaram) and the Coleroon river that drains into the Bay of Bengal were severely damaged claiming more than a 1,000 lives" (Ministry of Environments and Forests, Ministry of Earth Science 2005). Parangipettai village (adjacent to the Vellar river) witnessed maximum inundation upto 2.5 km as the initial terrain slope (from the coast) is very gentle and the far reaching areas are low lying. The villages further south of the Vellar inlet and its backwaters with mangrove vegetation acted as a barricade reducing the wave energy, otherwise the intensity of the damage could have been much more severe. The Vellar backwaters are a part of the Vellar-Pichavaram-Coleroon forest area (Ministry of Environment and Forests, Ministry of Earth Science 2005). Appreciation by the local community of the protection provided by the mangroves has lead to local acceptance of the need to conserve them. Before December 2004 the level of exploitation of the Pichavaram mangroves far exceeded the sustainable levels, which lead to rapid degradation of the habitat (Ministry of Environment and Forests, Ministry of Earth Science 2005) Therku Payyanallur looks like any other coastal area in Tamil Nadu. But on 26 December 2004 the people living there discovered a special feature of their ecology: sand dunes, 30-ft high, and stretch for 5 km acted like a protective fortress against the tsunami. The water did not enter the villages because of this natural shield. But erosion and human encroachment have taken their toll on the dunes. Only after the Tsunami did the community realise how important it was to restore and safeguard them (United Nations 2008).

21.1.2 Coastal Areas-Living on the Edge

Within Nature there is inherent order. Violating it has inherent consequences. Nature has established boundaries for human behavior. Yet not a day goes by without countless violations of nature's law, its values and its opportunities for human use with certain limitations and prohibition resulting in disastrous physical consequences. Amazingly, humans often argue that human's have the "right" to violate these boundaries (MacHarg 1971). As humans we are trespassing boundaries that are safe for humanity, we need to identify the safe operating limits for the planet, we need to understand what those planetary boundaries are (Sach 2011). For long coastal ecosystems have a standing connection to and benefits derived from its services and protection. But coastal ecosystems have not always fared well at human hands. Changes in land use and pollution from land has degraded critical habitat functions of estuaries, lagoons, mangroves, coral reefs, sea grass beds, backwaters, salt marshes, wetlands, rocky and sandy beaches. While commercial overexploitation of marine life has driven out the rich tapestry of marine life into loss of biodiversity and eventually into extinction (Sach 2011).

21.1.3 One Coastal Ecosystem: Many Services

The tsunami had illustrated powerfully the vulnerability of human populations that choose areas close to the shoreline to establish their settlements. However, for some communities, this choice is somewhat pre-determined by the nature of their occupations and sources of livelihood. The coastal ecosystem of India plays a vital role in the local economy by virtue of its valuable resources, rich biological diversity and productive habitats. There are 77 coastal districts out of 593 and is home to 17 % of the total population of India. It is estimated that nearly 25 % of people live within a distance of 50 km from the coastline. There are large cities and townships along the coasts of India, including some of the largest—Kolkata, Mumbai, Chennai, Kochi and Vishakhapatnam which support a large section of people in poverty, the section most vulnerable (UNISDR and UNDP 2012). Fishermen mainly depend on coastal resources for their survival. The coast line includes the Economic Exclusive Zone (EEZ) extending up to 2.02 million km^2 with several activities such as tourism, industries-oil, mining, fisheries, aquaculture, harbors, ports, marine protected areas, breading and nesting sites for marine fauna, and structures of archaeological importance and heritage sites that dote the coastal stretch. Major rivers such Ganges, Brahmaputra, Krishna, Godavari, and Cauvery, Mahanadi on the east coast, and Narmada and Tapti rivers on the north-west coast discharges its river water into the east and west coasts and marked by deltas, estuarine, beaches, island habitats, mangroves (e.g. Sundarban is about 1,430 km^2one of the largest single block of mangrove) and coral reef and sea grass beds (e.g. mainland coast of Gulf of Kachchh and Andaman and Nicobar Islands, Atolls make up the coral reefs of Lakshadweep) and sand dunes along the entire Indian coast. (Venkataraman 2008) Seagrass beds occur along the southeast coast (Gulf of Mannar and Palk Bay) and in the lagoons of islands from Lakshadweep (Arabian Sea) and Andaman and Nicobar (Bay of Bengal). Marine Protected Areas covers an area of 18.5 % of the islands and 6.16 % of the coastal bio-geographic zone in India (Marale and Mishra 2011).

21.2 Coast at Risk-Multitude of Factors

The December 2004 tsunami reaffirmed the importance of maintaining healthy coastal and marine ecosystems for disaster risk management and post-disaster recovery, as well as for human well-being. Threat to conserve coastal zones of India existed even prior to the devastating tsunami. The convergence of risk factors prior to tsunami was of the direct runoff of pollution from industrial sources and oil refineries, sedimentation due to industrial and port activities, eutrophication from sewage, expansion of dredging of wetland and shore line, destructive fishing practices, excessive harvesting of varied marine life and mining, rise in population density and agglomeration of settlements change the land uses upstream of rivers basins and within coastal zones (Venkataraman 2008). Also the combination of nearly 13 years of poor implementation of the Coastal Regulation Zone[1] (CRZ) Notification 1991 in recognizing coastal regulation for sensitivity coastal zones such as beaches, estuaries, creeks, rivers and backwaters, influenced by tidal action towards landward side and along the water bodies upto 500 m from the high and low tide line (Ministry of Environment and Forests 2009). The notification has been a mix of success and failures as the efficacy of CRZ notification[2] was assessed differently by key influencers. NGOs proactive in protecting coastal environment with local communities felt that the notification is an effective tool and stringent enforcement mechanism must be implemented. While industries, tourism sector and central as well state Government observed the notification to be a means of breach rather than in adherence to support the notification. (Coastal Management Programmes Parliament of India 2009). Though amendments were made, it introduced several relaxations for industrial and construction activities and number of settlements increased along the coast, particularly close to the shoreline. (Coastal Management Programmes

[1] Coastal stretches of seas, bays, estuaries, creeks, rivers and backwaters which are influenced by tidal action upto 500 m from the High Tide Line (HTL) and the land between the Low Tide Line (LTL) and the HTL are categorised as CRZ. HTL means the line on the land upto which the highest water line reaches during the spring tide and shall be demarcated uniformly in all parts of the country by the competent authority.

[2] The CRZ Notification, 1991 prohibits and regulates certain activities within 500 m from the High Tide Line (HTL) on the landward side and within 100–150 m from the HTL on the landward side along water bodies influenced by tidal activities—such as creeks, estuaries, bays, etc. The land use is regulated by classifying the 500-m regulated zone into four categories: CRZ-I (ecologically sensitive and inter-tidal areas), CRZ-II (urban or developed areas), CRZ-III (rural areas) and CRZ-IV (Andaman & Nicobar and Lakshadweep Islands).

Parliament of India 2009) It has been vocally expressed by key influencers including fish-worker associations, environment groups and concerned citizens, that the provisions of the CRZ Notification 1991 must be amended. Coupled with 25 major amendments to the notification, countless petitions pending in courts on non-implementation and the Supreme Court of India explicit orders, highlights violations of the law only to reaffirm the need to conserve the coastal ecosystem with better protection measures outside coastal habitats. As coastal habitats themselves occur as a mosaic of interconnected units of habitats, and the landscape. (Venkataraman 2008) It also made the prospect of rehabilitation phase according to the present provisions of the CRZ Notification a daunting task to protect sensitive zones of the coast (Coastal Management Programme Parliament of India 2009).

21.3 Ramification of Recovery Agenda on Coastal Protection Measures

Government of India learned from this terrible calamity. It left in its wake an opportunity to put in place a level of infrastructure and services in the affected areas which are of higher standards than those destroyed by the tsunami. After the tsunami, efforts were made to shift focus from immediate relief work to long-term rehabilitation (Asian Development Bank 2006). Initially post recovery measures was divided into three phases- of Immediate Relief and Response which was done by the states of Tamil Nadu, Kerala, Andhra Pradesh, the Union Territory of Pondicherry, and the Andaman and Nicobar Islands. Financed mostly from the central assistance-National Calamity Contingence Fund. Temporary rehabilitation and restoration of infrastructure for tsunami affected areas was carried out after assessing sector-wise damages in January 2005 under the Rajiv Gandhi Rehabilitation Package. The long-term tsunami reconstruction programme was monitored by the Planning Commission with a financial outlay of US$ 2,193 million. The source of funding for the rehabilitation and reconstruction programme was from the Rajiv Gandhi Rehabilitation package with the central government budgetary provision and support from external multilateral agencies (Central Audit General report 2006). Sector-wise utilization for the programme included 34 % on housing and internal infrastructure (water distribution, sewerage, roads, power distribution), 15 % on livelihood including agriculture, fisheries and social welfare programmes, 38 % on medium and long-term reconstruction covering ports and jetties, roads and bridges, power and communication, tourism and social infrastructure, and only 9 % on environmental and coastal protection measures (Ministry of Home Affairs 2007). The Recovery Framework in India was the 'one' programme that outlined the expected results of the combined efforts by the respective state governments of Tamil Nadu, Kerala, Andhra Pradesh and the Union Territory administration of Andaman and Nicobar Islands and Pondicherry, and the Central Ministries Departments and external agencies supports in tsunami recovery between January 2005 and end of 2008 (Ministry of Home Affairs 2007). The initial response of State governments, along with national and international agencies focused on rebuilding shelters for communities living along the coast and restoring their livelihoods. Environmental impacts of the 2004 tsunami, was addressed after 2 years. What was not so visible at first was the damage caused to the coastal ecology, become more and more important at later stage of the rehabilitation phase. Very little was done to understand the impact of this event on coastal ecosystem and their services. There was the inevitable and unprecedented consequence of tsunami on the fragile natural ecosystem along the coast and islands. Post tsunami scenario had exposed the relevance of the existing coastal management plans, trends of land use change by industries and construction activity within 500 m of the High Tide Line in the case of the sea, and lesser distances in the case of rivers, creeks and backwaters which are subject to tidal action and the environmental changes in more than one way (Ministry of Environment and Forests, Government of India 2009). The recovering lessons showed that plans should have sufficient and accurate baseline data on coastal ecology, socio-economic status, value of natural resources, hazard history of coastal settlements. Coastal community participation must be ensured and supported by strong research and scientific knowledge. And recovering interventions emphasized the need to adopt better plans for land management, locating of infrastructure services and development activity. And the vulnerabilities of coastal ecosystems must be checked and accounted inorder to minimize conflicting trade-off between development needs and conservation measures (United Nations 2008).

21.3.1 Pathway to Recover-Strengthening Coastal Area Management

A two-phase management framework was adopted for the restoration of coastal environment. In Phase-I (2006) the impacts of the tsunami was explored, many issues that affected the coastal populations and the ecosystems were identified at preliminary level. This was done keeping in mind the rapid coastal development, need of resource and changing legislative mechanisms. Key findings of Phase–I (2006) showed long-term impacts on the unique coastal zones of coral reef area decreases in coral cover and increase in turbidity after the tsunami (Indian Satellite Remote Sensing Organization 2005). Short-term impacts on sensitive zones of mangroves that reduced in area in Andaman and Nicobar Islands from 658 to 637 km^2 and in West Bengal from 2,120 to 2,118 km^2 mainly because of the tsunami (Ministry of Environment and Forests 2008; Ministry of Environment and Forests 2009). Fishing communities along the coast showed resilience and revealed the need to reform the system of local governance, autonomy to assume responsibility, recognition of rights and self-sufficiency to livelihood. There was a clear need for greater community participation and responsibility for managing and protecting the coastal zones. Since the central and state governments had different goals with respect to the implementation of the Coastal Regulation Zone (CRZ) notification 1991, it resulted in its non-implementation and acceptance by stakeholders of the coastal community. Phase-II (2007 and 2008) focused on the entire tsunami-affected coastline of mainland India. The intervention ensured long-term sustainability of the coastal ecosystem services—coastal habitats, river basin and landscape. In phase–II the ecological assessments done in phase–I helped to identify the knowledge gaps on critical coastal ecosystem services of the integrated rivers basins, coastal zones and landscapes, followed by the environmental change of coastal habitats and landscape outside coastal zones. The assessment provided feedback to the state government to reform existing coastal management plans and conservation policies to ensure long-term security and sustainability of the coastal ecosystem and its services to local communities. Policies related to implementation of legislations relevant to the coast such as regulation of economic zones for settlements, industrial, livelihood and tourism activities within the limits of coastal regulation zones along the tsunami-affected coast, were evaluated for its parameters on sustainability, efficiency and governance. As part of the suggestive amendments to the Coastal Regulation Zone notification 1991, the habitat development approach was conceived to integrate proper planning and beneficiary consultation process for unifying the physical and socio-economic changes with environmental sustainability of permitting protection of ecologically sensitive zones of Marine Protected Areas (MPA) and coastal habitats of mangroves, coral reef wetlands, and prohibiting or regulating urban and rural livelihood and development activities along coastal stretches and islands (United Nations 2008).

21.3.2 Learning to Sustain Coastal Zones

Post tsunami offered an opportunity to assess and monitor the resilience of coastal ecosystems to such extreme events. In turn the disaster has helped plan a suite of mitigating measures that reduces the impacts from natural risk of extreme wave action, which affect coastal areas periodically (Asian Development Bank et al. 2006). Impact from tsunami was an eye opener on the set back lines decided for protection, conservation of coastal sensitive area-mangroves, coral reefs, sea grasses, sand dunes, mudflats, salt marshes, estuaries, lagoons (Ministry of Environment and Forests 2009). Setback lines for regulating development activities are decided in terms of distance from high and low tide line as done under the Coastal Regulation Zone Notification 1991 (Asian Development Bank 2006). Though there was a notification (1991) by the Ministry of Environment and Forests, which sought to impose restrictions on setting up and expansion of industries or related operations through a regulation that is consistent within well established scientific principles of coastal zone management, showed trends that had allowed unplanned commercial and industrial expansion in the coastal zones (United Nations 2008). The notification failed to take into account that the Indian coastline is highly diverse ecology in terms of biodiversity, hydrodynamic conditions, demographic patterns, natural resources, geomorphology and geographic features. No clear procedures were laid down and no time lines stipulated for obtaining CRZ clearance.

Furthermore, there was no format given for submission of clearance applications, no post clearance monitoring mechanism nor a clear cut enforcement mechanism to check violations. Though the 1991 Notification sought to regulate developmental activities within the coastal regulation zone area of high and low tidal area and within 500 m on the landward side along the tidal influenced water bodies, there were no concrete steps to keep a check on the increasing anthropogenic pressure for development particularly for housing, tourism, hotel and oil industry, mining, and rise in coastal pollution affecting fragile coastal regulation zones. In compliance with the Supreme Court's orders of 1993, the National Coastal Zone Management Authority was constituted at the central level and in the states and union territories. The authorities had to prepare the Coastal Zone Management Plans (CZMP) by identifying and classifying the CRZ areas. However, none of the authorities had revised Coastal Zone Management Plan for the coastal state as per the procedure laid down in the CRZ notification. Restrictive nature of the 1991 Notification caused hardships to the communities living in certain ecologically sensitive coastal stretches, including traditional coastal inhabitants, fisher-folk, slum dwellers and other persons living in dilapidated and unsafe buildings (Sridhar 2005).

21.3.3 Contain Development for Coastal Zones

The run-up levels due to increase of sea level during tsunami as well as during storm surges added another dimension to land elevation which is considered as the principle for setback lines in coastal zones. It is necessary to incorporate the elevation levels for new and expanded settlement areas under the Town and Country Planning Act, so that people lives and property are saved. The low lying coastal areas and islands needed immediate attention in this regard. Proposals were submitted in December 2004 to the government for amendments to Andhra Pradesh Town Planning Act, 1920 and Andhra Pradesh Municipalities Act, 1965 by suggesting model amendments to the town and country planning, land use zone regulations and building regulations for hazard zones. Similarly the proposals were pending with the government in Tamil Nadu, for necessary amendments to Town and Country Planning Act, Building byelaws which were yet to be proposed to the Government by the Revenue Administration, Disaster Management and Mitigation Department. In Pondicherry, no amendments to Building bye-laws and Zoning Regulations were issued and a comprehensive review of the Building bye-laws and Zoning Regulation, 1972 was being finalized (Central Audit General 2006). There was the inability in the past to assess the ecological value of coastal ecosystem. Maps demarcating the extent of land areas vulnerable to seawater inundation, sensitive or protected areas and safe locations for settlement and vital infrastructure were to be prepared for all coastal zones in the country. Where possible, the settlements and other human gathering locations needed to be located at safer locations in a phased manner. The fact of less damage to coastal villages located in elevated areas with wide beach front, mangrove shields have laid great importance for the need to protect beaches, mangroves, coral reefs, offshore shoals, etc. as they act as excellent natural barriers. It is necessary all developmental activities both existing and planned in the future need to adopt the construction codes stipulated for their coastal region and adopt best environmental management practices to minimize loss or damage of natural barriers that ensures protection of human life. The huge loss of human life in villages located close to the coast amply demonstrates the vulnerability of human settlement located or occupied close to the coast, despite the areas have a beach front (Asian Development Bank 2006).

21.4 Post-Tsunami Learning and Improvements: Over 10 Years

Ten years have past since the recovery interventions were aided by a reduction in or elimination of the principal human threat. The recovery factors has communicated the order of nature's process so that community of people can avoid those devastating consequences. Accepting the proposition that Nature (ecosystem) is the process, it is interacting, it responds to laws, it represents values and opportunities for human use, with certain limitations and even prohibition. Taking this proposition to confront and resolve many problems, the nature is the arena of life and that a modicum of knowledge of nature's process is indispensable for survival, rather more for existence, health and delight. It is amazing how many apparently difficult problems present ready resolutions (MacHarg 1971). The tsunami reconstruction programme was an opportunity to rebuild at higher standards of safety and enhancing wide range of coastal area management. People are wise to know and to stay within Nature's prescribed boundaries. India identified two critical needs to improve in the management of coastal environment, both geographically, and across administrative departments and management targets. First is to tackle the many reasons for failure of management effort through institutional and regulatory framework. The second is to build better understanding of the true value of ecological goods and services and sustainably manage coastal environment. So that communities recognize what is at stake when their coastal ecosystem degrades. Based on the learnings from the post tsunami rehabilitation programme, the national government has given more priority to integrate the portfolio on conservation of natural resource management and coastal zone management supported through the national planning instrument. The integration is based on the management framework of ecosystem based adaptation approach to sustain the coastal ecosystem and its services. Through better planning, exploring reformed policy measures and good governance with participation of local people, the integration process is being done in a federal system collectively by understanding and obtaining consensus on conservation of natural resource policy, environmental legislation and management of governance, in order to address the risk reduction needs in coastal zones.

21.4.1 Towards Integrated Management: Prioritize Coastal Ecosystem Services

From recovery intervention the important lesson learnt for managing coastal ecosystem in India, has been a need to restrict setting up and expansion of industries, operations or processes in order to provide livelihood security to local communities, including fisher-folk and tribals and promote conservation and protection of unique marine area and costal zones. Ecosystem management has been initiated through national planning instruments at national and sectoral levels and policy and legislation reforms based on scientific assessment and lesson learned from recovery intervention. The national development planning process has attempted to consider ecosystem services as a component of development. Comprehensive amendments to the notification of 1991 (Center for Environment Education 2008) was declared in 2011. An integrated management approach was outlined to account for the vulnerability of the coasts to natural hazards, categorize the CRZ areas, prohibit and regulate permissible activities, for use of land within 500–100 m of the coast, ensure correct procedures for clearance and preparation of Coastal Zone Management Plans, demarcation of hazard line[3] along the coastline including shoreline changes and mapping, requires special consideration. (Society of Integrated Coastal Management 2010). Through this legislative amendment there has been an improvement in compliance with coastal management at state level. In view of the risks posed to Andaman and Nicobar and Lakshadweep Islands by natural disasters, the management of these islands has been delinked from the Coastal Regulation Zone notification, in order to manage and maintain the islands ecosystem services. A separate Island Protection Zone (IPZ) Notification in 2011 was issued. As per this notification, an Integrated Island Management Plan would be prepared for 340 islands of Andaman & Nicobar and 32 islands of Lakshadweep. The notification provides special dispensation for Andaman and Nicobar and Lakshadweep Islands. The coastal stretches of Middle Andaman, North Andaman, South Andaman and Greater Nicobar and the Lakshadweep island and their water area upto territorial water limit have been declared as the Islands Protection Zone (IPZ) with imposed restriction. Environmental management of the Islands of Andaman and Nicobar and Lakshadweep will be managed as per the Island Coastal Regulation Zone (ICRZ). Islands of Andaman and Nicobar and Lakshadweep is to be managed based on the Integrated Island Management Plans (IIMP), *interalia*, specifying existing and proposed developments, conservation and preservation schemes, dwelling units including infrastructure projects such as, schools, markets, hospitals, public facilities. Developmental activities in the Islands and ecologically sensitive areas shall be included in the IIMPs, in accordance with rules, regulations and building bye-laws for town and country planning in the Islands (Ministry of Environment and Forests 2009).

[3]The hazard line is a composite line of the shoreline (including sea level rise) due to climate change, tides and waves. This initiative forms a critical part of its responsibilities towards the planned management of the country's coastal zone.

21.4.2 Utilize Ecosystem Management Tools

In the background of tsunami tragedy concern have been raised on the safety of critical infrastructure projects located along the coast for instance power plants, oil storage depots, refineries. More such projects are likely to come up along coastal locations. Having capacity to utilize ecosystem management tools an exercise was commissioned to Survey of India to identify, delineate and demarcate the hazard line along the main coastline of India. The hazard line mapping exercise assumes special importance in view of tsunami disaster. Mapping of the hazard line is important to deal with impacts of coastal disasters. An appraisal committee constituted by the Ministry of Environment and Forests, relating to industry, infrastructure, thermal power, and nuclear power, had deliberated on tsunami-related risks and examined how the potential risks can be included in the terms of reference for Environment Impact Assessment (EIA) for future projects. In addition, it was necessary to review the current systems for assessing tsunami-type risks and other environmental risks, and post disaster phase as part of environmental impact assessment of such projects and evaluate what additional safeguards are needed for existing projects as well as suggest measures to be included in environmental impact assessment for projects to come up in future (Ministry of Environment and Forests 2011). Based on the Coastal Regulation Notification in 1991 and subsequent new CRZ notification in 2011 the activities in the coast zones are monitored by Space Application Centre in India. Earlier mapped the entire coastal zones of India have been extensively used in implementing the CRZ during the past 20 years. Recently Space Application Centre in 2011 in collaboration with scientific organizations and universities under the National Coastal Studies project supported by the National Natural Resource Management System Programme, detailed study and inventory of the entire coastal zones of India was published in 2012 as a coastal zone atlas to strengthen the knowledge gap on coastal zone information system. The study included the inventory of the coastal land use along with the demarcation of high and low tide lines, Ecological Sensitive Areas, mapping and monitoring of coral reefs and mangroves, sea rise on Indian coasts and (Ministry of Environment and Forests 2012) Survey of India (Department of Science and Technology) commissioned in 2010 to map, delineate and demarcate the hazard line along India's wide coastal belt. The hazard line for the mainland coast of India will be mapped, delineated and demarcated on the ground over a period of 5 years. This will include the collection and presentation of data, identifying flood lines over the last 40 years (which includes sea level rise impacts), and a prediction of erosions, (Ministry of Environment and Forests 2011).

21.4.3 Realign National Programmes

Post tsunami, the national planning process in India had considered coastal ecosystem as a component to improve environment management and sustainable use of natural resources in coastal zones. In doing so the programmes will promote cross sectoral policy integration, participatory decision making and innovations in linking environment management and development process. Settlers near the coast have been fisher folk who are dependent on the coast for their livelihoods and survival. Rights of the large section of fishing community must be protected from existing pressures on scarce coastal and marine resources. In consultation with the National Fish workers Federation (NFF) the draft on traditional coastal and marine fisher-folk (protection of rights) Act 2009 recognized traditional rights and occupation of fishing by fisher folk who have been living in coastal areas and carry on regular fishing activity (Sridhar 2005). The act provides a framework for protecting the rights keeping in mind sustainable use, conservation of marine biodiversity and maintenance of ecological balance along the coastal stretches. Thereby strengthening the conservation regime of the coasts and oceans while ensuring livelihood and security of the traditional fisher-folk (Sridhar 2005) The national planning instrument in India realigned it environmental programme on mangroves and coral reefs to address degradation of mangroves and corals due to developmental activities (Ministry of Environment and Forests 2009). Mangroves provide habitats for marine species and a resource base for sustainable tourism. Supported through the Integrated Coastal Zone Management approach, the scheme on conservation and management of mangroves[4] includes survey and demarcation, afforestation and restoration of mangroves, regeneration of corals, alternate and supplementary livelihoods, protection measures, and education and awareness. The scheme funded with central assistance, coastal states and Union Territories are to implement the approved Management Action Plans. As a result of this scheme two mangrove sites in the country are internationally recognized, viz. RAMSAR recognition for Bhitarkanika and UNESCO-MAB recognition for Sunderbans. Both Bhitarkanika and Sunderbans are rich in mangrove forests and the recognition is based on the richness of their biodiversity (Ministry of Environment and Forests 2004, 2005) Sundarbans is a key component of the Integrated Coastal Zone Management Project, that includes ecotourism, coastal erosion, and protection, livelihood improvement of communities. Indo-Bangladesh Forum on the Sundarbans jointly addresses issues of the region, recognizing that the entire Sundarbans region is one ecosystem (40 % of which is in India and the remaining 60 % in Bangladesh). For managing mangroves[5]

[4]Coastal Regulation Zone Notification (1991), recognizes the mangrove and coral reef areas as ecologically sensitive and categorizes them as CRZ-I (i). This implies that these areas are accorded protection of the highest order. The National Environment Policy, 2006 recognizes mangroves and coral reefs as important coastal natural resource. The national government support scheme on conservation and management of mangroves and coral reefs was initiated in 1987. The abundant coastal zone is endowed with a wide range of mangroves, coral reefs, sea grasses, salt marshes, sand dunes, estuaries, lagoons, and unique marine and terrestrial wildlife The offshore marine ecosystems include 6,740 km^2 of mangroves, including part of the Sundarbans and the Bhitarkanika, which are among the largest mangroves in the world.

[5]As per the latest 'India State of Forest Report 2009', mangrove cover in India is spread over an area of 4,639 km^2 in the coastal states/UTs of the country. Compared with the 2005 assessment, there has been an increase of 58 km^2 in mangrove cover because of plantations and protection measures in Gujarat, Orissa, Tamil Nadu and West Bengal mainly due to activities sponsored by

more effectively, the National Institute for Research in Mangroves and Coastal Bio-resources is set up in West Bengal. In order to address the coral reef issue, a National Coral Reef Research Centre at Port Blair, is established under the International Coral Reef Initiative (ICRI), which is a partnership among governments, international organizations and NGOs. The project on Mangroves for Future a regional initiative covering tsunami-affected countries-India, Indonesia, Maldives, Seychelles, Sri Lanka and Thailand, promotes investment in coastal ecosystem conservation and strengthen the environmental sustainability of coastal development. Coordinated by IUCN and Ministry of Environment and Forests the project address the continuing challenges of management of coastal ecosystems and livelihoods in four states—Gujarat, West Bengal, Andhra Pradesh and Odisha. The tsunami did not create the current problems faced by coastal ecosystems at present, but rather brought them sharply in focus. The project looks at the national management framework with view of revision in the coastal zone legislation to determine coastal area according to vulnerability lines, and to promote a more participatory approach to decision making and planning and link the national coastal zone management framework for regional collaboration in coastal management and sharing of transboundary coastal areas. (International Union for the Conservation of Nature 2008) After the 2004 tsunami, many of the actors in coastal management quickly became involved in the reconstruction process. Many of these efforts at environmental restoration and rehabilitation were guided by a short-term planning perspective. Currently, the valuable work to restore livelihoods, reducing pressures on the natural resource base, and minimising negative environmental impacts of settlement and infrastructure and in incorporating environmental concerns into coastal planning and development, had been sparked off by the immediate needs of post-tsunami recovery. Marine Protected Area[6] in India have been used as a tool to protect marine resources for biodiversity conservation and for well-being of people dependent on these resources. Scientific monitoring and traditional observations confirm that depleted marine resources are getting restored and pristine ecological conditions have been sustained in well-managed MPAs, in cooperation with local people (H B Singh 2003; Ministry of Environment and Forests 2011). India has identified 12 protected areas as trans-boundary protected areas under the framework of the IUCN Trans-boundary Protected Area Programme. Among these sites, two are MPAs, namely the Sundarbans Tiger Reserve and Gulf of Mannar Biosphere Reserve (Sivakumar et al. 2012) National programme on Integrated Coastal Zone Management was launched with

---

the scheme on Conservation and Management of Mangroves. Decrease in mangrove cover in Andaman and Nicobar Islands are attributable to the after-effects of the tsunami. (MoEF 2009)

[6] 18 marine protected areas (MPAs) in peninsular India and more than 100 MPAs in its Islands. Of the 18 MPAs, the Gulf of Mannar Marine National Park, Sundarbans National Park, Gulf of Kutchch National Park, Bhitrakanika National Park, Coringa and Chilika Wildlife Sanctuaries have unique marine biodiversity and provide a range of services to local communities. There are 105 protected areas in the Andaman and Nicobar Islands, and of these about 100 are MPAs. These MPAs cover more than 30 % of the terrestrial area of the islands and still protect more than 40 % of available coastal habitats of the islands (Sivakumar et al. 2012). In the Lakshadweep group of islands, Pitti Island (0.01 km^2) is the only island having the status of an MPA.

the support of the World Bank, over a period of 5 years. Society of Integrated Coastal Management (SICOM) has been established as a registered body to implement and responsible for mapping the national coastal hazard line, setting up of a National Centre for Sustainable Coastal Management and for piloting the Coastal Zone Management in three states namely along the coastal stretch in Gujarat (Gulf of Kachchh), Odisha (Paradip-Dhamra and Gopalpur-Chilka) and West Bengal (Digha-Shankarpur and Sagar Islands), to address the issues relating to coastal management improving livelihood of local communities, and disaster mitigation (Society of Integrated Coastal Management 2010).

21.5 Time to Bridge the Gap

Natural disasters show forth the power of nature. And in a similar way, all the cycles of nature, with their vivid colors and changing patterns, tell of Nature's process. In conclusion such rehabilitation programme after the devastating disaster was the key to identify environmental damages to ecosystem, prioritize restoration of degraded ecosystem and need for a comprehensive coastal zone management strategy. Such a strategy adapted reflects the dynamic nature of coastal environment and multiple-use objectives for livelihood needs, reduce vulnerability to natural hazards, conserve biodiversity and ecological services, without compromising the sustainable supply of ecological goods and services. Post tsunami factored in much more proactively the ecological dimension with opportunities for ecosystem restoration and management for better coastal zone protection and biodiversity conservation. It has lead to mangrove restoration, marine protected areas along coastal stretches which is indigenous to coastal biodiversity. Developing a coastal zone information system has improved management of coastal ecosystem and generate multiple benefits for different natural resource user groups.

21.5.1 Challenges to Integrate the Gaps

From tsunami recovery experience the key take-way learning has been to manage healthier coastal ecosystem through an integrated policy and management approach. More effective integration of disaster risk reduction and coastal zone management with development planning is unquestionably required to reform the present coastal zone management and adaptation practices at community level. At present the governance arrangements for national and sub national planning instruments do not facilitate integrated management of risk drivers, especially when responsibilities are entrusted to different governmental entities for critical issues such as disaster risk reduction, ecosystem based adaptation, coastal management, land tenure and development policy. The existing institutional framework must be enhanced to implement a actionable approach giving due regards to underscore importance for inter-linkages among relevant cross cutting issues of the environment and natural disasters. National and State Governments have been challenged to factor in disaster risk management considerations with ecosystem management, as a driver for security and sustainability. There have been repeated post disasters recovery issues at disasters sites that demonstrated lack of environment management and lack of joint resilience strategies for inter-linkages with river basin, watershed, coastal zones, flood prone land, wastelands and forest land. This calls for a holistic approach to have joint action plans that embraces disaster risk reduction and ecosystem management also fundamental for poverty reduction and sustainable development. The integration still remains to be model to be put into practice and communicated to rest of the people and stakeholders. Environmental degradation, loss of biodiversity and sensitivity to natural resource limits and environmental tipping points of ecologically sensitivity areas along coastal stretches continue to be neglected or skipped during post–disaster assessments and planning for transition from recovery to resilience. Identified to be a pressing concerns in context of the integration process of design, planning and execution of post disaster recovery plans. Moreover the national planning instruments must enhance capacity to improve environmental management for safety boundaries such as coastal zones, habitats, landscapes including sustainable use of natural resources by looking at long term post disaster recovery needs. More over effective participation and engagement of the influencers must address mismanagement of the ecosystems, enhancement of social and environmental vulnerability assessments and accounting of the ecosystem services. Voluntary sharing of sustainable experiences, lesson learned and partnership must be practiced to integrate both aspects of disaster risk reduction and ecosystem management for long term development. It also recommended to strengthen technical and human resource capacity, involving interdisciplinary stakeholders to understand complexity of interaction between ecological, social, and development process. There is need to effectively administer, manage and govern the integrated strategy for managing India's most sensitive natural resource-the coastal zones. Though presently the coastal zones is purely regulatory under the CRZ amendment regulatory framework post the tsunami, a process has definitely begun, fitting well to be integrated and consistent within the established principles of the ecosystem management approach. This approach that has come of years with number of corrections, reforms, pilots and up-scaled interventions and evaluation, post the tsunami 2004, facilitates protection and conservation of coastal ecosystem and giving due priority to people's well-being needs and reducing risks of community in coastal India.

References

Asian Development Bank, United Nations, World Bank (2006) India post tsunami recovery programme preliminary damage and needs assessment. Asian Development Bank, United Nations, World Bank

ATREE-UNDP (2005) Statement on the CRZ notification and Post-tsunami rehabilitation in Tamil Nadu

Centre for Environment Education (2008) Report on the public consultation on coastal management zone notification

Coastal Management Programmes, Parliament of India (2009) Rajya Sabha Report No 202, Rajya Sabha Secretariat pp 1–40
Government of India (2006) Report no 20 2006, performance audit report on tsunami relief and rehabilitation, Comptroller and Auditor General of India
Government of India (2008) Annual Report Ministry of Environment and Forests
Government of India (2008 and 2009) Annual Reports, Ministry of Environment and Forests
Indian satellite Remote Sensing Organization (2005) Rapid assessment of damage to ecological resources due to tsunami, pp 1–9
International Union for the Conservation of Nature (2008) Mangroves for the future-a strategy for promoting investment in coastal ecosystem conservation. pp 23–25
Marale SM, Mishra RK (2011) Status of coastal habitats and its management in India. Int J Environ Protect 1(1)
MacHarg I (1971) Design with nature. In: Sea and survival. Natural History Press, New York, pp 7–17
Ministry of Environment and Forests (2009) Final frontier, agenda to protect the ecosystem and habitat of India's coast for conservation and livelihood, security. Committee on the draft Coastal Management Zone notification, constituted by the Ministry of Environment and Forests
Ministry of Environment and Forests, Ministry of Earth Science (2005) Preliminary assessment of impact of tsunami in selected coastal areas of India, Department of Ocean Development ICMAM-Project Directorate, Chennai for Andaman and Nicobar Islands and Tamilnadu. Ministry of Environment and Forests, Ministry of Earth Science
Ministry of Environment and Forests (2004–2005) Annual report, Government of India, pp 1–28
Ministry of Environment and Forests, Government of India (2009–2010, 2010–2011) Report to the people
Ministry of Home Affairs, Government of India (2007) Report to the Nation, tsunami disaster management-the strategy. pp 1–5
Sach J (2011) The planetary boundaries, age of sustainable development
Sach J (2011) Ocean and fisheries age of sustainable development
Singh HB (2003) Marine protected areas in India. Int J Environ Protect 23(3):1
Sivakumar et al (2012) Marine biodiversity conservation in India, Wildlife Institute. Go4 Bio Div Newsletter 2(2):10–12
Society of Integrated Coastal Management (2010) Implementation of integrated coastal zone management project, pp 1–8
Sridhar A (2005) Pre-and post-tsunami coastal planning and land-use policies and issues in India. ATREE
India two years after tsunami- a joint report by United Nations, the World Bank, the Asian Development Bank (2006), pp 1–74
Tsunami-Three years after, a report by United Nations (2008), pp 1–75
UNISDR, UNDP (2012) Review paper – status of coastal and marine ecosystem management in south Asia
Venkataraman K (2008) Coastal and Marine Wetland in India. In: 12th World lake conference proceedings, pp 393–400

Chapter 22
Listening to the Communities

Hari Krishna Nibanupudi and Parnasri Ray Choudhury

Abstract The communities in the Indian sub-continent have learnt to live with disasters. The Indian subcontinent has been stuck by massive natural disasters in the recent decades. The catastrophic supper cyclone in Odisha in 1999 was followed by mega earth quakes in 2001 and 2005 in Gujarat and Kashmir respectively. The visual impact of the live coverage of tragedy caused by Indian Ocean Tsunami in 2004, was fresh in the memory of people in the sub continent until they witnessed the Indus river floods in Pakistan in 2010 that devastated lives and livelihoods of over 20 million people. The memories of tragic Indus floods were taken over by multi-hazard disaster in Uttarakhand in the summer of 2013 at about 3,583 m (11,755 ft) above sea level, dubbed by the media as "Himalayan Tsunami".

From Indian Ocean Tsunami to Himalayan Tsunami, the constant feature in the aftermath of all disasters is that communities have been the first responders and communities are the ones to stay back in the area to deal with future hazards with new wisdom and enhanced resilience. But the question is, are we (non-government organizations, donors, government agencies, research institutions, etc.) listening to the communities or are we interested only in telling what communities should do to deal with disasters?

The chapters discusses the comparison and difference between the concept of participatory appraisal and listening to communities. The chapter suggests that, listening to communities is much more than mere consultation and participation and it implies looking at the issues from community's perspective and having pure & objective understanding of issues that are context specific. While development professionals can be trained to become 'experts in participation', listening to communities requires development workers to undergo a mentoring to improve thinking and behavior and it must be promoted as an attitude alongside with participatory tools and approaches.

Taking Indian Ocean Tsunami as an example, this chapter discusses the role of communities from first response to the last action in the aftermath of massive disasters (like Indian Ocean Tsunami). Based on field experiences and a set of researches that inquired the issues of gender and social inclusiveness, participation, HIV AIDS,

---

H.K. Nibanupudi (✉) • P.R. Choudhury
Safe Citizen.Org, Kathmandu, Nepal
e-mail: harikrishna@safecitizen.org; parnasri@safecitizen.org

R. Shaw (ed.), *Recovery from the Indian Ocean Tsunami: A Ten-Year Journey*, Disaster Risk Reduction, DOI 10.1007/978-4-431-55117-1_22

accountability, shelter and risk reduction, in the Tsunami affected areas of India and Sri Lanka, this chapter analyze the impacts of Tsunami response as a result of listening to communities or lack of it. The chapter will high light the role of local knowledge, capacity, and priorities of communities in effective disaster response and its sustainability and calls for greater attention to empathetic listening among humanitarian workers.

Keywords Community participation • Empathy • Gender • HIV-AIDS & humanitarian response • Public health • Resilience

22.1 Introduction: Consultation, Participation and Listening to Communities

Listening to communities in humanitarian assistance programs implies involving community voices, community wisdom and community choices in the humanitarian response design and implementation strategies. The spirit and wisdom of listening to communities is closely related to popular development approaches, such as "rapid rural appraisal (RRA) which developed and spread especially in the 1980s and its further evolution into participatory rural appraisal (PRA) which has developed and spread fast in the 1990s" (Chambers 1994). While participatory approaches were considered as best practice approaches for many years since 1980s, today they become the ways of practicing development among local non-governmental organizations and the donor organizations that support them. However, the transition of participatory approach to humanitarian operations has not been automatic, for, a very long time, humanitarian assistance was seen from military or missionary perspectives. Humanitarian organizations thrust for quality and equality since 1990s can be attributed to their adaptation of participatory approaches in humanitarian action.

Minear and Weiss's survey in 1993, found that there has been an increasing realization among the humanitarian agencies on the need to embrace new age principles accountability and the need for appropriateness and contextualization (Minear and Weiss 1993), with a concern for quality and effectiveness of humanitarian aid. According to Dorothea Hilhorst, this transition has been triggered partly by public pressure, but mainly followed from lessons learned from the humanitarian experience (Hilhorst 2002). A number of humanitarian evaluations, impact assessment, lessons documentation reports do repeatedly reflect the ground level feedback on the need for enhancing community's role in humanitarian programming.

Analyzing the rehabilitation processes post Kobe Earth Quake of 1995, Shaw and Goda identified the need for focusing on community needs and priorities with a specific emphasis on socially vulnerable groups, need to bring communities into the decision-making process for reconstruction and to reach collective opinions with communities and suggest that to realize a safer society, it is essential that more people share fundamental ideas in common (Shaw and Goda 2004). Akinci also argue that, Failure to include community perspectives can lead to responses which are not relevant or acceptable to the people for whom they are intended (Akinci 2004). Collecting vital lessons from past earth quake humanitarian responses in different parts of the world, ALNAP and Provention consortium recommend that communities should be asked to prioritize their needs at least once a month and needs-assessment teams should have prior knowledge of local context. As Bourbé (2010) and Sanderson (2012) suggest, communities can provide invaluable information on which areas and households are most in need, and what is needed. Key informants and/or focus groups are essential to determining specific needs assessment criteria that reflect the needs of the community and the context (Bourbé 2010; Sanderson 2012).

Similarly, Nakagawa and Shaw inform that in the aftermath of Kobe Earthquake, It was individuals and their neighbors, who saved most of the victims right after the earthquake and it was also the community which determined whether each member was satisfied by the rehabilitation. Therefore, they aptly suggest that, Disaster recovery is not only about building houses but the reconstruction of the whole community as a safer place. To mobilize each member of the community in this collective action (community development), social capital is a crucial need (Nakagawa and Shaw 2004). Communities, past experience, contextual knowledge and their ideas for future are also essential components of social capital that is vital for successful rehabilitation programming. Brenda L Murphy too echoes this wisdom and strongly recommends that the role of the local authorities should prioritize providing opportunities for community members to learn more about the hazards prevalent in that locality and in fostering the social capital bonds that contribute to resiliency (Murphy 2007).

The importance of consulting communities, listening to them and giving them a say in disaster response planning and decision making has been articulated and recommended by a plethora of evaluations, impact and lessons learnt documents. Fritz Institute for instance, after a comprehensive documentation of lessons from Tsunami relief strongly recommends that, those affected must have a voice in the relief plan. The report further states that, affected communities must be consulted about their priorities and preferences about the type of assistance that is most needed, because, community leaders are most familiar with the needs of their communities, and incorporating these into relief plans can ensure that the help is targeted and appropriate. Further, consulting local communities before a disaster, during the assessment phase, and in the relief phase will ensure that those made vulnerable have a voice in describing their needs. The report laments that, during the Tsunami, the rush to help in any way possible resulted in mountains of used clothing and inadequate medical help and notes that, in those communities where relief agencies were able to perform consultations with the local communities during the assessment phase, the damage estimates tended to be more comprehensive and accurate (Fritz Institute 2005).

The references quoted above and many other references that couldn't be quoted here, equally articulate the wisdom of consulting communities and making them participants of humanitarian response programming. The non-governmental practice of development gained much respect in public perception for embracing participatory approaches for last few decades. Community consultations, participatory appraisals and more recently gender inclusive participation processes have become non-negotiable practices of development among non-governmental organizations, for good. However, at the same time, these approaches have also become mere rituals for some organizations who demonstrate them to please donors and observers, without really meaning them or committed to the spirit of participation. The result is under the gloss of participation, some organizations try and push their own agenda through communities and actually undermine communities own choices. If unchecked such practices actually erode the credibility of development work of entire non-governmental sector in the long run.

Bill Cooke and Uma Kothari (2001) discuss many limitations of the paradigm of participation. They argue that, participatory approaches fail to recognize individual's identity, context and conditions in their choice of participation and participatory approaches have the danger of overlooking the potential links between inclusion and subordination. Robert Chambers who originally championed the participatory approaches also warn against complacency and high expectations from participatory processes. He clarifies that the aim of participatory approaches is not to extract local knowledge for analysis elsewhere, but to mobilize indigenous capacities for the self-management of development projects (Chambers 1994).

The concept of listening to communities resonates the underlying philosophy of participatory approaches. However listening to communities is much more than mere consultation and participation. Listening to communities implies looking at the issues from community's perspective, respecting the voices of communities, shedding pre-conceived ideas and notions and having pure and objective understanding of issues that are context specific. While, participatory approaches can be mastered by practice, listening to communities requires internalization of the core philosophy of participation. While participation can be practiced with the help of tools, listening to communities needs to be practiced through behavior. While development professionals can be trained to become 'experts in participation', listening to communities requires development workers to undergo a mentoring to improve thinking and behavior. Therefore, it may not be possible to develop social tools for listening that every organization can adapt to overnight. Listening to communities must be promoted as an attitude and behavior alongside with participatory tools and approaches. Following chapters discuss some of the experiences of listening to communities.

22.2 The Context: Major Disaster and Greater Resilience

The Indian Ocean Tsunami affected India's Andaman and Nicobar Islands in the Bay of Bengal, and the southeastern provinces of Tamil Nadu, Andhra Pradesh, and Odisha, killing an estimated 11,000 people and displacing, by some estimates almost 650,000 (Indurfurth et al. 2005). In Sri Lanka, the tsunami resulted in 31,187 deaths, 4,280 missing people, 23,189 injured people, and the displacement of

545,715 people (Hollifield et al. 2008). The tsunami killed 167,000 people in Indonesia's Aceh, and displaced over 600,000. Three months later, a massive aftershock devastated the neighboring island of Nias (Tsunami 2004).

For coastal communities in Sri Lanka and Indonesia, tsunami was an added disaster. Before the tsunami of 2004, Sri Lanka was mired in a civil conflict between the government of Sri Lanka and the Liberation Tigers of Tamil Eelam, which had been ongoing since 1983, displacing over 200,000 people as of 2006 (Oxfam 2006). Similarly, before the tsunami of 2004, the Indonesian province of Aceh had suffered three decades of civil war (Oxfam 2006). The communities in these three countries that are most affected by tsunami have already been less fortunate and victims of economic disparity and massive environmental destruction going on in the coast for many decades.

The tsunami affected communities have shown a remarkable resilience, spirit of survival and gritty recovery after this massive disaster, with the assistance of aid agencies and respective Governments. Dozens of donor organizations, hundreds of aid agencies and thousands of local non-government organizations and the local government agencies have rendered admirable assistance to the communities affected by Tsunami. However, most of those agencies, as evident from their self congratulatory reports have failed to adequately acknowledge the remarkable resilience of affected communities who refused to break down after losing their families and livelihoods and the dignity they have shown in the process of recovery. Most importantly, these humanitarian agencies have missed the opportunity of listening to the communities and learning lessons in resilience and recovery from them.

22.3 Listening to Dignity

Hundreds of local, national and international humanitarian agencies thronged to the tsunami affected coast with various kinds' relief supplies and resources. Some of them stayed back to build back the lives of the affected communities. Most of these humanitarian agencies have long experience in disaster response and have established standards, operation procedures and protocols. Some of the well-intended but mis-informed staff of humanitarian agencies also went with the notion that, they were god sent to help the communities and it was an opportunity to correct all "ills" in the community's cultures and lives through un-invited social engineering processes. The participatory approaches of the humanitarian agencies and their professional approach to assessing needs have been appreciated by the communities as well as the local administration. However, the highly standardized approaches of these agencies had driven them to "listen to what they already assumed" and ignore what people had really to say. The notion of the humanitarian agencies that, surviving, living, food and shelter are the only needs of the communities driven their investments and energies to duplicate food supplies, houses, boats, fish nets, etc. during relief and rehabilitation phases. In the processes, humanitarian agencies have (probably) unintentionally created a hierarchy of givers and receivers.

Further, these well intended but less-informed social missions have at times undermined the capacities, resilience and dignity of communities who were thankful for the support received, but were not necessarily desperate for the support of external aid and the unwritten conditions that came with it. Many of these communities did already pick up the threads of life and livelihoods much before the arrival of the humanitarian aid. As can be seen in the personal account of the authors, the tsunami affected communities acknowledged the money and resources brought in by the aid agencies. But, what valued more was the audience to what they had to say, an equal space for discussion and a warm and friendly presence around in times of great emotional loss. For instances, listen to what Kamla a tsunami affected person in Cuddalore district of Tamilnadu state in India says to a humanitarian worker;

> I learnt that millions of rupees in donation are made by people from many countries. I'm deeply touched by the fact that so many people, who don't even know me, want to help me!. At the same time, I hate my situation of having to take help from people I don't know, because, I can't return their favor. If possible, I would like to know individuals whose contributions are helping me survive with basic food, water and shelter. I want to cook for them, feed them and honor them. I want to be able to help them, if ever they face this kind of situation. But I pray god that they should never face this situation, in fact no one should ever be in a situation of needing others help for survival.! (Krishna 2005a)

This statement of Kamla in middle of colossal tragedy is a clear indication that disasters don't necessarily turn human beings in to desperate seekers of charity. There are other things like dignity that many disaster affected people possibly value more than mere physical survival. Kamla's statement also reveals that she is thankful for the help, but taking help doesn't mean that she is inferior. She sees herself as equal and help back those helped her.

Similar, powerful expression of dignity was made by another woman in the neighboring state of Andhra Pradesh. In a public meeting in a village of East Godavari district, addressing a coastal fishing community, a Government officer said, "Tsunami didn't give adequate time to warn you and evacuate you to safer places, hence many people died. However, the hurricanes that hit the coast every year do give adequate time for early warning and evacuation. But you people are adamant to be evacuated; it will not be good for you". A woman from the community became furious at this statement; she got up and said;

> Your staffs evacuate us and take us to your (meaning Government's) cyclone shelter. It's better to die in a cyclone than to rot in that dingy cyclone shelter.

The local administration emphasizes on early evacuation of people from the coast on receiving cyclone warning from the met department. In the process of evacuation, the administration mobilizes local trucks, pushes people in to them and transports them like cattle to a nearby cyclone shelter. They are virtually locked in it until the threat of cyclone is subsided. These cylindrical shaped cyclone shelters are just closed walls devoid of bare minimum facilities. No toilets, no electricity, no water and no facility for cooking. Situation of women would be worst in these shelters as they are forced to share space with men from their villages and other villages for a few days and nights. Their sense of shame doesn't let them respond to natural urges for long periods and they have no place to hide when they menstruate. Therefore, as the woman said in her response "people are willing to die, but refuse to face ill treatment". The response of the government officer was—"*we are saving their lives, what else they want? If you give them a chance, they will ask you for the moon*" (Krishna 2005b).

22.4 Is Dignity as Important as it is Made Out to be?

It's not necessarily an over statement, when someone says that, dignity for them as important as or more important than life. For instance, over a quarter million farmers committed suicide between 1995 and 2010 in India, the worst-ever recorded wave of suicides of this kind in human history (Sainath 2012). All most all of them have the same reason, to escape humiliation in the hands of money lenders, whose loan they can't repay due crop failure. They found dignity in death instead of losing their face in the community. Chaman Pincha, who conducted a research for Oxfam on Gender issues in Humanitarian Response in the aftermath of Indian Ocean Tsunami, reported the experiences of women who told her that some of their friends preferred to be taken away by the tides.

> The waves were so violent that women were stripped of their cloths. Some women refused to climb naked in to the rescue boats because of an internalized sense of shame and honor (Pincha 2008).

We are saving their lives, what else they want? Says the Government, we are trying to ensure minimum standards of food, water, shelter for them in camps, says a bit more responsible non-government organizations. The humanitarian services rendered by the government and the non-government organizations can't be undermined. However, this good work need not limit the ability of these agencies to listen to the communities, whose loud and clear message is that their dignity, identity, personal and cultural values are as important as physical survival.

22.5 Listening to Gender, Class and the Cultural Sensitivities in Disaster Relief

The state and the non-state humanitarian actors were appreciably quick to move in with relief operations in India, Sri Lanka and Indonesia in the aftermath of Indian Ocean Tsunami. They were endowed with financial and logistical resources to reach out to the communities after Tsunami. The Government systems lacked the empathy to appreciate social, cultural and gender specific needs of the Tsunami affected people, non-state actors lacked the context specific understanding. For instance, in the Indian state of Tamilnadu, the state didn't recognize the existence of *Aravanis* (Transgender), the non-state actors, or non-government organizations, didn't even know about their existence and the losses they faced in tsunami. According to Chaman Pincha and Hari Krishna, *Aravanis* lost dancing costumes, small savings in cash, make-up kits, jewels and tools. But these losses suffered by Aravanis did not feature in damage assessments carried out by the government and non-government organizations, therefore they were deprived of post disaster assistance even after 4 years of tsunami disaster. "The exclusion of *Aravanis* from access to temporary shelter, housing and livelihood support points to a degree of gender blindness even among agencies genuinely committed to gender equality and human rights" (Pincha and Krishna 2008). Despite growing articulation of the rights of trans-gender under the umbrella of LGBTs (lesbian, gay, bisexual, and transgender), the gender and development discourse is yet to embrace recognize their distinct gender presence and identity in the world.

While Aravanis were almost completely excluded from the relief process, the choice of relief material distributed by humanitarian agencies revealed that they didn't pay enough attention to women's practical gender needs. For instance, the clothing supplied by the state and humanitarian agencies in the initial days of tsunami response in India, did not meet women's needs across different ages and different communities. Inner garments were not included in the clothes' packages for both men/boys as well as women/girls and *burqas* (a cloak worn by Muslim women to cover the entire body, when in public place) for Muslim women (Pincha 2008). Similarly, the relief and rehabilitation across communities was delivered to and in the name of men, with an s stereotype national that, men are the only bread winners and they only take care of the needs of the families. On the contrary, as Pincha reports, the cash compensation (ex-gratia) given in many cases was wasted on alcohol addiction and gambling, further aggravating vulnerabilities of the households and creating more hardships for women, children, and the elderly (Pincha and Krishna 2009).

Lack of space and opportunity for women to voice their needs and concerns made them vulnerable to increased gender based violence in Indonesia. For instance, gender un-informed and poor design of the camps for internally displaced people in Ache in the after math of tsunami, where five to eight people (both men and women) were housed a mere 4 × 5 m room created tremendous inconvenience and personal safety issues for women. As a result, young boys and girls preferred to stay out of the camps until mid-night, exposing themselves to other risks. Apart from the safety issues, women (tsunami affected women in Ache like their counter parts in India and Sri Lanka) suffered the absence of feminine hygiene products; lack of proper latrines and bathing facilities; lack of clean water in the camps; an increased burden of domestic work; limited access to humanitarian aid; lack of reproductive health services; and increased vulnerability to trafficking, sexual violence and sexual harassment (UNFPA 2005).

The inadequate awareness and understanding of the state and the humanitarian agencies in the tsunami affected countries was a result of not listening to communities and going ahead with set formats of needs assessment and deep routed stereotypes. Despite their best efforts to induce gender sensitive culture, most humanitarian organizations still carry the burden of patriarchal inheritance that make them inherently pay special attention to men, listen to them more and think of humanitarian needs from men's perspective. Decades of participatory approaches and gender sensitivity training in the development sector resulted in numerous tools and methodologies but failed to develop listening capacity of professionals. Humanitarian and development professionals need to develop the capacity to listen carefully to communities, their feelings and learning to respond to and validate their feelings. The big lesson to take away from tsunami response for the humanitarian agencies was the need for them to listen to communities without gender, cultural and social stereotypes so as to be able to internalize inclusiveness rather than trying to integrate without necessarily believing in it.

22.6 Listening to Vulnerabilities Beyond Food, Water and Shelter

As discussed in the previous chapters, approaching disaster relief and rehabilitation with set standards and experiences borrowed from different cultural settings had limited the ability of the humanitarian agencies to listen to, understand and assimilate local, context specific and dynamic vulnerabilities of the communities affected by Tsunami. For instance, a study commissioned by Oxfam in India 2 years after Tsunami has established the complex relationship between HIV and AIDS and disasters.

The Oxfam study worked through the operational definition of vulnerability as characteristics of a person or group or community, and their situation that influence their behavior and practices which lead or exposes them through a causal chain of events resulting in increased likelihood of acquiring HIV-AIDS. The results of the study reveal high levels of vulnerability of coastal communities to contracting HIV and AIDS in the tsunami affected areas studied. Of the four main modes of transmission of HIV, the sexual route was found to be the major mode that has contributed to increased vulnerabilities to HIV and AIDS. The study also revealed that about, 17 % of the population was found to be vulnerable to HIV and AIDS in Tsunami affected districts of India, of which 60 % were found to be 'directly' vulnerable. Of various sub-groups within the community, unmarried men followed by married men were found to be the most vulnerable groups, and unmarried women were the least vulnerable. High prevalence of unprotected multi-partner sex combined with low awareness on STI and HIV were found to be the main causes for vulnerabilities of communities. Alcohol consumption, access to commercial sex and temporary separation of spouses due to migration were the main factors (Oxfam and Swasti Health resource center 2008).

Though natural calamities do not transmit HIV, however, some of the post disaster situations may provide a fertile environment, which would enhance vulnerability of individuals to HIV. The tsunami disaster led to inadequate safe blood, shortage of clean injecting equipment for injecting drug users, an insufficient supply of condoms and health care facilities; limited facilities for treating STIs; and the vulnerability of displaced people, especially women and children to sexual abuse and violence. In addition, during the periods of population displacement, HIV and AIDS prevention and care is often disrupted (Oxfam and Swasti Health resource center 2008). Further, most of the relief shelters after Tsunami were over populated (Oxfam and Loyola 2006). This situation had led to absence of any privacy for the families and had probably led to 'no sex' between the married couples. Indirectly this situation has also contributed to escalate the number of men accessing sex workers, exploitation of women and adolescent girls at relief camps as post tsunami trauma affected the ability of many parents to provide adequate attention to their adolescent children (Swasti and Oxfam International 2009).

As World Disasters report suggests, the impact of HIV on people living in disaster zones is still not fully understood. An understanding of HIV is vital in disaster preparedness, especially in an age when all types of disasters appear to be on the increase (World Disasters Report (WDR) 2008). The process of moving towards AIDS competence in Humanitarian operations needs to evolve through a spiral of listening, learning, action and reflection, challenging problems and making the best use of available opportunities. Supporting community-led change allows the move towards AIDS competence to be at a pace that is locally acceptable, allows issues to be discussed and acted upon in an open manner and ultimately means any change is sustainable and holistic.

22.7 Listening to Local Resilience

Humanitarian agencies lay enormous emphasis on learning and improving from each disaster response experience. Some who don't agree with this observation say, humanitarian agencies learn same lessons from each disaster!. The certainty however is that communities don't miss their lessons and they don't repeat their mistakes in dealing with disasters. Contrary to what many humanitarian agencies may believe, communities have clear ideas as to who is responsible to help their recovery, what to expect from those sources and how to negotiate with them. Communities also know the role of non-government organizations, their capacities and limitations. Therefore, community wisdom tells them to be independently resilient first and build on the assistance that may come from the state or external agencies.

A study commissioned by Oxfam to understand local capacities in disaster resilience in the after math of Tsunami in Tamilnadu state of India, learnt that, after tsunami, communities enhanced their capacities to demand their dues form the government and how to guide external NGOs in their post disaster reconstruction activities. Their major source of support in their perception remained to be local traditional *panchayats* (traditional village councils) who are available to them locally all the time. Most importantly, the Oxfam study established the fact that, forcing blueprint-based approach to relief, rehabilitation and disaster resilience on communities were unsuccessful as compared to flexible, participatory and consultative interventions. Communities welcomed those initiatives that provide them with a platform to engage with the government and other service providers to help them have direct access to the humanitarian assistance. There were also instances, where communities asked the competing humanitarian agencies coordinate to provide joint assistance (Oxfam and BEDROCK 2009).

22.8 The Barriers: Why, We Don't Listen to Communities?

It has been said that Green revolution in 1970s was responsible for ensuring sustained and sufficient food production in India. In the eagerness to promote modern agricultural techniques and technologies for enhanced food production, the communication and education messages by India's agricultural institutions sought undermine the traditional and experiential knowledge of farmers. In protest, there was this light hearted cartoon, published in a magazine, that couldn't be traced by the authors for proper acknowledgement. The cartoon as narrated here, highlights the limitations of academic knowledge and the arrogance one carries with it.

An agricultural scientist, sporting a French beard was in conversation with a humble farmer. Standing in the shade of a tree on the farm, the scientists lectures the farmer on the importance of modern agricultural techniques. He concludes by saying, see, because, no one gave you such knowledge so far, you remained a poor farmer!. Had you known these techniques, earlier, this mango tree would have given you bumper yield this summer. I really feel sad for you that all mango tress on your farm are fruit less in this peek mango season! The farmer with his typical humility responds, "sir, I'm so fortunate that such a highly learned man like you stepped in to this small farm and generously shared so much knowledge with a poor farmer like me. But, sir, by using all those techniques, will I really be able to produce mangoes from a *Neem* Tree?".

This cartoon is a pointer to the unintended negative effects of professional education as it has been practiced in our University systems. The idea here is not to paint education system in a bad light, but to introspect, whether current practice of education is making pupils more arrogant than educated?

Typically, in the 'aid industry' the key programmers are drawn from top educational institutes, who takes decisions and control purse strings. How much ever committed they may be, their ability to have objective grasping of the ground situation remain inadequate due to their upbringing that is alien to the communities that they sought to serve. But they become confident and complacent after familiarizing with a few academic concepts of development. Interestingly, they preach participatory tools and approaches to the humble community organizers from local NGOs, who actually live with and work with local communities. As Vaux (2001) says *"developmentalism and professionalism caused particular problems and failures in the humanitarian response(to Ethiopian famine). These were ideological faults of minds shaped too firmly by experiences and perceptions that had their origins outside of the immediate situation and didn't relate to the person in need"*.

Industrialization of aid resulted in creating development agents whose capacity to serve the industry in place of development workers with an aptitude to work with communities. The International organizations and academic institutions who have been churning out intelligent, sophisticated and articulate development professionals must ensure that these pupils are groomed with right attitude, humility and listening skills instead of just mastering the skills that are useful only for the aid industry. Most importantly, these pupils must safe guard against the danger of our concern and ability to serve the needy slipping in to an arrogance of altruism and complacency of doing good that impede our listening attitude and skills.

22.9 And, What Happens as a Result of Not Listening to Communities?

As discussed extensively in the previous chapters, despite the good intentions of donors and of responding agencies, many post-tsunami relief and rehabilitation projects did not have the positive impact that agencies intended. Instead, many such interventions had actually compounded the suffering of the communities. These negative impacts could have been avoided, only if these agencies cared to listen to the communities. A report by Center for Environment Education (CEE), for instance, revealed that a Tsunami affected fishing in Galle district of Sri Lanka was resettled 15 km inland, "without proper consultation," At the new settlement, there were no public transportation facilities, schools or local employment opportunities. As a result, the community abandoned the new settlement and went back to its original habitat. The resulting outcome was a conflict with the local administration and the humanitarian agency that made investments (CEE 2008).

In another example, a study by University of Colombo and Oxfam in Sri Lanka, which has also been the experience in other countries, found that the rush of humanitarian agencies to provide boats to fishing community resulted in excess boats, duplication of resources and wastage of aid. At the same, the humanitarian agencies had no money to assist farmers who lost their crops and fertile soils and cottage entrepreneurs who lost their weaving equipment. Because aid agencies didn't listen to the needs of the communities, but went ahead with their standard and mechanical needs assessment process. According to the report of Colombo University, some humanitarian agencies, to avoid the difficult task of hard need assessment and negotiations, adapted a lottery system to allocate houses for the communities. As a result, some people whose homes had minor damage were granted new homes while others with much more heavily damaged housing were left in the lurch, leading to serious conflicts within the community, which has otherwise been living in harmony for decades.

In Pondicherry state of India, one organization consulted the community on home building design but ignored community requests to have relatives and friends as their neighbors. The organization felt that requiring people to live next to strangers would foster a sense of unity and minimize discriminatory behavior. As a result, the communities opted to stay where they were instead of moving to newly constructed houses as they wanted to live together than living in the strange neighborhoods (Oxfam and Loyola 2006). In a separate study on vulnerability to HIV following the tsunami, researchers found that the risk of contracting sexually transmitted diseases (STD) including, HIV/AIDS increased when shelter policies did not seek to preserve community networks. Researchers found that most shelters offered little privacy to girls and women, thereby jeopardizing their safety. Only a few of the temporary communities had TV halls, an important source of information and recreation. In a number of communities, some homes were built far away from schools against the advise of the communities, there by increasing student dropout rates. And most temporary villages lacked public health clinics. The study concluded that closer consultation with shelter residents might have averted many of these problems. Communities know best what they need to restore their livelihoods, rebuild their lives, maintain their community networks, and avoid risks (Swasti and Oxfam 2009). Therefore, aid agencies and local NGOs should have a genuine partnerships with communities in every step to achieve the objective of humanitarian assistance with quality and on time.

22.10 Conclusion and Recommendations

22.10.1 Humanitarian Work Begins with Listening to Communities

The field experiences and research reports cited in the above chapters was not meant to conclude that, humanitarian response in the after math of Indian Ocean Tsunami was a big failure. The main objective of the analysis of these cases was to highlight the fact that humanitarian response is bound to fail, when it begins without listening to communities. Some of the key factors that limited the ability of humanitarian agencies to listening to communities were; overwhelming demand from donors and national governments on many International agencies with limited local presence and little experience in mounting major aid operations, pushing standard protocols and tools through the local NGOs and forgotten principles of transparency and accountability. On the other hand, as documented by Oxfam, there were plethora of interventions by humanitarian agencies that were effective and successful due to listening to communities and involving them in all stages of relief, rehabilitation and reconstruction processes.

In Sri Lanka

- A thorough research and a participatory needs assessment formed the basis for a program that helped coir (coconut fiber) workers, most of whom are poor, marginalized women, double and in some cases triple their incomes since before the tsunami.
- Awareness raised through research on disaster preparedness helped a community in Kalutara District respond quickly to flooding in neighboring villages in 2008.

In India

- A community led survey of conditions in temporary shelters helped bring about the release of $1.4 million in government funds for repairs in Tamilnadu state of India.
- A participatory research on recent rainfall patterns, which have confounded traditional methods of weather prediction, enabled farmers in several villages to adapt their planting schedules to changing climate conditions (Oxfam International 2009)

22.10.2 Reducing Disaster Circumstances is Real Humanitarian Work

The findings from multiple research studies, organizational reports and field notes reinforce the importance of listening to communities and participatory research in helping aid providers understand local strengths, needs, context, and priorities. The key message from all these studies is also that disaster-affected women and men want to play a central role in identifying priorities and designing, implementing, monitoring and evaluating the programs aimed at assisting them. Also, it is important to recognize that, there is a line that separates those who needs basic humanitarian support in order to pick up the threads and move on with life and those would like to garner humanitarian support to move up in life. The dividing line can be called as dignity. Therefore, the goal of humanitarian work should be to reduce the opportunities for humanitarian work by mitigating disaster risk and vulnerabilities and strengthening resilience of the people to be self-sufficient and manage disasters with greater capacity.

22.10.3 Humanitarian Workers Should Develop Empathetic Listening

Every body agrees that sympathy is a good emotion, but most of us don't want to be in a position to be a subjective others sympathy. As narrated through a few cases in previous chapters, the communities affected by disasters also don't want our sympathy. Ability to make an empathetic connection with disaster affected communities is a prerequisite qualification for a humanitarian worker. The advantage of university education is that it empowers us with intellectual framework to learn chosen area of knowledge and skills to articulate the knowledge in different ways. However, the real use and application of these knowledge and skills for public good will become a reality only when we develop the art of empathetic listening to communities. Covey (1989)'s prescription of *"Seek first to understand, then to be understood"*. Similarly, its important to *show our desire to understand and reflect the person's feelings or felt meanings* (Atwater 1992) for developing communication and trust with communities.

References

Akinci F (2004) The aftermath of disaster in urban areas: an evaluation of the 1999 earthquake in Turkey. Cities 21(6):527–536

Atwater E (1992) I hear you (Rev ed). Walker, Pacific Grove

Bourbé L (2010) Identification of vulnerable people in urban environments. In: Assessment of sustainable livelihoods and urban vulnerabilities. ACF International, Paris, Cap Impression Press, Palaiseau

CEE (2008) Threads of disaster management: weaving youth experiences from tsunami response in to a learning document. Center for Environment Education, p 30. Ahmadabad, India, ISBN 978-81-89587-38-3

Chambers R (1994) The origins and practice of participatory rural appraisal. World Dev 22(7): 953–969

Cooke B, Kothari U (2001) The case for participation as tyranny. In: Cooke B, Kothari U (eds) Participation: the new tyranny? Zed Books, London, pp 1–15

Covey S (1989) The 7 habits if highly effective people. Simon & Schuster Inc., New York

Fritz Institute (2005) Lessons from tsunami: top line findings. www.fritzinstitute.org. Accessed Dec 15 2013

Hilhorst D (2002) Being good at doing good? Quality and accountability of humanitarian NGOs. Disasters 26(3):193–212

Hollifield M, Hewage C, Gunawardena CN, Kodituwakku P, Bopagoda K, Weerarathnege K (2008) Symptoms and coping in Sri Lanka 20-21 months after the 2004 tsunami. Br J Psychiatry 192:39–44

Indurfurth AKF, Fabrycki D, Cohen S (2005) The 2004 Indian Ocean tsunami, six month report. The Sigur Center Asia Papers, Sigur Center for Asian Studies, vol 7, George Washington University

Krishna H (2005a) Field note, post Tsunami needs assessment, Oxfam, Chennai, India (2 January 2005)

Krishna H (2005b) Field note, post Tsunami needs assessment, Oxfam, Chennai, India (30 June, 2005)

Minear L, Weiss TG (1993) Humanitarian action in times of war: a handbook for practitioners. Boulder Lynne Rienner (Also available in Spanish and French from UNICEF, New York)

Murphy BL (2007) Locating social capital in resilient community-level emergency management. Nat Hazards 41:297–315

Nakagawa Y, Shaw R (2004) Social capital: a missing link to disaster recovery'. Int J Mass Emerg Disasters 22(1):5–34

Oxfam and Loyola (2006) Rapid assessment of temporary shelters for tsunami victims, an unpublished report. Oxfam International, Chennai, India

Oxfam International (2006) Oxfam International Tsunami fund, second year report, vol 17. December 2006, Oxford, UK

Oxfam International (2009) Lessons from the Oxfam International tsunami research program. Oxfam International, Oxford, p. 4

Oxfam and BEDROCK (2009) Building local capacities for disaster response and risk reduction. BEROCK for DFYWA, Chennai, India

Oxfam and Swasti Health Resource Center (2008) Reducing vulnerability to HIV before and after disasters: a research brief. Oxfam International, Oxford

Pincha C (2008) Indian Ocean tsunami through the gender lens. Oxfam International, p 21. Chennai, India ISBN 81-86945-21-0

Pincha C, Krishna H (2008) Aravani's: voiceless victims of the tsunami. Humanitarian Exchange Magazine, Issue 41, Humanitarian Practitioners Network, ODI, London, UK

Pincha C, Krishna HN (2009) Post disaster Ex Gratia payments and their gendered impacts. Regional Development Dialogue (RDD), vol 30, issue 1, Spring 2009, Gender and Disaster Management, p 95, Tokyo, Japan

Practitioners. Lynne Rienner, London
Sainath P (2012) In 16 years, farm suicides cross a quarter million. The Hindu, English Daily, 27 May, 2012
Sanderson D (2012) Building livelihoods of the most marginalised in urban areas: strategic approaches from Dhaka. In: Burnell J, Sanderson D (eds) Environmental hazards, vol 2, issue 11. Earthscan, London
Shaw R, Goda K (2004) From disaster to sustainable civil society: the Kobe experience. Disasters 28(1):16–40
Swasti and Oxfam International (2009) Did the tsunami increase the vulnerability of coastal communities to HIV? Catalyst Management Services, Bengaluru, India, ISBN 978-81-908417-1-9
Tsunami (2004) Tsunami, 2004, facts and figures. http://www.tsunami2004.net/tsunami-2004-facts/
UNFPA (2005) Gender based violence in Aceh, Indonesia. United Nations Population Fund, A Case Study, p 12, ISBN 0-8971-4-828-2
Vaux T (2001) The selfish Altruist: relief work in famine and war. Earthscan, London, ISBN-1849774307
WDR-World Disasters Report: Focus on HIV AIDS (2008) The International Federation of Red Cross and Red Crescent Societies, Geneva, ISBN-92-9139-134-4

Chapter 23
Empowering Communities Through Disaster Management Strategies: Are We on the Right Track?

Asharose and Izuru Saizen

Abstract The Indian Ocean Tsunami spurred the growth of disaster management activities in India. Obviously, all of these efforts were intended to reduce the risk of future disasters, through empowering communities and increasing their resilience to disasters. An array of questions that hails in, at this point of time are: did these efforts achieve their ultimate goal; have communities been empowered? If so, to what extent? By further examining certain communities, gaps can be found which needs yet to be filled. A study on the coastal communities of Cuddalore District, Tamil Nadu, reflects the gaps and pitfalls that further underline the lack of sustainability in the implemented disaster risk reduction programs. The results clearly underpins that the communities striving even for their basic necessities gives least priority to disaster risk awareness and preparedness comparing to physical resilience as well as economic resilience. Thus implementation of disaster risk reduction programs in such communities will be ineffective. This chapter will focus on such gaps, as well as the challenges in molding a disaster-resilient community.

Keywords Coastal communities • Community empowerment • Disaster resilience

23.1 Introduction

Disasters have struck mankind throughout its long history. While each disaster may vary in its type, magnitude, and severity, the final result is always the same: devastation. Natural disasters can be devastating, particularly when they affect vulnerable, low-income households (Clarke and Grenham 2013). People below the poverty line are likely to suffer higher risks to life and livelihood (Ferdinand et al. 2012). The poor are particularly hard-hit because injury, disability, and loss of life directly affect their main assets: food production for self-consumption and real labor income (Shaw and Krishnamurthy 2009). Therefore, the empowerment of such people and

Asharose (✉) • I. Saizen
Graduate School of Global Environmental Studies, Kyoto University, Kyoto, Japan
e-mail: asharosevictor@gmail.com

R. Shaw (ed.), *Recovery from the Indian Ocean Tsunami: A Ten-Year Journey*,
Disaster Risk Reduction, DOI 10.1007/978-4-431-55117-1_23

communities is imperative in every sense. The United Nations International Strategy for Disaster Reduction (UN/ISDR) places considerable emphasis on building the resilience of communities as a necessary component of DRR (UNISDR 2009).

Every disaster provides an opportunity to strengthen the affected community's capacity to appropriately respond to the next disaster. To make that happen it has to be ensured that those opportunities are being used in its true sense. To accomplish that, it is essential to find the pitfalls of the pre and post-disaster management strategies that were undertaken in previous disasters.

This chapter intends to have a look into the disaster management strategies undertaken after the Indian Ocean Tsunami in India as well as the disaster management issues in selected coastal villages of Tamil Nadu (the worst affected state in the country) and its implications on the community. This chapter also discusses about the required improvements in the strategy level and community response level.

23.2 The 2004 Indian Ocean Tsunami

The December 26, 2004, tsunami significantly affected the coastal regions of India's southern peninsula. The death toll in India was 10,273, while the number of missing persons was 5,832 (Ministry of Home Affairs 2005). The tsunami affected 2,260 km of the mainland India's coastline, as well as all of the Nicobar Islands and some of the Andaman Islands (Murty et al. 2006).

In India, the states most severely affected by the tsunami were Tamil Nadu, Kerala, and Andhra Pradesh. The state of Tamil Nadu was the most-affected on the mainland, with a death toll of 7,793. Nagapattinam District, Kanyakumari District, Cuddalore District, Chennai, and Kancheepuram District all fall within the severely affected regions of Tamil Nadu. The state of Kerala experienced damage in three southern districts, named Kollam (131 dead), Alappuzha (32 dead), and Ernakulam (5 dead).

The December 2004 tsunami in the Indian Ocean, with its devastating toll on human suffering and destruction in the affected areas, was a painful reminder that coastal communities are vulnerable to unforeseen events.

23.3 Disaster Management in India Before and After the 2004 Indian Ocean Tsunami

A strategic approach to disaster management strategies, as well as well-coordinated action, is an essential factor for ensuring the safety of communities, property, the environment, and the sustainable growth of a country. When it comes to India, its geographic position and diverse geographic and climatic conditions make it prone to almost all types of disasters, and thus one of the countries that is worst affected by disasters. Furthermore, its rich cultural diversity makes acting and planning accordingly a requisite.

Droughts, floods, earthquakes, and cyclones have, with grim regularity, devastated the country year after year (Metri 2006). Almost all of India's regions have experienced one or more of these events (Gupta 2000). As much as 60 % of India's land is prone to earthquakes of varying intensities, over 40 million hectares are prone to floods, about 8 % of its total area is prone to cyclones, and 68 % of its area is susceptible to drought (Sharma and Khanna 2013). These data show the need for a systematic and strategic approach to reducing vulnerabilities to, and the risks posed by, hazards, as mentioned in the Hyogo Framework for Action.

India has its own disaster management system and approach, established decades ago, to deal with the aftermath of each and every disaster it has faced. In earlier times, it had a post-disaster centric approach that gave emphasis to the allocation of compensation funds, relief, and rehabilitation. To achieve these targets, the Government established the National Crisis Management Committee, the Crisis Management group, the Control Room, the Contingency Action Plan, State Relief Manuals, and a well-planned Funding Mechanism through the formation of a Calamity Relief Fund in each state (Menon, NDMA).

With major steps taken at the international level—such as the declaration of the Decade for the Reduction of Natural disasters (1990) in August of 1999—India, too succeeded in bringing changes to its disaster management policy priorities through the formation of the High Powered Committee, which prioritized pre-disaster actions for disaster preparedness, prevention, and mitigation, along with post-disaster relief and rehabilitation activities. Eventhough, the Orissa Super Cyclone (October 1999) and Bhuj Earthquake (January 2001), revealed the inefficiency and pitfalls of the built institutional and policy framework. Later, the National Committee on Disaster Management reviewed the High Powered Committee Report and approved some of the recommendations, one of the most significant of which being the shift of the primary responsibility for disaster management from the Ministry of Agriculture to the Ministry of Home Affairs in Government of India (Menon, NDMA).

After the mass devastation brought about by the Indian Ocean tsunami in 2004, the Indian Government further strengthened its Disaster Management Framework to ensure more effective management at national, state, and district levels through the enforcement of the Disaster Management Act in 2005. As per the act, the government formed a hierarchy of authorities (Fig. 23.1). Among the hierarchy, the National Disaster Management Authority is the apex body, chaired by Prime Minister; it is followed by the State Disaster Management Authority and then the District Disaster Management Authority for each state district.

Going further, under the DM Act 2005, the Government established the National Institute of Disaster Management for planning and promoting training and research in the area of disaster management, as well as the development of a national-level information base relating to disaster management policies, prevention mechanisms, and mitigation measures, and a National Disaster Response Force that provides specialist responses to disasters (IFRC 2005). Each of these institutions has their

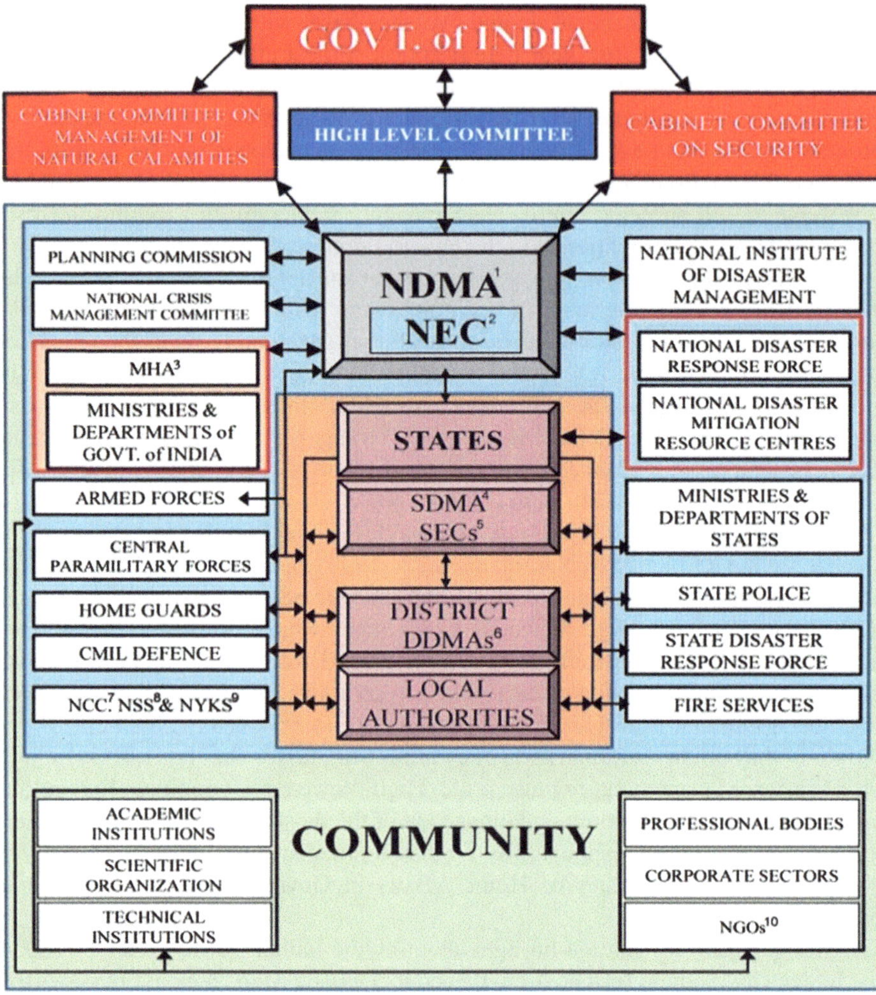

Fig. 23.1 Present disaster management structure in India. (*Source*: Pune Division Disaster Management System (partly edited by authors)). *Notes*: This diagram reflects the interactive linkages for synergized management of disasters, rather than a hierarchical structure. Backward and forward linkages, especially at the functional level, are made with a view to optimize efficiency. 1: National Disaster Management Authority; 2: National Executive Committee; 3: Ministry of Home Affairs; 4: State Disaster Management Authority; 5: State Executive Committee; 6: District Disaster Management Authorities; 7: National Cadet Corps; 8: National Service Scheme; 9: Nehru Yuva Kendra Sangathan; 10: Non Government Organizations

own assigned duties and functions, to be rendered at concerned authority limits. The integration and mainstreaming of disaster management with development planning was the other major venture after the 2004 Indian Ocean Tsunami.

The growth of the nation's disaster management perspectives clearly demonstrates how the 2004 Indian Ocean Tsunami influenced and spurred disaster management initiatives, policies, framework, and activities in the nation. In any county, the prime beneficiaries of all of the formulated policies, programs, frameworks are supposed to be, and should be, the communities. It is crucial to confirm that these actions positively influence communities, and that their objectives can be considered achieved, only if they have contributed in the capacity building and empowerment of communities towards a resilient one.

Changes brought in the nation's disaster management strategies reflected positively in the strategic disaster management approaches of all its states. As shown in Fig. 23.1 disaster management authorities at various hierarchical level formulated plans which helped states to focus more on the issues to prepare, mitigate and respond in a much better way than in the past. For materializing these strategies in its real sense, communities requires to play a pivotal role. Taking this into consideration, National Disaster Management Authority has brought a shift in its emphasis from a government-centered approach to decentralized community participation (Rahman 2012).

23.4 Empowerment of Communities Towards a Resilient One: Review

A community refers to a group of individuals and households living in the same location and having the same hazard exposure, who can share the same objectives and goals in disaster risk reduction (Ferdinand et al. 2012). In other context, community can also be interpreted as heterogeneous individuals and groups who share common interests and needs, and who are able to mobilize and organize themselves towards social and political change (Laverack and Wallerstein 2001).

Disasters have major impacts on vulnerable, generally less-developed, societies, and make achieving sustainable development exceedingly difficult (McBean 2012). Therefore, empowering such societies and communities towards resilience is imperative, in every sense. The significance of resilience is indispensable, as it has strong bonding with sustainable development. Meanwhile, community empowerment is usually considered as a process in the form of a dynamic continuum involving—personal empowerment: (i) development of small mutual groups, (ii) community organizations, and (iii) social and political action (Jackson et al. 1989; Labonte 1994; Rissel 1994).

Developing and evolving a standardized module can be seen essential, which should be specific for that specific area (region); have to ensure that the module is actually in sync with the community's culture and tradition. Interventions are often proven to be insufficient, as they tend to ignore the local preceptions, needs, and potential value of local resources, as well as their inherent capacities. The inherent differences and stratification of the community has to be noticed and taken into consideration from the start of planning through to the time of implementation. Therefore, it is always better to involve communities in work along, inviting them to contribute ideas and to participate in meetings and discussions while strategies are being shaped.

It will always be better to implement strategies through, or in conjunction with, local bodies, making community the core group. Whoever is working at the grass root level—which may vary with place and region in different regions of same country, or across the country—has a greater chance of better knowing the geographical, economic, and social background. Coordination has a major role in such an implementation. Carefully examining and comparing the implementation of former program and projects should be carried out in order to note and correct points of disagreement or to find suitable alternatives. Elements of the plan that impede the development of the community must be pinpointed by the community (either in the community's view or in the expert's view).

Empowerment promotes the capacity building of heterogeneous individuals, which can be seen in their ability to move towards small group activities, organizational structures, and links with people outside the community, along with an increased awareness of the broader social and political causes of their disempowerment (Laverack and Wallerstein 2001). Community empowerment and Disaster Risk Reduction both are linked one another and that link can be magnified by betterment collaborative mechanism between the local authorities, the local communities and other stakeholders. One of many benefits of community based disaster risk reduction (CBDRR) addressed by the United Nations International Decade for Disaster Reduction (UN-IDNDR) is that community participation will positively address the local socio-economic concerns in disaster reduction, by empowering the community with knowledge and skills and develop the leadership capability of the community members and further strengthen their capacity to contribute to development initiatives (Pribady and Mariany). Empowering the community by internalizing the tools and methods of disaster risk reduction is a good way to deal with the future potential risks (Safer Cities 21 2008).

To sustain empowerment, a consensus approach to decision making is recommended, which assures participation in identifying shared problems, as well as developing and implementing solutions that in turn facilitates the development of problem-focused coping, a sense of community, and a commitment to action (Paton and Johnston 2001). To enhance the community's abilities to identify the hazards based on the severity of damage and then to take proper actions in response, in Chittagong city, the PROMISE- Bangladesh project has adopted the following steps (i) Participatory Vulnerability and capacity Assessment (PVCA), (ii) Workshop on Community Risk Assessment Findings, (iii) Formation of Wrad Disaster Risk Management Committee, (iv) Preparation of IEC materials etc. (Safer Cities 21 2008). These activities have the capability to bring in a sustainable empowerment into that community.

Evaluation of the situation after a disaster and evaluation during frequent interval can help in designing new plans and modifying and thereby disasters can be taken as an opportunity that can result/bring in developments. To know how the impacts and footprints laid by 2004 Indian Ocean Tsunami in the affected communities resilience status as well as to know the present situation in selected coastal communities in Tamil Nadu focus group discussion was conducted.

23.5 Study Area

Cuddalore District is one of the 32 districts in the state of Tamil Nadu that lies on the southeastern coast of the Indian mainland. Cuddalore is a district prone to multiple hazards, including cyclonic depressions and floods, and it also falls within the Zone-3 with respect to earthquakes. Cuddalore District falls within the geographical constraints of 11°45 N and 79°45 E. The district has a coastline of 57.5 km. Considering its disaster significance as a multi-hazard prone district with annual cyclonic depressions, four coastal villages were selected from Cuddalore to conduct the Focus Group Discussions.

23.5.1 Methodology

In order to know how the so-called paradigm shift from a post-disaster management centric approach to a pre-disaster management approach has affected or influenced communities that are supposed to be the beneficiaries of these ventures—as well as how the 2004 Tsunami spurred disaster management policies, strategies, programs, and actions—Focus Group Discussions (FGDs) were conducted in four coastal villages. Focus Group Discussions was used as the information collection method to solicit views, insights, recommendations of the community. Its flexible format helped exploring unanticipated issues and provided checks and balances, thus minimizing false or extreme views. Sotikuppam, Devanampattinam, Mudasalodai, and Samiyarpettai, all in Cuddalore District. Among these, Sotikuppam and Devanampattinam are coastal villages of Cuddalore Taluk, and Mudasalodai and Samiyarpettai are coastal villages of Chidambaram Taluk (Fig. 23.2).

In addition, these discussions were aimed at determining the community's present problems, issues, and immediate needs. Discussions were conducted with a focus group consisting of 10–15 people in each village. To avoid bias, care was taken to include members from both genders and ensured participation from various age groups, village level representatives (such as the Village President, Village Assistant), and leaders (usually coastal villages in this region have a village leader (non official) elected by the community).

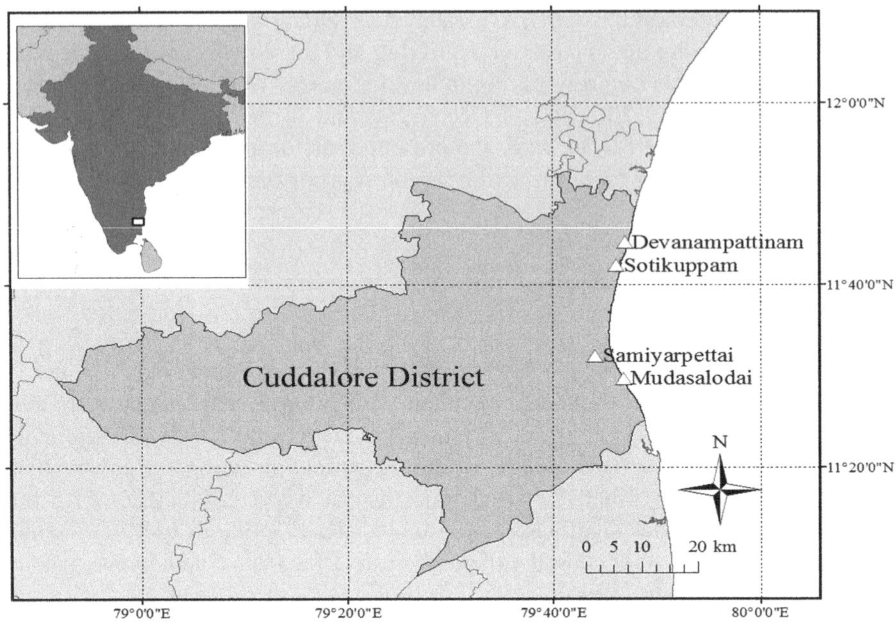

Fig. 23.2 The location of the selected coastal villages

23.5.2 *FGD Results*

The identified problems and issues were classified under different categories: Economic Resilience, Social Resilience, Physical Resilience, and Risk Preparedness and Awareness (Table 23.1).

In general, the major issues noticed were

i. The elements that ensure physical resilience are considered the most important by the four villages. Next, they placed importance on elements for economic resilience. They gave the lowest priority to risk preparedness and awareness, even though they have been severely affected by various types of previous disasters.
ii. The results also show that basic requirements like water, proper housing, and toilet facilities are all things that the communities regard as most important. Therefore, finding a solution that meets their basic requirements is of much importance, along with making the community aware of and prepared for risks.
iii. It can also be stated that they are neither particularly aware nor concerned about future disasters and the significance of the positive changes that risk preparedness can bring into their lives and into the community as a whole.
iv. The results underline the need for communities (i) to make themselves realize their potential, (ii) to increase awareness of the risks and vulnerabilities that they face, and how they can better prevent these adversities through risk preparedness and community empowerment, thereby making their community more resilient.

Table 23.1 Summary of village level responses and the related resilience type

Resilience type[a]	Response/solution
Sothikuppam	
1	Water facility, hospital facility, diseases like fever, malaria, dengue (drainage facility, mosquito repellent fumigation), garbage facility, toilets for houses and schools, transportation facility.
2	No warnings (warning/alarm for disasters), the sea has advanced further up the shoreline, releases from industries have polluted the air and environment leading to vomiting, skin diseases (itching), and breathing problems ought to occur
3	Need alternative livelihood.
Devanampattinam	
1	Sometimes brownish, mud-colored water is found in public supply system, private water supply is expensive (clean and quality water from Govt.), Dengue, Chikungunya (mosquito repellent fumigation, hospital).
2	Proper warning before disasters (should be provided by authorities), first aid training needed.
3	Soil quality decreased, so they cannot rely on main business of flower production (soil testing/treatment by Govt.), alternative livelihood needed (small-scale industries), sea erosion (planting trees).
Samiyarpettai	
1	Drinking water facility (quality water by Govt./public supply system), maintenance of tsunami and cyclone affected houses still pending, shortage of bus services (improved transportation facility needed).
2	Sea erosion (planting of trees), locker facility for storing documents during cyclone/disaster prone periods, proper warning facility.
3	Fish preservation facilities, livelihood options for women.
Mudasalodai	
1	Private water supply expensive (public water supply needed), hospital facility needed, improvements to transportation facility.
2	Cyclone shelter needed, sea erosion (jetty to be extended).
3	Not able to conduct fishing because of sediment deposition in river mouth (deepening to be done by Govt.), alternative livelihood option during non-working hours, fish preservation methods needed (cold storage facility by Govt.).

[a]*1* Physical Resilience, *2* Risk awareness and Preparedness, *3* Economic Resilience

Poor people without having much access to resources are more vulnerable to disasters. Here, the results clearly underpins that the communities that are still striving for their basic necessities have a lesser chance of showing interest in, and placing importance on, protecting themselves against disasters through risk preparedness well in advance. Lacking even the basic necessities for life, the implementation of risk preparedness programs in these communities will be ineffective in a way and bring an element of dissatisfaction into the community.

None of the four coastal villages expressed any issues that had the potential to negatively affect social resilience. This is mainly due to the management system they have at the village level. They usually elect a village head, as well as some additional members to serve as his assistants. They usually compose the decision-making body for any issues that face these coastal villages. It has been noticed, based on the solutions they have suggested for the problems they are facing, that they usually are of the opinion that disaster management and issued related to its risks are absolutely the job of the Government. Accordingly, a sense of their own potentiality needs to be encouraged among the communities, along with an increase in their awareness.

The identified issues underpin the gaps that need to be filled and the challenges faced when molding a resilient community within these villages. The lack of sustainability seen in the implementation of the Disaster Risk Reduction programs is the major factor for yet to be met challenges in the concerned villages.

23.6 Discussion

Despite the type of disaster, its community's capacity and will to do act accordingly which brings functioning of the community to normalcy. After the earthquake in Gujarat, India, the role of community leaders was prominent in using social capital in the recovery process and facilitating collective decision making (Nakagawa and Shaw 2004). Thus, successful recovery means empowering those impacted by a disaster in a way that preserves their dignity, embraces their right of choice, and demonstrates respect for their experience. Recovery task force members serve as advocates for those impacted by a disaster and provide assistance to them in accessing aid. (Community Disaster Recovery Guide 2006) In Sri Lanka international aid agencies and national NGOs focused heavily on replacing the assets that Indian Ocean Tsunami survivors had lost. Such an 'asset replacement' strategy cannot deal with pre-existing vulnerabilities—such as poverty or exposure to risks—and it can never give substance to the much-repeated mantra of 'build back better'. Whilst, Indian chapter showed that the Indian government took the opportunity to enforce and ensure the compliance of building codes and introduced a scheme to provide houses to the people who were willing to stay 200 m away from the sea (ADRC 2009). Government launched a policy that permanent houses within 200 m from the coastlines will principally be relocated to safer places.

Another important aspect is sharing of the successful, innovative strategies and lessons learned. In Mudasalvodai hamlet, MSSRF (MS Swaminathan Research Foundation) initiated a alert information program named FFMA (Fisher Friendly Mobile Application) which provides bad weather alert and warning text messages to mobile phones provided to fishermen. This can be called much of a people centered warning system which can be adopted in other nearby villages too. But even the nearby coastal village (Sothikuppam) doesn't have the facility. Thus, local/community level strategies as well as Government level strategies and risk reduction approaches implemented and the constraints in the implementation has to be made known to other places (communities) by knowledge sharing so as to help other communities to make use of innovative and effective methods to get benefited from.

Mainstreaming gender, ensuring participation from all sectors and age groups, and giving more importance to youth efforts are some of the main area of concentration when DRR programs are implemented. Apart from these, the other areas that need to be addressed in order to lead to an increased sustainability include:

23.6.1 Gaining Knowledge About Local Needs and Constraints

Taking time to understand a community's needs and constraints can be beneficial in providing the actual necessities by preventing the misuse of time, unwanted efforts, and the usage of inappropriate resources. It is advisable to do this well in advance of possible disasters, during normal (pre-disaster) periods, and as soon as possible after a disaster (the early stages of the post-disaster period). This can be accomplished by concerned local authorities, and could be forwarded to higher levels in order to ensure the integrated management of issues. Therefore, before the formulation of policies, programs, decisions, and their implementation, the local needs and constraints should be assessed to determine the effectiveness of all of the aforementioned stages.

23.6.2 Prevailing and Possible Livelihood Options

Community empowerment, and through it resilience, always has strong binding with a secured income source. Livelihood plays a big role in this. Furthermore, livelihood diversity in the community is essential for ensuring the generation of stable income in the community. Prevailing and possible livelihood options that have yet to be utilized should be identified. Additionally, plans to make use of possible livelihoods, to bring maximum livelihood diversity, and to improve the efficiency of prevailing livelihood have to be well conceived and implemented. At the same time, the sustainable use of resources should be assured.

23.6.3 Socio-Economic Situation

The socio-economic situation, or status, reflects the capacity of a community to cope with a disaster or any other unexpected event that may affect the community. It should always be borne in mind, while designing action plans (for making communities aware of risks and risk preparedness and reduction measures), that their socio economic uplift is equally important. Both the government and non-governmental authorities should take establishing action plans towards these ends seriously.

23.6.4 Cultural Background

Any management strategy can be well executed if and only if it is in tune with the cultural background and traditions of the concerned community. A well-planned approach to cultural issues therefore becomes a requisite in this case too. This should not be considered a simple step in a country like India, which is rich in cultural diversity.

23.6.5 Correct Selection of the Approach

The correct selection of an approach that implements the plan, ensures community participation, and increases community understanding is pivotal to the success of action plans. The approach can be based on discussions, awareness programs, Participatory Rural Appraisal tools, presentations, posters, pamphlets, booklets, workshops, and any other innovative ideas that assure the community is completely reached while making them involved and interested. The approach should vary depending on the target groups and objectives such as increasing awareness among students, acquiring ingenious knowledge or local wisdom from elders, etc.

Even after implementing well-planned and accurate action plans, strategies, and similar activities, two major problems are usually observed following implementation: (i) a lack of sustainability in the implemented activities, and (ii) a lack of assessment. Following each disaster, an enormous inflow of resources and money can be seen. What is required most at this point is to properly channel those inflows. Similarly, soon after the event, numerous programs and projects sprout up, of various dimensions and implemented by various authorities or organizations, be they government or non-governmental. These projects and programs usually completes within 3–4 years. Outside of the work done during the project period, follow-up programs are rarely seen. At a minimum, there should be an assurance that the community has reached the level needed to maintain the activities and processes implemented that can be regarded as sustainable before the completion of these projects. There is always the need to assess whether a project promotes a continuous process of development. Thus, assessment should be an integrated part of disaster management strategies, and assessments should be conducted at pre-determined time intervals. This time interval can be determined based on the purpose of the assessment. It will be better to include community members in these assessments to assure sustainability

23.7 Conclusion

It is true that disasters, especially the 2004 tsunami, have spurred disaster management activities in India. Still, there are communities that strive for even basic necessities, which in turn makes them vulnerable to disasters. There are pitfalls and gaps yet to be filled. Each disaster provides an opportunity to strengthen our capacity to respond appropriately the next time. We have to ensure that we are capable of making use of these opportunities in their true sense. To accomplish that, it is essential in finding out the pitfalls in the post-disaster management strategies that were undertaken in response to previous disasters. Government decisions to shift the focus to pre-disaster planning has been well-established through the formulation of policies, strategies, and actions, but now its high time to ensure the communities concern more to be shifted towards pre-disaster paradigm. This requires a tiresome amount of effort in educating and making them aware of the significance of predisaster planning and management. Management strategies should be designed focusing on the community, their culture, their need and vulnerabilities where they have to be more strengthened. Hence, multidisciplinary approaches to disaster management strategies are indispensable, as they can ensure community empowerment, in its true sense, which can lead to resilient societies that can, in the long run, make the entire nation resilient.

References

ADRC (2009) Multi-national mission to the tsunami affected areas in India. Final Report

Safer Cities 21 (2008) Community empowerment and disaster risk reduction in Chittagong city. Asia Disaster Preparedness Center, Bankok, Thailand

Clarke DJ, Grenham D (2013) Microinsurance and natural disasters: challenges and options. Environ Sci Pol 27s:s89–s98

Community Disaster Recovery Guide (2006) A guide for BC local Authorities and First nation. Provincial Emergency Program, Ministry of Public Safety and Solicitor General, British Columbia, Canada

Ferdinand I, O'Brien G, O'Keefe P, Jayawickrama J (2012) The double bind of poverty and community disaster risk reduction: a case study from the Caribbean. Int J Disast Risk Reduction 2:89–94

Gupta A (2000) Vulnerability and disaster management in India. Prehosp Disaster Med 15(3):s98

IFRC (2005) Disaster management Bill, 2005. http://www.ifrc.org/Docs/idrl/512EN.pdf#search=' Bill+2005+disaster+management'. Accessed 26 November 2013

Jackson T, Mitchell S, Wright M (1989) The community development Continuum. Community Health Stud 8:66–73

Labonte R (1994) Health promotion and empowerment: reflections on professional practice. Health Educ Q 21:253–268

Laverack G, Wallerstein N (2001) Measuring community empowerment: a fresh look at organizational domains. Health Promot Int 16(2):179–185

McBean GA (2012) Integrating disaster risk reduction towards sustainable development. Curr Opin Environ Sustain 4:122–127

Metri BA (2006) Disaster mitigation framework for India using quality circle approach. Disaster Prev Manag 15(4):621–635

Ministry of Home Affairs (2005) Annual Report 2004–2005. Government of India, New Delhi

Murty CVR, Jain SK, Sheth AR, Jaiswal A, Dash SR (2006) Response and recovery in India after the December 2004 Great Sumatra earthquake and Indian Ocean tsunami. Earthquake Spectra 22(S3):S731–S758

Nakagawa Y, Shaw R (2004) Social capital: a missing link to disaster recovery. Int J Mass Emerg Disasters 22(1):5–34

Paton D, Johnston D (2001) Disasters and communities: vulnerability, resilience and preparedness. Disaster Prev Manag 10(4):270–277
Rahman HA (2012) Community and tsunami disaster. International conference on environment science and engineering, p 32
Rissel C (1994) Empowerment: the Holy Grail of health promotion? Health Promot Int 9:39–47
Sharma RK, Khanna M (2013) Disaster Management: An Overview. Asian J of Adv Basic Sci 2(1):46–52
Shaw R, Krishnamurthy RR (2009) Disaster management global challenges and local solutions. Universities Press, India
UNISDR (2009) Disaster risk reduction in the United Nations: roles, mandates and areas of work of key area of United Nations entities. Geneva, Switzerland

Chapter 24
Impact of Higher Education in Enhancing the Resilience of Disaster Prone Coastal Communities: A Case Study in Nemmeli Panchayat, Tamil Nadu, India

R.R. Krishnamurthy and K. Kamala

Abstract The Government of Tamil Nadu has launched constituent colleges in the State during 2011 and 2012 in rural areas where the Gross Enrolment Ratio is low, which are managed by the State Universities. Nemmeli is an important coastal "hot spot" on the north-eastern part of Kanchipuram District, which has a cluster of about 15 coastal hamlets with majority of fishing community. Coastal beauty is endowed with lush green plantations bordering the shoreline and low-lying wetlands with agricultural lands on the landward side. Before establishing permanent infrastructures the University of Madras Constituent College is temporarily running in a renovated Cyclone Shelter and in Elementary School blocks. The Government of Tamil Nadu has chosen this region to establish a "Model Degree College", with zeal to produce under-graduates in Literature, Commerce and Computer Applications of the highest quality to cater the changing needs of the society and industry. About 400 students have enrolled for 2011–2012, 2012–2013 and 2013–2014 academic year courses and all of them are first generation learners from a family background of Below Poverty Line (BPL). As the area is prone to cyclone almost every northeast monsoon and devastated by 2004 Indian Ocean tsunami the students and the local community have realized about the importance of DRR through preparatory measures. An elective course on Disaster Management has been introduced in their curriculum and also conducted awareness campaigns through National Social Service (NSS) programmes. Local Residents' Alliance (LRA) has also formed with the support and guidance from SEEDS India to mobilise and sensitize the local community in DRR. Additionally, international collaboration with Australia and Japan has been established to carry out joint academic

R.R. Krishnamurthy (✉)
University of Madras Arts and Science College (University Constituent College), East Coast Road, Nemmeli Campus, 603104 Kanchipuram Dist., Tamil Nadu, India
e-mail: rrkrishnamurthy@gmail.com

K. Kamala
Benchmark Electronic Systems, Chennai, India

R. Shaw (ed.), *Recovery from the Indian Ocean Tsunami: A Ten-Year Journey*,
Disaster Risk Reduction, DOI 10.1007/978-4-431-55117-1_24

activities in disaster management. This chapter is devoted to document the establishment of a community college and to evaluate the impact of training and capacity building carried out in the field of disaster management, which helped the communities resilient to disasters apart from creating impact on societal transformation in this region.

Keywords Community based disaster management • Community college • Disaster higher education • Local residents' alliance • Nemmeli Panchayat

24.1 Introduction: Higher Education Scenario

India is considered as third largest country in the World as far as the Gross Enrolment Ratio (GER) of students in higher education stream. Few decades before the GER was recorded as less than 6 %, which is steadily growing and doubled during the last decade with a total number of about 613 universities which includes State, Central, Deemed and Private universities and around 33,000 colleges apart from 39 Institutes of National Importance (INI). However, this is far below than global GER of 26 %. Hence the Government of India is set a target of 30 % GER by the year 2020, which may need additional 500 universities and 31,000 colleges (www.reformingeducation.in; ASHE – Annual Status of Higher Education in States and UTs 2012). Higher education in India has been considered as a silent disaster/quiet crisis/sick child by the academics and stressed the decision-making body towards revamping higher education in the country. Additionally, the rural talents are being forced to migrate to cities to pursue higher studies and employment. The importance of higher education is also paved way for the initiative on National Commission for Higher Education and Research (NCHER) in 2011 by the Government of India. Hence the twelfth five year plan initiatives of the University Grants Commission (UGC), which is the apex body of higher education include not only revamping of curriculum but also encourages equity in education, multi-disciplinary studies, bilateral research, international collaborations, establishment of new campuses and centres of excellence, private institutions and many more.

However, the State Governments in the country have played a vital role in providing affordable and accessible higher education to the rural population through the establishment of constituent colleges, which are directly managed and maintained by the State and Central Universities in India offering undergraduate as well as post graduate courses in arts, law, physical education, science, engineering, technology and medical disciplines. The important criteria being considered for selecting the location is based on the Gross Enrolment Ratio (GER), especially rural areas of backward districts spreading from all over the country where GER is very low. Since the main objective of these constituent colleges is to bring the rural population in to the main stream of higher education, the Constituent Colleges are being considered as well as called as "Community Colleges" or "Model Degree Colleges" in general. Being Government funded the constituent colleges always aim to attract rural talents by liberalizing fee and by providing other incentives. Here is the full cover story on the genesis of a community college and how the motto and vision are helpful towards coastal rural development.

24.1.1 Constituent College at Nemmeli: Its Genesis

Soon after forming government the Hon'ble Chief Minister of Tamil Nadu has launched 11 constituent colleges in the State in July 2011 by choosing the rural places where the GER is low. The colleges are attached to state universities based on its location. Out of 11, one college is attached to the University of Madras, which is one of the first three academic institutions established in the year 1857 by the British India and considered as a mother university for several universities in South India. Normally the colleges in its jurisdiction will be affiliated to the University and this is the first time in the history of 155 old university directly run an undergraduate college as part of the university like various departments in the university system. This is a new experience for the university administration and initiated the process in a war footing stage in coordination with the state government in identifying temporary building, admission process, recruitment of faculty and staff etc. in order to start the academic year courses. Till the establishment of permanent infrastructures the government has provided an unused cyclone shelter to renovate and to do internal modifications to start the college (Fig. 24.1). Nemmeli is an important coastal "hot spot" on the northeastern part of Kanchipuram District, which has a cluster of about 15 coastal hamlets with majority of fishing community. Coastal beauty is endowed with lush green plantations on the seaward side and low-lying wetlands with agricultural lands on the landward side. University Constituent College at Nemmeli (UCC Nemmeli) is located on the slope of a stabilized sand dune is an ideal location to have sea breeze from the Bay of Bengal throughout the day.

24.1.2 Nemmeli Panchayat

Nemmeli is one of the 53 panchayats in Kanchipuram District with a majority of fishermen community live in villages consist of cluster of houses (Fig. 24.2). Each village normally has a population of about 500 with approximate number of about 80–90 houses. The East Coast Road (ECR) which connects the States capital Chennai with Puducherry was formed about a decade before in order to attract tourists (both national and international) to the picnic parts and amusement parks along the coast which also has a UNESCO's designated World heritage site called Mamallapuram. The ECR is formed on the stabilized sand dunes, which clearly divides the coastal villages on the sea front and the landward side. The local community live in the sea front are mainly engaged in sea fishing and in the land ward side are in agriculture, aquaculture, salt pan, etc. New Kalapakkam is one of

Fig. 24.1 Converted cyclone shelter for the college

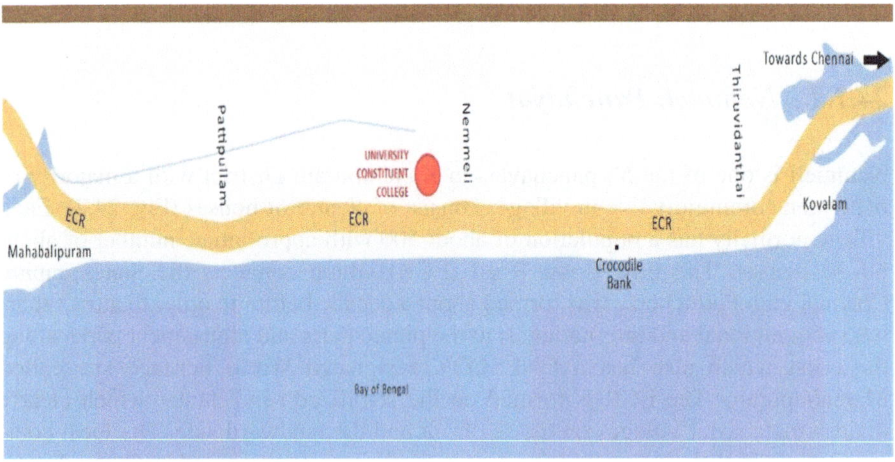

Fig. 24.2 Location of the Nemmeli Panchayat

villages under Nemmeli Panchayat has about 500 population and there are relocated from the place called Kalpakkam nearly three decades before at the time of establishment of Nuclear Power plant at Kalpakkam. Being very conservative orthodox coastal community mainly depending on fishing and other related coastal activities. Their connectivity to outside their Taluk and District is improved only after the establishment of ECR. Most of the older generation in this coastal region have not completed their schooling and the younger generation will normally complete their schooling and then follow their parent's occupation. The main reason for the younger generation not going to college education is due to poor economic condition as well as the distance to go to few private colleges and the course fee may not be affordable to them. Hence this coastal rural region has been not received much attention to provide them opportunity for affordable and accessible higher education and was considered as a neglected region by the local community. Hence the coastal population in this region has no other options for their livelihood expect following their traditional way since independence of the country in 1947.

Considering the natural disasters this region was severely affected by 2004 tsunami and cyclonic storms "Thane" and "Nelam" in 2011 and 2012. Though the life loss is very limited the infrastructure and environment are severely damaged during the disaster times. People living in the shore front villages are shifted to a temporary shelter that was located about 10 km away from their place. Most of the houses were re-build with the support from various donor agencies, which took about nearly 2 years to become normalcy in this region after 2004 tsunami. Many drop outs from the school due to remote location of temporary shelter and hence both the elementary and higher secondary schools run by the Government of Tamil Nadu have low enrolment ratio from 2004 onwards. The infrastructure built for accommodating expected strength is not being used and hence many school blocks are kept unused for several years after 2004 tsunami. After 7 years of 2004 tsunami the establishment of constituent college gave a new scope for bring this rural population in to main stream of higher education.

24.1.3 Vision and Mission of the Constituent College

Given opportunity to establish the university constituent college in a neglected place of higher education the motto of the college is framed as *"Dedication to the Rural Communities we serve and to produce Graduates to become Strong Leaders, World Class Professionals and Good Citizens with a strong sense of Personal Accountability and Social Responsibility"*. Hence the vision is defined by the founder principal as *"To be the premier source for education, training and capacity-building for coastal rural population of this region"*. The mission is to provide a foundation for a lifetime of learning through innovative and user-friendly educational environments, opportunities, and experiences that enable students to grow, thrive and prosper. Learning at UCC Nemmeli extends beyond the classroom to encompass a broad range of field based, professional and extracurricular opportunities. Ultimately we

Fig. 24.3 College logo

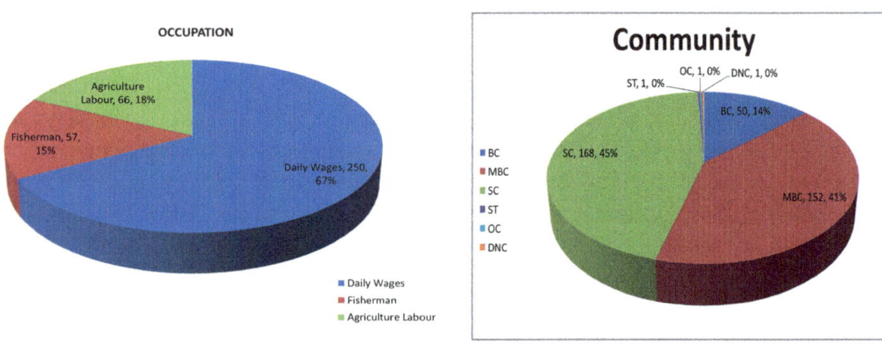

Fig. 24.4 Status of students' parents occupation and the caste they belong

need to produce skilled younger generations with an empowerment to move upward in their career path self-sufficiently. After a competition conducted to the students to finalize the logo designed for the college highlights the local landscape and the importance of life time learning (Fig. 24.3).

With a small dedicated team of staff and faculty the first academic year courses are started with 67 students and all of them have studied in Tamil medium till their schooling. The Government of Tamil Nadu has advised to start with four undergraduate courses Viz. Bachelor of Arts in Tamil Literature, Bachelor of Commerce (General), Bachelor of Commerce (Computer Applications) and Bachelor of Computer Applications (BCA). Tamil literature students also have to study English course till their final year and all other courses are being offered in English medium. This is the greatest challenge to us to transform these rural and economically backward students (Fig. 24.4) in to University curriculum. Nearly 45 % of the students are belonging to schedule caste followed by most backward and backward community and all the students are first generation learners. Considering the students' parents

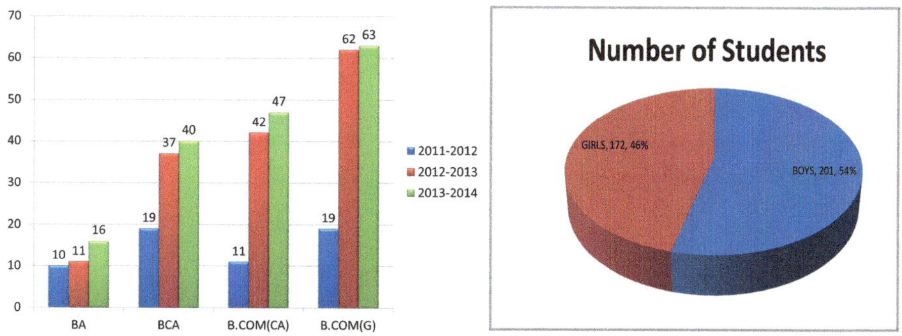

Fig. 24.5 Students' enrolment status course-wise and gender-wise

occupation majority of them are engaged as daily wages in salt pan, fishing, agriculture, construction industries etc. (Fig. 24.4). Only 15 % directly own boat and engaged in sea fishing and marketing. Our faculty team efforts initially spent mainly for confidence building and enhancing their life skills before training them in core subjects. Technology based teaching methods (ICT tools) and interactive sessions helped the students to multiplying their learning opportunities. The first academic year (2011–2012) there was much demand for computer application course and subsequently there was a higher demand for commerce courses (Fig. 24.5). Also we kept in mind the quality while meeting the basic criteria of affordability, accessibility and equality. The university has fixed a special fee structure, which is equivalent to 45–50 USD approximate for full academic year. All enrolled students are entitled to receive government scholarship, which is equivalent to 100 USD approximate with other students' amenities like free bus pass for their travel from home to college and back including holidays period. Most interestingly the enrolment of female students has increased considerably and as of now they are on the higher side when compare to male students (Fig. 24.5).

Boom of multinational IT industries, expansion of Chennai city limits in the form of new townships, gated communities and the expansion of coastal tourism and other industries like sweater desalination plants etc. gave confidence to the younger generation to enrol for their undergraduate degree courses in the college. From the academic year 2013–2014 the college has operates in two shifts 8:00 am to 1:00 pm and 1:00 pm to 6:00 pm with total number of 400 students. The Government of Tamil Nadu has earmarked 10 acres of land, which is just adjacent to the temporary building to construct permanent infrastructures such as modern class rooms, administrative blocks, conference hall, playground, students' hostel and staff quarters. Witnessing the fast expansion of the college the real estate business has also picked up very well and many private parties are interested to buy lands in this region. Apart from enhancing the literacy level the local community are having the feeling that the growth of the constituent college will make a societal transformation in this region due to the increase in land values, more new establishments, employment opportunities etc.

Fig. 24.6 Questionnaire survey and group activities on hazard mapping of villages

24.1.4 Introduction of Disaster Management in the Curriculum

Fundamentals of disaster management as an elective course with five units covering all important aspects have been introduced from the first academic year (2011–2012). The three main objectives of this 2 credit course are (1) to develop an understanding on the process of disaster management, (2) to understand the mitigation programmes including technology applications and (3) to understand the disaster management policies and legislations. All theoretical aspects will be taught in the class room with lecturing, interactive sessions, group discussions and take the students to the nearby villages to prepare hazard maps, collect household survey data through interactions with the local community (Fig. 24.6). There are several advantages in introducing this course (1) students have faced fury of 2004 tsunami devastation, (2) their practical experience with tsunami and coastal floods in monsoon timing helped to understand the importance of this course and (3) easy to visit field sites to gain more practical knowledge. Whatever they learned in the class room can be linked to the field visit experience on the same day. Though they have some lacking in English language skills for their presentation their understanding on the subject has been observed as highly commendable. It was also observed that majority of the students have a sense of feeling that their class room knowledge should be used by their community and their home towns should be prepared for disasters. The students' skills are also demonstrated by conducting poster competitions on world disaster day and debates on disaster preparedness. The importance of disaster preparedness through various extension activities of the college has emerged very strongly, which paved way for organizing field camps, awareness campaigns in all the neighbourhood villages from the subsequent academic years.

From second academic year (2012–2013) with a sufficient number of trained students through National Social Service (NSS) programmes two field camps are conducted to create awareness on the importance of disaster preparedness including emergency services, identifying environmental issues, health hazards etc.

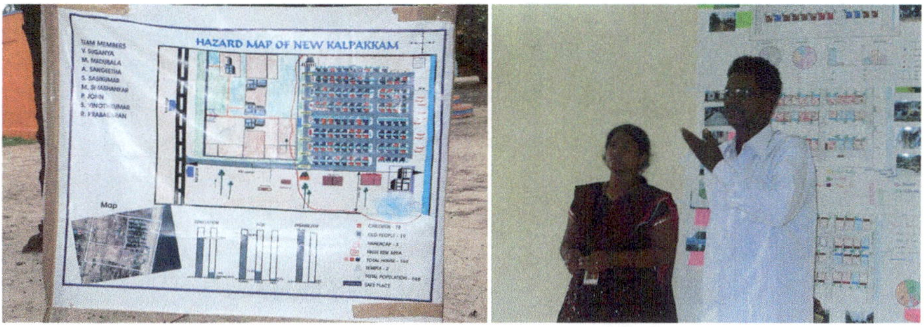

Fig. 24.7 Display and demonstration of hazard maps to the local community

A two-day disaster preparedness for Nemmeli Panchayat training programme was conducted in July 2013 in which about 60 students from the constituent college and about 60 students and NSS officers from city colleges are participated and camped at New Kalpakkam Kuppam—a fishermen village just a km away from the college. This is an another important occasion in which both rural and city students interacted, stayed together and jointly prepared hazard map and disaster management strategies and demonstrated to the local community, which received much appreciation by the stakeholders i.e. village leaders, media etc. (Fig. 24.7).

Apart from core courses now the students have the feeling that proper learning of disaster management course is for their life saving skills. Faculty are trained on specialized disaster management course and training programmes conducted by the Academic Staff College in order to handle this course more effectively.

24.1.5 Initiation of International Collaborations in Disaster Management

It is felt that the new institution should enrich its skills in disaster management through international collaborations and hence the process of experts exchange visits and joint academic events are planned under Indo-Japan bilateral cooperation with Kyoto University and Indo-Australian bilateral cooperation with University of Melbourne and Victoria University. An initiative by Kyoto University, Japan, in networking of Universities/Institutions in Asia under the title "Asian University Network for Environment and Disaster Management" www.auedm.net has helped in enriching curriculum in the University of Madras (Fig. 24.8). It is very much experienced and realized that the international collaborations will be immensely helpful not only enriching the disaster management experience from Australia and Japan but also to learn the people life style, environment etc. through interactive session by the visiting experts.

The Department of Science and Technology, Government of India under the Australia (Fig. 24.9)—India Strategic Research Fund (AISRF) has sanctioned a

Fig. 24.8 Consul from Consulate General of Japan and researcher from Japan are conducting interactive sessions on the culture and environment of Japan

Fig. 24.9 Australian delegation interaction with local community in Nemmeli village

grant to conduct an international workshop entitled "Spatially enabled nations: Workshop for a meta integrative framework towards improving disaster management practices in India and Australia" during 21–24 January 2013 in which about 40 key experts from the following organizations have participated and deliberated on the importance of improving disaster management practices:

Department of Science and Technology, New Delhi
University of Madras Chennai
Indian Institute of Technology Madras in Chennai
National Institute of Technology, Karnataka
National Maritime Foundation, Chennai
Department of Revenue, Health and Family Welfare, Government of Tamil Nadu
Commissioner of Information, Government of Gujarat
Anna University Chennai
Coordinates Magazine and
SEEDS India

This international workshop analysed current practices and outlined best practice measures for both governments. A review of national technological capacity has also undertaken based on the state-of-the-art research in natural and unnatural disasters. The presentations from government and non-government officials cantered on current practices with the view to identifying real challenges; presentations from academics cantered on high level engineering inputs for the benefit of decision makers and the discussions cantered on the way forward for improving current practice. Apart from technical presentations, panel discussion, a visit to neighbourhood villages of the constituent college is an eye opener to the international participants to know the ground reality. This event is just a beginning of our bilateral collaboration with Australia and is expecting more long term academic and research activities in the field of disaster management. Certainly Nemmeli Panchayat is an ideal location to conduct field based projects in order to enhance the resilience of the local community.

International collaboration with Australia and Japan helped to understand how best the modern technology tools like satellite remote sensing, telecommunication, medical emergency services, solar and biogas energies etc. are important to the local community and how to proceed with the preparation of community resilient to disasters. Planning is underway to document indigenous knowledge and to train the community to the use of modern application tools like mobile navigation etc. under international collaboration. Awareness on disaster preparedness among the local community varies considerably among Australia, Japan and India and also the impact of disaster recovery and the experience in institutional cooperation mechanisms. It is felt that more focused research for vulnerable sites are warranted, which will be helpful in the following ways towards making resilient communities:

1. address the lacking in institutional cooperation mechanisms,
2. self-reliance of local community to prepare with emergency food, medicines etc.,
3. building rapport among local community, government machineries towards effective implementation of bottom-up approach, and
4. transfer of modern technology tools to the field for DRR.

24.1.6 Local Residents' Alliance in DRR

Realizing that the more focused research through international collaboration will take time, requires funds and manpower, and hence there should be another way to reach the community to do field based activities with limited resources and available expertise locally. Under this circumstance the SEEDS India http://www.seedsindia.org came forward with an innovative idea of establishing a Local Citizens' Alliance (LRA) for Disaster Risk Reduction in the beginning of 2013 with the following objectives:

- Raise awareness on local disaster and climate change induced risks at local level,
- Promote mainstreaming of risk reduction and disaster resilience in local development programmes,

Fig. 24.10 LRA meeting in the college and Members undergoing DRR training at SEEDS New Delhi

- Demonstrate through community led actions to reduce risks at local and neighbourhood level, and
- Advocate for improvement in local risk governance and emergency preparedness for long lasting change.

After the brainstorming sessions and interaction with the village leaders it has been agreed to experiment LRA in Nemmeli Panchayat (Fig. 24.10). As of now there are about 15 members representing all villages in this alliance and many of them are students' parents of the constituent college. This is really very unique way of maximising participation of the community by keeping the college as a secretariat for LRA. More interestingly the student learnt out of class room teaching on disaster management will be exchanged with their parents who are members in the alliance. The Alliance supports the local area in preparing development strategies that link disaster risk reduction and climate change adaptation with their economic growth and poverty reduction objectives and activities are planned in a collaborative manner. Under this Alliance activity two local fishermen had undergone training at SEEDS New Delhi during March 2013 (Fig. 24.10) along with similar participants from different coastal states.

After building considerable awareness through the students and alliance members in the neighbourhood villages, the students are engaged in conducting demonstrations to the local community and efforts are underway to carry out the following activities during the next 2 years in all others 52 Panchayats in Thiruporur Taluk in Kanchipuram District:

- Awareness raising among local community on the importance of disaster preparedness,
- Exercise on hazard and vulnerability maps preparation and utilization,
- Importance of evacuation routes, methods and safety measures,
- Linkage between College curriculum and the field, and
- Prepare the local community resilient to future disasters.

Parallely efforts are underway to prepare community based disaster management (CBDM) plan for Nemmeli Panchayat in order to project this Panchayat as a model to the entire State. Few Television channels have also highlighted the efforts that have taken towards this on the mark of 10th anniversary of 2004 Indian ocean tsunami.

24.2 Integrated Approaches Towards Implementable CBDM

In order to make the vulnerable community resilient to disasters, the following integrated approach is being adopted by keeping the constituent college curriculum as its starting point (Fig. 24.11). In order to reach the end point i.e. CBDM the vital role is being played by the Local Residents' Alliance, which directly support the college curriculum and also enhancing stakeholders participation in collective decision-making towards implementable CBDM. International collaboration is also having an important role both in enriching the college curriculum as well as CBDM preparation and implementation.

Thanks to the Government of Tamil Nadu for providing free laptop to all students of the college, which immensely helpful to the students to carry out students project work by taking various issues on environment and disasters in the Nemmeli Panchayat. Project on "Web based Nemmeli Panchayat Information System" and "Village Information System" are involves the preparation of scientific database, which will support the decision-making on CBDM. Students are using satellite remote sensing data, integration of spatial and non-spatial data towards identification of vulnerable sites etc. are progressing well. Students' project outcomes and LRA activities are linked to strengthen the preparation of CBDM as a science based people centered approach. Few more silent disasters like shore erosion (Fig. 24.12) in few pockets of coastal villages in the aftermath of new engineering structures that

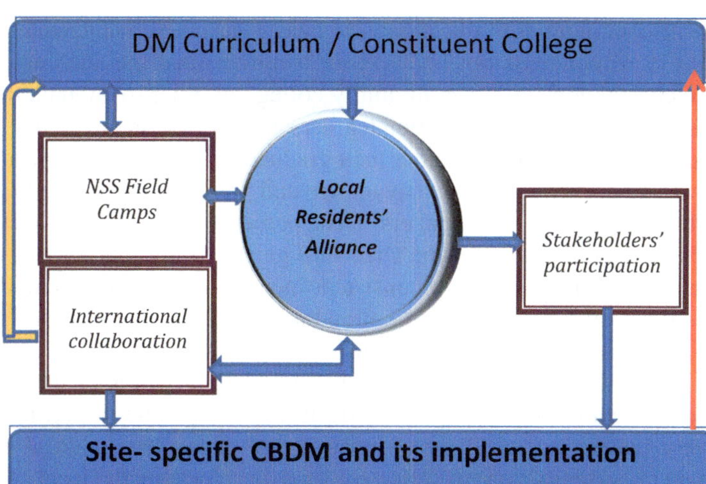

Fig. 24.11 Integrated approach towards site-specific CBDM

Fig. 24.12 Few pockets of coastal villages are undergoing shore erosion recently

all constructed just few years before, free movement of cattle's from the villages which leads to road accidents on the East Coast Road etc. are being given importance to be taken up under the joint activity of the project work and LRA.

24.3 Discussion

Based on the above highlights it is evident that the piecemeal of activities in DRR through class room teaching linking with community participation including the extension activities like NSS and the field based activity like LRA are integrated towards the preparation and implementation of site-specific CBDM in Nemmeli Panchayat, which is focused as a model site in the Tamil Nadu State by the media and other stakeholders. This is one of the salient achievements in the field of higher education in disaster management, which is witnessed with enhancing the resilience of vulnerable communities. A questionnaire survey has been conducted by choosing 1:10 ratio (39 respondents out of 390 total strength) and the students are from vulnerable coastal areas of the community college in order to validate the above. All the respondents are first generation learners and 85 % of them are in immediate need to earn for their family soon after their graduation. Remaining 15 % are female students who do not have the necessity to earn and they may follow their traditional way of life style i.e. as home makers. Cent percent of the students enrolled for undergraduate degree course for the chance of getting better job and this might be due to their change in mind set, inconsistent income, frequent disasters etc. More than 50 % of the students are on the opinion that a college degree will not only provide better job opportunity but also the chance of relocation to safer places (Fig. 24.13). In the event of getting an employment 72 % will not have any second thought for relocation and only 26 % will decide at the time of getting an offer. This is very interesting result which proved that the college education and the employment opportunities are the effective ways in relocating the local community from most vulnerable places. The federal government is struggling to relocate the

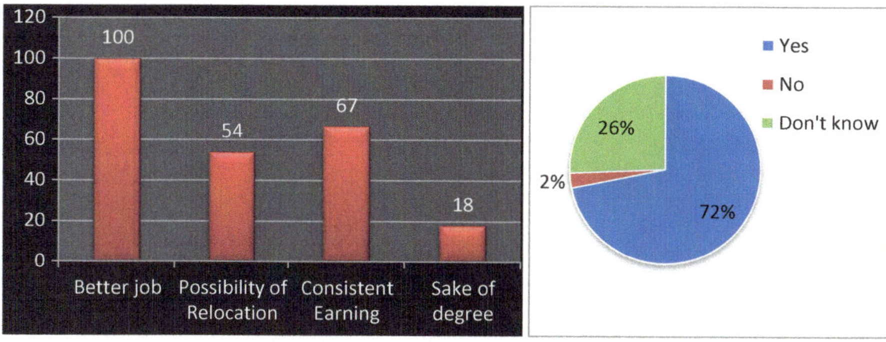

Fig. 24.13 Need of higher education and willingness to relocate

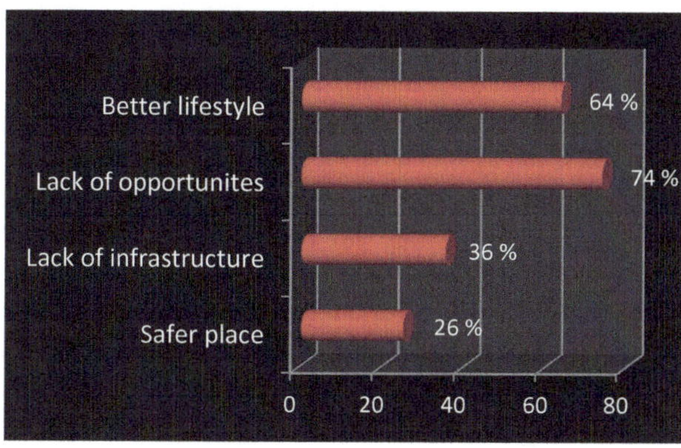

Fig. 24.14 Reasons for relocation

vulnerable local community through the introduction of Coastal Regulation Zone (CRZ) during the last two decades and more. Only a very narrow percentage of students are coming to higher education for sake of getting a degree and not willing to relocate (Fig. 24.13). Considering the nature of job 87 % are ready to engage in any employment, which comes in their way and only 13 % are particular that the job should be relevant to their field of specialization. It is proved that the coastal rural areas, which are lacking resources, alternate livelihood and vulnerable to disasters are reformed very effectively through the introduction of higher education and employment opportunities. This fact is also supported by opinion on the importance of the community college by the respondents. All the respondents have opined that the establishment of a community college is for their betterment and will have highest usefulness to the local community.

Considering the important causes for relocation lack of opportunities and resource potential are the prime factor followed by the ambition to have better life style. Apart from lack of infrastructure and about 26 % of the population consider their place of stay is not safe with reference to natural disasters, which also influences them to relocate (Fig. 24.14). The establishment of the college is found to be

Fig. 24.15 Feedback on the college environment

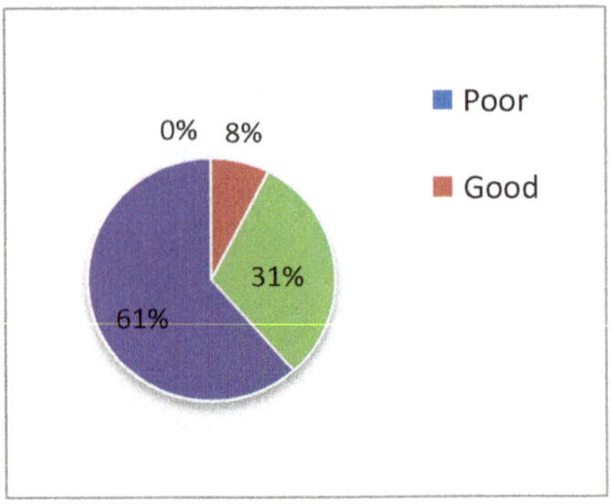

right time for the local community, which facilitate in getting employment opportunity and there by relocation and better life style for the younger population.

The dream of younger population is almost fulfilled by the community college by the sincere and consistent efforts been taken during the last three academic years. Though the college is being operated in temporary buildings due to the delay in constructing permanent infrastructures the dedicated team of faculty have infused the knowledge skills including considerable extracurricular activities to the students (Fig. 24.15). Hence majority of the respondents have rated that the teaching quality is high followed by extracurricular activities and practical skills among all other parameters. The impact of lack of infrastructure such as modern class rooms, conference hall including insufficient library and sports facilities are also reflected in this survey results.

Offering an elective course on disaster management to undergraduate students especially in Arts and Science Colleges is very limited in the State, which might have due to lack of expertise as well as students interest. Also teaching this inter-disciplinary subject requires strong field experience, research experience and adequate teaching materials like videos, usage of ICT, engaging students in practical aspects etc. The experience gained from various national programmes on coastal zone management, tsunami inundation mapping, establishment of national warning system for tsunami and storm surges, training and capacity building in DRR etc. helped to instil the importance of disaster management to the students of this college. More than 60 % of the respondents have rated that the effective of the disaster management course as "Excellent" and 31 % as "Very Good" and 8 % as "Good" (Fig. 24.16).

The course not only helped them to learn theoretical aspects but also the practical skills through the extension activities like National Service Scheme (NSS) field camps and DRR training programmes conducted for the local community. The concept of LRA introduced in the field also provided an opportunity to the students to interact with other stakeholders. The expertise gained in various aspects of disaster management has helped towards better preparedness, educating the family and local community and also to establish a rapport with local bodies as shown in Fig. 24.17.

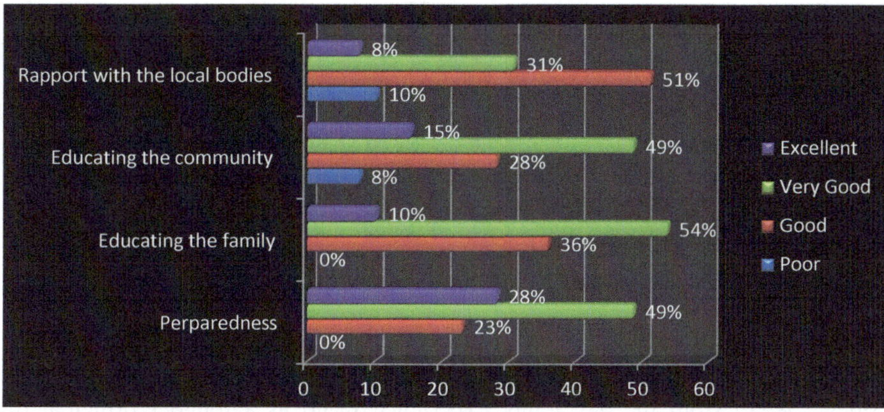

Fig. 24.16 Effectiveness of disaster management course

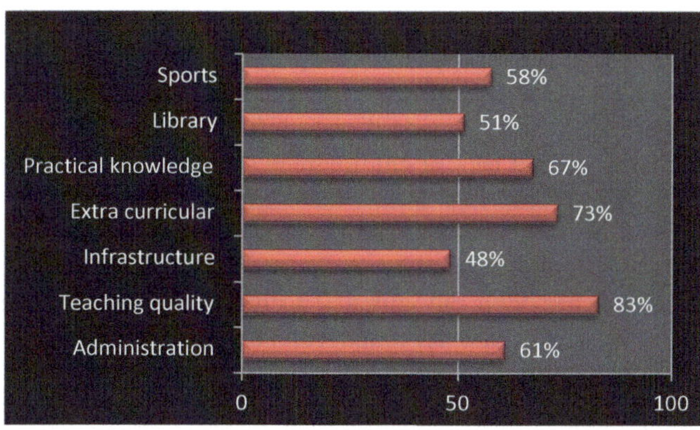

Fig. 24.17 Impact of disaster education

24.4 Conclusions

Higher education in disaster management is not just the curriculum but it is essentially the mode of delivery and the suitable approaches. Ultimately, the role of higher education should lead to societal and economic development and fulfil the expectations of the society (Shaw et al. 2011). Disaster-related education can enhance personal preparedness, which is crucial in mitigating the disaster risks as witnessed in the Andaman coast in Phang Nga province, Thailand (Muttarak and Pothisri 2013). It is clearly evident that the efforts taken in imparting disaster education in a community college helped not to the local community in getting better employment and alternate consistent livelihood but also to enhance their resilience to disasters. The support and contribution from all primary stakeholders

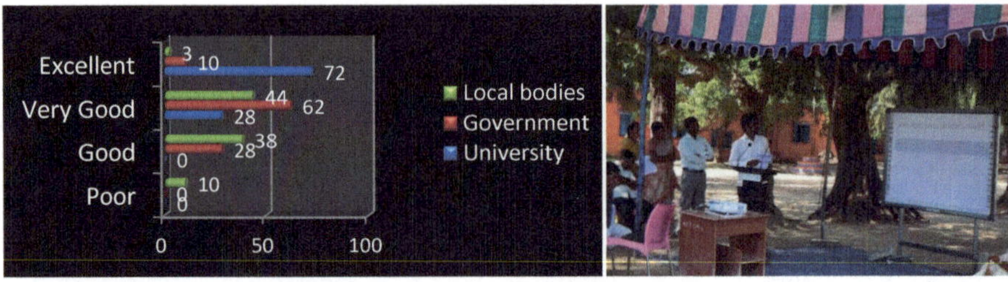

Fig. 24.18 Contribution of stakeholders in the establishment of community college

Viz. University, Government and Local bodies is vital in establishing a community college. Apart from the day today experience faced as a founder principal the respondents of this questionnaire survey have also reflected that the support from local bodies is still below the expectation level. The University's contribution is very high followed by the State Government towards achieving the societal transformation of coastal rural area through higher education projects as shown in Fig. 24.18.

As highlighted by Shaw et al. (2012) it true that the disaster higher education is given less importance when compared with school or community education in India. This is mainly due to lack of employment opportunities in higher education sectors and few educational institutions have winded up the courses due to the limited number of takers. There was lot of debate and discussions in the aftermath of 2004 Indian Ocean tsunami and how to do innovations in higher education. The University of Madras has come out with a new concept of introducing elective subjects on different aspects of disaster management and environmental studies in the existing postgraduate curriculum and few mandatory (core) subjects to the postgraduate courses in earth sciences. In that way the essentials of disaster management are injected in to the higher education.

Shaw et al. (2011) have also commented that having appropriate academic freedom, scope of innovation and flexibility in research, setting up of long term goals, promoting good educational governance through regional and international networks, and ensuring quality control of teachers and students are essentials to achieve tangible results. Normally, all these essentials may not be provided by the institutions. Also the quality control of teachers in rural educational institutions is again a question mark. However, the experiences that are gained from this study confirm that the will and skill of the person heading such institutions has higher role in achieving it. During the academic year a quantitative assessment on the academic potential of the college is assessed based on the performance of the teachers and students. All the stakeholders Viz., students, teachers, students' parents and local leaders have participated in brainstorming sessions in which educational experts will comment on the academic performance and suggest remedial measures (Fig. 24.19). This is found to be an important innovative approach to maintain the quality of education and made everyone involved is accountable.

Fig. 24.19 Presentation and discussion on performance of students and teachers in brainstorming sessions

24.5 The Way Forward

Efforts are underway in establishing permanent infrastructures including students' amenities in a 10 acre land area earmarked by the Government of Tamil Nadu. Proposals are under consideration to introduce more inter-disciplinary and integrated 5 year courses under the 12th five year plan period will also further leads to total transformation of this coastal region by 2020. Disaster education may be resulted in disaster preparedness. But in this project we made a unique attempt to link these two components. This is an ideal example for the usefulness of rural higher education projects that helped the vulnerable communities to resilient to disasters and also have created impact in their mind-set to opt for alternate livelihood and relocation opportunities. Hence the establishment of community college has been considered an innovation in DRR, which has happened nearly a decade after the 2004 Indian Ocean tsunami.

Acknowledgements The authors are thankful to the University of Madras, Government of Tamil Nadu, SEEDS India, NSS Regional Centre in Chennai, Consulate General of Japan at Chennai, Kyoto University, Department of Science and Technology, AISRF and the faculty and students' team at the University Constitute College for their support and cooperation in successfully carry out the tasks mentioned in this chapter. The authors are also thankful to the Nemmeli Panchayat including village leaders and members of Local Residents' Alliance for their support and cooperation. Special thanks to Mr. V. Senthil Kumar and his team in the Computer Science Department for their help in analysis of questionnaire data.

References

ASHE – Annual Status of Higher Education in States and UTs (2012) Planning Commission. Government of India, Summary Report, p 16

Muttarak R, Pothisri M (2013) The role of education on disaster preparedness: case study of 2012 Indian Ocean earthquakes on Thailand's Andaman Coast. Ecol Soc 18(4):51

Shaw R, Mallick F, Takeuchi Y (2011) Essentials of higher education in disaster risk reduction: prospectus and challenges. Community Environ Disast Risk Manag 7:95–113

Shaw R, Takeuchi Y, Krishnamurthy RR, Pereira JJ, Mallick F (2012) Universities and community-based disaster risk reduction. Community Environ Disast Risk Manag 10:55–66

Part IV
Sri Lanka

Chapter 25
Reviewing Indian Ocean Tsunami Lessons Learnt Practices of Sri Lanka: In Order to Emphasize Disaster Risk Reduction Endeavours

Poorna Yahampath

Abstract Most devastated single event disaster of the country is Indian Ocean tsunami 2004. Post-tsunami recovery and reconstruction attempts have been mainstreamed by Government and other organizations. Beginning of this chapter shows some facts and figures on tsunami damage and briefed before-tsunami disaster management structure and functionality of the country and then emphasised planning and implementation of recovery and reconstruction. The main body of the chapter is on tsunami recovery and reconstruction lessons learnt practices which was leading the disaster risk reduction endeavours that has been elaborated through, institutional arrangements and coordination mechanisms; government involvement in recovery and reconstruction; collaborating for working with partners, and mechanisms; coordination and involvement of CBOs; strategies for institutional and coordination mechanism; institutional and coordination mechanism at central and local level; integration of environment and ecological aspects and by case-studies on changing of tsunami housing policy; declaration of tsunami buffer zone linked to coastal zone.

Keywords RADA • Recovery and reconstruction • Risk reduction • Sri Lanka • Tsunami

25.1 Introduction

Issues, challenges as well as developments in post-disaster recovery and reconstruction being experienced in both developed and less developed countries. It's highlighted especially in large scale disasters. Standing government agencies have been tasked

P. Yahampath (✉)
Consultant – Integrated Environmental Disaster Risk Management, Environmental Disaster Risk Management Group, No 112, Piliyandala Road, Maharagama, Colombo, Sri Lanka
e-mail: poorna.yahampath@gmail.com

R. Shaw (ed.), *Recovery from the Indian Ocean Tsunami: A Ten-Year Journey*, Disaster Risk Reduction, DOI 10.1007/978-4-431-55117-1_25

Table 25.1 Deadliest tsunamis of the world

Rank	Death toll	Event	Location	Date
1.	230,000	2004 Indian Ocean tsunami	Indonesia	December 26, 2004
2.	123,000	1908 Messina earthquake	Italy	December 28, 1908
3.	36,417–120,000	1883 eruption of Krakatoa	Indonesia	August 26, 1883
4.	40,000–50,000	1755 Lisbon earthquake	Portugal	November 1, 1755
5.	30,000–100,000	Minoan eruption	Greece	12nd Millennium BC
	40,000	Possible tsunami—a cyclone origin	Taiwan Strait	1782
6.	31,000	1498 Meiō Nankaidō earthquake	Japan	September 20, 1498
7.	30,000	1707 Hōei earthquake	Japan	October 28, 1707
8.	27,122	1896 Meiji-Sanriku earthquake	Japan	June 15, 1896
9.	25,674	1868 Arica earthquake	Chile	August 13, 1868
10.	23,024	1293 Kamakura earthquake	Japan	May 27, 1293

Source: Wikipedia

with additional responsibilities of managing recovery and reconstruction or if exceed the capacity new agencies have been established with a strong coordination role for support the various line agencies and other organizations. But these agencies faced with many encounters, e.g. managing multiple donors and their resources intended to implement in the field with limited time periods. Mainly these agencies established in national and sub-national level, but in locally additional responsibilities imposed to existing administrative bodies and local organizations to operate in grass-root level with less supplementary facilities.

Experiences from recent large scale disasters like Pakistan floods 2011; East Japan earthquake and tsunami 2011; Haiti earthquake 2010; China-Sichunan earthquake 2008; Myanmar cyclone Nargis 2008; Pakistan earthquake 2005; Indian ocean tsunami 2004; India Gujarat earthquake 2001…etc., can be analysed to understand lessons learnt of post-disaster recovery and reconstruction practices under disaster risk reduction framework (Table 25.1). The Chapter is focused on Sri Lankan experience in 24th Dec, 2004 Indian Ocean tsunami, it's the most deadly tsunami in the world history; killed 230,000 people in 14 countries. Please refer following table for world tsunami history (Tsunami Global Lessons Learnt Project 2009).

25.2 Damage

Most devastated single disaster event of Sri Lanka is the 2004 Indian Ocean tsunami; followed by extreme winds and cyclone, animal attacks and landslides. Highest number of individuals affected by floods (42 %) and droughts (43 %), but tsunami takes only 3 %. Figure 25.1 showing analysed percentage of affected individuals and percentage of killed individuals by disasters recorded since 1965 to present.

Some of the key human, economic and social impacts of the tsunami can be described as follows. Nearly 35,322 people killed and more missing; 21,441 number

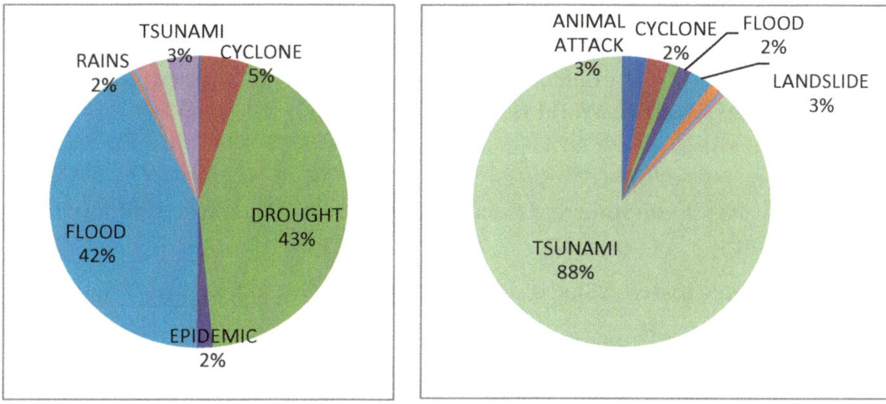

Fig. 25.1 Affected individuals and loss of lives by disasters. *Source*: Poorna Yahampath, Plan International Sri Lanka, DPP National Workshop (2012); reproduced based on http://www.desinventar.lk

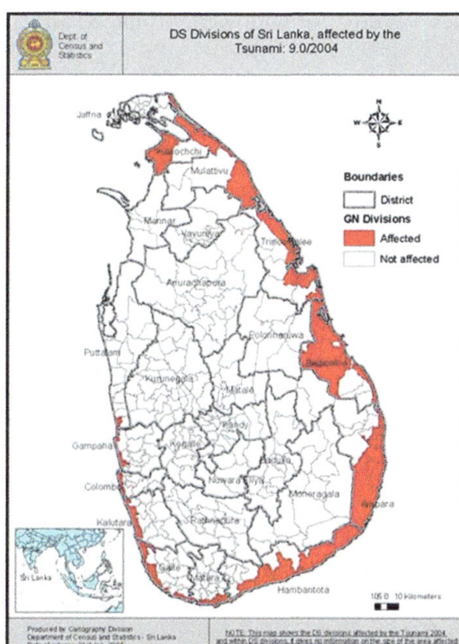

Fig. 25.2 Tsunami affected Divisional Secretariat Divisions of Sri Lanka. *Source*: TAFREN Central Bank, MoF (2005)

of people injured; 516,150 IDPs recorded; 105,293 Houses damaged or destroyed; about one million people affected; over two thirds of the county's coastline areas devastated (Fig. 25.2); 1,500 children left without parents; social networks were disrupted; assets, water and electricity supplies and infrastructure severely affected;

sanitary conditions deteriorated; lack of clean drinking water and shelter observed; delayed access to medical aid was strong; psychological trauma was observed; short term and long term environmental damages occurred in affected coastal areas (ADBI 2007; ADB, JIBC, World Bank 2005).

Box 25.1 Key Economic and Social Losses due to Dec, 2004 Tsunami in Sri Lanka [1]

Key economic losses Value of lost assets Million 900 US $

- Number of lost livelihoods 150,000
- Number of houses damaged 98,000
- Proportion of fishing fleet destroyed 75 %
- Extent of salinated agricultural land 23,449 acres
- Damage to tourism infrastructure; Large hotels 53 out of 242 and Small hotels 248
- Affected related small enterprises 210

Key social losses Number of widowed, orphaned and affected elderly and disabled 40,000

- Number of health facilities damaged 97
- Education facilities damaged
    - Schools 182
    - Universities 4
    - Vocational Training Centres 15
    - Schools used a camps for IDPs 446
    - School children affected 200,000

25.3 Before-Tsunami Disaster Management Structure and Functionality

When tsunami hits, the country was very fresh for such an experience, also the word "tsunami" was totally new. There were couple of historic evidences of Sea Surge had been recorded but has limited to folk story and there was no particular lessons learnt brought forward. Although, the country was being used traditional sustainable practices embedded in the society. E.g. Deforestation is not a practice in coastal areas to protect from sea erosion. Basically, executive government/political, central government and local government systems established in the country in order to govern national, provincial, district, divisional, local and village level. Ministry of Social Welfare was operated in national level with responsibilities of disaster

[1] *Source*: TAFREN, Central Bank, MoF (2005).

response with coordination and support of line ministries/departments and forces. National Disaster Management Center (NDMC) was the centralized body to cover disaster preparedness activities. NDMC worked closely with District Secretary who was acting as the leading focal person for overall disaster response and preparedness activities at district level with coordination of Divisional Secretaries who has the decentralized implementation authority. Grama Sewa Niladari (GN) worked under DSs and considered as the grass-root level and community level representatives. Extensive support has been taken from other agencies like Forces, Police, NHDA, NBRO...etc.

District Secretary has to maintain favourable coordination and cooperation with Provincial Chief Secretary and provincial line departments such as provincial education authorities, engineering services. It was mandatory to maintain respectable relationship with all types of executive/political governing bodies to have a smooth disaster response and recovery operations. NGOs worked at national, sub-national and community levels within unsustainable coordination systems, even though best practices were highlighted and brought forward to originate further disaster management arrangements (UNDP/BCPR 2004).

Some organizations were experienced with response systems for serious disasters such as 2002 drought, 2003 floods and landslide which had been a reinforced capacity for kick-off the tsunami response. As an example, Sri Lank Red Cross (SLRCS) included some established institutional setup, trained and equipped human resources and internationally supported systems. Following figure shows Ratnapura branch of SLRCS deployed to heavily damaged Ampara district for tsunami emergency response (Fig. 25.3).

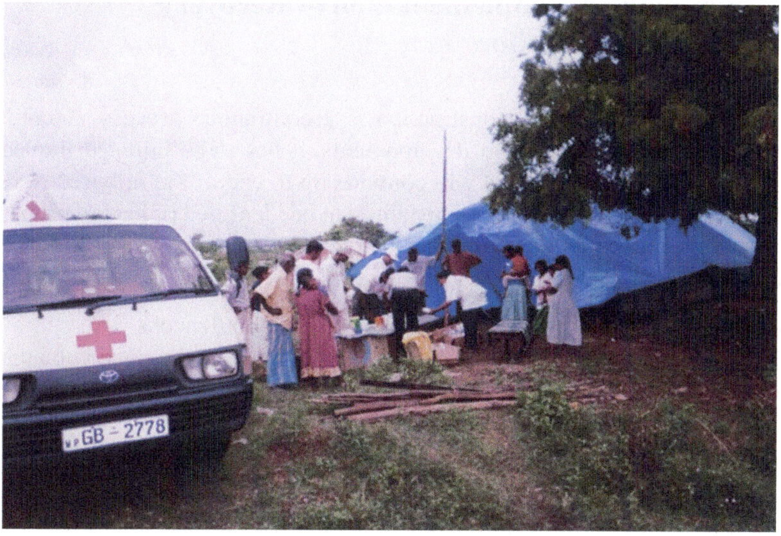

Fig. 25.3 SLRCS tsunami emergency response Pottuvil, Ampara. *Source*: Poorna Yahampath and SLRCS (2004)

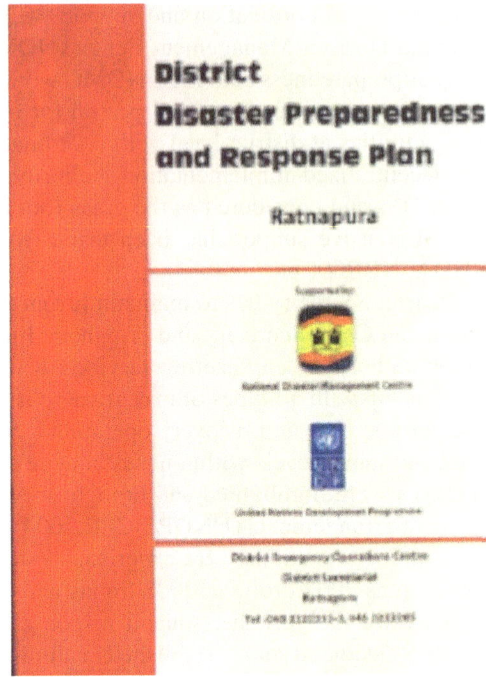

Fig. 25.4 NDMC and UNDP at developed DM plans for districts. *Source*: Poorna Yahampath (2004), Ratnapura District Secretariat (2005), UNDP/NDMC (2004/5)

25.4 Planning and Implementation of Recovery and Reconstruction

Post-disaster recovery and reconstruction is a continuous process. According to recent disaster risk management developments, it has to be initiated immediately after the disaster occurrence and will continue up to end of the fully recovered status. As mentioned below number of benefits can be identified of having proper planning for post-disaster implementation.[2] Some are experienced in post-tsunami recovery operations in Sri Lanka.

- Reduce risk of uncertainties, delays, overlaps and bottlenecks
- Limits poor decisions taken under post-disaster stress and reduce vulnerability
- Clear delegation of roles and responsibilities and facilitates business continuity
- Allows good governance and accountability, which helps in mobilization of resources
- Provides legal basis for effective recovery planning
- Ensures that decision-makers have policies and procedures to be followed
- Establishes the modalities for financing recovery processes

[2] Haas et al. (1977), and UNDP/BCPR (2011).

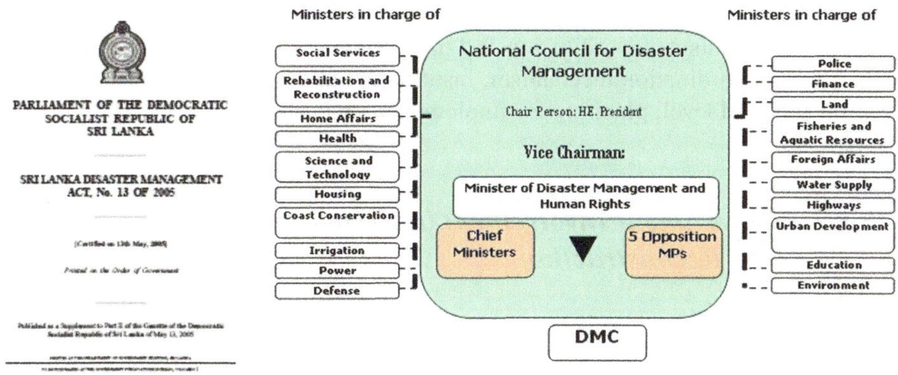

Fig. 25.5 Cover page of Disaster Management Act and overall system of disaster management under the National Council for Disaster Management in Sri Lanka. *Source*: Disaster Management Centre, Sri Lanka

- Prepares stakeholders for their full participation in recovery including prior arrangement for cooperation and assistance from neighbouring countries and humanitarian agencies
- Facilitates humanitarian to development transitions

Recovery and reconstruction policies have been urged overall planning process for implementation and monitoring. Sri Lanka, soon after the 2004 tsunami has led to initiate whole process of recovery and reconstruction policy development through a Parliamentary Select Committee to establishment of a coastal buffer zone, provide a legal basis and institutional arrangements for recovery initiatives through the endorsement of a new Disaster Management Act and through that development of national disaster management plan and creation of national policy on disaster management and recovery were highlighted.[3] According to the Disaster Management Act No. 13 of 2005, National Council for Disaster Management and Disaster Management Centre (DMC) established at nationally (Fig. 25.5). District Disaster Management Coordination Units (DDMCUs) established under the leadership of District Secretary in every District to work on wide range of disaster management aspects.

25.5 Tsunami Recovery and Reconstruction Lessons Learnt Practices Which Lead to Disaster Risk Reduction

25.5.1 Institutional Arrangements and Coordination Mechanisms

Institutional arrangements and coordination mechanisms has been provided the main strategy for tsunami lessons learnt practices which was leading to address disaster risk reduction in recovery and reconstruction processes. Government

[3] ISDR/IRP (2007).

involvement in recovery and reconstruction; communicating for working with partners, and their mechanisms; coordination and involvement of CBOs; strategies for institutional and coordination mechanism; institutional and coordination mechanism at central and field level; planning for implementation has been described.

25.5.2 Government Involvement in Recovery and Reconstruction

To address the needs of the emergency response, early recovery, recovery and reconstruction period, the focal agency has to be worked in the very complex, challenging and dynamic environment to cater the future risk reduction. It is very important to make the best decisions during post-disaster periods on institutional arrangements and coordination mechanisms. Such decisions should have a broad spectrum which may lead to safer communities, their resilience and sustainable development. In that case the leadership and involvement of the government is vital to have an overall supervision.

Sri Lanka Reconstruction and Development Agency (RADA) formed as the single government agency for recovery issues after the tsunami. Prior to the RADA, the agency called TAFREN (Task Force for Rebuilding the Nation), formed by the Government and the structure and composition of TAFREN was not able to be fully effective in that role. TAFREN was dominated by private sector representatives, and lacked links to line ministries and clear lines of authority. This greatly hampered its ability to efficiently coordinate activities among government agencies. RADA was acted as the main disaster risk reduction agency during tsunami operations to reduce future risk and vulnerability of the affected people in coastal region of the country.

Sri Lanka has coordinated major recovery sectors such as Transitional Shelter, Permanent Houses, Restoring Livelihoods, Health and Education, Infrastructure (water and sanitation, roads, bridges, power and energy) and Cross-cutting issues. The progresses reflected in period coordination meetings and other official representations via single documentation and, this effort has minimised duplication and miscommunication. As the main government apex body for recovery and reconstruction RADA's role reflected best practices for disaster risk reduction in recovery and reconstruction, this leads to repeat the same structure and functionality in future endeavours.

Sri Lanka 2004 tsunami is showing another best practice by formation of the Centre for National Operation (CNO) aftermath of tsunami, it was essential to deal with risk reduction in emergency operations and early recovery. Also, evolution and progressive development of institutional arrangements according to the needs and address the issues are important in risk reduction framework. The best example in Sri Lanka was for rehabilitation period CNO dissolved and created TAFREN, as a successor of that RADA was created for Recovery and Reconstruction. RADA is placed under the Presidential Secretariat and given the statutory authority by an Act of Parliament. Furthermore with implementation progress, the sectors that were coordinated by RADA were gradually handed over to relevant ministries/departments and RADA was closed in June 2007.

25.5.3 Collaborating for Working with Partners, and Mechanisms

Essentially partnerships and stakeholders are meaningful to have a success of post-disaster recovery and reconstruction in all sectors. Other than the particular government, the key players in recovery and reconstruction include UN agencies, regional agencies, Red Cross and Red Crescent Movement (RC/RC Movement) and Non-Government Organisations. UN Organizations and other partner organisations being delivered their extended expertise services to the government directly and indirectly, also behave as auxiliary to the government. This was same in tsunami operations in Sri Lanka, reflecting inter-agency coordination mechanism is vital in post-disaster recovery and reconstruction in any atmosphere when focusing in future risk reduction applications and practise.

Box 25.2 Inter-Agency Coordination Model (IASC & UN Coordination Mechanisms)

Example: Inter-Agency Standing Committee (IASC) is a unique inter-agency forum for coordination, policy development and decision making involving the key UN and non- humanitarian partners. The IASC has Cluster lead mechanism i.e. for nine areas of activities identified as follows:

* Camp coordination/management: UNHCR (conflict situation) and IOM (disaster situation) * Early recovery: UNDP * Emergency shelter: UNHCR (conflict situation) and IFRC (disaster) Convenor * Emergency Telecommunication: OCHA/UNICEF/WFP * Health: WHO * Logistics: WFP * WASH: UNICEF * Nutrition: UNICEF * Protection: UNHCR (conflict) and UNHCR/OHCHR/UNICEF (disaster)

IASC has been applied in Sri Lanka for disaster/conflict situations, it was chaired by the UN Resident and Humanitarian Coordinator and comprises the Executive Heads of the following organizations: OCHA, UNHCR, UNICEF, WFP, UNDP, UNFPA, FAO, WHO, IOM, OHCHR, CHA, FCE, Sarvodaya, Sewa Lanka, Oxfam, NRC, CARE, World Vision, ACF, ZOA, Solidar, Save the Children, Merlin. In addition the ICRC, IFRC, World Bank and ECHO attend as observers. Such inter-agency coordination for disaster response will lead automatically for pre-disaster preparedness in all kind of disaster response operations too. Hence addressing risk reduction, the inter-agency coordination mechanisms played significant role by strengthening communication, coordination and collaboration between the organizations, further it will help to avoid duplication and replication of humanitarian recovery deliverables also by reduce the vulnerabilities.

Such best practices can be identified in HFA Priority Action 5 for risk reduction said: Strengthen disaster preparedness for effective response at all levels "At times of disaster, impacts and losses can be substantially reduced if authorities, individuals and communities in hazard-prone areas are well prepared and ready to act and are equipped with the knowledge and capacities for effective disaster management" (DMC 2010).

25.5.4 Coordination and Involvement of CBOs

Community Based Organizations (CBOs) and Civil Society Organisations (CSOs) played important role not only in tsunami emergency response, also in tsunami recovery and reconstruction which was supported by understanding and trust of community apart from the technical, human and financial resources. CBOs and CSOs, including their community leaders acted as the first level tsunami responders in mass evacuation, search and rescue, emergency relief supply…etc. Such a capable CBOs and CSOs were able to reduce impact of tsunami, further well establishment and properly functionality of those organizations will reduce social and institutional vulnerabilities in a community to reduce multi-hazards risk.

As a response in 2003 floods and landslides of Sri Lanka, there were successful CBO's coordination systems supported by district and divisional level administration. Government and CBO leaders were represented at the monthly coordination meetings. Information shared and updated status of their different programmes and activities; thus helped to identify gaps and needs of the affected communities. Accordingly lessons learnt brought forward in tsunami response in 2004.

"Village Welfare Societies" as most sustain CBOs in Sri Lankan context extended their services for post-tsunami emergency response, early recovery as well as recovery and reconstruction implementation processes. Grama Seva Neladari (GN) represented the lowest level central government agents in grass-root level, with coordination with other government and NGOs bodies have been delivered their services to Village Welfare Societies for effective tsunami recovery. Village Disaster Management Committees (VDMC) chaired by GNs and their Sub-Committees acted as leading as well as monitoring bodies in recovery and reconstruction processes, further they have been trained for disaster management activities with designated roles and responsibilities. As an example VDMC trained for Camps/Shelter Management with collaboration of DDMCU by Red Cross (Fig. 25.6). [4]

[4] GN level Disaster Preparedness & Response Plans, DMC & Poorna Yahampath, Red Cross, 2007

Fig. 25.6 Camps/Shelter management training for VDMC

Box 25.3 Designated Sub-Committees of Village Disaster Management Committees

- Early Warning Dissemination
- Search, Rescue and Evacuation
- Camps and Relief Management
- Health and First Aid
- Drinking water, Sanitation and Dead body removal
- Patrol, Vigilant and Coordination

Establishment of such Committees and Sub-Committees operated under Community/Village Disaster Preparedness and Response Plan in a community is essential to achieve future risk reduction measures. E.g. Vulnerable groups in the community such as women, children, economically poor, mentally and physically disable people...etc. are sometimes left out in the recovery process. As a role of CBOs and related NGOs are the key players to track such matters, if the government service not inclusive due to many reasons.

Coordination and financial and technical support of NGOs is being facilitated the role of CBOs always. Sri Lanka Women's Coalition for Disaster Management (WCDM) played important role on gender issues in Tsunami recovery.

In Batticaloa and Ampara districts in Sri Lanka, the Women's Coalition for Disaster Management (WCDM), which was initiated by Suriya, a local women's organisation in Batticaloa, in mid-January, played an important role in post-Tsunami relief and reconstruction work. The WCDM initially lobbied for a women's committee to be set up in every camp. The committees then identified the basic needs of women, such as private space, appropriate facilities (such as private bathing and toilet facilities) and access to supplies.

The WCDM also formed an action group called Gender Watch, involving local and international non-government organisations. The initiative enabled women to report domestic violence, sexual harassment and discrimination to the group. Gender Watch documented violations in the camps and distributed the information to international agencies and the government. Remedial action taken included: suspending a government officer for violations; protecting five orphaned children; ensuring women have access to oral contraceptives; facilitating access to the police in the case of domestic violence; providing temporary shelters to single women who were originally excluded because they did not possess the right papers; and registering women for the provision of ration cards to provide them with access to goods.[5]

25.5.5 *Strategies for Institutional and Coordination Mechanism*

Based on extracted lessons learnt practices in tsunami affected countries, following strategies have been developed for institutional and coordination mechanism in recovery and reconstruction process.[6]

Standing Strategy—Recovery and reconstruction is coordinated and implemented by the existing line ministries and departments. For 2004 tsunami recovery and reconstruction some Indian Local Governments applied this strategy by devolving implementation authority to existing local authorities.

Representative Strategy—Existing government may designate representative body to manage post-disaster recovery and reconstruction processes (E.g. Task Force or Commission) In Sri Lanka tsunami 2004 established Task Forces were later upgraded into specialized organizations. 2003 floods and landslides in Sri Lanka, the executive government of Sri Lanka appointed a Cabinet Subcommittee to oversight the housing and reconstruction which was implemented through the National Disaster Management Centre under the Ministry of Social Welfare.

Interim Strategy—Interim or permanent dedicated agency is created to coordinate and/or implement the recovery and reconstruction. It consolidates recovery and reconstruction in one agency that provides oversight, a single point of coordination for external partners and additional capacity to implement and expedite the projects. Best example is RADA and it was created in Sri Lanka for focus on reconstruction and development issues after the tsunami 2004.

---

[5] *Source*: SWS Fact Sheet: Women and Disaster, UNDP/IRP (2007)

[6] *Source*: Poorna Yahampath, ADPC (2011)

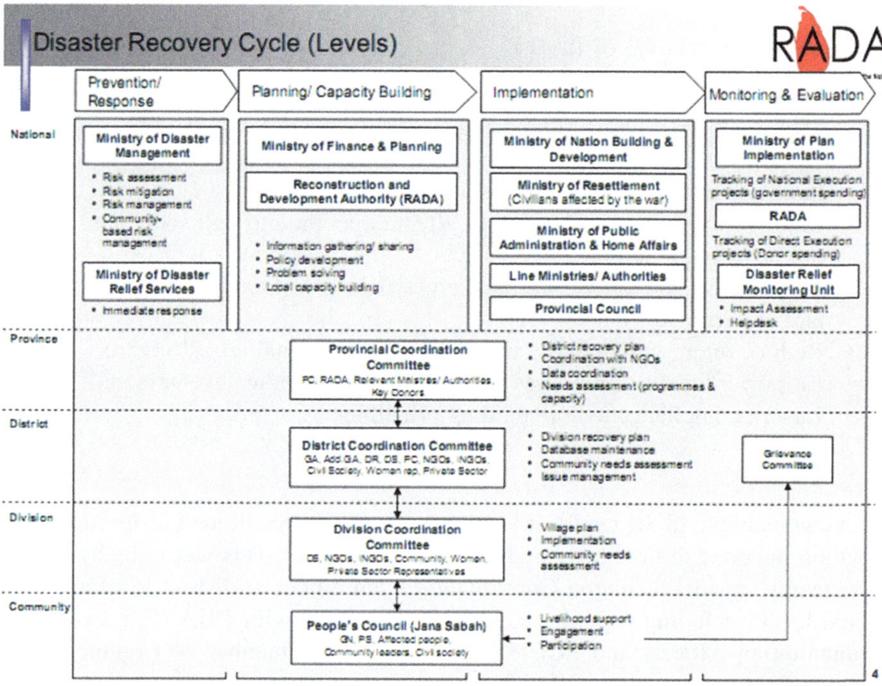

Fig. 25.7 Tsunami recovery and coordination mechanism of RADA. RADA Mission, Objectives, Organisation Structure and Progress. *Source*: RADA (2006a, b, c)

25.5.6 *Institutional and Coordination Mechanism at Central and Local Level*

Under the vertical and horizontal coordination systems, the overall institutional setup required at central as well as local level. In Sri Lanka, RADA operated centrally with link to local level which was established with following objectives, to achieve the Vision of *"Ensure rapid recovery of affected areas and foster long-term social and economic development to achieve Sri Lanka's 21st century growth plan and Millennium Development Goals through empowerment of local communities and strategic coordination of existing government and non-government stakeholders"* (Fig. 25.7). Eventually, RADA was acted to create risk reduction in tsunami affected individual/parties in longer term aspect. Hence, this covers HFA Priority Action 4: Reduce the underlying risk factors, which described as "Disaster risks related to changing social, economic, environmental conditions and land use, and the impact of hazards associated with geological events, weather, water, climate variability and climate change, are addressed in sector development planning and programmes as well as in post-disaster situations".

Box 25.4 Objectives of RADA

- Accelerate sustainable reconstruction and development through information gathering, programme monitoring and project coordination
- Optimize utilisation of resources, strengthen existing government structures and provide policy leadership
- Improve long-term well-being of affected people through social and economic development
- Ensure short and long-term social protection for vulnerable groups, women and children
- Reduce future vulnerability through improved social infrastructure
- Empower local communities to foster local economic development
- Ensure compliance to core guiding principles

As an example in Sri Lanka at locally; post-tsunami livelihood restoration coordination operated at field level by building partnerships. This was done by created nine teams called Livelihood Coordination Committees at district level and divisional level, including tripartite partner arrangements with CHA (Consortium for Humanitarian Affairs) and FCCISL (Federation of Chamber of Commerce and Industry Sri Lanka) and key I/NGOs, Private Sector and Public Sector.

It is important to have decentralized approach and that will lead institutional arrangements near the project site at locally. In Sri Lanka the responsibility of drive the process of housing construction for tsunami recovery has been delegated to the District Secretaries, by providing an operating framework and the discretion to make decisions in Secretaries, with support of RADA and THRU.[7]

Another best practice highlighted in Sri Lanka prior to tsunami; in the case of 2003 May floods and landslides recovery programmes, the housing reconstruction programme were over-looked by the Cabinet Sub-committee and coordinated with various actors such as; beneficiaries, NGOs, village leaders, divisional and district administrations, professional institutions like Sri Lanka Institute of Architects (SLIA), technical agencies like the National Housing Development Authority (NHDA), the National Building Research Organisation (NBRO) and National Disaster Management Centre (NDMC).

25.5.7 Integration of Environment and Ecological Aspects

Overall, the tsunami affected two-thirds of the coastline of Sri Lanka width from tens to hundreds of metres inland including wetlands of lagoons, estuaries, fringing and offshore reefs, mangrove, swamps, sea-grass beds, salt marshes, beaches, sand

[7] Tsunami Housing Reconstruction Unit, Under the Ministry of Housing and Urban Development

spits, rocky shores and dunes and ecosystems affected by seawater inundation, penetration and hits of damaged materials and debris. The Government of Sri Lanka, through its Ministry of Environment and Natural Resources quickly requested urgent assistance from UNEP to work with national experts in conducting a Rapid Environmental Assessment (REA) of the damage caused by the tsunami. Environmental issues specified for assessment included coral reefs, shore erosion, coastal land use planning, water pollution, and soil contamination. The Government further specified the UNEP after the assessment that was in developing a forward-looking plan for coastal zone management, in which lessons learned from the disaster would be adequately incorporated. The assessment team involved relevant international experts, national experts, university, government line departments, INGOs, NGOs members.

Different methods were applied in field assessments, as an example tsunami impacts assessed in "Green" for Ecosystem and Biodiversity impacts and "Brown" for Pollution, Debris and impacts on Human Settlements and Infrastructure. [8]As a part of the study, identified environmental damages in seven Special Area Management (SAM) sites affected by tsunami; Negombo, Lunawa, Maduganga, Hikkaduwa, Habaraduwa, Mawella and Kalametiya.[9]

Box 25.5 Green and Brown Assessments

Green Assessment—Described at 1-km intervals over more than 800 km of affected coastline, transects perpendicular to the shore and running inland from the high tide line. Data were collected on vulnerability, physical, ecological and social damage, land use, constraints on and options for land use, and on the precise pattern of tsunami inundation.

Brown Assessment—Concentrated on contamination at over 750 sites where particular risks were known to exist because the tsunami affected facilities for storage or processing of potentially hazardous materials. These included both established solid waste dumps and new ones used to dispose of tsunami debris, as well as storage and processing facilities associated with the commercial, health, security, transport, tourism, agriculture, fisheries, mining and other sectors.

Further, the assessment recommended strengthening measures to build resilience of the coastal zone. Reconstruction of defensive ecosystems around the coast was considered as single highest priority for medium and long term risk reduction. Although some locations were not easy to defend like many bays suffered devastating property damage because of tsunami was funnelled and strengthened by the shape

[8] Post-tsunami research initiatives in the coastal zone of Sri Lanka, NSF (2005).

[9] National Rapid Environmental Assessment Report- Sri Lanka, GoSL (2005).

of the sea bed and surrounding land, while river outlets through dune systems inevitably allowed the wave access to lagoons and inland areas. Broad, flat beaches and hinterlands of the north-east offered an easy path deep inland for the tsunami surge and hard to enhance natural capacity for building resilience entirely. As the coast is very heterogeneous in terms of terrain, natural ecosystems, and human pressures, there will be no one solution that can be applied in all places; instead a number of options are available; Relocating settlements, Redesigning new construction, Building artificial breakwaters, Establishing sand dunes, Planting mangroves, Using other vegetation types, Establishing artificial coral reefs, Establishing marine protected areas…etc.

25.6 A Case-Study "Changing of Tsunami Housing Policy"

Changes in tsunami housing policy of Sri Lanka has described for better planning in housing reconstruction implementation strategy; different options for hosing assistances; applicability housing assistances according to coastal zones definition; administrative guidelines; definitions of terminologies; technical specifications for housing construction; specimens for MoU between Government and the Donor… etc. Tsunami recovery programme had been identified to building of 98,525 units of shelter facilities at the beginning. This was planned by combining the strategies of "Donor-driven", "Owner-driven" and "Private Sector/Private individuals and groups housing programmes" (Fig. 25.8).

The IDPs were moved from emergency shelters to transitional shelter and finally to the permanent houses. The re-settlement housing programme (Owner driven) supported by the government and re-location programme (Donor driven) funded by the donors. 2006 May, the government revised the tsunami housing policy on the principle of *"a house for a house, regardless of land ownership"*, and the policy describes;

- Government land + donor-built house under the donor-driven housing programme (an already existing scheme) primarily for all those who lived within the previous buffer zone
- Government land + government cash grant (Rs. 250,000) to construct a new house + regulated donor assistance provided to complete the houses (not less that Rs. 250,000 depending on costs to meet the minimum standard house) through co-financing agreement
- Government cash grant (Rs.150,000 for Ampara three divisions and Rs.250,000 for Colombo) to purchase land + government cash grant (Rs.250,000) to construct a house + regulated donor assistance provided to complete (not less than Rs.250,000 depending on costs to meet the minimum standard house) through co-financing agreement (GoSL 2006; TRIAMS 2009; Government of Sri Lanka 2006).

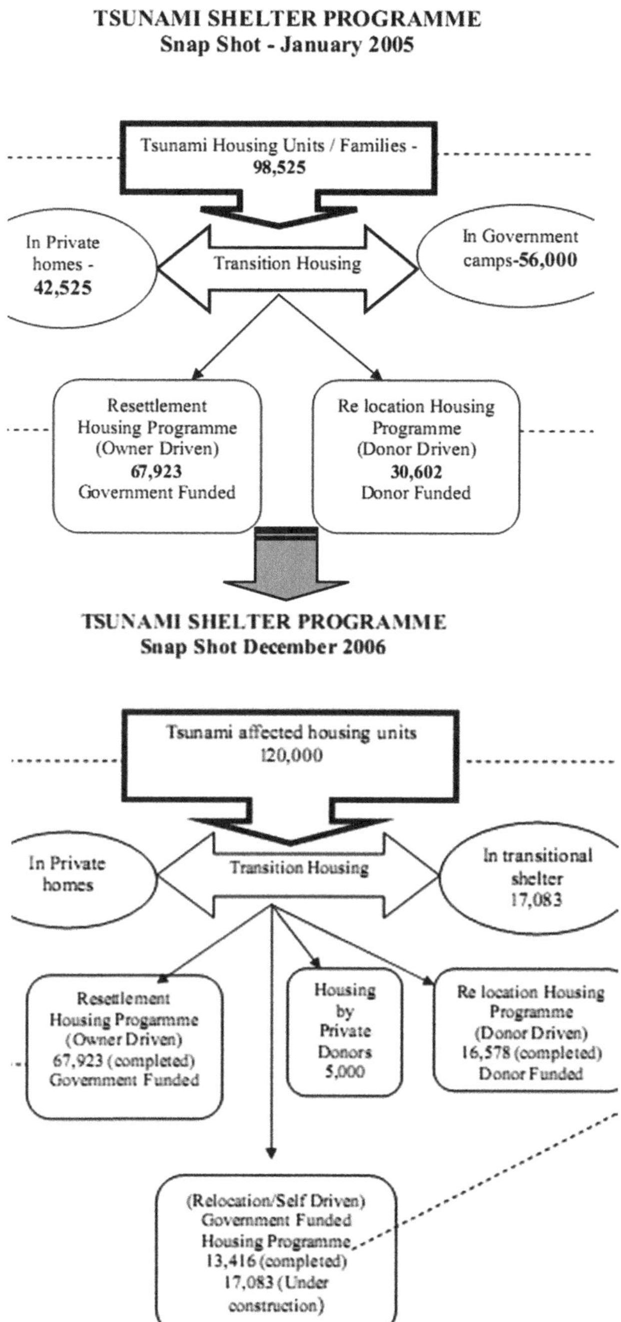

Fig. 25.8 Snap-shot of tsunami shelter programmes and changes in housing policy

- Housing reconstruction grant (Rs. 250,000 for fully damaged homes and Rs. 100,000 for partially damaged homes) + regulated donor assistance provided to complete only fully damaged houses as required for meeting minimum standard house through co-financing agreement.

Thereby the tsunami housing policy was pushing the total housing needs from 98,525 to approximately 120,000 and, it led to changing of overall programme of housing reconstruction in Sri Lanka.[10] Such a policy changes to solve the post disaster shelter issues always supported for reduce future disaster risk and provide best practice and lessons learnt for forthcoming disaster recovery programmes.

25.7 A Case-Study "Declaration of Tsunami Buffer Zone Linked to Coastal Zone"

Due to severe tsunami devastation in coastal zone in Sri Lanka; considering future tsunami and disaster risk, the government declared a 100/200 m strip of land as a "no build zone" along the coastal belt. The Cabinet approved policy in January 2005 prohibited any new construction of buildings (permanent or temporary), reconstruction of completely or partially damaged buildings, and additions and alterations to existing undamaged buildings within the buffer zone.

The Coastal Zone is defined in the Coast Conservation Act of Sri Lanka as the area lying within a limit of three hundred meters landwards of MHWL and a limit of 2 km seaward of MLWL. In the case of other water bodies the landward boundary extends to a limit of 2 km (Fig. 25.9).

Later the factors understood that coastal areas affected by tsunami revealed that the spatial distribution of its impacts from the beach towards the interior was at moderate and low levels. Same time discussions and requests made by stakeholders such as District and Divisional Secretaries, Line Ministries and Departments, affected families…etc., made influence to revise the coastal buffer zone policy of 2005. Hence, with the recommendations of Coastal Conservation Advisory Council of Coast Conservation Department, revised setback standards for coastal buffer zone was released in 2006. Revised buffer zone included minimum 25 m to maximum 100 m depend on the coastal area of the each locality (Fig. 25.10). (Example minimum 25 m at Galle district Closenberg, Rumassala and Matara district Browns Hill, Eliyakanda; maximum 100 m at Jaffna district Maruthankemy, Point Pedro)[11] (Samaranayake RADB 2006).

[10] *Source*: RADA, Post Tsunami Recovery and Reconstruction Mid-Year Review, June 2006

[11] *Source*: Urban Development Authority, Press Release Revision of Setback Standards (2005).

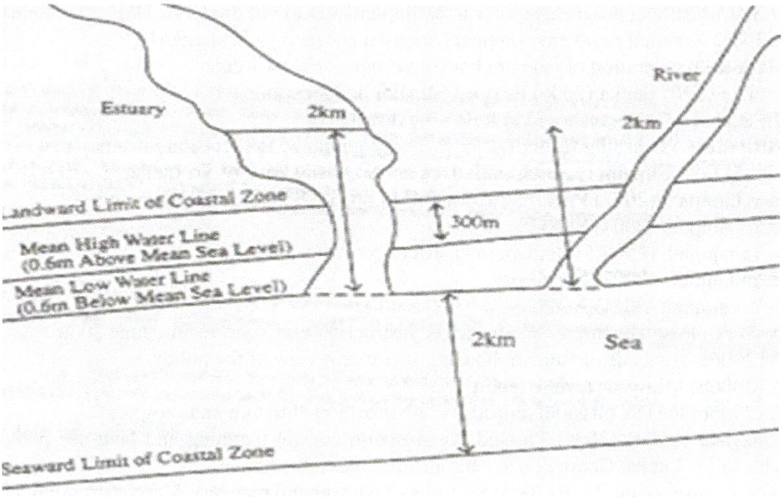

Fig. 25.9 Coastal Zone demarcation in Sri Lanka according to Coast Conservation Act

Fig. 25.10 Revised buffer zone and new setback standards *Zone 1*: Inside new setback limit. *Zone 2*: between 1997 Setback and former buffer zone (100/200 m). *Zone 3*: Outside former buffer zone. *Source*: Yahampath P (2011)

References

ADB, JIBC, World Bank (2005) Preliminary damage & needs assessment. ADB, JIBC, World Bank, Sri Lanka

ADBI (2007) Economic challenges of post-tsunami reconstruction in Sri Lanka

DMC (2010) National progress report on the implementation of the HFA. DMC, Sri Lanka
GoSL (2005) National rapid environmental assessment report - Sri Lanka
GoSL (2006) Acceleration of tsunami housing Programme – Circular
Government of Sri Lanka (2006) Revised tsunami housing policy
Haas JE et al (1977) Reconstruction following disaster
ISDR/IRP (2007) Learning from disaster recovery-guidance for decision makers
NSF (2005) Post-tsunami research initiatives in the coastal zone of Sri Lanka
Poorna Yahampath (2012) Plan International Sri Lanka, DPP National Workshop
Poorna Yahampath (2004) SLRCS
Poorna Yahampath (2004/5) Ratnapura District Secretariat
Poorna Yahampath (2007) Red Cross
Poorna Yahampath (2011) ADPC
RADA (2006) Post tsunami recovery and reconstruction mid-year review, June 2006
RADA (2006a) Analysis of tsunami housing sector & review of the policy
RADA (2006b) Mid-year review report
RADA (2006c) RADA mission, objectives, urbanization structure and progress
Samaranayake RADB (2006) Pre-and post-tsunami coastal planning and land-use policies and issues in Sri Lanka. Coast Conservation Department, Sri Lanka
TAFREN, Central Bank, MoF (2005) Sri Lanka-Post tsunami recovery & reconstruction: progress, challenges, way forward. National Planning Department, Sectoral Reports
TRIAMS (2009) Lessons learned in post-crisis recovery monitoring
Tsunami Global Lessons Learnt Project (2009) The tsunami legacy: innovation, breakthroughs and change
UNDP/BCPR (2004) Flood recovery: a case study from Sri Lanka
UNDP/BCPR (2011) Methodological guide for pre-disaster recovery planning processes guidelines and actions for national, regional and local governments
UNDP/IRP (2007) Review of post disaster recovery needs assessment and methodologies
UNDP/NDMC (2004/5) District disaster preparedness & response plan Ratnapura
Urban Development Authority (2005) Revision of Setback Standards. Sri Lanka
Yahampath P (2011) Handbook on disaster recovery practices – draft

Chapter 26
Capacity Gaps in Post Disaster Waste Management: Case Study in Sri Lanka

Gayani Karunasena and Dilanthi Amaratunga

Abstract Disaster waste is one of the major consequences aftermath of any disaster, impacts on public and environment, rescue and emergency services, provision of lifeline support and socio-economic recovery of affected areas. Thus, management of wastes created by disasters has become an increasingly important issue to be addressed in responding to a disaster. This chapter intends to present the prevailing gaps in disaster waste management and approaches to minimize the impacts on disaster management at developing countries with special emphasis to Sri Lankan context. Findings revealed that, unavailability of single point responsibility and provisions for disaster waste in existing policies and capacity constraints of the prevailing peace time solid waste management practices as major capacity gaps. Establishment of a regulatory body and enforceable rules and regulations with necessary levels of capacities were identified with seven areas for capacity building for post disaster waste management. The research enabled to attain sustainable post disaster waste management for future resilience.

Keywords Capacity gaps • Disaster waste • Post disaster

26.1 Disaster Waste and It's Importance

Increasing nature of impacts from disasters has made post disaster management a key area of concern. Management of disaster waste is identified as an area of least concern yet it presents momentous challenges for disaster management those with inadequate capacities due to the large volume and hazardous constituents created, in particular in developing countries. Shibata et al. (2012) highlighted that Great East Japan earthquake and tsunami occurred in 2011, estimated generated waste in Fukushima prefecture is 16 million tons which was equivalent to 14 years of waste

G. Karunasena (✉)
Department of Building Economics, University of Moratuwa, Moratuwa, Sri Lanka
e-mail: gayanik@uom.lk

D. Amaratunga
University of Salford, Manchester, UK

R. Shaw (ed.), *Recovery from the Indian Ocean Tsunami: A Ten-Year Journey*, Disaster Risk Reduction, DOI 10.1007/978-4-431-55117-1_26

generation. Brown et al. (2011a) state that disaster debris impacts not only public and environment but also on rescue and emergency services, provision of lifeline support and socio-economic recovery of affected areas. Haiti earthquake in 2010, Victorian bushfires in 2009, hurricane Katrina in 2005 and Indian Ocean tsunami in 2004 area few disasters that generated large volumes of waste overwhelming existing solid waste management capacities, requiring special approaches (Basnayake et al. 2005; Luther 2008; Booth 2010). Srinivas and Nakagawa (2008) mention of risks caused to public and environment by prolonged exposure to disaster waste after the Indian Ocean tsunami in 2004. Thus, management of wastes created by disasters has become an increasingly important issue to be addressed in responding to a disaster (Thummerukudy 2012).

Importance of focusing on long term ecological and economic sustainable waste management strategies for resilience to future disasters is emphasized by Blakely (2007). Further, many researchers have emphasized the importance of designing early-stage strategies to be managed in the most environmentally sound manner possible, maximizing source reduction and recycling options and minimizing land disposal (Baycan and Petersen 2002; Brown et al. 2010, 2011a). Brown et al. 2011a further argued that most exiting plans on disaster waste management provide more technical guide on how to manage with little guidance on decision-making and option consideration in different disaster situations.

In this context, this chapter focuses on gaps in post disaster waste management and presents the approaches to minimise the identified gaps. The scope of the chapter limited to the gaps created from natural disasters covering all types of waste with special emphasis to Sri Lankan context.

26.2 Capacities Needs for Post Disaster Waste Management

Recent decades has placed more focus on capacity building to increase resilience to natural hazards due to associated economic, social and environmental challenges. Capacity building dominates disaster management policies and practices in developing countries more vulnerable to disasters, particularly to impacts of climate change due to poverty, weak governance and ecosystem degradation (Webb and Rogers 2003). Coping with disasters and enhancing capabilities of communities are termed as prior targets of vulnerable countries (Ozden 2007). Hartwig et al. (2008) identify it as a key concept facilitating sustainability in developing countries. Boyd and Juhola (2009) explained that it provides an opportunity to understand strengths, weaknesses, threats and opportunities towards a resilient future through identification of broader issues around sustainable development of a particular program, project or process, including unique cultural, social and ecological characteristics. Capacity building is necessary due to lack of financial, institutional and technological capacities and access to knowledge to deal with risks and benefits (Ayele and Wield 2005).

It is impractical to initiate capacity building programs solely for disaster waste management, as it is necessary to enhance sustainable post disaster waste management systems. According to Brown et al. (2011a) sustainability of disaster waste management systems depend not only on required technologies or guides but also on development of institutional and human capacities that enhance preparedness and responses to future disasters. Institutional capacities need to be built to prevent, prepare and respond to disasters, enhancing resilience of disaster-affected communities (Baycan and Petersen 2002; Tadele and Siambabala 2009). Intervention of communities can be more successful, leading to genuinely positive impacts on human well-being, building on local knowledge and existing capacities (Allen 2006). Many highlighted that capacity building of local level governments, particularly in developing countries is also essential (Petersen 2004; United Nations Development Programme (UNDP) (2006); Bjerregaard 2007). Apart from that Milke (2011) pointed out the important processes of capacity building such as development of educational modules for processing, storage and disposal of post disaster waste and development of a free database and information source for disaster waste management.

26.3 Study Background and It's Post Disaster Waste Management

Sri Lanka prone to natural disasters such as floods, windstorms, landslides and droughts. A cyclone in 1978, floods and landslides in 2003 and tsunami in 2004 were major disasters that caused immense damage, interrupting economic and social activities of affected areas (Disaster Management Centre (DMC) 2005a). In addition, various human-induced hazards are caused by deforestation, indiscriminate coral, sand and gem mining and industrial pollutants (Disaster Management Centre (DMC) 2005b). Three decades of ethnic war also has cast huge economic and human impacts. Indian Ocean tsunami in 2004 is widely acknowledged as the largest, most devastating natural catastrophe reported in the history of Sri Lanka.

Floods, droughts and landslides are frequent natural disasters managed by local government authorities except in case of critical disasters. Subsequent to Indian Ocean tsunami in 2004, recognizing the magnitude and urgency of disaster situation, Sri Lanka established three task forces of Task Force for Rescue and Relief (TAFRER), Task Force for Law and Order and Logistics (TAFLOL) and Task Force for Rebuilding the Nation (TAFREN), for effective co-ordination (Task Force for Rebuilding the Nation (TAFREN) 2005a, b; Jayawardane 2006). National Council for Disaster Management (NCDM) established under Disaster Management Act No. 13 of 2005 is a high-level inter-ministerial body that provides direction to disaster risk management work in the country (Disaster Management Centre (DMC) 2005a, b, 2006a, b; Jayawardane 2006; Karunasena et al. 2009, 2012a). Disaster Management Centre (DMC) is the lead agency implementing activities related to all phases of Disaster Risk Management (DRM) in the country within the Ministry of Disaster Management and NCDM. These activates are carried out in coordination with relevant stakeholder ministries, national and provincial level government and private entities, civil society, non government organizations, community based organizations and communities.

Literature revealed disaster waste removal programs implemented in Sri Lanka along with occurrence of Indian Ocean tsunami in 2004 highlighting many failures incurred due to non-existence of pre-planned disaster waste management strategies and enforceable or mandatory rules and regulations (Basnayake et al. 2005; European Commission (EC) 2006; United Nations Environmental Programme (UNEP) 2005). Review of national polices on disaster management (refer, Sri Lanka Disaster Management Act, No 13 of 2005) and waste management (refer National Environmental Act, no 47 of 1981) disclose that no specific provisions on disaster waste management exist. Rules and regulations related to peace time solid waste management processes are imposed on management of disaster waste (refer: National Environment Act, 1981; National Environment (Amendment) Act, 1988). National Disaster Management Plan and National Emergency Operation Plan expected to be implemented in future contains inadequate provisions on disaster waste management. Thus, it is pertinent that Sri Lanka as a country prepare sustainable post disaster waste management strategies. Brown et al. (2011a) reveal that most developing countries do not have plans prepared in advance for disaster waste management.

As most other developing countries Sri Lanka also disposed its disaster waste during the post Indian Ocean Tsunami with the assistance of international aid organisations and United Nations agencies (Petersen 2004; United Nations Development Programme (UNDP) 2006; Brown et al. 2011a). Evidence of large scale processing of disaster waste in Sri Lanka is non existent as most of waste is disposed by land filling (Basnayake et al. 2005). Disaster waste generated after the Indian Ocean tsunami in 2004 at Telwatte (Hikkaduwa) was used to fill coral mined pits and lands with Central Enviornment Authority's (CEA) permission (Basnayake et al. 2005). The only recycling plant for construction waste established in Galle for processing post tsunami construction waste was subjected to operational delays and transportation costs of waste for recycling was costly, significantly reducing benefits of recycling (Raufdeen 2009; Karunasena et al. 2012b).

Next presents research findings identified through case studies and verified through expert interviews for post disaster waste management in Sri Lankan context. Three case studies which included fifteen individuals and six experts representing government, non government and others were selected for data collection. Face to face semi-structured interviews were held as the main data collection method and code based content analysis and cognitive mapping were used to analyse collected data. Approaches presented for capacity building was further valided through the three experts representing waste management in Sri Lanka.

26.4 Capacity Gaps in Post Disaster Waste Management

Capacity gaps identified are summarised namely in seven areas such as skills and confidence building, organizational implementation, linkages and collaborations, continuity and sustainability, investments in infrastructure, research and development and communication and coordinationat as illustrated at Table 26.1.

Table 26.1 Capacity gaps in post disaster waste management

Capacity gaps
Skills and confidence building
• Fewer opportunities for personal development—training/workshops
• Unavailability of formal procedure for preparation, conducting, monitoring and evaluation of training and awareness programmes
• Unavailability of strategies to retain valuable human resource
• Unavailability of provisions for disaster waste management in existing policies
Organization implementation
• Unavailability of single point responsibility at national level for post disaster waste management
• Inefficiency and ineffectiveness of prevailing peace time solid waste management practices, policies and responsible authorities
• Non-revision of existing waste management systems/procedures
Linkages and collaborations
• Unavailability of formal procedures to establish linkages and collaborations
• Availability of projects with complete proposals without implementation
• Reduced active participation of NGOs and INGOs
Continuity and sustainability
• Less consideration of incorporation of sustainable concepts into disaster waste management practices
• Loopholes in prevailing solid waste management practices, policies and with responsible authorities
• Unavailability of formal procedures for monitoring and evaluation of implemented projects
Investments in infrastructure
• Loopholes in government rules and regulations on fund raising and procurement
• Less consideration for environmental protection
Research and development
• Reduced interest on research and development -government sector
• Inadequate opportunities for collaborative research programmes
• Inadequate transfer/sharing of knowledge and technical know-how
Communication and coordination
• Uniformity of prevailing centralised framework
• Inadequate efficiency and effectiveness of existing systems
Other
• Vacuum between relief and early rehabilitation
• Policy issues, such as enforceability and wider scope
• Lack of awareness about peoples' needs
• Overlapping functions among institutions

Unavailability of formal procedure for preparation, conducting, monitoring and evaluation of training and awareness programmes is a major capacity gap as evidenced by a lesser number of programmes conducted on soft skills development as against many programmes conducted for technical skills development at local authority level focusing more on peace time solid waste management with less emphasis to disaster generated waste such as Construction and Demolition waste Lesser opportunities for personal development such as training, workshops and scholarships and inadequate strategies to retain valuable human resources are identified as the other main capacity gaps prevalent in skills and confidence building.

Unavailability of a single point responsibility at national level for post disaster waste management and absence of provisions on disaster waste management in existing policies are major capacity gaps of organisation implementation. Inefficiencies and ineffectiveness of prevailing peace time solid waste management practices, policies and of responsible authorities is another capacity gap that impacts on disaster waste management. Non-revision, retraining or monitoring of existing solid waste management systems at frequent intervals further aggravates above.

Unavailability of formal procedures to establish linkages and collaborations is a major capacity gap, impacting on transparency and accountability. It is revealed that projects with complete proposals and documentation exist without proper implementation. There is also a noted reduction in active participation of NGOs and INGOs when compared with the period immediately after the Indian Ocean Tsunami in 2004.

As mentioned loopholes in prevailing peace time solid waste management practices, policies and also with responsible authorities and unavailability of formal procedures for monitoring and evaluation of implemented projects exist as capacity gaps impacting on continuity and sustainability of post disaster waste management in Sri Lanka. Less consideration at national level for incorporation of sustainable concepts into disaster management practices including disaster waste management is also a prevailing capacity gap of this area.

Loopholes in rules and regulations on fund raising and procurement procedures are major capacity gaps impacting on investments in infrastructure at government sector entities. Less consideration for environmental protection and conservation by donors is another capacity gap.

There is inadequate interest on research and development, specially in the government sector. This is further aggravated by traditional government practices that do not facilitate new approaches in the long run. Inadequate opportunities for collaborative research programmes and lack of transferring and sharing of knowledge and technical know-how are also prevalent as capacity gaps in R&D.

Identified capacity gaps of communication and coordination include tightly formal approaches established for communication and coordination during emergency situations, uniformity of prevailing centralised framework, lack of efficiency and effectiveness and less transparency and accountability of established communication and coordination system.

Apart from those capacity gaps identified within aforementioned seven areas, findings further revealed of capacity gaps influencing post disaster waste management in a general context. An example is the vacuum between relief and early rehabilitation which leave disaster waste unattended. Lack of awareness on peoples' needs and overlapping functions are also identified as prevailing capacity gaps. Lack of political will, inadeqaute funds, lack of awaernass and attitudes of the public towards waste management are identified as the key influencing factors contributing for aforementioned capacity gaps. A study conducted on disaster waste management a developing country, Samoan Tsunami in 2009 by Brown et al. (2011b) also revealed similar capacity gaps such as unavailability of responsible authorities, lesser synergy among ministries, lack of strategy for coordination, unavailability of disaster funds and formal procedure to monitor funds.

Next presents proposed apparoches for enhance capacities of post disaster waste management to address the above mentioned capacity gaps.

26.5 Approaches for Enhance Capacities in Post Disaster Waste Management

Approaches proposed for enhance capacities in post disaster waste management are also summarised in seven areas aforementioned as illustrated at Table 26.2.

Table 26.2 Approaches for enhance capacities in post disaster waste management

Approaches
Skills and confidence building
• Provide more opportunities for career development—local and international exposure
• Establish procedures to prepare, conduct, monitor and evaluate local and foreign programmes
• Enhance capacities of the government sector to promote interactive working
• Promote holistic approach for capacity building with more focus on local and sustainable approaches
• Develop an expert knowledge database on disaster management
Organizational implementation
• Incorporate disaster waste mgt. into existing solid waste management practices, policies and authorities
• Restructure institutional practices allocating specific functions to each with single point responsibility
• Increase collection of recyclable items, provide incentives to recyclers
• Develop enforceable rules and regulations for prevailing solid waste management/disaster waste
• Introduce cash paying programmes for waste management
• Change procedures to facilitate quick and easy payment of compensation to affected parties
Linkages and collaborations
• Develop formal and transparent procedures to establish linkages
• Enhance capacities of government sector to promote interactive working, at local levels
• Provide more opportunities for collaborative projects
• Promote diversification
• Enhance active participation of NGOs and INGOs

(continued)

Table 26.2 (continued)

Approaches
Continuity and sustainability
• Train general public and officials on sustainable techniques with special emphasis on environmentally friendly, culturally supported mechanisms
• Introduce procedures to obtain permission for projects on quality, operational, maintenance and environmental impacts to ensure continuity and sustainability
• Establish formal procedures for monitoring and evaluation of implemented projects
• Promote holistic approaches for implementing waste projects
Investments in infrastructure
• Enhance capacities of staff to obtain funds through project proposals
• Establish transparent and accountable formal procedures for project selection
• Provide incentives to recyclers and mobilization of peoples
Research and development
• Establish resource centres with knowledge on new developments
• Organize open discussion forums for sharing research knowledge
• Provide opportunities and incentives for collaborative research
• Establish transparent systems in providing opportunities for career development
Communication and coordination
• Decentralize the system within established rules and regulations
• Provide adequate resources for communication systems
• Appoint responsible persons at each level of the communication
Other
• Prepare orders enforceable by law that clearly define responsibilities and functions of each institution
• Capacity building and needs identification from bottom to top
• Design framework for disaster waste management through District Coordinating Committees
• Provide provisions for disaster waste management when preparing urban development plans

Provide more opportunities for career development of responsible persons, with local and international exposure through seminars, workshops and scholarships to enhance capacities of officials at national level is identified. Parallel to that provide opportunities for self training through field activities, specifically in disaster waste management is emphasised Provide incentives to attract and retain staff such as life insurance/pension schemes and sufficient grants for career development, specially for government employees are proposed. To avoid repetition or duplication of programmes and unethical practices, establish formal procedures to prepare, conduct, monitor and evaluate local and foreign programmes. Implement a national level project to build technical support, assigning Disaster Management Centre (DMC) Sri Lanka with responsibility of training and awareness. Additionally introduce monitoring and evaluation methods such as beneficiary evaluations, statistical and non-statistical measures and progress reports. Sharing and disseminating knowledge among respective parties can enhance personal interests. Also enhance capacities of the government sector to promote interactive working, such as collaborative projects.

Enhancement of soft skills is proposed as an approach to eliminate traditional bureaucratic red tape. Alignment of capacity development at each level with existing policies for real time implementation, such as individual capacity building programmes with master plans is suggested. These will eventually align capacity development with economic development of the country. Promote holistic approach for training and development focussing on indigenous and sustainable approaches for skills and confidence building, giving consideration to new aspects as good governance, livelihood development and resilience. Gupta and Sharma (2006, p67) pointed out that good governance and social capital as important elements for ensuring equitable recovery processes, as well as ensuring appropriate capacity building for marginalised and highly vulnerable communities. Development of an expert knowledge database consisting of experience of experts on typical disasters is also proposed as a propose approach for long term resilience. Further, researcher proposes a specific database on disaster waste management comprising of technical information on safe waste handling, disposal options, facilities, regulations and contact information of those involved in disaster waste management, similar to a decision support tool called EPA's Suite of Disaster Debris Management and Disposal (DDMD) in the USA (Thorneloe et al. 2007).

Incorporate disaster waste management into existing peace time solid waste management practices and policies including environmental and wet land protection policies to improve disaster waste management, reinforced with disaster waste management guidelines prepared specifically for developing countries with little or no existing infrastructure and expertise by United Nations Joint Environmental Unit (Joint UNEP/OCHA Environment Unit (JEU) 2010). Expansion of existing peace time solid waste management practices such as zoning and seven steps process is proposed to promote sharing of resources and collaborations among local authorities. Establishment of enforceable rules and regulations for peace time solid waste management as well as disaster waste is necessary for long term sustainability. Restructuring of institutional processes allocating specific functions with single point responsibility is also a key approach to improve institutional structures for better disaster waste management. This will overcome non-functioning of important and necessary committees on waste management. These changes need to be incorporated into activities of entities as modes of mitigation, adaptation and recovery. Continuation with awareness and training programmes for public and local authorities on native and sustainable approaches for waste management emphasising on environmental protection and conservation will reduce inefficiencies and ineffectiveness prevailing in existing solid waste management processes. Examples are conducting of awareness programmes for domestic solid waste management at urban areas and enhancing of capacities of local authorities on composting and bio gas generation. Parallely, increase collection of recyclable items, provide incentives to recyclers and mobilize peoples' support for recycling. Enhance existing procedures to facilitate quick and easy payment of compensation to affected parties from polluters, together with an effective spot fining system.

Developing formal and transparent procedures to establish linkages and collaborations with local and international entities is an important approach to enhance capacities by exchange of skills and practical knowledge. An example is to produce reports on benefits gained through established linkages and collaborations. This will enhance transparency and accountability of linkages leading to commitment of parties. Gupta and Sharma (2006) indicated networking among governments, NGOs, academia and communities is crucial for informed decisions and improved practices based on lessons learnt. Enhance capacities of government entities to promote interactive working, specially at local levels to gain effective and efficient outcomes from partnerships. Also, promote diversification to build new relationships and collaborations among entities. Enhanced active participation of NGOs and INGOs in disaster management also create opportunities for collaborations.

As an approach to enhance continuity and sustainability, create awareness among general public and train officials on sustainable, environmentally friendly and culturally supportive techniques on disaster waste management. Create awareness on how to convert waste into profitable businesses. As an example, promote holistic approaches in initiating projects such as composting and recycling, together. Further, it is important to change rules and regulations to facilitate sustainability concepts in disaster waste management practices. Establish formal procedures for monitoring and evaluation of implemented projects to avoid duplication of work and illegal projects with increased government intervention at regular intervals. Introduce procedures to obtain prior permission for projects on such aspects as quality, operational maintenance and environmental impacts to ensure continuity and sustainability. Additionally, at the end of a project, a certificate can be issued on achievement of sustainability standards.

Establishment of formal, transparent and accountable procedures for project selection and evaluation is important to improve investments in infrastructure, increasing confidence among investors. An example is to share financial reports at the end of a project among all involved parties. Create awareness among investors on how to make waste a profitable business while providing incentives to investors. Enhancing capacities of staff to obtain funds through project proposals and implementing policies, rules and regulations that facilitate self financing are important steps for the government sector.

Establishment of a transparent system to provide opportunities for career development, such as foreign training, workshops and scholarships is important to enhance research and development capacities. This is specially relevant to ignite and enhance interest on research and development within government entities. It can further be improved by allocation of sufficient funds for staff development and grant of promotions based on research performances, publications presented at recognised conferences, symposiums and papers published in academic journals. Further, it is necessary to provide opportunities and incentives for collaborative research work. Also, establish resource centres with data on new developments and adequate facilities. Conduct awareness programmes to develop a research culture in government entities, changing attitudes and traditional practices. Documentation of project outcomes for future reference and organisation of open discussion forums to share research interests at regular intervals will also enhance research and development.

Appointment of responsible persons at each level of communication and coordination process is necessary to enhance transparency and accountability of existing systems. This can prevent lack of responsibility in the existing system. Provision of adequate resources and new technology, such as wireless and online communication facilities can improve effectiveness and efficiency of existing systems. Conduct awareness and training programmes for officials and general public on new technologies in communication and coordination at regular intervals. Decentralisation within existing rules and regulations can minimise adverse effects of inadequate implementational powers of Disaster Management Centre.

Preparation of lawfully enforceable provisions with clearly defined responsibilities and functions of each institution involved in disaster management is one such approach. It can minimise duplication of work and non-functioning of important committees and entities. Establishment of an institutional framework for disaster waste management with single point responsibility and adequate implementational powers is an important aspect of such an approach. Experts propose designing of a framework for disaster waste management through district coordinating committees and adequate provisions for disaster waste management shall be made when preparing urban development plans. It is further proposed to enhance capacities related to certain specific disasters, such as earthquakes that generate large quantities of waste.

26.6 Conclusions

Disaster waste has become a crucial issue not only in Sri Lanka but also in worldwide due to the increase of number of disasters. Thus, it is a challenge, which has to be faced by local governments in order to minimise or manage waste following a disaster. In Sri Lankan context, findings revealed that contribution for disaster waste management is very less and number of gaps which are prevailing in disaster waste management process lead to an improper management of waste after a disaster. Disaster waste management gaps such as unawareness, policy issues such as unenforceability, inadequate government support and unavailability of institutional arrangement are identified as capacity gaps in post disaster waste management. Further various approaches to minimise the prevailing gaps in disaster waste management are also discussed.

References

Allen KM (2006) Community-based disaster preparedness and climate adaptation: local capacity building in the Philippines. Disasters 30(1):81–101

Ayele S, Wield D (2005) Science and capacity building and partnership in African agriculture: perspectives on Mali & Egypt. J Int Dev 17:631–646

Basnayake BFA, Chiemchaisri C, Mowjood MIM (2005) Solid wastes arise from the Asian tsunami disaster and their rehabilitation activities: case study of affected coastal belts in Sri Lanka and Thailand. In: Tenth international waste management and landfill symposium, Sardinia

Baycan F, Petersen M (2002) Disaster waste management-C&D waste. In: ISWA (ed) Annual conference of the international solid waste association, Istanbul, pp 117–125
Bjerregaard M (2007) MSB/UNDP debris management guidelines. disaster waste recovery. Swedish International Development Cooperated Agency, Lebanon
Blakely E (2007) Collaborating to build communities of opportunity. Paper presented at Roosevelt institute symposium, New Orleans
Booth W (2010) Haiti faces colossal and costly cleanup before it can rebuild. The Washington Post, Haiti
Boyd E, Juhola S (2009) Stepping up to the climate change: opportunities in re-conceptualising development futures. J Int Dev 21:792–804
Brown C, Mike M, Seville E (2010) Waste management as a life line?' A New Zealand case study analysis. Int J Disast Resilience Built Environ 1(2):192–206
Brown C, Mike M, Seville E (2011a) Disaster waste management: a review articles. Waste Manag 31:1085–1098
Brown C, Milke M, Seville E (2011b) Disaster waste management for the 2009 Samoan tsunami. In: International conference on building resilience, Kandalama, Sri Lanka
Disaster Management Centre (DMC) (2005a) Annual report 2005. Disaster Management Centre, Colombo
Disaster Management Centre (DMC) (2005b) Towards a safer Sri Lanka – road map for disaster risk management. Disaster Management Centre, Colombo
Disaster Management Centre (DMC) (2006a) Annual report 2006. Disaster Management Centre, Colombo
Disaster Management Centre (DMC) (2006b) Towards a safer Sri Lanka – road map for disaster risk management. Disaster Management Centre, Colombo
Disaster Management Act, No 13 of 2005 (2005) Published as a supplement to Part II of the gazette of the Democratic Socialist Republic of Sri Lanka of May 13, 2005. Government publications bureau, Colombo
European Commission (EC) (2006) Progress report on post tsunami rehabilitation and reconstruction program [Online], European Commission (EC). http://ec.europa.eu/comm/world/tsunami/index.html. Accessed 5 August 2008
Gupta M, Sharma A (2006) Compounded loss: the post tsunami recovery experience of Indian island communities. Disaster Prev Manag 15(1):67–78
Hartwig K, Pashman J, Cherlin E, Dale M, Callaway M, Czaplinski C, Wood EW, Abebe Y, Dentry T, Bradley EH (2008) Hospital management in the context of health sector reform: a planning model in Ethiopia. Int J Health Plann Manag 23:203–218
Jayawardane AKW (2006) Disaster mitigation initiatives in Sri Lanka. In: International symposium on management systems for disaster prevention, 9th–11th March 2006, Kochi
Joint UNEP/OCHA Environment Unit (JEU) (2010) Disaster waste management guidelines. Joint UNEP/OCHA Environment Unit, Geneva
Karunasena G, Amaratunga D, Haigh R, Lill I (2009) Post disaster waste management strategies in developing countries: case of Sri Lanka. Int J Strat Property Manag 13(2):171–190
Karunasena G, Amaratunga D, Haigh R (2012a) Post disaster construction & demolition waste management: a Sri Lanka case study. Int J Civ Eng Manag 18(4):457–468
Karunasena G, Rameezdeen R, Amaratunga D (2012b) Post-disaster construction & demolition waste management: the case of COWAM project in the city of Galle, Sri Lanka. 6th International conference and workshop on the Built Environment in Developing Countries (ICBEDC), 4–5 Dec. 2012, University of South Australia, Adelaide
Luther L (2008) Managing disaster waste: overview of regulatory requirements, agency roles, and selected challenges, congressional research service. http://wikileaks.org/wiki/CRS-RL34576. Accessed February 25
Milke M (2011) Disaster waste management research needs. Waste Manag 31:1
Ozden AT (2007) Constituting a sustainable after disasters: the role of architecture. J Facul Built Environ 9(3):1–17
Petersen M (2004) Restoring waste management following disasters. In: IF (ed) International conference on post disaster reconstruction, UK. IF Research group, Coventry

Raufdeen R (2009) Construction waste management: current status and challenges in Sri Lanka. COWAM publication, Sri Lanka

Shibata T, Solo-Gabriele H, Hata T (2012) Disaster waste characteristics and radiation distribution as a result of the Great East Japan earthquake. Environ Sci Technol 46:3618–3624

Srinivas H, Nakagawa Y (2008) Environmental implications for disaster preparedness: lessons learnt from the Indian Ocean tsunami. J Environ Manage 89(1):4–13

Tadele F, Siambabala MA (2009) Building disaster resilience through capacity building in Ethiopia. Disaster Prev Manag 18(3):317–326

Task Force for Rebuilding the Nation (TAFREN) (2005a) Recovering from tsunami: a report on the progress of the post-tsunami recovery of Sri Lanka and the role of TAFREN in the recovery effort. The Task Force for Rebuilding the Nation, Sri Lanka, November 2005

Task Force for Rebuilding the Nation (TAFREN) (2005b) District wise post tsunami rehabilitation and reconstruction projects: rebuilding Sri Lanka, Post tsunami recovery and reconstruction strategy update of Part II – Progress made and problems encountered. Task Force for Rebuilding the Nation, Colombo

Thorneloe S, Lemieux P, Rodgers M, Christman R, Nickel K (2007) Decision support tool for the management of debris from homeland security incidents. In: XI International waste management and landfill symposium, Sardinia

Thummerukudy M (2012) Chapter 11: disaster waste management: an overview. In: Shaw R, Tran P (1 eds) Environment disaster linkages, Emerald Group Publishing limited, Emerald

United Nations Development Programme (UNDP) (2006) Post-tsunami recovery and reconstruction strategy. United Nations Development Programme, Colombo

United Nations Environmental Programme (UNEP) (2005) Sri Lanka post tsunami environmental assessment, United Nation Environment Program (UNEP). UNEP (DEP/0758/GE), Geneva.

Webb P, Rogers B (2003) Addressing the “In” in food insecurity. Food and Nutrition Technical Assistance (FANTA) Project, Academy for Educational Development (AED), Washington, D.C.

Chapter 27
Housing and Resilience: Case Studies from Sri Lanka

Iftekhar Ahmed and Esther Charlesworth

Abstract Research on housing and disasters indicate a critical need for assessing the resilience of permanent housing built after disasters. A survey of current research in this field led to the development of an evaluation tool for that purpose. The analytical framework of the tool consisted of five main factors—Inputs, Output, Result, Impacts and Effects, and External Factors—which are explained in this chapter. The severe impact of the 2004 Indian Ocean Tsunami on housing in Sri Lanka and the consequent abundance of housing reconstruction projects allowed the testing of the evaluation tool there. Two case studies of post-tsunami housing reconstruction projects implemented by international NGOs were examined in Galle, southern Sri Lanka. Using the tool, it was found that both the projects had improved the resilience of the beneficiary communities. However a number of challenges were also evident, particularly in meeting the wider needs of the beneficiaries and coordinating with local government authorities. The importance of the evaluation tool is underscored in the global context of widespread disaster occurrence and their devastating impact on the housing sector, and the multiplicity of post-disaster housing reconstruction projects by humanitarian agencies.

Keywords Evaluation tool • Housing • Resilience • Sri Lanka • Tsunami

27.1 Background

Housing is often the most valuable asset for many people and in disasters it is usually the most visible component that is damaged or lost. Displacement or loss of housing makes people vulnerable to aftershocks, as well as to the climate—rain, snow, heat, etc.—thus compounding the effects of the disaster with significant impacts on household and community health. Therefore housing that is resilient and adequate can safeguard people from such disaster impacts. This has been emphasised in a number

---

I. Ahmed (✉) • E. Charlesworth
Humanitarian Architecture Research Bureau (HARB), RMIT University, Melbourne, Australia
e-mail: ifte.ahmed@rmit.edu.au

R. Shaw (ed.), *Recovery from the Indian Ocean Tsunami: A Ten-Year Journey*, Disaster Risk Reduction, DOI 10.1007/978-4-431-55117-1_27

of disaster and shelter related publications (ADPC 2002; Coburn et al. 1995; HFHA undated; IFRC 2011; Seraj and Ahmed 2004; UNNATI 2006).

In places where disasters are frequent and recurrent, such as in the case of floods in Bangladesh, or typhoons in Vietnam, resilient housing initiatives are often on the agenda (Ahmed 2005; Tro 2011). However, more often the need for resilient housing assumes significance, tragically, after devastation by a disaster. Most guidelines and initiatives for safer buildings have arisen after major disasters such as earthquakes and tsunamis (for example ERRA 2006; NHDA 2005) and in many post-disaster recovery programs maximum resources and priority is allocated to shelter and infrastructure reconstruction compared to other sectors (Lang 2008). Thus during reconstruction there is the opportunity to understand and thereby address and overcome the underlying vulnerabilities that had previously prevented resilient housing construction and the risks that threaten the durability and sustainability of housing. Building housing back to a better standard that is less vulnerable to hazards can contribute to reduced disaster risks in the long term, especially if based on local knowledge and participation (Lyons et al. 2010). Reconstructed or rehabilitated housing with future risk in mind will prove more resilient and sustainable.

This chapter is concerned with this significance and need for housing resilience in the context of increasingly intense and frequent disaster impacts in this era of climate change, and the potential for post-disaster reconstruction to enable resilience, explored in Sri Lanka after the 2004 Indian Ocean Tsunami, the most massive disaster in recent history. The chapter is derived from a research project entitled *Scoping study: shelter and disaster risk reduction in the Asia-Pacific region* (Charlesworth and Ahmed 2012) undertaken by the authors for the Australian Shelter Reference Group (SRG).

In the scoping study, firstly a review of literature was undertaken on key themes relating including housing and disasters (for example ADRC 2012; Lyons et al. 2010), lessons from past projects (for example Aquilino 2011; UNHCR 2009, 2010, 2012) and evaluation frameworks (for example AusAID 2005; Twigg 2007; Lizarralde 2002). The review highlighted, (a) The importance of understanding resilience[1] in permanent shelter given the limited literature on the subject in the Asia-Pacific; (b) The need for an evaluation tool for assessing resilience of housing that encompassed a range of physical and social dimensions; and (c) An evaluation framework adapted from the most relevant evaluation approaches.

Secondly, following from the literature review, an evaluation tool for assessing resilience in housing projects was developed. The tool consisted of three main stages—Pre-Assessment, Assessment and Consolidation—with each stage including a set of guided activities (Fig. 27.1). To understand its utility, the tool was tested in actual housing projects, where it was found that the Key Informant Interviews was one of the most significant activities.

---

[1] 'Resilience' of housing here is viewed in a broader sense beyond only its physical aspects. While the focus here is on housing and related aspects, it is in line with wide-ranging community resilience frameworks such as that proposed by IFRC (2012) based on its work in Sri Lanka.

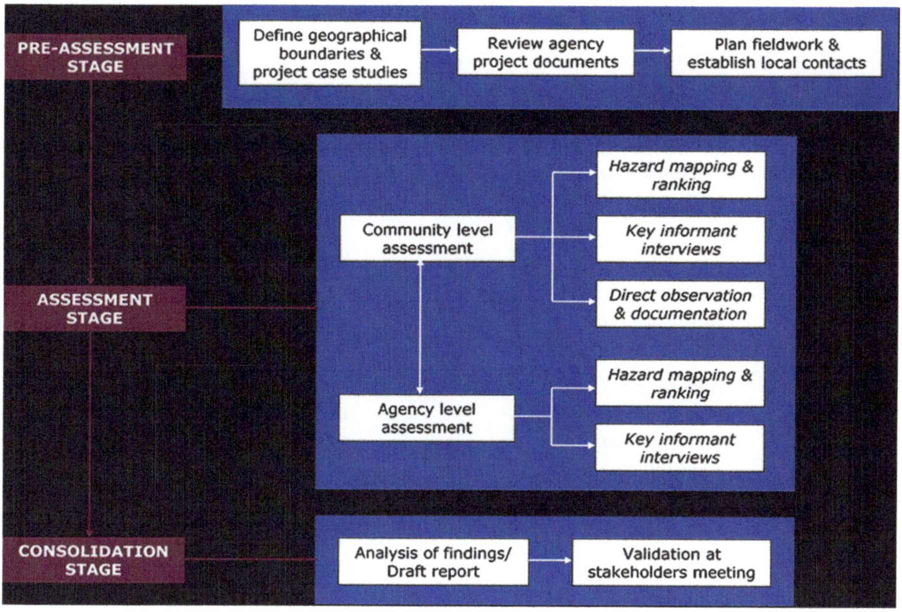

Fig. 27.1 Key aspects of the assessment process

To test and refine the evaluation tool developed, two countries in the Asia-Pacific region—the Cook Islands and Sri Lanka—were selected in consultation with SRG member agencies. Within each country, a number of case study housing projects were selected, through which the evaluation tool was tested via extensive in-country fieldwork. This chapter discusses some of the case studies in Sri Lanka.

An Analytical Framework consisting of five main factors—Inputs, Output, Result, Impacts and Effects, and External Factors—established through the literature review, was followed in the evaluation tool and utilised during the test assessment interviews in the field (Table 27.1).

27.2 Disasters and Housing in Sri Lanka

The Indian Ocean Earthquake and Tsunami struck on the morning of 26 December 2004, battering coastal communities in fourteen countries with 30-m high waves and killing almost 230,000 people (USGS 2013). The initial earthquake of an average 9.2 magnitude was the second largest recorded so far, with its epicentre in the Indian Ocean about 150 km from the northern tip of Indonesia. The seabed rose by up to five metres, generating gigantic tsunamis across the Indian Ocean that travelled more than 500 km/h and within a couple of hours heavily impacted countries fringing the Indian Ocean, such as Sri Lanka. Aftershocks were felt as far away as Alaska and the impact was similar to an atomic bomb explosion (ADB 2005).

Table 27.1 Analytical framework of the evaluation tool (adopted from Lizarralde 2002)

Factors	Definition	Aspects	Key questions
(a) Inputs	Human, material and financial resources required to incorporate resilience in shelter	Efficiency	• Were the local and external resources optimised (cost-effectiveness)?
			• Was the community specifically engaged in design/construction?
			• Was there a dedicated skills transfer/training component?
(b) Output	Articulation of resilience options before applying it	Results	• Were the resilience options realised?
		Timing	• Were they available at the right time?
		Quality	• Are the resilience options 'good' in the local context?
(c) Result	Direct consequence for the beneficiary of applying the resilience options	Pertinence	• Were the resilience options available to the most vulnerable people?
		Acceptability	• Did the local community use the resilience options?
			• Were they pre-determined/required, or optional?
			• Were they replicated outside the project?
			• Are they easy to maintain?
(d) Impacts and Effects	Indirect or later consequences for the beneficiary of using resilience options (or the situation originating from the project)	Strategy	• Did the resilience options correspond to the needs of the community?
		Scope	• What proportion of vulnerable people was covered?
		Ultimate objective	• Did the project reduce the disaster risks of the community?
			• Do the community/households feel a greater sense of security?
(e) External Factors	Factors beyond the control of the implementing agency	External aspects	• How did the context and environment affect the results of the project?

Following the disaster, large and generous donations for relief and recovery amounting to nearly US$14 billion were made from around the world (Cosgrave 2007).

The island country of Sri Lanka with extensive coastal communities was among the most severely impacted countries, second only to Indonesia (PreventionWeb 2008); more than a million people were affected, more than 35,000 people were killed and almost 150,000 houses were destroyed or damaged. A long stretch of coastline of more than 1,000 km along the eastern and southern seaboard was affected, from Jaffna in the north to Colombo in the southwest (DCS 2005; UNEP 2005). The widespread damage to housing and consequent displacement of many coastal communities spurred many housing reconstruction programs by various international and national agencies, coupled with the government's 'buffer zone' policy of resettling the affected people away from the coast. Sri Lanka was selected in the SRG study because of the abundance of housing projects implemented there after the tsunami.

Sri Lanka also faced other hazards such as floods and cyclones, as well as the challenges of rebuilding a country after 30 years of civil war, making it particularly appropriate to examine housing and disaster resilience there. The two housing case studies discussed in this chapter were located in tsunami-affected Galle district in southern Sri Lanka. They consisted of permanent shelter in resettlement projects built after the tsunami. The case study projects also addressed local hazards such as flooding and water-logging, which were key hazards in the project locations. Thus the projects allowed gaining insights on various aspects of resilience.

27.3 Basis of Selection of Housing Case Studies

The case study housing projects were selected through extensive consultation with SRG member agencies. Two prominent international non-governmental organisations (NGOs), Habitat for Humanity and Caritas, offered their post-tsunami resettlement and reconstruction projects in Galle for the test assessments using the evaluation tool. Both the projects incorporated disaster resilience features and hence they were considered suitable for assessment to understand the extent resilience had been achieved. The large scale intervention by many agencies, particularly in southern Sri Lanka, also allowed visiting projects implemented by agencies other than SRG members, enabling a broader insight.

27.4 Case Study 1: Habitat for Humanity

This project in Galle was implemented by Habitat for Humanity Sri Lanka (HfH-SL) in coordination with several government stakeholders and service providers including Galle Municipal Council, Urban Development Authority (UDA), Reconstruction and Development Authority (RADA), National Water Supply and Drainage Board (NWSDB) and Ceylon Electricity Board (CEB). Thus it relied on the support of these stakeholders, which had a bearing on its effectiveness as assessed in the study.

27.4.1 Context of the Project

In addition to massive destruction on the Galle coast, water from the 2004 Indian Ocean Tsunami travelled 3–4 km inland through rivers and canals, causing further devastation. The settlement of Samagiwatte, located about a kilometre from the coast, was inundated by water flowing through an adjacent canal (Fig. 27.2) and connected channels. This was largely a low-income settlement where most houses had earthen floors and rough timber walls, and hence damage to shelter here was extensive.

Fig. 27.2 Canal adjacent to Samagiwatte through which water from the tsunami flooded the settlement

27.4.2 Key Housing Features

In this project 22 damaged houses were rebuilt on existing sites during 2006–2007 (Fig. 27.3). Each single-storey house was about 350 ft^2 (32.5 m^2) with 2 bedrooms, a living room, kitchen and bathroom. The cost of each house was Sri Lankan Rupees 350,000 (about US$3,500). HfH-SL provided masons and skilled construction workers and beneficiaries provided sweat equity by helping the workers.

Because the area was low-lying marshy land, each beneficiary household had to fill the housing site at own cost by bringing soil from outside and raising the level of the land before the houses were built. Houses were built on pad foundations for structural stability on the soft soil.

Because the area was low-lying, flooding/water-logging was the main hazard. Thus the key DRR feature involved raising the land. Additionally by replacing the previously flimsy wooden houses with houses built of durable materials, vulnerability was addressed.

27.4.3 Main Findings

As mentioned above in Sect. 27.1 and shown in Table 27.1, the evaluation framework consisted of field investigations with regards to five main factors—Inputs, Output, Result, Impacts and Effects and External Factors. Accordingly the findings of the investigations are discussed below.

Fig. 27.3 Houses built by HfH-SL in Galle

27.4.3.1 Inputs: Resources for Incorporating Resilience

A list of potential beneficiaries was prepared after the tsunami by the office of the Divisional Secretary of Galle and HfH-SL screened it to make the final selection of beneficiaries. The beneficiaries were living without tenure on government land before the tsunami. Before the reconstruction project was implemented, HfH-SL assisted the beneficiaries to secure land tenure by obtaining land titles and ownership. Each beneficiary paid SL Rs. 10,000 (about $900) as fees to the Municipal Council for a plot of land of 5 perches (1 perch = 25 m^2).

The process of acquisition of land ownership involved extensive engagement and support from HfH-SL, and consultation with beneficiaries. Although there was some consultation on positioning of the house on the site, there was hardly any consultation on developing the house design, which was designed by HfH-SL based on the Urban Development Authority's (UDA) guidelines.

There were limited inputs for livelihoods in the form of support for home-gardening. There was no direct input for DRR in terms of development of an early warning system, disaster preparedness and community awareness raising and capacity building.

27.4.3.2 Output: Articulation of Resilience Options

Key resilience features included raised land elevation to avoid flooding, pad foundations for structural stability on soft soil, about 30 cm high plinth and durable construction. Such construction included a concrete floor instead of the former typical earthen

Fig. 27.4 Flimsy houses such as this was typical in the area before the tsunami. Note in the background the much better type of house built by HfH-SL

floor, brick walls instead of rough timber (previously common) and clay tiles instead of corrugated iron (CI) sheet to avoid lifting off in strong wind. The land was filled and the soil was compacted to prevent settlement of structures.

To support overall resilience, through negotiations with local government authorities tenure security was legalised, and municipal water supply and electricity was formalised from previous illegal connections. Also septic tanks were built by HfH-SL to ensure sanitation.

27.4.3.3 Result: Direct Consequences of Application of Resilience Options

Overall, most people who benefited from the project were poor and vulnerable. Previously they lived in flimsy houses in a low-lying area that was often flooded and water-logged (Fig. 27.4), and these conditions had improved. The quality of construction and building materials was generally good. However laminated plywood doors at the rear of houses that were exposed to the outside had deteriorated and generally were not of good quality.

Standard design guidelines of the Urban Development Authority (UDA) were followed. However the UDA guidelines suggested a house of minimum 500 ft^2 (46.5 m^2), but the HfH-SL houses were 350 ft^2 (32.5 m^2). This was done for cost-effectiveness and to support more beneficiaries, but for large households it presented space constraints. In general, the bedrooms were rather small. Nonetheless the houses had a high roof with the roof slope more than 30° to prevent damage by strong wind. Thus, inside, the houses felt spacious and cool. From the outside, the houses seemed large and gave their owners a sense of pride. The roofs were well-built and provided protection from rain adequately.

Although a kitchen was provided inside the house, most households had converted it for other functions (small bedroom, store, dining, pantry, etc.) and built an extension at the rear to serve as a kitchen. Gas fuel was unaffordable for most of them, and using wood fuel in the indoor kitchen would have made the whole house smoky.

27.4.3.4 Impacts and Effects: Indirect/Later Consequences of Application of Resilience Options

Despite raising the land by 1–1.5 m, because of the very low original elevation, flooding and consequent water-logging persisted. Particularly in the area close to the canal, regular flooding, sometimes up to 1 m height, was experienced. However houses further away from the canal experienced less flooding and here water did not enter into the houses. Nonetheless, although the area still suffered from flooding, on the whole the community was in a better and improved situation than before and the efforts of HfH-SL had succeeded in reducing the overall vulnerability of the community.

Most beneficiaries were generally satisfied with the quality and durability of the houses, and mentioned that it was a great improvement over the type of house they had before. One beneficiary commented, "Previously we lived like animals in a shed, now it's so much better—we are thankful." Particularly because tenure security had been legalised, beneficiaries felt more secure. As one beneficiary mentioned, "I can now sleep better."

No professional contractors were appointed; rather local head masons were appointed who organised their construction worker teams. This allowed local workers from Galle to obtain employment and thereby the project contributed to the local economy and livelihoods.

27.4.3.5 External Factors: Beyond Control of Implementing Agency

In Sri Lanka, constructing roads and roadside drains was the responsibility of the Municipal Council. Internal roads within the settlement were constructed several years after the houses had been built. In the interim residents moved about with difficulty especially in the rainy season when the unpaved roads became muddy. The roads were constructed with a high elevation to avoid water-logging. Thus, although the roads remained free from water, because they were higher than adjacent house plots, water flowed from the road into the plots, aggravating the flood problem.

More than 5 years after the houses had been built, during the field visit in October 2012, it was observed that no drains had been constructed. Open unpaved drains throughout the settlement presented a health hazard; in addition to overflowing during floods, in places where water was stagnant mosquitoes bred and added to the health hazard.

27.5 Case Study 2: Caritas

This project in Galle was implemented by Caritas Social and Economic Development Centre (Caritas SED) of the Catholic Diocese of Galle. The housing and settlement design was done by Avant Garde Urban Design Partnership, a consulting firm from the capital city Colombo. As in the previous case study, the same government stakeholders and service providers were involved: Galle Municipal Council, Urban Development Authority (UDA), Reconstruction and Development Authority (RADA), National Water Supply and Drainage Board (NWSDB) and Ceylon Electricity Board (CEB).

27.5.1 Context of the Project

This project was implemented in line with the Sri Lankan government's 'buffer zone policy' of resettling victims of the 2004 Indian Ocean Tsunami away from the coast. Low-lying vacant land in an area called Waluawatta about half a kilometre inland from the coast was provided by the government to establish this settlement of 76 houses with a community centre and playground.

27.5.2 Key Housing Features

In this project, 76 two-storey house units in duplexes were built on reclaimed marshy land in 2006 (Fig. 27.5). Each house was about 600 ft^2 (55.8 m^2), 100 ft^2 (9.3 m^2) larger than the UDA-prescribed house size. The houses included a living-dining area, a kitchen and bathroom on the ground floor and two bedrooms with a balcony on the first floor. Each duplex was built on 5 perches (1 perch = 25 m^2) of land of which ownership was transferred to the beneficiaries. The cost of each duplex was Sri Lankan Rupees 1,400,000 (about $14,000).

The houses were built by contractors selected through an open tender system. Because the area was low-lying marshy land, Caritas SED had to raise the elevation by landfill, prepare the site for construction and build special foundations. Community facilities including a community centre, pre-school, children's playground and shopping centre were incorporated into the settlement plan.

Fig. 27.5 Houses built by Caritas SED in Galle

By raising and improving the land, and building drains, the risk of flooding and water-logging was overcome. By establishing the settlement in this area located away from the coast, the risk of coastal hazards such as tsunamis and cyclones was reduced.

27.5.3 Main Findings

As in the preceding case study, the assessment was done following the evaluation framework consisting of the same five main factors—Inputs, Output, Result, Impacts and Effects and External Factors—as discussed below.

27.5.3.1 Inputs: Resources for Incorporating Resilience

A list of potential beneficiaries was prepared after the tsunami by the office of the Divisional Secretary of Galle and Caritas SED screened it to make the final selection. Houses were designed by a consulting firm in Colombo, built by contractors and then allocated to beneficiaries by lottery. There was no beneficiary consultation or participation in the selection, housing design and construction processes.

There were no specific direct inputs for livelihoods. Because of the inland location, beneficiaries found it difficult to pursue their previous coastal livelihoods.

Fig. 27.6 Production of compressed soil-cement blocks for building walls in Caritas shelter projects created livelihood opportunities (*Photo*: courtesy of I. Anda)

Nonetheless compressed soil-cement blocks were used for wall construction, produced through a labour-intensive process that created local job opportunities during project implementation (Fig. 27.6).

During the early stage after beneficiaries had moved into the houses, some disaster preparedness awareness-building activities were conducted with the support of the governmental Disaster Management Centre (DMC) focusing on first aid and search-and-rescue. However this was not part of any program on enabling long-term community resilience.

27.5.3.2 Output: Articulation of Resilience Options

Key resilience features included site improvement and raised land elevation to avoid flooding, drainage to prevent water-logging (drains from back of houses connected to a main drain), deep foundations for structural stability (foundation trench filled with 1.5–2.0 m of sand with 1.5–2.0 m deep foundations resting on the sand-filling), 30 cm high plinth and durable building materials (concrete floor and reinforced concrete frame; stabilised soil-cement block walls; roofing of clay tiles on good quality timber frame instead of CI sheet to avoid lifting off in strong wind). The land was filled and the soil was compacted to prevent settlement of structures.

The area although inland was still close to the coast and experienced the effects of cyclones, even if not direct impacts. There was a cyclone in early 2012, but there was no damage to the houses, demonstrating their sturdy construction.

Municipal water supply and electricity was provided after 3 months and internal roads were constructed by the Municipal Council about 4 months after beneficiaries moved into the houses. In the short interim, inevitably, they experienced walking through muddy roads with many potholes and using candles or lanterns at night.

Roadside drains were built about a year after beneficiaries moved in, and in the interim there was water-logging and overflow from septic tanks. Because of strong complaints by the beneficiaries, the drainage system was provided by the Municipal Council. However, as observed during fieldwork in October 2012, the drains had not yet been extended to the last four houses on the edge of the settlement; it was expected that this would eventually done, but in the interim the residents had to live without drainage. Caritas SED provided overhead water tanks and constructed septic tanks to ensure sanitation.

This project thus represented a gradual process of consolidating the resilience of the beneficiary community.

27.5.3.3 Result: Direct Consequences of Application of Resilience Options

The project beneficiaries were all tsunami victims who lost their houses, although they were of varying social and economic backgrounds. It was reported that in this community a mix of people from different ethnic backgrounds lived in harmony.

The quality of construction and building materials was generally good. However laminated plywood doors at the rear of houses that were exposed to the outside had deteriorated and generally were not of good quality.

The living-dining space in the houses had a double height at one end and hence felt spacious and cool. However households with elderly members faced a problem of climbing the stairs to go to the bedrooms above and hence were compelled to sleep on the ground floor in the living-dining area. This led to a privacy problem, especially when guests were received inside and also during the regular household activities.

Most beneficiaries used the indoor kitchen as it had a chimney and so smoke from wood fuel could escape. Building two units on 5 perches land resulted in very little open space on the plot and therefore little scope for extension, though some beneficiaries would have preferred to build an outdoor kitchen in keeping with local tradition. Some large households were planning to build such a kitchen so as to get extra space inside the house.

Because there was no extended eave on the gable ends, there was water penetration from the side during heavy rainfall. Thus dampness resulted in the gable wall. In some cases, tiles were not laid properly and allowed rainwater to enter, and beneficiaries had to repair and replace defective tiles.

Nonetheless, overall, beneficiaries reported the houses and building materials to be of good quality contributing to their resilience. Some of them mentioned that the

Caritas SED houses were better than the houses built by other organisations in the area, and also much bigger. One beneficiary said, "This house is luxurious compared to other [agency] houses. The tsunami was actually a blessing for us to be able to get this house."

27.5.3.4 Impacts and Effects: Indirect/Later Consequences of Application of Resilience Options

Despite some initial problems the community gradually became more secure and resilient. Most beneficiaries were generally satisfied with the quality, size and durability of the houses, and mentioned that they were happy to receive such a house. The housing seemed to serve its purpose well even after more than five years since being built.

27.5.3.5 External Factors: Beyond Control of Implementing Agency

The government created pressure on Caritas SED to build the houses quickly so as to move the tsunami victims out of the transitional camps. There were various problems of coordination between different stakeholders and thus Caritas had to implement the project under very trying circumstances.

Additionally the prices of building materials had increased after the tsunami because of the great demand created by the large number of reconstruction projects; it was also difficult to find skilled construction workers —many had died in the tsunami—and thus there was great demand for them. This impacted on the project implementation and created many challenges for Caritas SED.

27.6 Summary of Main Findings from the Case Studies

The key aspects and findings of the two case study projects are shown in Table 27.2, and summarised below.

In the case of the HfH-SL project, although the project area continued to suffer somewhat from flooding from the adjacent canal, on the whole the situation of the beneficiary community was much improved by the project. Its overall vulnerability had been reduced and thereby its resilience increased. However external factors such as uncoordinated road construction by local authorities and the lack of drainage undermined the efforts of HfH-SL and contributed to the community's vulnerability.

In the Caritas SED project, inland location, adequate site preparation, construction of good quality houses and provision of drainage had led to a resilient community. This project represented a gradual process of consolidation of community resilience.

Table 27.2 Key project aspects and findings

Implementing agency	No. of houses	Key resilience features	Key findings
Habitat for Humanity Sri Lanka (HfH-SL), Galle, Sri Lanka	22	• Inland location to avoid coastal hazards • Elevated land to avoid flooding	• Although the area still floods, the overall situation of the community was much improved and the community's vulnerability reduced
		• Durable building materials	• External factors—uncoordinated road construction and lack of drainage by local authorities—undermined HfH's efforts and increased the vulnerability
Caritas Social and Economic Development Centre (SED), Galle, Sri Lanka	76	• Inland location to avoid coastal hazards • Elevated land to avoid flooding	• Inland location, adequate site preparation, construction of good quality houses and provision of drainage resulted in a resilient community
		• Durable building materials	• Gradual process of consolidation of community resilience

Despite the overall positive findings some challenges became evident in the test assessments, especially with regard to the Result factor concerning housing design issues. A key example was the one-size-fits-all approach followed in the projects; given the diversity of beneficiary households, providing the same sized house to all the households resulted in the obvious problem of lack of space for large households and too much space for small ones. Other challenges as noted above, such as poor quality doors, water leakage, cultural inappropriateness of indoor kitchens and difficulty experienced by elderly people in two-storey houses, were due to the bearing of the Inputs factor on the Result: Although the houses were designed and built professionally following government (UDA) guidelines, there was no community consultation and participation in those processes, hence arose the challenges.

That other factor that was found to have played a problematic role was External Factors. While good intentions and efforts were demonstrated by the implementing agencies, slow and inadequate support from the government undermined community resilience. On one hand communities were sheltered in well-built disaster resilient houses built by the NGOs (HfH-SL and Caritas SED) and on the other hand effective service provision (drainage, roads, etc.) by the government was lacking and thereby contributing to the vulnerability of the communities, thus undoing the efforts of the NGOs.

It should be noted here that the exact nature of initiatives for housing resilience depends on the type of hazard. For example, in areas affected by strong wind, strengthening the house roof structure is important (Mayo 1988); in earthquake-prone areas attention is given to the walls (Arya et al. 2010); in flood-prone areas attention is given to the base of the shelter (Ahmed 2005). In the context of tsunami risk, the key resilience strategy is to avoid building housing on the coast, reflected in the Sri Lankan government's 'buffer zone policy'. This policy, while reducing disaster risk, resulted in various problems, particularly in pursuing coastal livelihoods (Amnesty International 2006; CPA 2006; Hettige 2007), which was evident in the Caritas SED project. Nonetheless, given the satisfaction of its beneficiaries with the wide range of benefits the project provided, this factor did not appear to have much bearing on the overall resilience of the community. On the other hand, the HfH-SL project was built in-situ in an inland location where flooding was the key hazard; although there was also some risk of indirect tsunami impact from the adjacent canal, this was essentially a form of flooding. Thus the approach to raise the land and build durable houses contributed to the beneficiary community's resilience.

27.7 Conclusion

This chapter presents the findings from two case studies of housing reconstruction in Sri Lanka, and it would be difficult to draw generalisations only from them. However the authors have undertaken extensive studies on the subject; in addition to the larger study from which this paper has been derived (see Sect. 27.1), various other research projects (Ahmed and McEvoy 2010; Shaw and Ahmed 2010) on post-tsunami housing reconstruction in Sri Lanka has allowed validating the findings presented here at a broader level. Two main aspects were evident. Firstly, because these projects were implemented after a major disaster, and hence were not regular housing projects, special attention was given to resilience against future disasters; the projects were found to have reduced disaster risk to varying levels, and had thereby contributed to beneficiaries' resilience. Secondly, these projects achieved a marked improvement of previous living conditions and thereby increased the overall resilience of most of the people belonging to the target group; quite often, such people lived in inadequate housing or in locations exposed to hazards, a key reason for their housing to be impacted, and the projects generally provided more robust and durable housing.

However challenges in terms of Result and External Factors affected the overall performance of the projects. While resilience features were applied in the design and construction of housing, social, cultural and contextual issues relating to housing design and settlement planning were overlooked. 'One-size-fits-all' approaches to house design and regimental, barrack-type and unimaginative settlement layouts were common.

Having grossly insufficient financial and other resources for recovery from a massive disaster like the 2004 Indian Ocean Tsunami, the Sri Lankan government allowed a plethora of humanitarian agencies to undertake the colossal reconstruction effort, but did not have a well-formulated mechanism for coordinating it with its own programs. Thus a mismatch between the GO and NGO actors emerged, as evident from project beneficiaries living in well-built houses without adequate provision of public services that make a settlement habitable.

To add an endnote: The evaluation tool that allowed arriving at the above findings and conclusion has been found productive in a variety of contexts and projects and would therefore serve as a useful resource for agencies interested in evaluating whether and how resilience has been achieved in their housing projects. It is being recommended that such evaluations using purpose-built tools should be a standard procedure, given the widespread and global occurrence of disasters, their devastating impact of the housing sector and the multiplicity of post-disaster housing reconstruction projects by humanitarian agencies.

Acknowledgements Shelter Reference Group (SRG), Australia for commissioning and facilitating the research project *Scoping study: shelter and disaster risk reduction in the Asia-Pacific region*, from which this paper is derived. Caritas-Australia for allocating funding for the above study. AusAID for providing the funds to Caritas-Australia under the Humanitarian Partnership Agreement, a component of which related specifically to disaster risk reduction and management.

References

ADB (2005) From disaster to reconstruction: a report on ADB's response to the Asian tsunami. ADB (Asian Development Bank), Manila

ADPC (2002) Proceedings of the regional workshop on best practices in disaster mitigation: lessons learned from the Asian Urban Disaster Mitigation Program and other initiatives. ADPC (Asian Disaster Preparedness Center), Bangkok

ADRC (2012) Natural disaster data book 2010: an analytical overview. ADRC (Asian Disaster Reduction Center), Kobe

Ahmed I (2005) Handbook on design and construction of housing in flood-prone rural areas of Bangladesh. ADPC (Asian Disaster Preparedness Center), Bangkok

Ahmed I, McEvoy D (2010) Post-disaster housing reconstruction: post-occupancy case studies from Sri Lanka. In: Proceedings of ANZAScA 44th annual conference, Auckland

Amnesty International (2006) Sri Lanka: waiting to go home – the plight of the internally displaced. http://www.amnesty.org/en/library/asset/ASA37/004/2006/en/f5d52b8c-d460-11dd-8743-d305bea2b2c7/asa370042006en.pdf. Accessed 21 February 2014

Aquilino MJ (2011) Beyond shelter: architecture and human dignity. Metropolis, New York

Arya AS et al (2010) Guidelines for earthquake-resistant non-engineered construction. International Association for Earthquake Engineering, Tokyo

AusAID (2005) AusGuideline: the logical framework approach. AusAID, Canberra

Charlesworth E, Ahmed I (2012) Scoping study: shelter and disaster risk reduction in the Asia-Pacific region. Report, Shelter Reference Group, Australia

Coburn A et al (1995) Technical principles of building for safety. Intermediate Technology Publications, London

Cosgrave J (2007) Synthesis report: expanded summary. Joint evaluation of the international response to the Indian Ocean tsunami. TEC, London

CPA (2006) Reflections on tsunami one year on: lessons to be learnt. CPA (Centre for Policy Alternatives), Colombo

DCS (2005) Tsunami census 2004/2005. DCS, Colombo

Earthquake Reconstruction & Rehabilitation Authority (ERRA) (2006) Guidelines for earthquake resistant construction of non-engineered rural and suburban masonry houses in cement sand mortar in earthquake affected areas. ERRA (Earthquake Reconstruction & Rehabilitation Authority), Pakistan

Hettige ST (2007) Tsunami recovery in Sri Lanka: retrospect and prospect. SPARC and ActionAid International, Colombo

HFHA (undated) Housing & health (factsheet). Habitat for Humanity Australia (HFHA), Sydney. http://www.habitat.org.au/document.doc?id=487

IFRC (2011) Shelter safety handbook. IFRC (International Federation of Red Cross & Red Crescent Societies), Geneva

IFRC (2012) Understanding community resilience and program factors that strengthen them. IFRC, Geneva

Lang H (2008) Community housing in post disaster area on Nias islands, Indonesia: responding to community needs. In: Proceedings of the 4th international i-Rec conference, University of Canterbury, Christchurch

Lizarralde G (2002) Organizational design, performance and evaluation of post-disaster reconstruction projects. In: Proceedings of conference on improving post-disaster reconstruction in developing countries, Montreal

Lyons M et al (eds) (2010) Building back better: delivering people-centred housing reconstruction at scale. Practical Action Publishing, Rugby, UK

Mayo A (1988) Cyclone-resistant houses for developing countries. Building Research Establishment, Watford, UK

NHDA (National Housing Development Authority) (2005) Guidelines for housing development in coastal Sri Lanka. Ministry of Housing and Construction, Colombo

PreventionWeb (2008) Tsunami – data and statistics. http://www.preventionweb.net/english/hazards/statistics/?hid=71. Accessed 23 October 2013

Seraj SM, Ahmed I (2004) Building safer houses in rural Bangladesh. Bangladesh University of Engineering & Technology, Dhaka and University of Exeter, Exeter, UK

Shaw J, Ahmed I (2010) Design and delivery of post-disaster housing resettlement programs: case studies from Sri Lanka and India. Report, ARC Linkage

Tro PTT (2011) Atlas/Compendium of housing vulnerability and measures for safe housing in Vietnam. Development Workshop France, Hue, Vietnam

Twigg J (2007) Evaluating disaster risk reduction initiatives. In: Benson C, Twigg J (eds) Tools for mainstreaming disaster risk reduction: guidance notes for development organisations. IFRC and ProVention, Geneva

UNEP (2005) After the tsunami: rapid environmental assessment. UNEP, Nairobi

UNHCR, IFRC, UN-Habitat (2009, 2010, 2012) Shelter projects. UNHCR, IFRC, UN-Habitat, Geneva

UNNATI (2006) A people's guide to building damages and disaster safe construction. UNNATI (Organisation for Development Education), Ahmedabad

USGS (2013) Magnitude 9.1: off the west coast of northern Sumatra. http://earthquake.usgs.gov/earthquakes/eqinthenews/2004/us2004slav/#summary. Accessed: 23 October 2013

Chapter 28
Ten Years of Resettlement in Eco-Village, Sri Lanka

Miwa Abe and Rajib Shaw

Abstract As part of Indian Ocean Tsunami recovery projects, an "eco-village" was established in Lagoswatta, located in the Kalutara district near Colombo in Sri Lanka. There are 55 tsunami-affected families relocated from eight different communities. It was supported by local NGO. UNEP and UN-HABITAT initiated the Eco-housing project in 2004 to build capacity and to increase awareness on this issue. This case study will show the impacts of thematic resettlement as eco-village through field survey between 2006 and 2012. Questionnaire and interview survey were conducted in eco-village named 'Danmiyangama'. Observation survey involved understanding the changing of the community structure and physical life condition in the resettlement area.

Keywords Disaster recovery • Eco-village • Resettlement • Tsunami

28.1 Introduction

Following the magnitude 9.0 earthquake on December 26, 2004, the tsunami affected the coastal regions of Indonesia, India, Thailand, Maldives, Sri Lanka and other countries in coastal Africa. Sri Lanka was among the several countries that was greatly devastated by the 2004 Indian Ocean Tsunami. About 60 % of its coastal region was damaged by the tidal wave and more than 38,000 lives were lost. At the aftermath of the disaster, people were urged to withdraw from their damaged houses. Many found it hard to leave a place where memories ran deep. However, they were left with no choice. Not only the houses along the coastline collapsed but there were other considerations such as government policy, fear of another tsunami and so on.

M. Abe
Center for Policy Studies, Kumamoto University,
2-39-1 Kurokami, Chuo-ku, Kumamoto 860-8555, Japan

R. Shaw (✉)
Graduate School of Global Environmental Studies (GSGES), Kyoto University,
Yoshida-Honmachi, Sakyo-ku, Kyoto 606-8501, Japan
e-mail: shaw.rajib.5u@kyoto-u.ac.jp

R. Shaw (ed.), *Recovery from the Indian Ocean Tsunami: A Ten-Year Journey*,
Disaster Risk Reduction, DOI 10.1007/978-4-431-55117-1_28

A buffer zone was implemented in the coastal region by the government after the disaster. People previously residing in the non-buffer zone were prohibited by law to return to their houses. The only option, therefore, was relocation of coastal communities. In the new relocation strategy in Sri Lanka, it is noteworthy that not a single villager has left the new community. Many have settled in quite well in the new setting. Problems encountered in the past such as distance to the working place and social relationships in the new community found solutions among the community members themselves and through supporting organizations. It was important to note that problems can be overcome when options are available in a resettlement of people from different villages or in one located a good distance away.

For understanding resettlement life and sustainability of eco-activities in eco-village, questionnaire survey and interview survey was conducted. In the previous research, Imura et al. (2011) showed that importance of common activities for resettlement after disaster. Especially, if some community cannot relocate with same community members (it means people should relocate with different community members), common purpose will be a one of trigger to improve bonding with each other in new community. Therefore, the purpose of this study are (1) to understanding resettlement people's life in relocated land during past 10 years and (2) to consider with adequacy of eco-activities for community recovery.

28.2 Disaster Recovery and Community Resettlement

Previous study identified several the relocation issues after disaster namely settlement planning and land ownership, livelihood options, human and social networks and so on (Imura and Shaw 2009). Dikmen (2008) concluded that new settlements were unsuccessful due to the following failures in post-disaster reconstruction projects:

- Quick decisions,
- Lack of user participation in early decision-making process,
- Inadequate site-selection criteria (distance from livelihood and social networks),
- Lack of interdisciplinary work during site-selection,
- Lack of consideration of the lifestyle of the beneficiaries,
- Lack of guidance to beneficiaries during the construction phase of houses
- Socio-culturally inappropriate settlement layouts

Success or failure judgment of resettlement is difficult, and it lacks critical indicators. Involuntary resettlement often involves removal from an environment in which the society has from evolved centuries of adaptation to their old patterns. This relationship is based on economic, political or social factors or a combination of any or all three. For these reasons, many researchers consider resettlement as something to be avoided or minimized in development projects if at all possible (Smith 1991). Partridge (1989) pointed out that "from the perspective of displaced people, forced resettlement is always a disaster." The World Bank's resettlement policy adopted in 1980, gave the message that it is essential to handle the difficult process of resettlement equitably and efficiently, if involuntary resettlement is mismanaged, the displaced people will suffer unjustifiably and will be impoverished (The World Bank 1996).

Usually evaluation of disaster recovery is conducted by support organization or government side. It is difficult to conclude people's opinion for evaluation and long term approach, because there are no common indicators. Tamura et al (2000) showed seven life recovery tasks from their research as recovery from Great Hanshin Awaji earthquake in Japan, 1995. This research was collected by workshop participants of local people. These recovery tasks are Housing, Social Support Network, Land Use Planning, Physical and Mental Health, Social Infrastructure and Preparedness, Government Assistance Policy, and Economic/Financial Situation. Kimura et al (2004) have continued the survey over a period of 10 years after Great Hanshin Awaji earthquake, and their group have introduced that the bigger impact to urban area, the longer it takes for social recovery. In addition, in terms of action pattern of affected people, it was noticed that recovery time was seen as—10, 100, 1000 h after disaster, suggesting that the complex measures for evaluation and time for the recovery. Determination of who and when in disaster recovery is still unclear. Therefore evaluation of relocation and resettlement after disasters proves to be a challenging issue.

28.3 What Is Eco-Village in Sri Lanka?

28.3.1 Process of Established Eco-Village

In 2004, the United Nations Environment Programme (UNEP) in Bangkok, Thailand and UN-HABITAT joined together for an eco-housing project in the Asian region. This was to promote eco-housing as a key disaster preventive measure in the Asia-Pacific region. The composite project addressed four key areas: (1) knowledge building, (2) educational initiative, (3) networking, and (4) implementation. The implementation strategy involved planning in Indonesia, Maldives, Sri Lanka, Thailand, China and Bhutan. In May 2005, at a national level workshop, the guidelines for Indonesia, Sri Lanka and Maldives was prepared. In Indonesia, the project was implemented in the 2004 tsunami-affected areas of Banda Aceh and Calang. An "Eco-village", one of the tsunami recovery projects, was implemented in Lagoswatta, in the Kalutara district near Colombo in Sri Lanka. The project considered social, ecology and economy and respective activities were derived. Kalutara is one of the districts in Sri Lanka (Fig. 28.1). Kalutara is a semi-urban town located approximately 50 km south of Colombo, Sri Lanka's capital. It lies between latitude 6°41′34″ and 6°43′10″ and longitude 80°02′53″ and 80°05′03″. The area was separated north and south Kalutara by Kalu River. The south side is more urban than north side. Floods occurred in the rainy season.

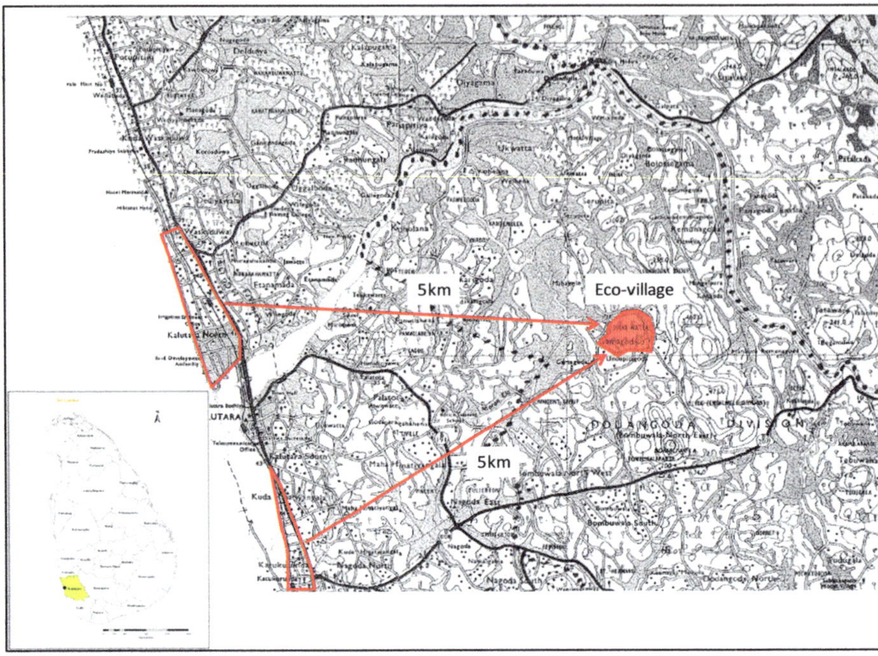

Fig. 28.1 Map of Danmiyangama, Kalutara district in Sri Lanka

28.3.2 New Tsunami Recovery Community as Eco-Village

The community named "Danmiyangama" has been newly formed in the Lagoswatta area in Kalutara district, which started on May 5, 2005. It is 5 km away from main national road (Galle road) with a land mass of 5 acres. The Danmiyangama resettlement is one of the projects initiated by Sarvodaya, the local NGO that supports tsunami-affected families. The building site is on a piece of land on a gentle slope bordering the rice field. It has a main road and two arterial roads dividing the site into four areas. The proposed site did not fall under the disaster control zone. The surrounding is covered with vegetation. The allotted plot for development is connected to the city through a main road, which runs across the site. At the beginning, the total number of people who settled down in the Eco-village was 245 and the number of households was 55. Income generating activities were fish merchants, carpenters, masons, labors, and drivers etc., with the majority being labors. Although they lost everything in the tsunami, they were willing to continue with the same activities. Sarvodaya, a local NGO, provided the assistance to improve their living conditions. The definition of Eco-village applied in 'Danmiyangama', Sri Lanka is as follows:

> Eco-villages are communities striving to create cooperative lifestyles in harmony with their local environments. Important features of the eco-village are developing and refining social and ecological tools such as consensus decision making, inter-generational care, alternative economic models, whole systems design, permaculture practices, renewable energy systems, and alternative modes of education that offer positive visions and real-life solutions for humanity and the planet. (Sarvodaya 2005)

Fig. 28.2 Characteristics activities in eco-village. (**a**) Waste separation centre. (**b**) Rain water tank. (**c**) Compost bin. (**d**) Solar power panel

Their project's concept was to integrate human activities into the natural world in a way that is supportive of healthy human development and can be successfully continued into the indefinite future. Especially, supporters request to achieve (a) Cyclic use of material resources, (b) Use of renewable energy sources, (c) Composting organic waste and (d) Roof top captured rain water will contribute to produce fruit and vegetables (Fig. 28.2). The Eco-village project in Sri Lanka was supported by Sarvodaya, a local NGO, which promotes the 'Shramadana' Movement which encourages the use of shared work, voluntary giving and sharing of resources to achieve the personal and social awakening of everyone from the individual level, to the village level, eventually to the international level.

28.4 Characteristics of the Study Areas

People were relocated 5 km inland from their original land by local government and Sarvodaya (See Fig. 28.1). The main income generation activities of these recipients were fish merchants, carpenters, masons, labors, and drivers etc. Especially, many people were labors.

Table 28.1 Survey schedule

	Survey period	Collected sample (Total households: 55)
First survey	9–23 January, 2007	54
Second survey	2–12 March, 2010	50
Third survey	29–30 November, 2012	37

New relocated people faced to eco-activities with unfamiliar people, which was mandatory in order to adapt to new community with rehabilitation from Tsunami. Before the Tsunami, villagers were not interested in environmental issues. The areas that they lived were facing issues such as no garbage separation, way of recycling activity, and direct flush wastewater to nearby rivers. After Tsunami, participants of the project agreed to make the eco-village using the UNEP guideline of eco-community. The first step was to select 55 families from eight tsunami-affected villages listed up by the Sri Lanka government. These families got a chance of new life and receiving environmental education on topics such as water management, appropriate garbage management, alternative energy sources and so on. They changed their old lifestyle to a new eco-friendly one. The project concept was "Human activities are harmlessly integrated into the natural world in a way that is supportive of healthy human development and can be successfully continued into the indefinite future". Supporters is required to achieve four subjects as use of renewable energy sources, cyclic use of material resources, composting organic waste and roof top captured rain water will contribute to produce fruit and vegetables.

28.5 Data Collection

For understanding the change in the activities and relationships in the eco-village, questionnaire survey and interview survey were conducted. Target households were 55. It was conducted in 2007, 2010 and 2012 (Table 28.1). All surveys used the same questionnaire sheet, and was answered by the same house owner from the previous (2007 and 2010) survey. If house owner was changed, the family heads was asked to respond instead. Collected sample was 54 in 2007, 50 in 2010, and 37 in 2012. All interviews were taken with the assistance of a translator who is a master student at the University or NGO staffs. Therefore, local people's language was translated from Sinhalese to English. Date of survey is as follow:

28.6 Changing Life Condition During Ten Years

28.6.1 Life Before Tsunami

Before the tsunami, all villagers lived near the sea in Kalutara. Their houses were close to the city center. Therefore they could easily go to the market, school and any institutions on foot. The traffic condition was good. People built their own house

Fig. 28.3 Previous community of eco-village people. (**a**) Damaged house by tsunami of Eco-villagers at Kartala South. (**b**) Damaged house in previous land and house owner

close to each other. Some families did not have their own toilet and had to use common toilets. Sometimes there are only one or two toilets for ten families. People lived near the sea but they were not fisherman. Some houses had a small shop and some families had a car that could be used to guide tourists. They were not poor as compared to the fisherman, and they could employ some house workers. Therefore their area had a mixture of low income families and middle class families. In some area there were human problems such as fighting, drunken alcohol and shouting, domestic violence, murder and drug, and police comes to keep things in order. In Sri Lanka, women of low income families tried to go abroad to work. Prior to the Tsunami, many women in the coastal area go to countries such as Saudi Arabia, Singapore, Lebanon, and Iran etc. to work. Some women came back from these countries and lead a better life before the tsunami, and some other women were called back by their family after tsunami. Many household electrical appliances bought by these women were affected by the tidal waves. The wave attacked not only their houses but also the assets from their migrant work and their hopes (Fig. 28.3).

28.6.2 After Relocation (from 2007 to 2010)

Along with relocation, people could get a new house and common/individual facilities by tsunami recovery support. Usually village people hoped for support at relocated land. Some families had called back their wife or daughters who are working abroad for living a new house. At the first survey, people asked for support to the researchers and Sarvodaya staffs. And they answered that they are happy and are enjoying their new found life and even welcomed another tsunami. One person answered he can hardly wait for next tsunami. However in 2010, responses changed. Relocated people and support organization have struggled because people do not like the interposition of Sarvodaya. Since the relocated site was 'Eco-village',

Fig. 28.4 Difference of apparent condition from 2007 to 2012. (**a**) Main road of village in 2007. (**b**) Main road of village in 2012

people had to keep their village, house clean as well as to continue doing other eco-friendly activities, which was the requirement for those affected people were relocated to the area. Further, they were unable to extend nor reconstruction the buildings in the eco-village since housing design was built with eco-friendly (e.g. well-ventilation) consideration. People had wanted to change the housing style, life styles to better adapt to their life but it was difficult. Once one family starts rebuilding and re-designing their house and life style, other will follow. Gradually, the amount of recovery support was reduced because of complex results such as budget and non-cooperation of people. On the other hand, it can also be considered that the relocated people have started to be more independent from the support groups.

28.6.3 Present Condition of Eco-Village (in 2012)

The eco village has undergo many changes over the 10 years, eco-village. At the beginning, village was new, houses were arranged in an orderly manner and gave a vibrant atmosphere. People received their houses, and started new life at relocated site (Fig. 28.4a). Every month community meeting was held and every house owner attended. Sarvodaya, which is local NGO supported to eco-village, was also attending at the community meeting. Many foreigner groups carried out inspections at the village and the village was busy. The appearance of the eco village has changed a lot since it was first build (Fig. 28.4b). During the past decade, people and Sarvodaya had plant many trees in community space, roadside and border of houses (Fig. 28.5a, b). In 2012, trees have embosk houses. Village people enjoy using green at own garden and talking with neighboring families. Community people were surprised to see foreigners, and even commented 'Until when will you be calling us tsunami victims?'. To the community people, they have already passed the post disaster recovery phase, however, whenever researchers or support organization staffs visits the village,

Fig. 28.5 Using of greens in the community. (**a**) Home garden in 2012. (**b**) Border of next house

Fig. 28.6 Changing of well using. (**a**) in 2007. (**b**) in 2010. (**c**) in 2012

they felt that they are seen as disaster victims. Tsunami recovery village which was relocated from coastal side had changed to a village.

In the second survey in 2010, however, the number of activities has reduced year by year because support organizations have withdrawn from affected site. Since the recovery activities were greatly dependant on the support organization, with the withdrawal of the support groups from the village, there was no one managing the activities. The management of the eco-village can be seen in Fig. 28.6a–c. At first survey, it was well-managed by community members with Sarvodaya's support. In addition, community has an environmental committee who are selected from villagers. Wells were kept clean and community people have used common wells and common baths in 2007 (Fig. 28.6a). In 2010, community leader and committee system was disbanded with the withdrawal of recovery support. Common water resources were no longer managed, wells and common baths became dirty (Fig. 28.6b). At the third survey in 2012, community leader showed a well which became clean again (Fig. 28.6c). The community leader explained that after the support groups withdrew from the eco-village, people started to talk about 'eco-village'. When eco-village was first build, there were many visitors to study on establishing environmental friendly village as disaster recovery support. However, the number of visitors reduced over the years. In 2011, community members re-organized their

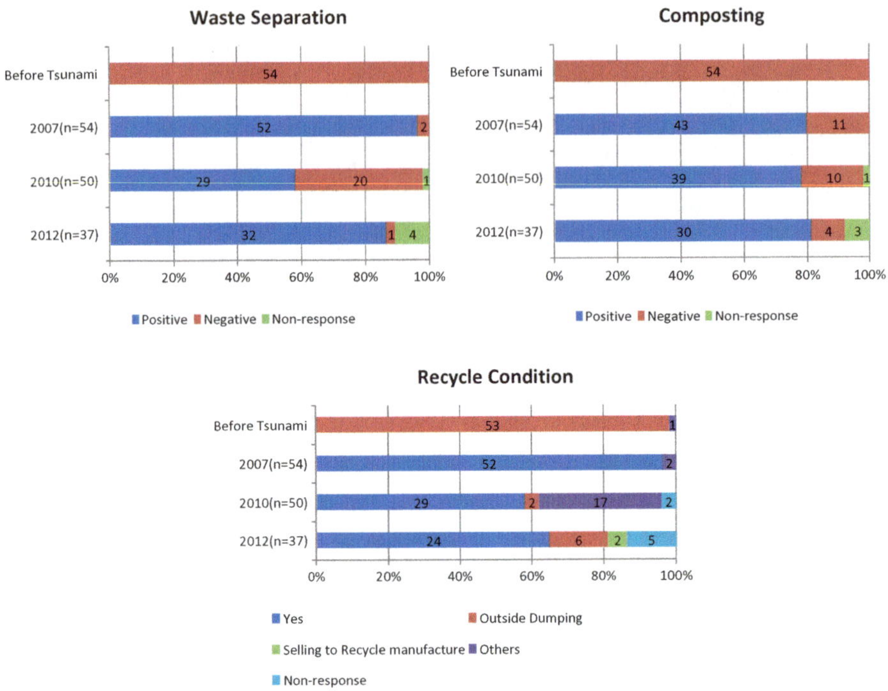

Fig. 28.7 Condition of each activity in 2007, 2010 and 2012. Actual status of waste separation (*upper left*), composting (*upper right*), recycle (*bottom*)

own committee, and elected a new leader. Now community members organize a cricket match, New Year festival and so on with neighboring villages. New interaction was initiated. These activities were triggered by a local government staff who visited to eco-village, and noted the deteriorating condition of the village such as dirty water resources and no using common facilities. Staff explained to the community people that this eco-village is special case in Sri Lanka, and the people there should appreciate the valuable facilities as other tsunami recovery villages does not have these things. Since then, the village people remembered about the 'eco-village' concept and the community lead felt the need to improve the current condition so as to attract the foreign researchers to visit the village again.

28.6.4 Data Analysis (Change of Environmental Activities and Human Relationships)

Figure 28.7 shows that condition of individual activities such as environmental in 2007, 2010 and 2012. For example, waste separation activity. Before the tsunami, no one separated their own waste. After relocation in 2007, over 52 households in 54

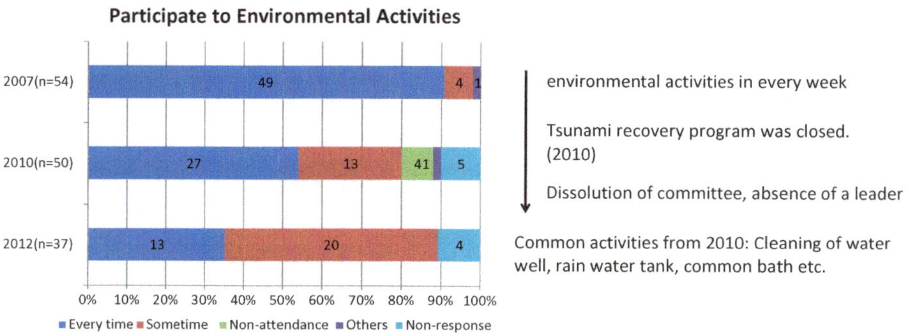

Fig. 28.8 Participation situation for environmental activities in eco-village

had started the activity (over 90 %). However, in 2010, 29 household in 50 was separating waste continuity. In 2012, 32 households resumed the activity. Composting is considered a continuity activity in eco-village with approximately 80 % participation. Similarly, with recycling, in 2007 people had tried to do recycle, the participation decreased in 2010, but has shown increment in 2012. And 6 households have returned to outside dumping. Before tsunami, all members dump their garbage as paper, glass and metal out of their houses. After relocation eco-village people learned about recycle from Sarvodaya, and people have tried to recycle. Before the tsunami, people dug a hole in their garden to dumpy their kitchen waste, a traditional way of wasting in there. After the tsunami, people were taught how to make compost. Some of ladies in the village responded that they do not know how to grow flowers and vegetables in own garden. They lack the skill and since they were located near the beach, their garden was mainly sand. Now people have made compost and started using it for their own garden for growing flower and vegetables. The flower and vegetables can be for self-use or sales. Although the amount earned is not huge but they have enjoyed growing and selling them. Some has taken composting as a hobby. Other environmental activities have not been conducive to continuity.

In the eco-village, there had took place in some others environmental activities such as cleaning rain water tanks and common baths, workshop by Sarvodaya, and so on. Figure 28.8 showed the participation to these kinds of activities. In 2010, most of all recovery support from support organizations ended. As such, the number of activities decreased. Despite people was interested to join the activities but there were no longer activities organized by third-party support. Some community member have voluntary started to clean and plan some of events from 2011.

Figure 28.9 shows that number of interaction with neighboring villages having increased in 2012. Just after relocation, people had a problem with nearby villages. Caste system still exists in Sri Lanka. Eco-village people who were relocated from coastal area did not share the same caste as the host community, As a result, some people who has daughter in own family did not like it. People had believed that they have different culture. Therefore, host community had opposed to having

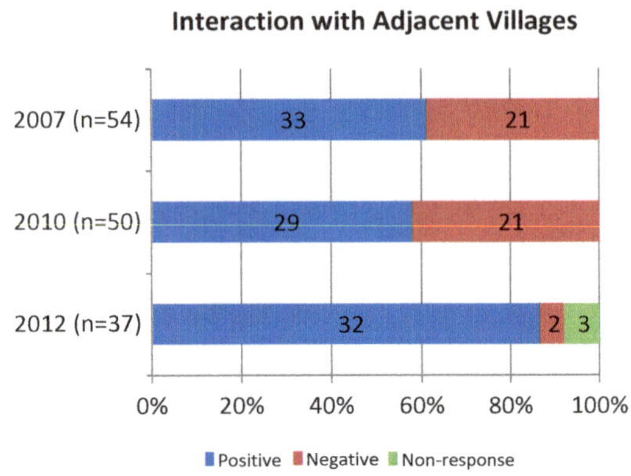

Fig. 28.9 Impact of Resettlement. Will of continuing living in eco-village

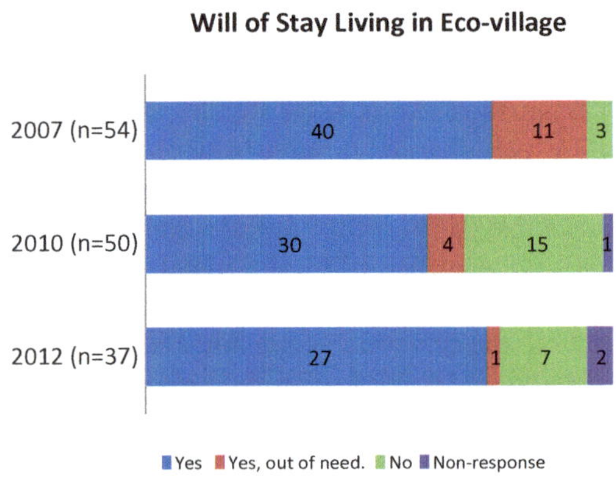

Fig. 28.10 Interaction with neighboring village families

the relocated people. Finally, support organization and local government had explained about relocation with common facilities as pre-school, community hall etc., and promised to set up infrastructure which covered their community. Sarvodaya put in much effort to improve their relationships and interactions in 2007. Survey at 2012, 32 households (about 90 %) answered they have an interaction, suggesting that Sarvoday's effort paid off and the communities was able to make new networks with each other. People have some acquaintance or relatives in there. And Fig. 28.10 shows that people hope to live in eco-village. The number was decreased on a temporary basis in 2010, but has been consistent since.

28.7 Summary of Major Changes

There are many process of relocated life, community activities, people's attendance and mind also was changed (Fig. 28.11). The 10 years process of relocation in eco-village can been seen as three stages namely carnival, conflict and renaissance.

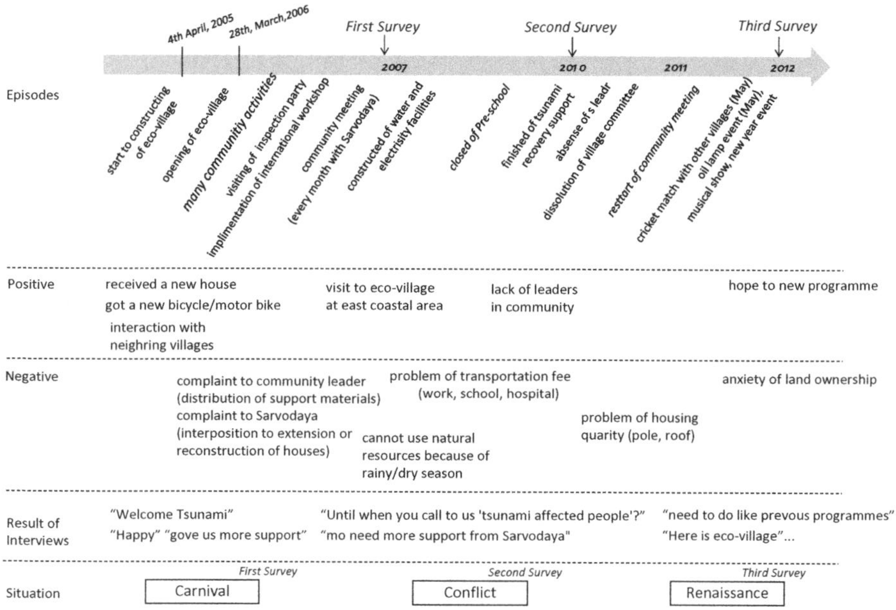

Fig. 28.11 Changing of condition in eco-village

The first stage is the carnival stage, the phase just after relocation. At the initial survey conducted in 2007, tsunami affected people was supported by government, domestic organization such as local NGOs as well as international organisation. Affected people could get new house, land, money, methods of transport and so on. This was unexpected to many of them. Some of people who lived in coastal area wanted to move to other place but they could not it because of finance reasons. For the relocated families, tsunami disaster and relocation became a chance for them to change their life. People relocated about 5 km from previous housing area and resulted in a move from coastal side to inland region. Despite the change in environment, they decided to relocate to start their new life. In eco-village, many government officers and foreigners who are staffs of international organizations had visited to eco-village because it was special recovery programme in Sri Lanka and explained to the local villager on eco-village, eco-life, and providing support goods. The villages were 'happy' and even looked forward to the next Tsunami, thinking it may help change their lives.

The second stage is the conflict stage. At the second survey conducted in 2010, it was observed that people were tired of the eco-friendly life, and preferred not to be controlled by Sarvodaya. It seems that there have some troubles between the villagers and support staffs. There could be two possible reasons. One, tiredness of continuing eco-friendly life. Many activities which are related with environment as cleaning well and bath, separating garbage and so on. People have not done these before tsunami. At the beginning, relocated people could enjoy because people could join many events and know each other. However, people are more accustomed to disposing their garbage and burning in their garden, which is more convenient, most importantly, it is their habit. Furthermore, eco-friendly facilities were limited by the seasons. Water tanks and wells could not use during dry season. And solar electricity was not available during rainy season. Second reason is the conflict within the village controlled by support organization. Eco-village was a pioneer project and was designed by the support groups based on special consideration. However, people wanted to change housing style such as color of wall, adding a room, making a park and so on was inhabited. Support organization have tried to keep original eco-village but relocated people wanted to their own life styles. They argued that they are no longer in post disaster recovery but has returned to their normal daily life and wishes to have more control of things, becoming more independent from the support groups.

In 2012, unexpectedly, community mind decided to keep the eco-village concept. Community leader and committee were set up and started organizing events. Over the past 10 years, eight house owners were changed. The reason of the change of owner was due to death of owner and inheritance by children, marriage and moved, rental business and so on. Moving to eco-village, people have used their networks (relatives or friends) for getting information about empty house or rental house. New comers know that it is an eco-village. Leader and committee member organize common events. The community leader commented that those who know previous activities such as environmental and cultural, has become a memory. People hope to resume the life of 'eco-village' again without tight restrictions of eco-work. This voluntary initiation amongst the villagers in wanting to resume eco-friendly activities is the renaissance stage.

Through the three stages—carnival, conflict and renaissance, the village has evolved to a 'new eco-village'. It is not guided, but a voluntary movement. The first common activity was to help connect the people from different communities in the eco-village. Over the period of 10 years, people start to appreciate these activities and its impact to others, and considers a new eco-village direction. This suggests the changes and development in resettlement process. If the policy makers try to evaluate relocation impact in 2007, it will be positive. If they evaluate it in 2010, resettlement condition will be negative. For understanding relocation and resettlement impact, policy maker should monitor over a long term period. This way, they would be able to observe the different stages in relocation and resettlement process.

Acknowledgment This work was supported by JSPS KAKENHI Grant Number 25870930. Grant-in-Aid for Young Scientists (B). The authors are thankful to Dr. Deepthi Wickramasinghe, in the University of Colombo, and Dr. Vinya Aryaratne and Mr. Nandana in Sarvodaya which one of the local NGO in Sri Lanka for their assistance and support in the field work, and to the people of the villages for their spontaneous response to the questionnaires.

References

Dikmen N (2008) Relocation or rebuilding in the same area: an important factor for decision making for post-disaster housing projects. http://www.grif.umontreal.ca/pages/DIKMEN_Nese.pdf. Accessed 27 Mar 2014

Imura M, Shaw R (2009) Challenges and potentials of post-disaster relocation. Asian J Environ Disaster Manage 1(2):197–219

Imura M, Shaw R, Tekeuchi Y (2011) Adaptive and sustainable post tsunami human resettlement in Sri Lanka and India. Research Publishing, Singapore, 220 pp

Kimura R, Hayashi H, Tatsuki S, Tamura K (2004) Psychologically defined life reconstruction process of disaster victims in the 1995 Hanshin-Awaji earthquake. Inst Social Safe Sci 6:241–250

Partridge W (1989) Involuntary resettlement in development projects. J Refugee Stud 2:373–384

Sarvodaya (2005) Eco-village Lagoswatte pilot project implementing. Lanka Jathika Sarvodaya Shramadana Movement in Sri Lanka

Smith OA (1991) Successes and failures in post-disaster resettlement. Disasters 15(1):12–23

Tamura K, Tatsuki S, Hayashi H (2000) The external validity of life recovery tasks and their structure among the disaster victims of the great Hanshin-Awaji earthquake. Inst Social Safe Sci 2:25–32

The World Bank (1996) Resettlement and development the bankwide review of projects involving involuntary resettlement 1986–1993. Environment Department Paper No. 032, 188 pp

Chapter 29
How the Tsunami Disaster Triggered a Change Process in the Education Sector of Sri Lanka: Lessons Learnt for Introducing Disaster Safety Education

Patrizia Bitter

Abstract The tsunami disaster clearly demonstrated that the people of Sri Lanka have not been prepared to face all the hazards threatening them. Many did not know how to protect themselves and responded inappropriately in the emergency situation. Therefore the education sector became a partner in the new Disaster Management strategy of the Government. The Ministry of Education and the German Agency for International Cooperation (GIZ) included DRR into their ongoing programme. In 2005, the "Disaster Management and Psycho-social Care" project was initiated and later integrated as "Disaster Safety Education" project into the "Education for Social Cohesion" Programme. This chapter will demonstrate the progress since the tsunami disaster in establishing comprehensive school disaster safety in the education system of Sri Lanka. The first step was to evaluate the current situation and identifiy the needs. Having done that, political commitment from education authorities for disaster safety education was achieved by meeting DRR/Education 'champions' in India. This was followed in Sri Lanka by the creation of partnerships with all relevant government and non-governmental stakeholders. In a joint effort, National Guidelines on School Disaster Safety were developed and are being implemented in a growing number of schools. The integration of DRR in school curricula and teacher training was undertaken by the National Institute of Education with support from GIZ in a step-by-step process over two curricular review cycles. Ten years after Tsunami the education sector has taken up disaster safety at all levels of education and has integrated knowledge on multi-hazards and development of practical safety skills.

Keywords Change process • Disaster prevention education • Disaster risk reduction education • International education cooperation • School Safety

---

P. Bitter (✉)
International Education Consultant, Gumpendorferstr. 88B, 1060 Vienna, Austria

R. Shaw (ed.), *Recovery from the Indian Ocean Tsunami: A Ten-Year Journey*, Disaster Risk Reduction, DOI 10.1007/978-4-431-55117-1_29

Abbreviations

ACSS	Asian Coalition for School Safety
ADPC	Asian Disaster Preparedness Center
COGSSDPE	Coalition for Global School Safety and Disaster Prevention Education
DMC	Disaster Management Centre
DRM	Disaster Risk Management
DRR	Disaster Risk Reduction
DSE	Disaster Safety Education
ESC	Education for Social Cohesion
ESDFP	Sector Development Framework Programme
GAR	Global Assessment Report
GIZ/GTZ	German Agency for International Cooperation
HFA	Hyogo Framework for Action
ICE	International Conference on Education
IDP	Internally Displaced Persons
IFRC	International Federation of Red Cross
INEE	International Network for Education in Emergencies
ISDR	International Strategy for Disaster Reduction
JAICA	Japan International Cooperation Agency
MoE	Ministry of Education
NCoE	National College of Education
NDMCC	National Disaster Management Coordination Committee
NIDM	National Institute of Disaster Management
NIE	National Institute of Education
TPKE	Thematic Platform for Knowledge and Education
WCESD	World Conference on Education for Sustainable Development)

29.1 Introduction

The chapter explores the impact of the Disaster Safety Education' project, which is conducted by the Ministry of Education (MoE) and the German Agency for International Cooperation (GIZ) in Sri Lanka. The divastating tsunami disaster 2004 made the education authorities aware of the imminent need for better preparedness. GIZ supported the education authorities in using this awareness for a change process towards a culture of safety. This case study covers 8 years of project work as collaboration of the european GIZ team with Srilankan GIZ experts and educational staff from the MoE and the National Institute of Education (NIE). The main objective is to describe the step-by-step process of integrating disaster safety into the whole education system and draw lessons for strengthening a culture of safety in future schools. The leading question for the case study was how to manage a sustainable implementation process.

Under the focus of education in conflict and emergencies contents of this chapter have been presented at the International Conference on Education and Development UKFIET—Post 2015: Reflecting, Reviewing, Re-visioning. Oxford, Sept 2013.

For writing the present chapter the following tools were used:

(a) Disaster Risk Reduction (DRR) in Education concepts and events: conference documentation, webresources and literature;
(b) GIZ DSE project: project documentation, interviews and feedback sessions with government partners and other stakeholders, monitoring visits, classroom observation.

The text starts with briefly reviewing the global context of DRR in education, giving an overview on ongoing initiatives and key documents to build a culture of safety through education. Also mentioned are international and regional organizations, campaigns and networks that created awareness on DRR, milestones in the process of promoting comprehensive school disaster safety.

Following is a description of the DRR framework in Sri Lanka with a focus on education. After tsunami, it became obvious that education authorities, teachers and students had not been prepared to face disasters and so the disaster triggered a change process in the education system. This opportunity was used by the ongoing GIZ supported, Disaster Safety Education' project. It aimes for integrating disaster safety into the whole education system through policy development, curricular reform, teacher training and school safety programmes. Sustainability of the project's concepts and methods could be achieved by integrating those into the existing system in key areas, e.g. national curriculum, national colleges, zonal evaluation teams. The present text gives an overview on partner structure, methodical approaches, implementation process and outcomes of the DSE project. For school disaster safety we can see the development since the tsunami 2004 until 2013:

- The education sector is a vehicle of central importance for achieving disaster safety. Children are taught in school how to protect themselves from hazards and how to respond correctly in an emergency.
- The MoE has established its role as the coordinating and steering body of comprehensive school safety in the education sector. Improved communication channels, increased trust and transparent decision-making have led to more efficient collaboration and the joint implementation of school safety programmes.
- The MoE, supported by UNICEF, has included Disaster Safety into the second Education Sector Development Framework Programme (ESDFP II).
- The Disaster Management Center (DMC) supports the education system at national, district and school level and has included school safety in its action plans.
- National School Disaster Safety Guidelines have been introduced in collaboration with DMC, provincial and zonal education authorities and international organizations.
- The NIE has linked basic concepts of climate change adaptation with DRR topics and integrated in the reviewed school curricula of the following subjects: Sciene, Geography, Civics and Physical/Health Education, Technical Education and Agriculture.

- Innovative methodological approaches—Project based and multi-media Learning—are used for disaster safety education and being integrated in Curricula, Teacher Guides and Textbooks.
- Common understanding on School Disaster Safety was achieved among the stakeholders, didactic materials were shared and planning and monitoring meetings established.
- 80 DSE coordinators work in teacher training institutions, provincial and zonal education offices all over the island, creating networks among them, using their capacities for reflection and implementation.

29.2 DRR in Education and School Safety: The Global and Regional Context

From the educational perspective it is important to register that whenever and wherever a disaster strikes, children and youth are among the most affected, because children are highly vulnerable in a disaster situation. They are physically less protected from death, disability and injury. They may become orphans, they often drop out of school and loose their option for a better life.

The *Hyogo Framework for Action* (ISDR 2005) has identified five main priorities for taking action, among them Priority 3 'Use knowledge, innovation and education to build a culture of safety and resilience at all levels.' People can develop their capacities through education and training to prevent disasters as well as respond to and recover from extreme natural events. Key activities are: (1) Information management and exchange; (2) Education and training; (3) Research and (4) Public awareness. Schools should become an important player as a community resource in DRR plannning and in post-disaster recovery.

Also in 2005, the *Coalition for Global School Safety & Disaster Prevention Education* (COGSSDPE) established an international network of advocates and activists. In 2006–2007 ISDR focused on Priority 3 and organized the worldwide campaign 'Disaster Risk Reduction Begins at School' (see ISDR 2006). It inspired a series of international, regional and national conferences and meetings and dynamized networking amongst education and DRR scholars and practicioners. A comprehensive review on DRR in education (Wisner 2006) provided the first guidelines and resources for the educational community on how to promote DRR in education and how to strengthen the active participation of children. Petal (2008) introduced the Guidance notes for education sector decision makers.

Further milestones for DRR in education have been the establishment of the 'Thematic Platform for Knowledge and Education' (TPKE) at the *Global Platform for Disaster Risk Reduction* in Geneva 2007 and the publication of the global baseline report (ISDR-TPKE 2012). In 2008, DRR was integrated in the 48th session of the *International Conference on Education* (ICE). A campaign for 'Safe Schools and Hospitals' as part of implementation of the Hyogo Action Plan helped to further promote educational aspects in global DRR strategies.

In 2009, at the *World Conference on Education for Sustainable Development* (WCESD) in Bonn, the TPKE advocated school safety and disaster prevention through education. Also in 2009, during the second session of the *Global Platform for Disaster Risk Reduction* the participating countries committed themselves to undertake national assessments of the safety of existing education facilities until 2011. During the third session in 2011 the commitment was further concretized to develop concrete action plans for safer schools and to include DRR in school curricula by 2015.

In order to support the education authorities in the implementation of these objectives, UNESCO-IIEP, the Global Education Cluster and UNICEF developed '*Guidance notes for educational planners on integrating conflict and disaster risk reduction into education sector planning'*. In 2012, UNESCO and UNICEF intensified their support on curriculum development for DRR through a case study from 30 countries (Kagawa and Selby 2012) and ISDR-TPKE published, *A Global Baseline Report'* which sets out to provide a baseline on school disaster safety by using existing initiatives from governments, civil society, UN, donors and other major stakeholders—over the past 7 years from 81 countries—that aim at assessing and improving school safety. The desk study defines the concept of comprehensive school safety and has become a guidance document for school safety.

In 2013, Education was a leading topic at the fourth Global Platform for Disaster Risk Reduction and included in the Communiqué aiming for improvement of school infrastructure and a global safe school campaign. Recommendations for the post-2015 global agreement on DRR (HFA2) were made by children and youth, who led a session on Children's Charter for DRR that became a highlight of the conference. Furthermore, key stakeholders in DRR and Education launched the, *Global Alliance on Risk Reduction in the Education Sector'*.[1] Stakeholders like UNESCO, UNICEF, UNISDR, Save the Children, Plan International, IFRC, World Vision, INEE and World Bank carry on to establish a culture of safety.

Due to the high vulnerabilty of Asia to natural hazards and the increasing impact of disasters in the region, the movement for school safety and DRR in education developed a strong dynamic in the Asian region. Sparked off by the ISDR School campaign, a lively network of technical DRR organisations, government bodies, donors and NGOs evolved and created a platform for sharing good practices, experiences and resources. Important milestones were the Bangkok and Ahmedabad Action Agendas (2007) as well as the following events: Regional Workshops on, *Education for Natural Disaster Preparedness in Asia Pacific in the context of Education for Sustainable Development (ESD)*, organized by UNESCO and ADPC in 2006 and 2008 in Bangkok; the *International Conference on School Safety* in Ahmedabad 2007 organized by SEEDS India; the *International Conference on School Safety* in Islamabad in 2008 organized by the Aga Khan Foundation; *Safe School National Conferences* in Indonesia (2010), Sri Lanka and Philippines (2012).

[1] See report of Transburg 2013.

In 2012 the *Asian Coalition for School Safety* (ACSS) was established to bring about comprehensive school safety based on three 'pillars': (1) Safe School Facilities, (2) School Disaster Management and (3) DRR/CC Education. The founders of the Coalition are ADPC, Plan International, Save the Children, UNESCO, UNICEF and IFRC. One of the key activities of ACSS is to foster Capacity Development for comprehensive school safety in the region.

In 2013 the *Global Assessment Report 2015* (GAR) was launched by UNISDR. It contributes to the achievement of the HFA through monitoring risk patterns, trends and progress in DRR and through providing guidance to governments and non-governmental actors. The GAR15 will be published prior to the World Conference on DRR in 2015, in which governments will adopt a successor of the HFA. One of the 22 core HFA indicators is looking into how school curricula, education material and relevant training include DRR and recovery concepts and practices. The research will be undertaken by Plan International in the following high-risk South Asian and South East Asian countries: Bangladesh, Cambodia, Indonesia, Pakistan, and Sri Lanka.

29.3 The Concept of Comprehensive School Safety

The main stakeholders for DRR in education together published a brief description of Comprehensive School Safety aiming to bring the various existing efforts and approaches into a clear and unified focus.

According to the leaflet from UNICEF et al. (2013) the goals are:

- To protect learners and education workers from death, injury, and harm in schools
- To plan for educational continuity in the face of expected hazards
- To safeguard education sector investments
- To strengthen climate-smart disaster resilience through education

Comprehensive School Safety is addressed by education policy, planning and practices in line with disaster management at national, district and local levels. Multi-hazard risk assessment is the foundation for planning. The concept consists of three pillars (Fig. 29.1):

In the context of this latest conceptual development, the strategic goals for the education sector in relation to the Hyogo Framework for Action are suggested as follows:

1. Integrate DRR into sustainable development policies and practices in the education sector.
2. Develop and strengthen educational institutions, mechanisms and capacities to build resilience to hazards at national, sub-national and local levels.
3. Systematically incorporate risk reduction approaches into the implementation of emergency preparedness, response and recovery programmes in the education sector.

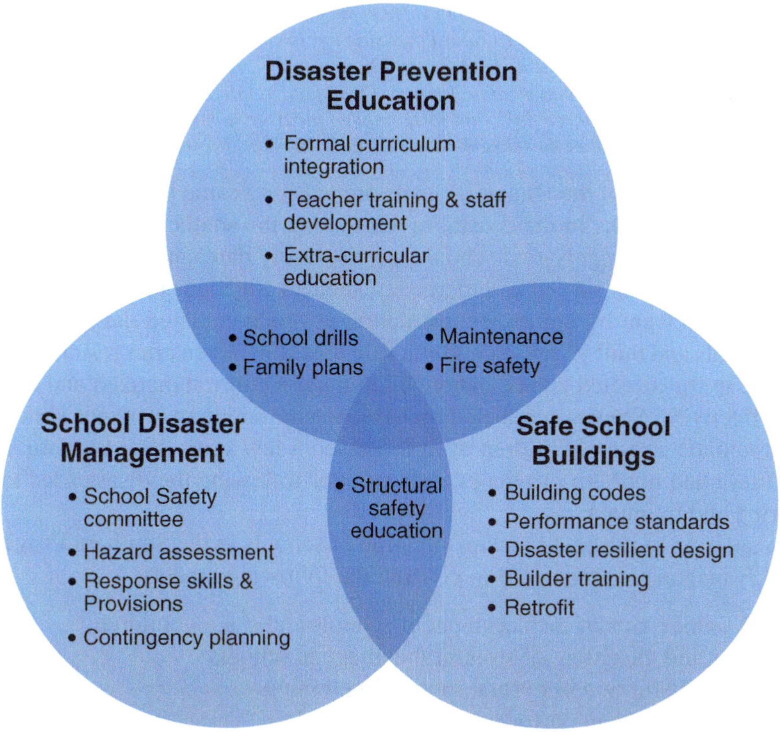

Fig. 29.1 Three Pillars of Comprehensive School Safety (adapted from UNICEF et al. 2013)

29.4 The Impact of Tsunami on the Education Sector in Sri Lanka

In Sri Lanka, the 2004 Indian Ocean tsunami took the lives of 35,000 people and left behind over one million surviving victims, among them some 200,000 school-age children and university students as well as 3,000 of their teachers. The number of victims would have been much higher if the disaster had struck on a school day instead of a holiday. Anyhow, the effects of the tsunami disaster clearly demonstrated that the people of Sri Lanka were not prepared to deal with the hazards threatening them. Teachers and children did not recognise the risks, did not know how to protect themselves and often responded inappropriately.

Tsunami report by 11-year-old Perinparaja Brindha from Kalmunai, student at Sri Mamanga College

My name is Perinparaja Brindha. I study in grade five in KM/Sri Mamanga Vidyalayam. I was caught in the Tsunami waves on the 26th Dec. 2004. That day I went to the school for a religion lesson. At that time, I could see a big wave coming from the school playground. We were afraid. We told the teacher that we wanted to go home. The teacher said that 'there is nothing to be afraid', because the water was only due to heavy rain. I was afraid and ran away from the school. At that time, a big wave came and I was pulled by that wave. One old lady saw me and asked me not to run to that side but to run to the other side. I listened to her and ran to the side she asked me to run. There was a van parked. Some police officers put me into the van. The van went to Ampara. One of our relations took me at Ampara and kept me with them for three days and later handed me over to my parents. All my friends who were in the school at that time died. (GTZ/ESC 2007)

In the aftermath of the tsunami, classroom teaching came to an almost complete halt for several weeks in the coastal regions along the southern and eastern shorelines. One hundred eighty-two schools were severely damaged or destroyed, while another 287 schools served as emergency shelters for thousands made homeless. Many months went by before reconstruction efforts completed the repair of damaged schools and built 95 new schools at safe locations. Even after resuming regular teaching in the affected zones, many children were still traumatized and unable to learn effectively. Studies conducted by JAICA in the southern province showed that post-traumatic stress of children even increased a few years after tsunami. And in the eastern and northern provinces, the civil war following the disaster led to more suffering and traumata.

An analysis of the post-tsunami situation of schools in the Southern Province by G.K. Liyanagunawardhana (2006) covered the following aspects:

- Areas and extents of displacement of schools and school children
- Number and the extent of structural damage in schools
- Extent of disruption in normal school functioning
- Teacher and student performance post tsunami
- Percentage of students and teachers who suffered or died in the tsunami
- List of schools used as camps and the time period
- Number of children attending schools from camps

The study showed that there was no uniformed strategy on how to deal with education post tsunami: Schools continued to function in their own distroyed premises, were relocated or attached to other schools, used as shelters (welfare camps), or begun re-construction. Mostly principals and teachers were unable to give information about the students of their school. Nearly 2 years after the tsunami disaster, no effort had been taken to evaluate the damage to physical properties and audit the equipments. Although NGOs had donated buildings and equipments, essential teaching/learning materials as syllabi and teachers' guides had not been provided by the education authorities.

The school principals of most of the tsunami affected schools in the southern province confirmed a decline in students' performance. Especially students coming from IDP camps suffered from post-traumatic symptoms. School attendance was reduced in all schools subjected to the study. Such factors led to severe disruption of school functioning. The real number of dropouts was not known, because no formal procedure of school entrance was adhered to after tsunami.

Teachers and students did not know how to prepare themselves to face disaster, and the education authorities were not prepared for response and relief. An urgent need for action in the field of disaster preparedness and management was evident.

The Government had to take systematic action to better mitigate disaster risks and create a more disaster-resilient society. The disaster triggered the sustainable establishment of national and local structures and processes for DM. In July 2005, the Sri Lanka Disaster Management Act No.13 of 2005 was enacted which provides the legal basis for instituting a DRM system in the country. The Ministry of DM is responsible for directing and coordinating inter-ministerial efforts. In a first step, it prepared a national plan of action for DM, termed the 'Road Map' (DMC 2005). It is based on an analysis of all sectors to determine their potential for reducing disaster risk. It documents the contributions of the various ministries involved, thereby establishing a binding framework. The Act also provides for establishing the Disaster Management Centre (DMC) to be the apex body for the purpose of planning, co-coordinating and implementing of natural and other disasters. The National Disaster Management Coordination Committe (NDMCC) as multi-stakeholder national coordination for DRR provides or mobilizes the combined knowledge, skills, and resources required for DRR and its mainstreaming into development policies, planning and programmes.

29.5 A Layout of the Disaster Safety Education Project of the Sri Lankan–German Development Cooperation[2]

The tsunami disaster has triggered the political committment to include disaster management into all relevant sectors' policies and plans, including the education sector. Real change is not easily achieved, people like to hold on to habits and to believe to be safe, but a tragic event can mobilize energies to drive a change process resulting in taking precautions and developing emergency skills.

DRR being a completely new approach for the education system, the Ministry of Education, which had already a long standing cooperation with the German Agency for International Cooperation, GIZ[3] (former GTZ), required technical support for introducing Disaster Risk Management (DRM) and for providing psycho-social care for tsunami traumatized students. In 2005, the 'Education for Social Cohesion' (ESC) programme was complemented by the 'Disaster Risk Management and Psycho-social Care' project, financed by tsunami funds.

It became clear that the entire education sector had to be involved in order to mainstream DRM in the education system in the long term:

- The Ministry of Education MoE assumes the lead role for directing the education process. It is responsible for issuing education policies and developing strategies, including cross-cutting issues as disaster safety.
- The National Institute of Education NIE is in charge of developing guidelines for implementation on the basis of policy provisions. It deliveres training of college

[2] Sources used for this chapter are Donga & Bitter (2008), Bitter & Edirisinghe (2003).

[3] Deutsche Gesellschaft für Internationale Zusammenarbeit GmbH.

lecturers and instructors, teacher training and development of syllabi and instruction materials—all key areas for integrating new topics into the curricula.
- In Sri Lanka teacher pre-service training is provided by 18 National Colleges of Education. In line with national standards, but with different areas of specialisation, these colleges train new teaching staff for the entire country.
- In-service training for teachers and school principals is structured as follows: There are about 100 Teachers Training Centres responsible for expanding the methodological knowledge and skills of teachers. School based in-service teacher trainers pass on what they have learned to their colleagues in the schools.
- In addition, the NIE maintains the Centre for Educational Leadership Development, to prepare teachers to assume leadership roles in education management and administration.

In 2006, none of the areas of Comprehensive School Safety were answered to by the education sector:

1. School buildings were located in hazardous zones, e.g. in the coastal belt or below flood level, and could not withstand possible disasters, e.g. cyclone.
2. Emergency planning and management was unknown at field level, except in few schools that had received first aid equipment and training by NGOs after the tsunami.
3. Disaster prevention education was not integrated into the national curricula and respective teaching/learning materials were not available.[4]

The objective of the DRM project was to integrate Disaster Safety into the whole education system, covering

- All levels—education managers, principals, teachers, students, school committees;
- All aspects of learning—knowledge, skills and attitudes;
- All areas of education—classroom teaching, co-curricular activities, education administration, including emergency plans for educational buildings.

Main obstacles for successful education in Sri Lanka are extremely exam-oriented education, overloaded and age-unadequate curricula, textbook based and teacher-centered teaching, highly competitive education culture and high social pressure, dependance on tuition classes (second income system for teachers), low teachers salaries, insufficient school buildings and equipments. Based on the experience with disaster safety education, we observed some key factors that can enhance education:

- Horizontally/vertically harmonized and child-adequate curricula and didactic materials
- More practical oriented (learning-by-doing) teacher training system
- Use of participatory learning methods (e.g. project based, multi-media learning, learning by doing) and development of life skills (e.g. fire safety)
- Change from hierarchical to more democratic structures in school management

---

[4] However, some topics related to natural hazards were already in the Science and Geography syllabi, e.g. tectonic plates and earthquakes.

- Cooperation between school and community
- Performance oriented career system for teachers
- Investment in school maintenance and equipment (e.g. disaster resilient retrofit)

As GIZ was the only donor active at all levels of intervention of the education sector, it was identified by the Sri Lankan Government to assume with the Ministry of Education the lead role for Priority 3 'Education' as lined out in the Road Map. World Vision Germany agreed to co-finance the project.

When the DRM project started in 2006, everybody had experienced the tragic of tsunami and was willing to learn and work for protection. Anyhow, still there was no conscience for multi-hazard vulnerabilty and preparedness. College and school staff often said: 'We don't need fire safety, because we never had a fire here.' Developing a culture of safety needs more than knowledge, it needs practical skills for prevention, protection and response. Comptencies need to be applied and extended by personal conviction and engagement for a safer environment in the community, at home, in school.

In 2006, the project started with a baseline survey on disaster preparedness and school safety (Dias-Wanigasekera 2006) in all National Colleges of Education (NCoEs) and in a sample of 54 schools. The results confirmed the preliminary assessment that schools, teachers and education authorities were not well prepared to face (post) disaster situations. The baseline survey in NCoEs[5] focused on risk factors, facilities for emergency cases, preparedness, training, curriculum and teaching on DRM. Generally, the survey showed that the NCoEs were not prepared to cope with hazards (no building safety, lack of emergency communication system and emergency equipment, gas/fire/electrical hazards) and DRM was not included in syllabi or co-curricular programmes. The results helped to define the assistance needed and allowed to monitor the progress closely.

Beeing a large-scale change project, to get started, the project needed the political support of decision makers within the education sector. Based on the experience that introducing change in a social system does not work through a purely hierarchical (top-bottom directive) or managerial (planning—implementing—evaluating) approach, the project aimed to win over key persons and to motivate them to act as change champions for the project. An initial training course in India was organized with the National Institute of Disaster Management (NIDM),[6] who tailored a programme on 'DRR in Schools' for the Sri Lankan delegation. Twenty-five executives from the Sri Lankan Ministry of Education, the NIE and NCoEs met up with passionate 'DRR champions' in India, participated in school events and experienced how DRR was integrated in the Indian education system.

The knowledge they gained through the course helped the Sri Lankan educators in the following years to apply disaster safety education in their own country, but even

[5] GIZ/ESC, 2006. Baseline Survey on DRM at NCoEs.

[6] NIDM was chosen for three reasons: several years of experience on DRR and education in India; the socio-cultural similarity between the two countries; existing links to Indian DRR/Education experts.

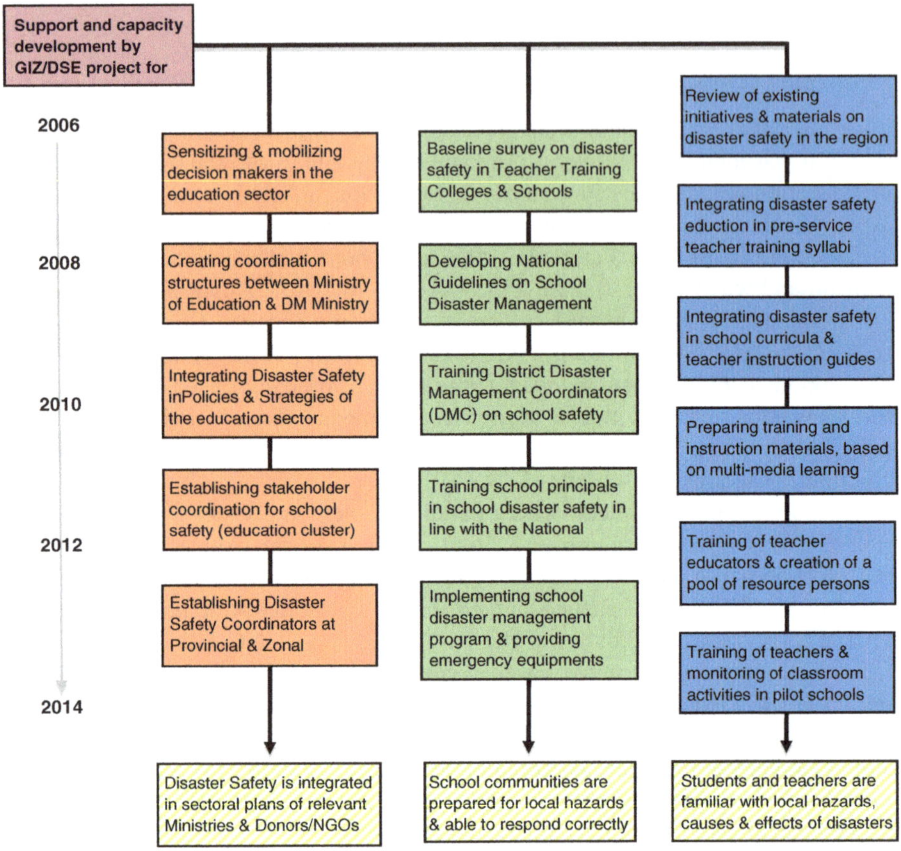

Fig. 29.2 Time line, main intervention areas and impact of the Disaster Safety Education Project

more important was the personal enthusiasm they brought back with them. Personal commitment became the driving force, releasing sufficient energy to successfully introduce the new cross cutting topic into the system. Following the course in India, a wide number of training events were held at the NCoEs, involving annually about 3,500 teacher students and their lecturers in DRR programmes. DRR was integrated into the national teacher training curricula and all future teachers learn the basic concepts of DRM and how to implement emergency management at schools. Figure 29.2 shows the implementation process of sector wide integration of disaster safety.

Ten years post tsunami all NCoEs are equiped with emergency equipment, staff is trained on first aid and fire safety, safety committees are established, mock drills are practiced, students are taught DRR and climate change concepts. Curricula and co-currcular programmes cover basic concepts and practical skills development.

29.6 Policy Development and Partnerships at National and International Level

In 2009 with the start of the second project period, a conceptual shift was done from DRM to Disaster Safety Education (DSE), taking up the understanding of the school safety community in India that in the education sector, a) a positive connotation and b) the focus on safety of the school community, create stronger motivation and a better understanding of the project. The DSE project aimes to foster disaster preparedness, prevention and mitigation among school children and in school communities.

The Asian Disaster Preparedness Center published a guideline on integration of DRR into school curricula (RCC 2007). A main issue pointed out is establishing coordination and partnerships among the stakeholders, starting from the responsible ministries and reaching out to other stakeholders, especially the education cluster or sector group. This approach has also been applied during the implementation of the DSE project from the beginning onwards. The project gave great importance to coordinating and agreeing the various activities both within and between the individual education authorities, as well as between the education sector and the Ministry of Disaster Management.

The same principle was pursued on a transnational basis. Using the Asia and Pacific Forum and Conferences on DRR and school safety,[7] the project contributed on the one hand its experience in Sri Lanka to the regional network and on the other hand streamlined the national activities in the global context and the related discussion. On issues of school disaster safety, the project collaborated with ADPC and NIDM.

Within the post-tsunami establishment of DM in the country, the Ministry of Education appointed a focal point for disaster preparedness. GIZ offered her DRR training in India and facilitated the creation of an inter-ministerial working group. The first task of the group was to develop National Guidelines on School Disaster Safety. Such a concrete undertaking, which involved the national Centre for Educational Leadership Development, helped to establish inter-institutional teamwork. Step by step, roles and responsibilities as well as areas of competence and collaboration of the governmental stakeholders were clarified within the setup of inter-ministerial cooperation.

With this partnership as foundation, all relevant international and national organizations involved in the field of disaster safety education were integrated in the working group. Since 2008 all stakeholders[8] support the Ministry of Education in the implementation of the National School Safety Guidelines. As the education cluster was dissolved after the end of civil war and emergency, the group was in 2011/2012 taking over the function of the cluster.

---

[7] See also milestones for DRR in education in Sect. 29.2.

[8] UNICEF/Emergency Education and Mine Risk Education, UNDP, UNESCO, Save the Children, Sri Lanka Red Cross, German Red Cross, International Federation of Red Cross.

Ten years after tsunami, the MoE has established its role as the coordinating and steering body of comprehensive school safety in the education sector. Improved communication channels, increased trust and transparent decision-making led to more efficient collaboration and the joint implementation of school safety programmes.
By 2013, a common understanding on School Disaster Safety has been achieved, a standardized model for training activities agreed upon, joint training programs conducted, didactic materials shared and joint planning and monitoring meetings established.
Visible impact at Colleges or schools could be shown to national level decision makers and motivated them to support the programme and include it into their overall plans. The MoE, supported by UNICEF, has included Disaster Safety into the second Education Sector Development Framework Programme (ESDFP II). The Disaster Management Center (DMC) supports the education system at national, district and school level and has school safety in its action plans.

29.7 National Guidelines on School Disaster Safety

A milestone towards establishing disaster safety in education are the National Guidelines on School Disaster Safety. The concept was adapted from India (seven steps of school disaster management) and ADPC (school emergency plans). Having piloted these models, the Sri Lankan Centre for Educational Leadership Development, the MoE and DMC, supported by GIZ, adapted the guidelines and published them. The introduction in 10,000 schools by MoE is done step by step. The joint effort with stakeholders like GIZ, UNICEF, UNDP, Save the Children and Red Cross has reached estimatively 4,000 schools by 2013.

All these schools create safety teams, disaster management plans and practical skills to cope with emergency situations. School communities become messengers of disaster safety, carrying Do's and Dont's home to the families and communities. Campaigns like Flood Protection; Fire Safety; Dengue, Bee sting[9] and Snake bite prevention show visible impact in the communities.

A vital element of this strategy was to establish about 80 DSE coordinators as change agents in teacher training institutions and provincial as well as zonal education offices all over the island, to create networks among them and use their capacities for reflection and implementation. Northern and eastern provinces have well functioning structures and processes in place, involving education and DM authorities and international organizations. About 4,000 schools are aware of disaster safety and most of them practice the programme. These school communities are now safer and better prepared for disasters.

[9] Cases reduced from monthly 68 to 2 after the campaign.

29.8 Integration of DRR in School Curricula and Teacher Training

During the first project period (2006–2009), DRR was integrated in school curricula and in all pre-service and in-service teacher training curricula and respective training modules. Teaching and learning materials were hardly available in the country and had to be collected, adapted and distributed. UNESCO organized material sharing and adaptation through '*Education for Natural Disaster Preparedness in Asia-Pacific in the context of Education for Sustainable Development (ESD)*' activities. GIZ supported the NIE in production of resource materials for lecturers and teachers and audio-visual materials for students on disaster safety (media kits).

The Guidelines for DRR in school curricula (UNICEF & UNESCO 2013) are based on good practices in about 40 countries. In Sri Lanka, integrating DRR into the national school curriculum was done mainly through Science and Social Science subject areas from grade 6 to 9 under topics like 'Preparedness for the management of natural disasters' with six periods of 40 min across the curriculum. In the primary curriculum, which follows the spiral approach, each theme can be worked out at different levels in each grade. DRR has been infused into a range of themes like 'The world around us', 'Our earth and the sky above it'.

The DSE project supported the introduction of project based and multi-media learning, including instructional video clips and interactive DVDs. School monitoring showed that these methods changed teacher-students relations towards more symmetry and cooperation. Teacher educators, trainers and education officers were trained on the new contents and methods. All training activities included practical and skill development activities, for example evacuation drill, fire safety, post-disaster food preparation. Resource person pools are created in the different institutional groups and training materials available.

In 2013 the basic concepts of climate change adaptation are linked up with DRR topics and integrated in the reviewed school curricula of the following subjects: Sciene, Geography, Civics and Physical/Health Education, Technical Education and Agriculture. Related teacher training modules have been developed and training is supposed to start in 2014.
Innovative methodological approaches—Project based and multi-media Learning—are used for disaster safety education and being integrated in Curricula, Teacher Guides and Textbooks.

29.9 Conclusions

At the start of the project, disaster preparedness was weak and safety no issue in school curricula or teacher training. The DSE baseline survey 2006 at National Colleges of Education and schools showed that awareness, knowledge and skills related to DRM were practically non-existent with exception of some tsunami affected schools, that had received external support and training.

In 2014, DSE is integrated in the education sector plan (ESDFP II) and the national plan of the Ministry of Disaster Management, in school and teacher training curricula as well as in co-curricular school programmes.

The project team identified the following Success Factors for achieving the sector wide integration of DSE:

- Motivation of decision makers and key persons, e.g. through meeting champions
- Partnerships among all stakeholders
- Skills orientated capacity building
- Piloting concepts in the field and establishing action-reflection-cycles at field and central levels
- Working through local coordinators
- Step-by-step infusion of new concepts and methods in curricula and teacher training
- Following a comprehensive school disaster safety approach

Disaster Safety Education has been well rooted in the Sri Lankan education sector and educationalists are aware that disaster preparedness of school communities is important. The challenge for the next years is to reach out to all schools in the country and turn the innovation into a rountine of day-to-day school safety activities.

References

Ahmedabad Action Agenda for School Safety (2007) International school disaster risk reduction conference, Ahmedabad, Jan 18–20 2007. SEEDS, Delhi

Bangkok Action Agenda (2007) Outcome of the Asia-Pacific regional workshop on school education and disaster risk reduction, Oct 8–10. UNISDR, UNICEF & UNESCO, Bangkok

Bitter, P, Edirisinghe S (2013) Introduction of disaster safety in the Sri Lankan education system: factors for success. UKFIET, Oxford. www.ukfiet.org/cop/wp-content/uploads/2013/11/Bitter-Edirisinghe_-Introduction-of-Disaster-Safety-in-the-Sri-Lankan-Education-system-Factors.pdf

Dias-Wanigasekera R (2006) Baseline survey on disaster risk management at national colleges of education (NCoE). GIZ/ESC, Colombo

DMC (Disaster Management Center) (2005) Towards a safer Sri Lanka. Roadmap for disaster risk management, vol 1–2. DMC & UNDP, Colombo

Donga M, Bitter P (2008) Teaching disaster risk management in Sri Lanka's schools. Experience Since the 2004 Tsunami. GTZ, Eschborn

GTZ (2007) A collection of Tsunami experience reports. Collected by GTZ/ESC, Colombo

ISDR (2006) 2006–2007 World disaster reduction campaign. Disaster risk reduction begins at school. ISDR, Geneva

ISDR (2005) Hyogo framework for action 2005–2015: Building the resilience of nations and communities to disasters. ISDR, Geneva

ISDR-TPKE (2012) Assessing school safety from disasters: A global baseline report. ISDR, Geneva

Kagawa F, Selby D (2012) Disaster risk reduction in school curriculum: Case studies from thirty countries. In: UNICEF/UNESCO (ed) Paris/Geneva

Liyanagunawardhana GK (2006) An over view of the impact of tsunami on schools & school children in the south. (Manuscript) Sri Lanka

Petal M (2008) Disaster prevention for schools. Guidance for education sector decision-makers. Consultation version, November 2008. ISDR, Geneva

RCC (Regional Consultative Committee on Disaster Management) (2007) RCC guideline 6.1 Integrating Disaster Risk Reduction into School Curriculum. Mainstreaming Disaster Risk Reduction into Education. Consultation version. RCC & ADPC, Bangkok

Transburg P (2013) Education and the global platform for disaster risk reduction. www.ineesite.org/. Accessed 28 May 2013

UNICEF (2012) Climate change adaptation and disaster risk reduction in the education sector: Resource manual. UNICEF, New York

UNICEF & UNESCO (2013) Towards a learning culture of safety and resilience: Technical guidance for integrating disaster risk reduction in the school curriculum. PILOT VERSION. Geneva and Paris

UNICEF et al. (2013) Comprehensive school safety. Working towards a global framework for climate-smart disaster risk reduction, bridging development and humanitarian action in the education sector (No editor and place of publication indicated)

Wisner B (2006) Let our children teach us! A review of the role of education and knowledge in disaster risk reduction. UNISDR, Geneva

Part V
Thailand

Chapter 30
Change of Livelihoods and Living Conditions After the 2004 Indian Ocean Tsunami: The Case of the Post-Disaster Rehabilitation of the Moklen Community in Tungwa Village, Southern Thailand

Monsinee Attavanich, Andreas Neef, Hirohide Kobayashi, and Terdsak Tachakitkachorn

Abstract The Moklen—often referred to as "sea gypsies"—are an ethnic minority with a long history of settlement in Southern Thailand. In contrast to the culturally and ethnically closely related Moken—who traditionally spent about 9 months at sea—, the Moklen have settled along the Andaman Coast for a long time and rely on coastal resources (e.g. mangrove ecosystems) and plantation work for their livelihoods. The Moklen do not have official land ownership titles due to their previous semi-nomadic lifestyle and restrictive government land policies. In this chapter, we first describe the impact of the 2004 Indian Ocean Tsunami on Moklen communities in Southern Thailand. We then analyze how a Moklen community in Phang Nga Province was affected by the various post-disaster recovery efforts. Our findings suggest that the Moklen people have become more vulnerable after the tsunami disaster due to hasty rebuilding of their settlement and persistent conflicts with expanding tourist resorts and related pressures on the natural resources they depend on.

Keywords Land conflict • Livelihoods • Moklen • Post-disaster recovery • Thailand • Tsunami

---

M. Attavanich • H. Kobayashi
Graduate School of Global Environmental Studies, Kyoto University, Yoshida Honmachi, Kyoto 606-8501, Japan

A. Neef (✉)
Faculty of Arts, Centre for Development Studies, School of Social Sciences, University of Auckland, Auckland 1142, New Zealand
e-mail: a.neef@auckland.ac.nz

T. Tachakitkachorn
Faculty of Architecture, Chulalongkorn University, Bangkok, Thailand

R. Shaw (ed.), *Recovery from the Indian Ocean Tsunami: A Ten-Year Journey*, Disaster Risk Reduction, DOI 10.1007/978-4-431-55117-1_30

30.1 Introduction

30.1.1 The Moklen of Southern Thailand

The "*Chao Lay*" (lit: *Sea People*) also known as *Sea Gypsies* comprise several groups of ethnic minorities in Southern Thailand, whose livelihoods have been closely tied with the sea. In the past, they mostly lived nomadically in the sea except during the monsoon season when they built temporary shelters on the islands and along the coast to avoid the rages on the sea (Ivanoff 2001; Ferrari et al. 2006). Many years ago they began to settle permanently in coastal areas, still maintaining a close connection with the sea. There are three groups of *Chao Lay* with their own name to distinguish their geographical location and their lifestyles: the "*Moken*", the "*Urak La Voy*" and the "*Moklen*".

The Moklen settle along the western coast of Phang Nga Province. Among the three groups the Moklen are the one with the longest tradition of settling on the land, having a relatively high degree of cultural assimilation with the Thai majority population (Larish 1992). There are an estimated 2,000–3,000 Moklen in several districts in Phang Nga Province, such as Kuraburi, Takuapa and Tai Muang district, living in around 20 villages scattered on the coasts and mangroves. They are also known as *Thai Mai* (New Thais), referring to the fact that the Moklen have already adopted a sedentary lifestyle and acquired Thai citizenship. In addition, they are the only group that is practicing swidden agriculture. As they changed their nomadic life to live permanently on land, some of them refer to themselves as "*Chao Bok*" or coastal people as opposed to the Moken which they refer to as "*Chao Kok*" or island people (Andaya 2008; Arunotai et al. 2008a). Some of them still keep their livelihood near the sea and the remaining patches of mangrove forest and live from fishing or diving for marketable sea products. Many of them, however, have abandoned their sea-based livelihoods and have engaged in land-based occupation, such as working in rubber or coconut plantations and as laborers for miscellaneous tasks. They still maintain their animistic beliefs, although many of them have adopted Buddhism as their official religion.

Until recently, the *Moklen* were not very well-known because their scattered settlements were located in areas that were not frequented by tourists and the assimilation process with the local culture made them appear not too different from local people (Arunotai et al. 2008b). After the Indian Ocean Tsunami in December 2004, however, they have become well known as one group of *Chao Lay* that had been heavily affected by the disaster.

30.1.2 The Indian Ocean Tsunami of 2004

The tsunami disaster in the Indian Ocean unfolded on 26 December 2004, when a magnitude 9.0 earthquake triggered massive tsunami waves that hit six provinces of Thailand: Rayong, Phang Nga, Phuket, Satul, Krabi and Trang. It caused the loss of life of a recorded 5,395 people and loss of property that was worth up to 1,132 million Thai Baht (The Director of Civil Defense of the Kingdom 2006, p. 13). The damaged area covered the coastline of Phang Nga Province which was the most heavily damaged province, with a recorded 1,739 deaths (Community Organization Development Institute (CODI) and Confederation of Fishing in Southern Region 2006). As the Moklen settled on or near the coast, they were directly affected by the tsunami. Many settlements were severely damaged, and villagers lost their family members and their property, especially on their residential land. In contrast to Thai communities, many Moklen could not set up their villages in the original places, because the landownership was claimed by others, which led to severe land conflicts.

30.1.3 Land Conflict Problems After the Tsunami

Because of their scenic landscapes and rich natural resources, the coasts and islands of Southern Thailand especially in the Andaman Sea have become popular tourist attractions. Following the Indian Ocean Tsunami, there have been many attempts by private sector actors to claim landownership—often in complicity with the local administration—to develop the tourist business, while several Thai government departments aimed to conserve the natural resources of the area. The issuance of restricted area policies also caused a number of land conflict cases after the tsunami, covering six provinces with a total of 412 villages being affected. Land conflicts were solved for 325 villages after 2 years, while 87 villages have seen a continuation of the conflicts that are still under legal consideration. Fourteen of the reported cases are conflicts among private sector actors, while 113 cases are classified as conflicts with government sectors in several departments. In 2012, 28 *Chao Lay* communities continued to have problems with residential insecurity and unresolved territorial claims (Chumchonthai Foundation 2012).

30.2 Study Rationale and Methodology

The Indian Ocean Tsunami of 2004 was the first natural disaster of such dramatic scale in Thailand. The limited experience in providing aid to tsunami victims and in handling the long-term recovery process makes the study particularly interesting and relevant, especially with regard to post-disaster management of residential issues. This research aims at studying changes of living conditions that occurred after the tsunami and at understanding the land conflict problem that affected the livelihoods of the most vulnerable and marginalized groups in the coastal society.

The *Chao Lay* were the community that was most affected by land conflicts, as a consequence of their being an ethnic minority with limited land tenure security which makes them particular prone to suffering from land grabs and displacement. In this research, the Tungwa Moklen community was chosen as a case study because the majority of its population belongs to the Moklen whose livelihoods and culture have been based on land much longer than other *Chao Lay* subgroups. The study of land conflicts in Tungwa community provides important insights into the multiple dimensions and social conflicts inherent in post-disaster residential recovery in non-egalitarian societies.

In this study we employed a questionnaire-based household survey and semi-structured interviews with key informants for collecting in-depth information to compare the living conditions and land tenure challenges before and after the Tsunami. We also used aerial photographs to triangulate the information gathered by surveys and interviews. The questions asked in household interviews and semi-structured interviews referred to changes of land use and living conditions to understand land conflicts that affected the residential area, the housing conditions and the economic livelihoods based on the use of natural resources. In addition, the patterns of houses and living conditions were elicited to analyze the livelihood challenges. The research results provide essential lessons for future post-disaster residential planning, for a better understanding of land conflicts in post-disaster recovery and for an improved residential management that provides long-term security of *Chao Lay* and other marginalized groups in the future.

30.3 History and Livelihoods of Tungwa Community

30.3.1 Brief History of Tungwa Community

Historically, there were four Moklen communities in this area, namely Tungwa Nok, Tungwa Nai, Bang Kaya and Bang Niang (Fig. 30.1). The history of settlement of Tungwa community in Tungwa Nok goes back to a Moklen woman who had claimed land ownership in this area. She was joined by four families that looked for work in the tin mining business around 50 years ago (Thongtaweewiwat 2006). They preferred the location because it was easy to reach by road and—at the same time—not far from the sea making it an ideal settlement place for the Moklen at the time.

During the pre-tsunami decade, there were migrations of Moklen and people from elsewhere to the village because of its convenient coastal location which allows fishing and working as day laborers. Villagers from Tungwa Nai moved to this community because landowners who were their relatives had sold the land that they were living on, so they needed to find a new place. Another group moved from Bang Niang area after they had been evicted by the official village headman. Thus, the community grew from an early settlement of only four families through the migration from other villages and relocation of nearby communities to a larger, culturally heterogeneous community of 40 families prior to the tsunami in 2004.

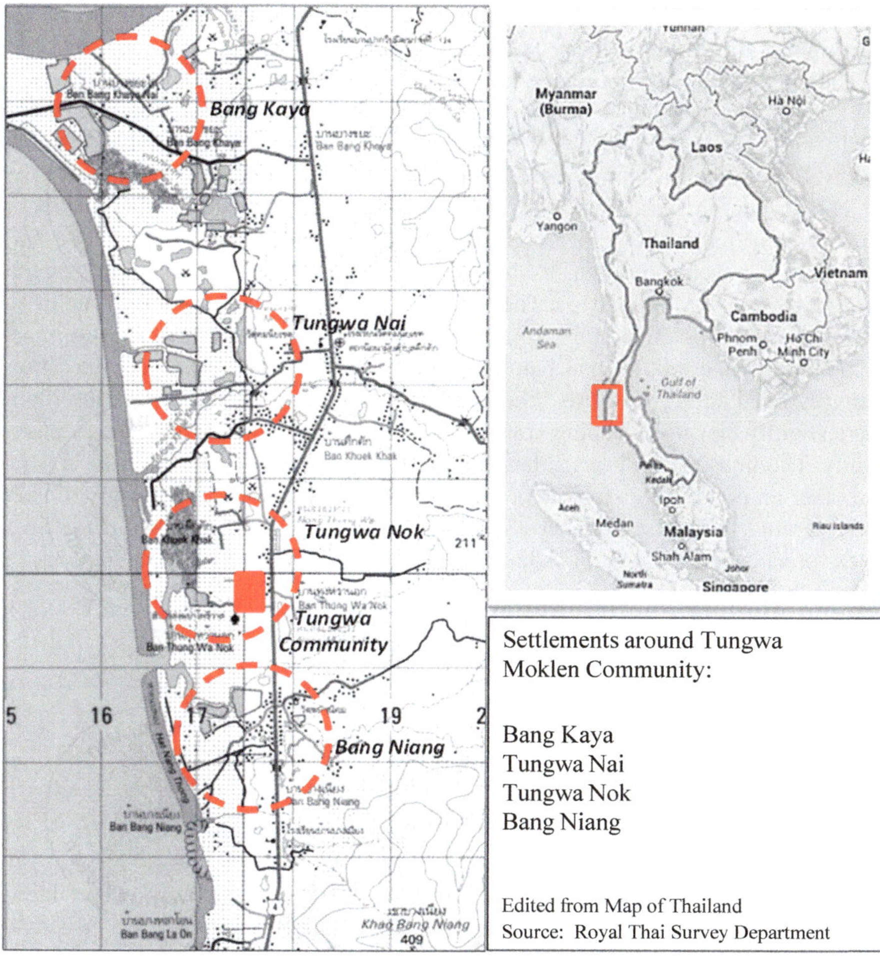

Fig. 30.1 Moklen communities in the area and location of Tungwa community

30.3.2 *Pre-Disaster Livelihoods and Living Conditions of Tungwa Community*

Like most Moklen communities, the Moklen community of Tungwa lived a simple life close to the sea, near a river mouth that provided freshwater supply. While most men had their primary occupation in the tin mines and in rubber and coconut plantations, collection of sea products, such as shellfish, and of natural materials from the surrounding jungle provided supplemental means of subsistence (cf. Larish 1992).

The people of Tungwa had a great sense of community and kinship relations, with the shaman (called *Por Ta*) acting as the spiritual center of their animist belief.

A shrine was used to hold regular community ceremonies, most of them related to worshipping the spirits of the sea (*Kwan*, interview, 2013 March).

The major areas of importance for Moklen people's livelihoods were the beachfront, the mangrove forests, the cemetery and the residential area. The beachfront, locally called *Nha Had*, was used to support their economic livelihood, and simple huts were built there as temporary shelters for fishing activities. The mangrove forest, called *Papang*, was another important natural resource that provided livelihood support. Spiritually, the cemetery[1] near the village (called *Pa Plew* or *Tung Hua Ting*) which covered an area of about 80 *rai* (12.8 ha) played an important role (*Hong*, interview, March 2013). The residential area was built along the road to get easy access to jobs outside the community. The bide-structure houses were made from natural materials, such as bamboo, rattan and *nipa* palm leaves, obtained from mangrove and adjacent forests. Each house consisted of a kitchen, a living room, and a slightly elevated sleeping area and provided sufficient space for the extended family. The houses had a raised basement to protect their inhabitants from rodents and other animals, with a terrace or *charn* in front of the house, and plenty of open space around the house. Prior to the tsunami, the Moklen of Tungwa raised livestock for sale, and most families had their own semi-commercial home gardens.

30.4 Post-Disaster Response and Recovery in Tungwa Community

30.4.1 Damages to the Tungwa Community and Evolving Land Conflict

The seaside location of the Tungwa community made it particularly vulnerable to the 2004 Indian Ocean Tsunami. The power of the tsunami waves killed 42 people, carried away or destroyed 18 fishing boats, destroyed all valuables and washed away around 40 houses. Among the people who died was the spiritual leader of the community, *Por Ta*. Hence, the disaster affected the community both materially and spiritually. After the tsunami had ravaged their community, Tungwa villagers had no officially acknowledged evidence or documents to claim their landownership rights, because they did not have prior knowledge of landownership issues, as they believed that the land was everyone's property for shared living. Rice (2005, p. 16) quotes the community leader as follows:

> *I have no idea who owns the land, but I have lived here since I was born. Our ancestors are buried here. My children were born here. It is our home. We are not intruders and will not be going anywhere. Assistance will soon end and we want to resume our livelihoods as soon as possible. We don't want anything, just our land.*

[1] In contrast to their Thai Buddhist neighbors who cremate their dead, the Moklen bury their deceased relatives and plant a coconut seedling on the burial place (Bristol 2010).

The Sub-district (*Tambon*) Administration Organization (TAO) at that time planned to move the Tungwa residents to Baan Bang Kaya, the residential project that had been set up for supporting tsunami victims. While the surviving Tungwa community members were living in temporary shelters, a sign was put up on their village site stating that a public hospital will be built there with funding from the German government (Bristol 2010). It later turned out that this was an attempt by district and provincial authorities to seize the public land (Rice 2005; Bristol 2010). In order to prevent the area from being taken over, community members—with support from NGOs—decided to get back on the land and start building new houses as quickly as possible. Following the involvement of the Board of Subcommittee on Land Issues, a land-sharing option was agreed upon by the conflicting parties. It was verified that the community had inhabited the area before the issuance of title deeds for public purposes in 1969. On this basis, the Tungwa community was allowed to return to their original residential area. However, the land was divided, with 16 *rai* (2.56 ha) going back to the community in the form of a renewable collective land lease, while ten *rai* (1.60 ha) was claimed by the local government for non-specified public use. In their effort to wrest the community's customary land from the control of the government and private sector interests, they were supported by the Chumchonthai Foundation, the Danish aid agency DANIDA, the United Nations Development Program (UNDP) and volunteer architects who helped in the design process. Villagers participated actively in the layout of the settlement and the design of the individual houses. The Community Organization Development Institute (CODI) and other agencies provided financial support to rebuild the houses on the original residential land. Overall, 21 organizations participated in this rehabilitation project.

The land-sharing agreement between the conflicting parties and the consultative reconstruction process was hailed as "Thailand's first post-tsunami, shoreline land-sharing scheme" (Rajah 2006, p. 30) and an "exemplary initiative" and a "model project for all the tsunami-affected communities involved in fighting for their rights over land rights disputes in Thailand" (Rice 2005, p. 16). However, the land conflict with the local administration and the private sector has never been fully resolved, as will be further discussed in Sect 30.5.1.

30.4.2 Post-Disaster Livelihoods in Tungwa Community

The Tsunami had a massive impact on people's livelihood. Most boats for making a living from fishing had been lost and—according to community members' accounts—there was no adequate help in terms of boat donations like in neighboring Thai communities. "*They didn't give us boats for every community member. And some of the boats they gave us had no engines*" (*Hong*, interview, November 2013). Many community members had to change to other ways of fishing, such as net-fishing close to the shore. Nearly 10 years after the Tsunami, most boats have broken down and several community members also could not afford the fuel costs when

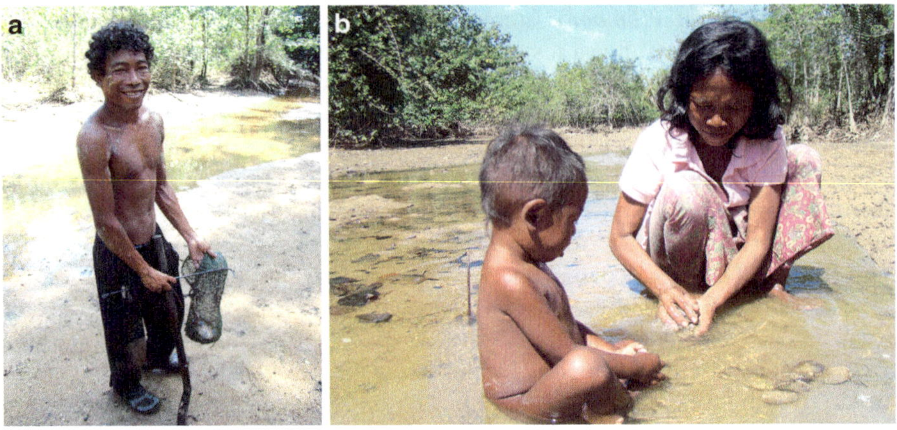

Fig. 30.2 (**a**) Moklen man from Tungwa collecting crabs. Photo by A. Neef. (**b**) Moklen woman from Tungwa collecting freshwater clams with her child. Photo by A. Neef

going fishing. "*Sometimes, if we want to go fishing by taking the boat to the sea, we have to share the costs of fuel with friends*" (*Nate*, interview, November 2013). The village survey found that the percentage of Tungwa community members who make a major livelihood from fishing decreased from 36 to 18 % during the post-tsunami recovery phase. Other activities, such as collection of crabs and freshwater clams in the small mangrove area, have become a primary means of subsistence instead (Fig. 30.2a, b). Many villagers have turned to be workers and engage in miscellaneous works both on a daily or monthly income basis, such as working as a gardener, cleaning maid, and construction workers, working for a hotel or being a municipal employee. However, getting job opportunities outside the village has become more difficult as it is necessary to have a motorbike to go the workplace. "*If you don't have a vehicle to go to work, there is no place to find a job*" (*Nate*, interview, November 2013). Around 10 % of the villagers were unemployed at the time of the survey because the job market has become increasingly difficult, especially for the Moklen people.

Nearly 10 years after the tsunami, the Tungwa community continues to rely on natural resources to make a living. The beachfront area is still used for fishing, but the ecology and structure of the shoreline has changed. "*In the past we could just walk into the water from the shore and it was not necessary to use a fishhook. Now the stream and the ocean floor have changed, thus we need different fishing gear*" (*Nate*, interview, November 2013). The Moklen also cannot build huts on the beach as in the past, due to the new tourist developments. The mangroves or *papang* continue to be a major place for collecting shellfish and crabs, but the villagers are afraid that adjacent tourist developments will constrain their access in the near future (*Hong*, interview, January 2013).

The area of the cemetery has been substantially reduced and the Ministry of Defense has notified the community that they can only use an area of 15 *rai* (2.4 ha) as their cemetery. On the other hand, the community has received a number of public facilities, such as a daycare center, a children playground, a football field, and the Chao Lay Cultural Center that serves as a museum and tourist attraction. While the Moklen people acknowledge the value of these public facilities established with the help of various national and international organizations in the post-tsunami rehabilitation process, they do not think these constitute an adequate compensation for the substantial loss of livelihood opportunities and the infringement on their cultural assets that they have suffered from after the tsunami.

30.5 Change of Living Conditions Following the Post-Disaster Recovery

30.5.1 Land Use Changes

Aerial photographs are used as a basis to analyze the livelihood changes in Tungwa community. The land use of Tungwa community is depicted at various times, namely in 1986, 2002, 2004 and 2009, with particular emphasis on three elements: the livelihood area of the community, the cemetery area and development projects, mainly for tourism purposes (Fig. 30.3).

The aerial photograph of 1986 shows the community settlement with a relatively intact livelihood area. The community was conveniently located at the main road and had full access to the mangroves and the beach front, which gave them nearly unlimited opportunities with regard to fishing and collection of marine life for subsistence. Most of the area along the beach consisted of coconut groves and forests, with very few tourism-related buildings. The area of the cemetery cannot be determined exactly, but is based on the narratives of community leaders.

The aerial photograph of 2002, taken 2 years prior to the tsunami, shows the substantial developments that had been undertaken between 1986 and 2002 in and around the community. Some houses in the north of the village had to move southwards because the ownership of adjacent land was claimed. Several beachfront resorts limited accessibility to the shoreline towards the south, as they wanted minimum interference from the Moklen in order not to have their guests disturbed.

The aerial photograph of 2004 shows the Tungwa community area just before the arrival of the tsunami. The livelihood space had been greatly reduced due to the development of various resorts. The large resort pictured in the center of the photograph—south of the temple—had carved out a significant portion of the mangrove area and further blocked community members' access to the sea (*Hong*, interview, March 2013). Further resort development along the shoreline caused additional strains on the livelihoods of the community. Most of the tourist resorts would be greatly damaged or completely destroyed by the tsunami.

The aerial photograph of 2009 shows the village area 5 years after the devastating tsunami. While livelihood opportunities along the shoreline were reduced due to

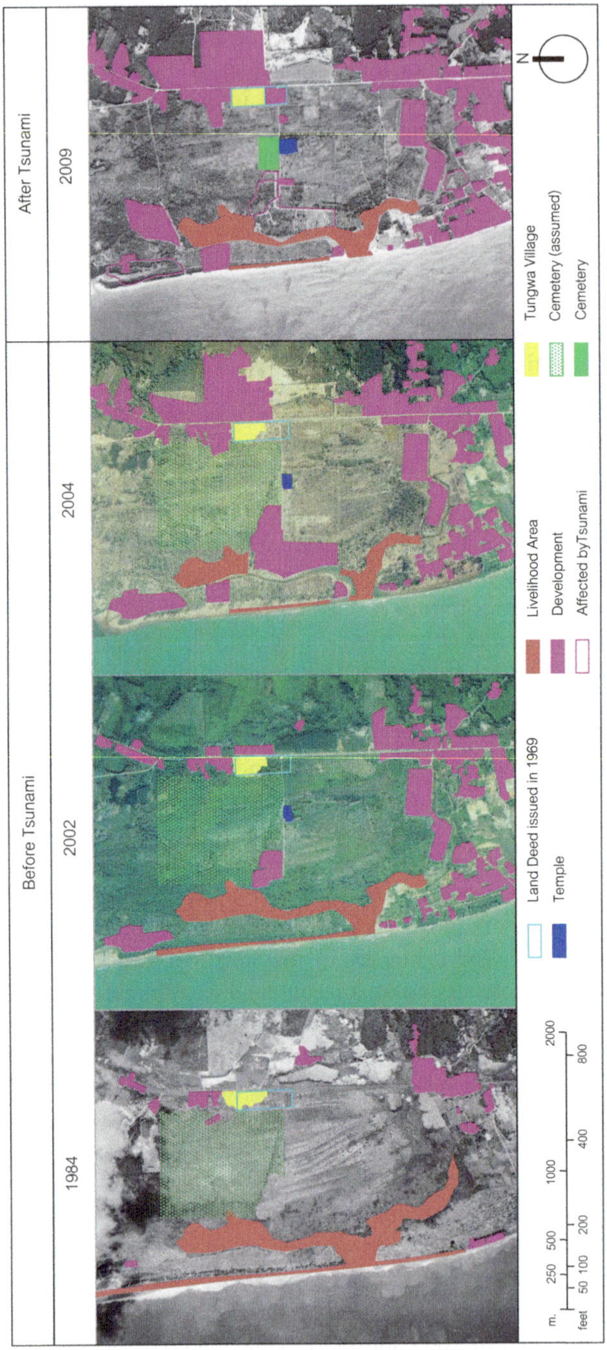

Fig. 30.3 Land use changes in Tungwa community. *Source*: Based on the aerial photograph of the Royal Thai survey department (RTSD)

beachfront resort development, the Tungwa community could still use the mangrove area, as the tourist resort that had substantially compromised access to the mangrove area was not rebuilt. The green area between the beach and the village remained undeveloped or was used as plantation land by private land owners. The Sub-district (*Tambon*) Administrative Organization (TAO) drastically reduced the cemetery area of the community in accordance with the notification of the Ministry of Defense. According to a local NGO that supports the community the cemetery has been regularly cleared and further reduced in size by the authorities (Chumchonthai Foundation, interview, 2013).

30.5.2 Changes to Village Facilities and Family Housing

We use another set of aerial photographs to analyze the changes in physical living conditions in the Tungwa community. The photographs depict the residential area of Tungwa community at various times, namely in 1986, 2002, 2004, and 2009 by considering three elements, including the village layout, the residential houses, and community facilities (Fig. 30.4).

The aerial photograph of 1986 shows the characteristics of the community and the distribution of housing. At that time, there were only seven to eight houses, with trails connecting them and with each house having a sizeable open space area and being built in the traditional Moklen style.

The aerial photograph of 2002 shows the condition of Tungwa village 2 years prior to the Tsunami. The number of houses had increased to about 45 houses due to the in-migration as described in Sect. 30.3.1. The settlement had become much denser, the area of open space around the houses reduced, but there were still large trees and extended home gardens. The settlement had become a mix of Moklen traditional houses and more contemporary houses that were constructed with industrial material.

The aerial photograph of 2004 shows the condition of the community a few weeks after the Tsunami. The water from the tsunami waves had already dried up, but the devastation is clearly visible. Only three houses were left standing, with all others completely destroyed, leaving 42 of their inhabitants dead. The vegetation around the houses was also severely affected, all home gardens had perished and nearly all domestic animals had been killed.

The aerial photograph of 2009 shows the village 5 years after the tsunami, the community's layout had changed drastically, the buildings had been neatly organized in rows, with each house sharing an equal area of land, but with considerably less space than in the pre-tsunami period. All houses display the same design with a high basement, gabled roof and modern construction materials used, resembling typical Thai rural houses in this coastal region (cf. Fig. 30.5).

In sum, the tsunami and the post-disaster events dramatically altered the physical living conditions of the community. Individual families' living space has been diminished, and the increased density affects the well-being of the community

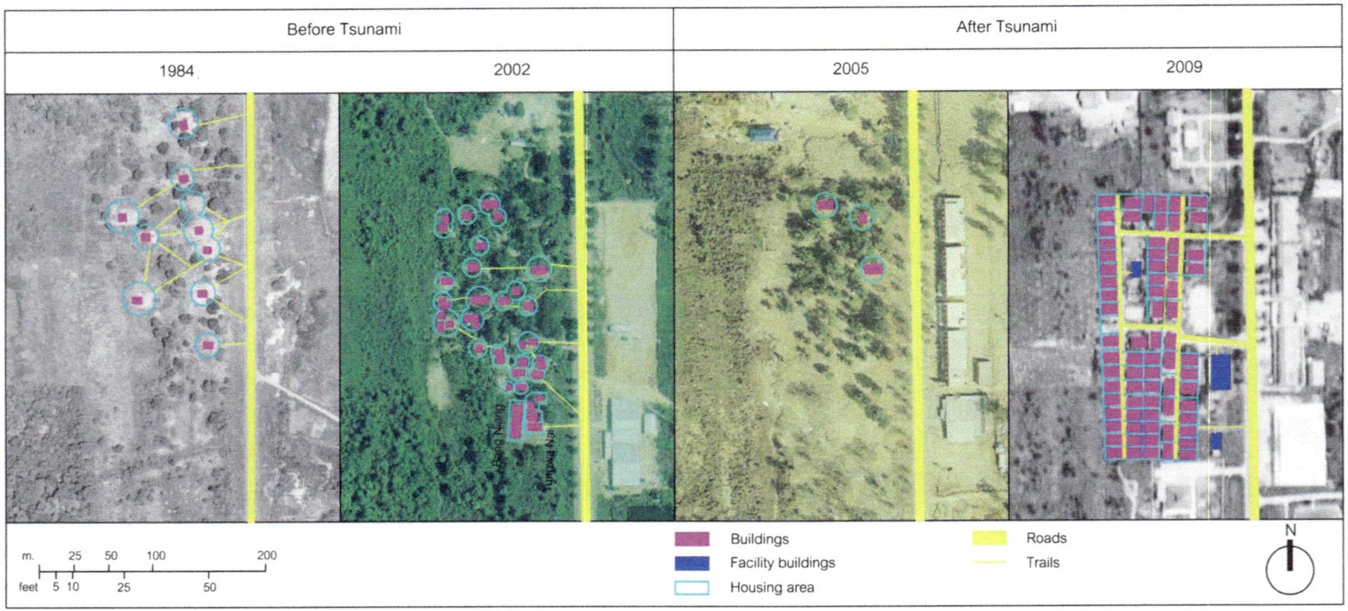

Fig. 30.4 Change of settlement area of Tungwa community. Based on the aerial photograph of the Royal Thai survey department (RTSD)

Fig. 30.5 The post-tsunami standard housing design in Tungwa community. Photo by A. Neef

members who now have less space to rear farm animals as they used to do in the pre-tsunami period. The dense village design also negatively affects the ventilation of the houses. In addition, changes in house characteristic demanded a considerable adjustment from the Moklen who have not been used to high basements.

Due to the reduction of livelihood opportunities for the Moklen, social problems have spread in the community; gambling, alcoholism and drug use—even by children—have become common phenomena.

30.5.3 Status of Land Conflicts Involving the Tungwa Community in 2013

During our inquiries at community and household level carried out in three field visits during 2013, we found that land conflicts continue to affect the Moklen people's livelihoods and tenure security. Although it has been proven that the residents of Tungwa community have lived in the area long before the issuance of title deeds, the land is still owned by the government and the community's residential rights depend on the regular renewal of a temporary communal land lease. The issuance of a permanent community land title deed—based on an initiative instigated by the government of former Prime Minister Abhisit Vejjajiva and adopted by subsequent administrations—could offer higher tenure security, but it is unclear whether this will happen any time soon, given the notoriously slow speed of implementing new

Fig. 30.6 (**a**) A sign board advertising land for sale on the customary cemetery grounds of Tungwa community. Photo by A. Neef. (**b**) Moklen community leader showing a demarcation pole set up by a tourist resort. Photo by A. Neef

land policies in the country. Meanwhile, Tungwa community members are worried about signboards advertising (public) land for sale to private developers and demarcation poles suggesting the further expansion of tourist resorts (Fig. 30.6a, b).

The Moklen community leader revealed his fears and frustrations about the ongoing land conflicts in an interview in January 2013:

> *I am worried about the new generation for they will surely not be able to access any resources in the sea and on the shores. Now we face many difficulties with regard to fishing and searching for food in the sea or on the beachfront because we are told by Thai resort workers not to enter the resort area as this would make their guests unhappy. [...] Our lives have ended after the [tourism-related] development and the tsunami. [...] We still have some land to live on, but we only have a permission which has to be renewed every five years. We were asked [by the local administration] to sign some documents related to the land, but we denied to do so, as we were afraid of being fooled.*

This statement is an expression of the Moklen community's frustration about rising pressures from tourism development on their socio-economic livelihoods and their distrust in the local government.

The conflict regarding the community cemetery area also continues to affect the community. According to information provided by Chumchonthai Foundation (interview, November 2013), the cemetery area has been decreasing every year (Fig. 30.7a). This could be witnessed during our visit in January 2013 when extensive land clearance with heavy machinery was taking place around the cemetery area (Fig. 30.7b).

Having lost their spiritual leader and the community shrine in the tsunami, the infringement on their customary cemetery area by local authorities and private developers constitutes another significant loss of cultural identity and heritage for the Moklen community.

Fig. 30.7 (**a**) Signboard allocating 15 *rai* (2.5 ha) as community cemetery land. Photo by A. Neef. (**b**) Land clearance on the communal cemetery land of the Moklen community. Photo by A. Neef

30.6 Conclusion

As the traditional livelihoods of the Moklen people of Tungwa community were dependent on close access to the sea, they were particularly negatively affected by the tsunami which claimed a high number of lives among them. Post-disaster reconstruction of the residential area had to be done hastily, as the Moklen risked being relocated to other areas by the local authorities. Although the design of the community rehabilitation scheme was done in a consultative manner, the Moklen could not resist subsequent tourism development that has threatened their resource-based livelihoods. After the withdrawal of external assistance from foreign donors and Thai NGOs and in the absence of a supportive legal and institutional framework, the Moklen find themselves exposed to increasing local pressures in this prime tourism area, where land prices have skyrocketed after the rehabilitation of infrastructure in the post-tsunami recovery process.

The major conclusion from our study is that consultative forms of participation in post-disaster reconstruction processes at the local level are no substitute for deeper empowerment of marginalized people through a legal framework that recognizes their long-established customary rights, respects their cultural heritage and considers both their immediate and long-term livelihood needs. Communal land ownership would give the community a stronger sense of security and cultural recognition, but political will at local and national level towards establishing permanent land rights for marginalized ethnic groups appears to be lacking.

Acknowledgements The authors are grateful for a grant-in-aid provided by the Japan Society for Promotion of Science (JSPS) under the research project "Integrated Assessment of Post-Disaster Recovery in Coastal Communities in Asia-Pacific" (JSPS KAKENHI Grant Number 24401007). Additional financial support was provided by Kyoto University's Global COE Program 'Sustainability/Survivability Science for a Resilient Society Adaptable to Extreme Weather Conditions'.

References

Andaya LY (2008) "Leaves of the same tree": trade and ethnicity in the straits of Melaka. University of Hawai'i Press, Honolulu

Arunotai N, Na Pompejara P, Buntaotuk J (2008a) Uraklavoy, Moklen and Moken: experts of the Andaman sea islands and coasts. Andaman Pilot Project, Social Research Institute, Chulalongkorn University, Bangkok, Thailand (in Thai)

Arunotai N, Na Pompejara P, Buntaotuk J (2008b) Human and community security. Andaman Pilot Project, Social Research Institute, Chulalongkorn University, Bangkok, Thailand (in Thai)

Bristol G (2010) Surviving the second tsunami: land rights in the face of buffer zones, land grabs, and development. In: Lizarrallde G, Johnson C, Davidson C (eds) Rebuilding after disasters: from emergency to sustainability. Spon Press, Abingdon, pp 133–148

Chumchonthai Foundation (2012) The crisis of sea gypsies' way of life crisis. Chumchonthai Foundation, Bangkok, Thailand (in Thai)

Community Organization Development Institute (CODI) and Confederation of Fishing in the Southern Region (2006) Summary of communities and villages affected by the Tsunami. CODI, Bangkok, Thailand (in Thai)

Ferrari O, Hinshiranan N, Utpuay K, Ivanoff J (2006) Turbulence on Ko Phra Thong: Phang Nga Province, Thailand. Anthropologie Maritime, Centre d'Ethno-Technologie en Milieux Aquatiques, Museum National d'Histoire Naturelle, Paris, France

Ivanoff J (2001) Rings of coral: Moken folktales. White Lotus, Bangkok, Thailand

Larish M (1992) Who are the Moken and Moklen on the islands and coasts of the Andaman Sea? http://sealang.net/sala/archives/pdf8/larish1992who.pdf. Accessed 21 Feb 2014

Rajah D (2006) Land not for the taking. In: SEAPA (ed) Disasters in southeast Asia: stories of struggle, resilience, greed. Southeast Asian Press Alliance (SEAPA), Bangkok, Thailand, pp 21–31

Rice A (2005) Post-tsunami reconstruction and tourism: a second disaster? Tourism Concern, Stapleton House, London

The Director of Civil Defense of the Kingdom of Thailand (2006) Assistance to earthquake and the Andaman ocean tsunami victims. Ministry of Defense: Bangkok, Thailand (in Thai)

Thongtaweewiwat K (2006) Community rights and interest in Thai society: a case study of Moklen land ownership after tsunami in Baan Tungwah Khukkuk Sub-district, Takuapa District, Phang Nga. Master Thesis of Arts Program in Political Economy, Faculty of Economics, Chulalongkorn University, Bangkok, Thailand (in Thai)

Chapter 31
Post-Tsunami Recovery and Rehabilitation of Small Enterprises in Phang Nga Province, Southern Thailand

Andreas Neef, Arusa Panyakotkaew, and Peter Elstner

Abstract In Southern Thailand, tourism-related businesses and small-scale fisheries were among the sectors that were hardest hit by the 2004 Indian Ocean Tsunami. Drawing on an extensive literature review, conversational-style interviews and an innovative recovery profiling methodology used in the most affected province Phang Nga, this chapter discusses the factors that enabled or constrained small enterprises in the tourism, fishery and farm sector in the process of recovering from the disastrous impact of the tsunami. We find that small businesses benefitted from individual support systems in their social networks and from a diversification of their business strategies rather than from government support and large-scale donations by international organizations. The second part of the chapter looks into the particular case of the expanding birds' nest business in one coastal community that had suffered from nearly complete destruction by the tsunami. Our results show that the production of birds' nests in specially designed buildings has primarily benefitted the wealthier and politically well-connected families in the village and outside investors who own more than half of the birds' nest houses. We conclude that post-disaster economic recovery can deepen income disparities and may disrupt the sense of community if local elites and absentee business owners are able to take advantage of the lack of sound governance structures in the aftermath of a major natural disaster.

Keywords Coastal communities • Disaster recovery • Fisheries • Southeast Asia • Tourism

---

A. Neef (✉)
Faculty of Arts, Center for Development Studies, School of Social Sciences, University of Auckland, Auckland 1142, New Zealand
e-mail: a.neef@auckland.ac.nz

A. Panyakotkaew
Independent Researchers, 101 Moo.3, Santisuk, Doilor, Chiang Mai 50160, Thailand

P. Elstner
Independent Researchers, 21/7 Moo 3, T. Mae Sa, A. Mae Rim, Chiang Mai 50180, Thailand

R. Shaw (ed.), *Recovery from the Indian Ocean Tsunami: A Ten-Year Journey*, Disaster Risk Reduction, DOI 10.1007/978-4-431-55117-1_31

31.1 Introduction: Impact of the 2004 Indian Ocean Tsunami on Coastal Communities and the Local Economy in Southern Thailand

The Andaman coast of Southern Thailand was one of the regions that were hardest hit by the 2004 Indian Ocean Tsunami. The worst affected province was Phang Nga which suffered most human casualties and vast damages to buildings, infrastructure and the local economy, dominated by agriculture, fisheries (incl. aquaculture) and tourism. The highest concentration of tourism-related businesses is found in and around Khao Lak, which had been an eco-tourism destination—predominantly targeting European tourists—prior to the tsunami disaster. This area was particularly affected by the disaster due to its relatively flat coastal topography, extensive development along the beaches and coastal corridors, and removal of natural barriers (e.g. mangrove forests) to enhance ocean views and provide access to the sea (Bell et al. 2005; Römer et al. 2012; Wong 2012). Eighty percent of the hotels suffered severe damages, and most beachfront hotels and infrastructure with a high share of small, family-owned businesses were completely destroyed (Alovisi et al. 2005; Wong 2012). The harbour and fishing village Ban Nam Khem was nearly wiped out by the tsunami, with more than a quarter of the population perishing in the waves.

Post-disaster reconstruction and recovery in this area started with a significant delay, as priority was given to the prime tourist destinations on Phuket Island[1] (Wong 2012). In some parts of Phang Nga province, the complete restoration of infrastructure took up to 30 months, causing substantial constraints to economic activity in the area (Willroth et al. 2011). Overall, it was estimated that more than 100,000 tourism-dependent livelihoods had been affected by the tsunami disaster (UNCTT 2005).

Theoretical and conceptual frameworks on post-disaster recovery suggest the existence of distinct phases of the disaster recovery cycle. Paul (2011), for instance, suggests a disaster recovery cycle that consists of (1) relief, (2) rehabilitation, (3) reconstruction, and (4) development. In an analysis of post-tsunami recovery efforts in Sri Lanka, Birkmann (2010) identified six distinct phases, ranging from (1) destabilization, (2) inflow of external aid and (3) government interventions to (4) residential reconstruction, (5) restoration of technical and social infrastructure, and (6) livelihood recovery and income stabilization. Yet Brown et al. (2008, p. 1) challenge this view and hold that "[t]he use of recovery phases is thought to oversimplify the recovery process and mask how in reality, the various roles overlap and interact with each other."

The importance of social networks, collective action and social capital for post-disaster recovery has been underscored by a number of studies and scholarly publications in recent years (e.g. Chamlee-Wright and Storr 2011; Aldrich 2012; Neef

[1]According to Nidhiprabha (2007; cited in Larsen et al. 2011), damages and losses resulting from the tsunami disaster in Phuket accounted for 90 % of the province's GDP, while in Phang Nga Province they equalled 70 % of GDP.

and Shaw 2013). Most studies, however, have focused on the impact of social capital on individual, household and community recovery, whereas the role of social networks in the recovery of small enterprises has received much less attention. Moreover, the socio-political dimension of business recovery, including issues of corruption, elitism and power, tends to be ignored by the majority of post-disaster recovery studies.

In this chapter, we look at individual recovery trajectories as embedded in local and translocal social, political and economic contexts. We describe the rationale for the study area selection and our research methodology in Sect. 31.2. The results in Sect. 31.3 are divided into two case studies, one describing the recovery of small- and medium-sized enterprises in the tourism, fisheries and farming sector and the other scrutinizing the economic transformations in one coastal village in the post-tsunami recovery process. Section 31.4 discusses the findings and concludes the chapter.

31.2 Methodology

31.2.1 Study Area

For this study we selected the northern part of Phang Nga Province that was most heavily affected by the 2004 Indian Ocean Tsunami, covering the coastal area stretching from Khao Lak town to Baan Nam Khem and the islands of Koh Kho Khao and Koh Prathong (Fig. 31.1).

The area of Khao Lak is mostly dominated by tourism, with large- and medium-sized tourist complexes dotting the coastline, in combination with clusters of small tourist-related businesses, such as restaurants, beauty and spa services, car and motorbike rentals, and souvenir shops. According to a representative survey of Willroth et al. (2011), 44 % of households in Khao Lak derived their income from tourism or related businesses in 2008.

31.2.2 Methods

In this study we used a mixed-method approach. About 25 small entrepreneurs in the tourism, fishery and farm sector were interviewed, using conversation-style, open-ended interviews. Following the interview, respondents were asked to complete a personal recovery profile on a score card (Fig. 31.2), indicating to which degree they had been satisfied with their socio-economic situation at five different points in time, namely in January 2001, November 2004 (1 month prior to the tsunami), January 2005 (1 month after the tsunami), January 2009 and January 2013. All respondents were asked to provide explanations for their scores, ranging from one (completely unsatisfied) to ten (completely satisfied). For analysis, we identified prototypical patterns in the recovery profiles, resulting in three types,

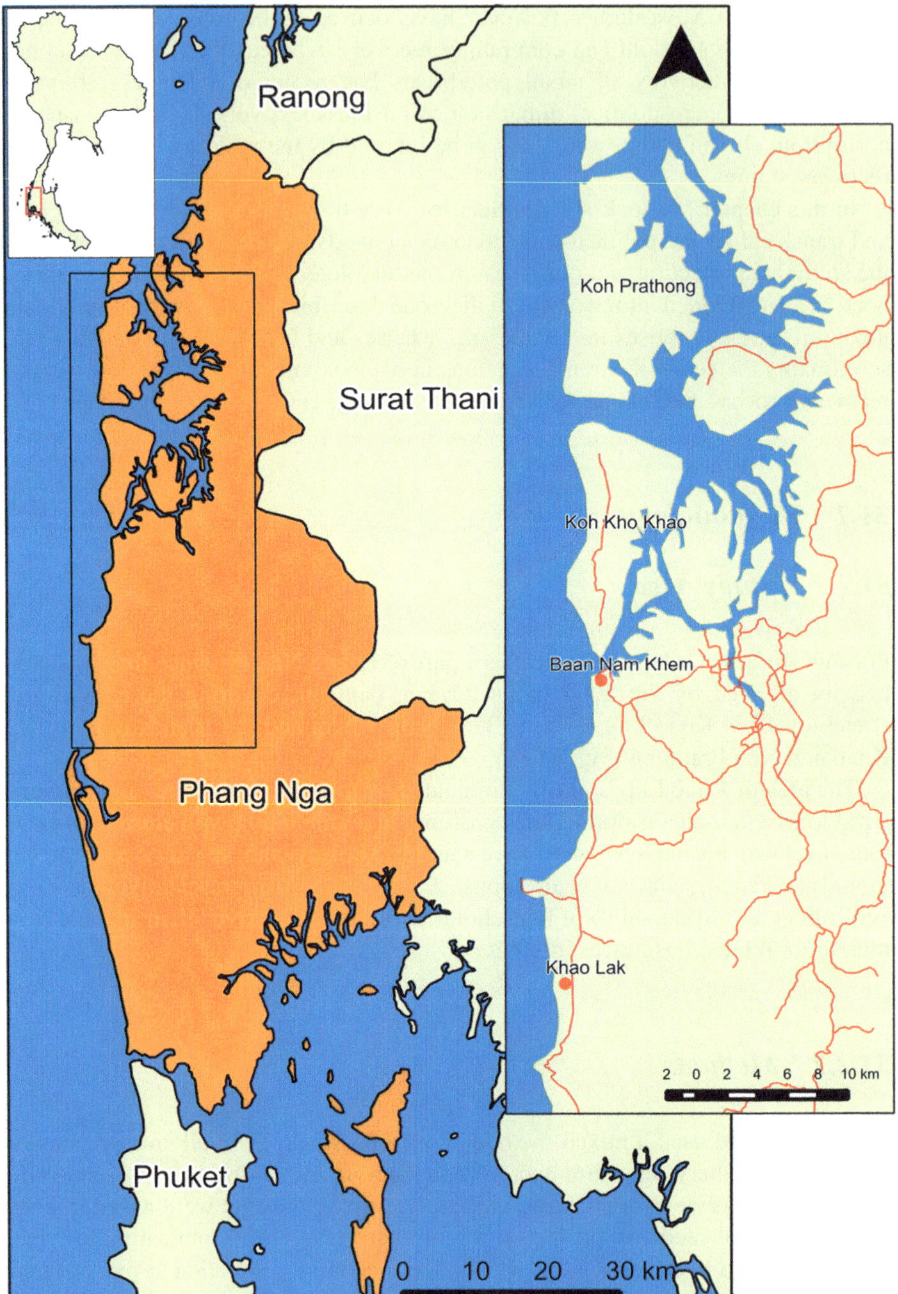

Fig. 31.1 Map of Phang Nga Province with location of main study sites

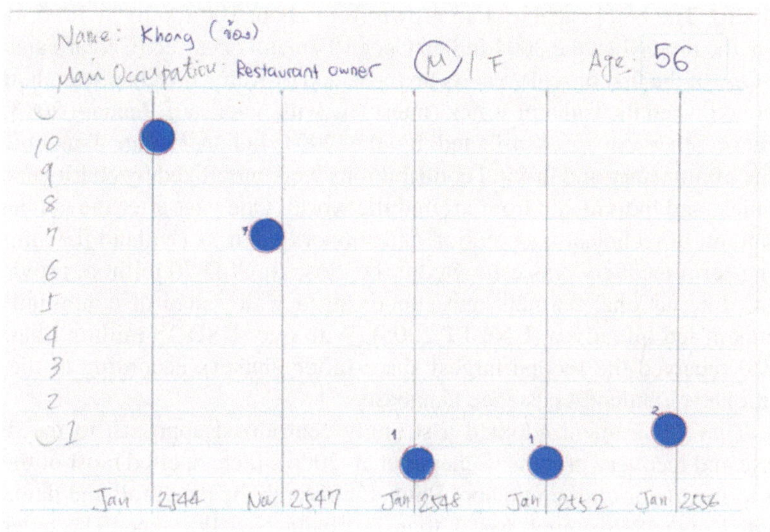

Fig. 31.2 Example of a score card for the recovery profile

i.e. (1) complete recovery; (2) partial recovery and (3) unsustainable recovery. Additional qualitative interviews were conducted with NGO representatives and employees in the tourism sector. All names of interviewees were anonymized to protect respondents' identities. Our primary data was complemented by an extensive review of the literature available on various periods of the recovery process.

In the case of the bird's nest business of Baan Nam Khem, we conducted a complete inventory of all bird's nest factories (or 'birdhouses'), including building size, location (identified with Global Positioning System—GPS), time of construction and ownership. Several birdhouse owners were interviewed through conversational-style interviews. We also analysed secondary data, e.g. a documentary video containing a number of interviews on the bird's nest business in Baan Nam Khem.

31.3 Results

31.3.1 Recovery of Small Businesses in the Tourism, Fisheries and Farming Sector of Khao Lak, Koh Kho Khao and Koh Prathong

Prior to the tsunami, the area of Khao Lak—formerly an insiders' tip among eco-tourists and backpackers and a gateway to the Similan and Surin islands in the Andaman Sea—had just been discovered by the mass tourism industry. At the time of the disaster more than 120 resorts and guesthouses were operating in the area (Hüger et al. 2006). The hotel capacity had grown from about 100 rooms in 1996 to 5,315 rooms at the time when the 2004 Indian Ocean Tsunami occurred (Calgaro and Lloyd 2008). One of the first upscale resorts on the island of Koh Kho Khao was about to be inaugurated when the tsunami struck (interview with hotel staff, January 2013).

Like in other countries affected by the 2004 Indian Ocean Tsunami, large amounts of monetary and in-kind contributions were mobilized by charities, private companies, and individuals from around the world. One year after the tsunami, the total amount of technical cooperation and support given to Thailand for immediate and long-term recovery was estimated to be close to USD 70 million, provided by some 45 international partners and supporting a wide range of community- and government-led initiatives (UNCTT 2005). With over USD 18 million Phang Nga Province received the second largest share (after Phuket), according to the TICA Tsunami Development Assistance Database.

The Thai government adopted a strongly centralized approach to the disaster response and recovery process (Scheper et al. 2006). It channelled most of the relief efforts to the major tourism hotspot Phuket, although the death toll and damages to residential areas were much lower than in Phang Nga Province. The emergency disaster relief was followed by medium- to long-term recovery efforts, expressed in the Andaman Tourism Recovery Plan and the Andaman Sub-Regional Development Plan, which included tourism marketing campaigns and provision of credit to support the recovery of the vital tourism business (Larsen et al. 2011). However, many small-scale tourist-related businesses—particularly unregistered entrepreneurs in the informal tourist sector and owners of small guesthouses—were apparently bypassed and received little or no assistance (Rice 2005). This was also confirmed in the interviews, as the following quotes show:

> *As business owners we did not receive any support from the government. We were not covered by any insurance like the big resorts. I had to take a loan from a French aid organization's fund to start up my business again. (Small resort and restaurant owner, Cape Pakarang, interview, January 2013).*

> *We did not get any subsidies or support from the government or other sectors [after the tsunami]. They might think we are wealthy because we ran a business. But actually we are not rich. In the future I may need to sell the land and the resort to get rid of my debt. (Small resort owner, Cape Pakarang, interview, January 2013).*

Small entrepreneurs did not just suffer from the lack of financial assistance. Several respondents mentioned post-traumatic syndromes that had affected either themselves or close relatives, which caused significant social and psychological burdens and also constrained the economic recovery process (cf. Willroth et al. 2011). Not surprisingly, all respondents characterized the immediate aftermath of the tsunami as the absolute low-point of their personal life and socio-economic situation.

Yet the picture becomes much more diverse when looking at individual recovery trajectories. We found three distinct types which we call (1) complete recovery, (2) partial recovery, and (3) unsustainable recovery. Figure 31.3 shows the recovery profiles of those respondents that reported a complete recovery. In this category are small entrepreneurs whose business was not directly affected by the tsunami, had a relatively diversified business prior to or after the tsunami or could rely on strong

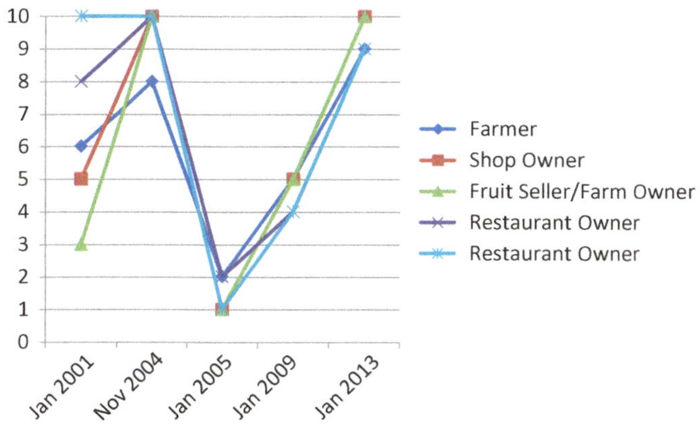

Fig. 31.3 Recovery profiles of small entrepreneurs (complete recovery)

family ties and social networks, including links to influential people, as evidenced by the following quotes.

> *A lot of people could recover [from the tsunami], reconstruct their houses and get money [to restore their business] because they are members of an influential family or have good connections to some local politicians. [...] We had a boat construction business, but it was all destroyed by the tsunami. We started life off again by renting out our land which had been affected by the tsunami. We also bribed military personnel, so we could get two reconstructed houses for our large family. (Restaurant owner, Baan Nam Khem, interview, January 2013).*

> *Five members of my family died in the tsunami and we went through a very difficult time. But my [rubber tree and oil palm] farm is located uphill, so it was not affected by the tsunami. I am now working in various jobs, as a farmer, a driver and a waiter. My younger brother and his wife are running a restaurant. (Farmer, Kuk Kak, interview, January 2013).*

Successful recovery of tourist-related small businesses was enhanced in those locations where the local administration made sustained efforts to revive the tourism industry, as evidenced in the case of the island of Koh Kho Khao (Restaurant owner, interview, January 2013).

The second category of recovery profiles relates to those entrepreneurs who reported only a partial recovery of their economic situation. Among the three representatives of this profile, two are fishermen (Fig. 31.4). Subsistence and small-scale fishery had been particularly adversely affected by the tsunami, with the majority of fishing boats destroyed and most fishing gear lost (Willroth et al. 2012). Therefore, boat rebuilding activities and supply of new equipment for fishermen were one of the major priorities of aid organizations. Hüger et al. (2006) refer to a survey by D-TRAC which reported that about 30 % of all aid agencies provided boats, yet often without a thorough demand analysis and without coordination among the various organizations. Apparently, this led to an oversupply of fishing boats, with some fishermen receiving more than one boat as they had applied for help from several organizations (Willroth et al. 2011; Hüger et al. 2006). Drawing on focus group discussions and personal

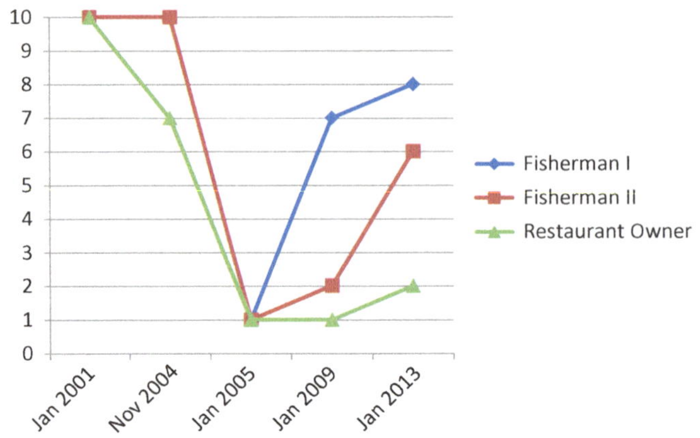

Fig. 31.4 Recovery profiles of small entrepreneurs (partial recovery)

Fig. 31.5 (**a**) Destroyed corals continue to litter the beach at Cape Pakarang (September 2012). Photo by A. Neef. (**b**) Signboard of a restaurant opened on public land by a tsunami survivor. Photo by A. Neef

communication from fishermen in Baan Nam Khem, Willroth et al. (2012) maintain that the overcompensation with boats and supply of more and better fishing equipment were responsible for overfishing and forced some fishermen to look for alternative income opportunities. However, this narrative may not provide the complete picture. Another factor that came into play was that the marine and coastal ecology was dramatically altered by the tsunami. At Cape Pakarang, a large band of coral reefs was completely destroyed and coastlines have been reshaped (Fig. 31.5a). Some mangrove areas had also been severely damaged, as reported by one of our respondents:

> *Large mangrove areas were destroyed by the tsunami which had provided habitat for fish and other sea life. It has become more difficult to find fish after the tsunami. (Fisherman on Koh Kho Khao, interview, January 2013)*

The destruction of reefs and alterations of coastlines also had a negative impact on some of the small businesses that depended on tourists who wanted to enjoy the beautiful scenery. The recovery profile of one restaurant owner—whose wife died

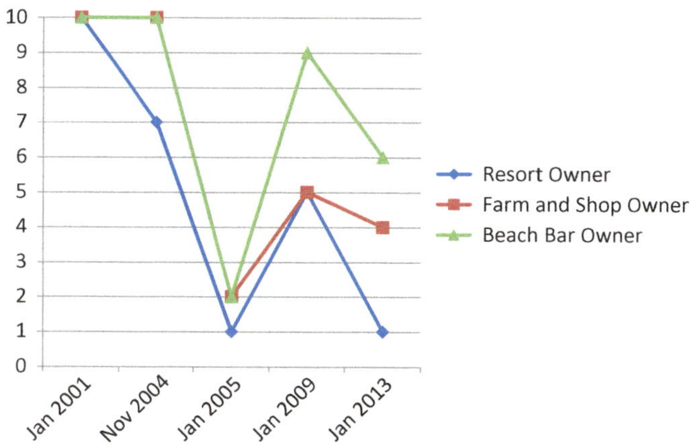

Fig. 31.6 Recovery profiles of small entrepreneurs (unsustainable recovery)

in the tsunami—reflects his very slow economic recovery. He used to be a fisherman before the tsunami, while his wife ran a small restaurant which was destroyed by the tsunami. He never fully recovered mentally from the death of his wife and had to sell his house to open a new restaurant on rented land, but had to abandon the business because it was not successful. He opened another restaurant on public land (Fig. 31.5b), as he needed money to repay the debts incurred from this failed business and expensive medical treatments for heart disease (which he attributed to the immense stress in the aftermath of the tsunami). He lives in constant anxiety because he is not certain whether the local administration will eventually expel him from the land (interview, January 2013).

Unsustainable recovery is a third category that has emerged from our recovery profiling method (Fig. 31.6). Several of our interviewees from the tourism sector reported that they had received personal financial support from overseas, e.g. from foreign tourists who had previously stayed in their resorts or visited their beachfront bars and restaurants and felt compelled to provide some form of personal assistance towards recovery. In some instances, this form of external financial assistance was very substantial and helped the beneficiaries to rebuild their small businesses, as the following excerpt from an interview shows:

> *I got 100,000 Baht [US$3,300] from a Swedish friend who was a long-time customer at the bar of my uncle [who died in the tsunami]. He told me to use the money to travel and start a new life. I started exploring other places. I also participated in a scuba diving project for 30 tsunami survivors. Eventually, my Swedish friend gave me another 400,000 Baht to restart the [beach bar] business on my uncle's land. But now the competition is getting tougher, there is another beach bar next door where a new investor has partnered with the owner after the tsunami. Now I have to be very concerned about money issues because I am managing the bar all by myself. (Beach bar owner, Cape Pakarang, interview, January 2013)*

The owner of a small resort further inland on Cape Pakarang had just started her first business season when the tsunami struck and caused severe damage to her property. She had taken a bank loan to construct the resort and—as she did not have insurance coverage—had to take another loan after the tsunami to pay for all the repairs which overstretched her financial situation:

> *The business has not been very good in recent years, as my resort is not located directly at the beach and tourists prefer beachfront accommodation. I am now thinking about selling the land and the business because I need to get rid of my debts. It will make me sad, but it's still better than having debts that have made me suffer. (Small resort owner, Cape Pakarang, interview, January 2013)*

The lack of insurance coverage of small businesses and uneven access to financial capital were important factors of unsustainable recovery processes. A study of Calgaro et al. (2009) found that only 9 % of interviewed small entrepreneurs in the Khao Lak area had insurance coverage prior to the tsunami and that this situation remained largely unchanged after the tsunami. High loan repayment commitments put particular financial burdens on small businesses, while medium- and large-scale business could rely on profits accumulated during the pre-tsunami period and had far better credit ratings, putting them onto a more sustainable recovery pathway (Calgaro et al. 2009).

According to some respondents, the farm sector—which was the major contributor to Phang Nga's provincial GDP prior to the tsunami—was neglected in the post-tsunami recovery process. Low and fluctuating prices for important agricultural and aquaculture commodities (e.g. palm oil, rubber, shrimps) were mentioned as a factor of unsustainable recovery processes and were to a major extent attributed to the lack of government support, particularly with the onset of political instability following the military coup of September 2006.

Aside from the economic and social sustainability dimensions of the post-disaster recovery process, the environmental sustainability of the recovery of the tourist sector also needs to be scrutinized. Prior to the tsunami, the tourism industry already had adverse effects on the environment, as mangrove areas have been reduced and coastal vegetation removed. Nearly 10 years after the tsunami, short-term profit maximization in the tourist sector still prevails over long-term, ecologically sustainable planning. It is evident that official requirements of setting back hotels and resorts from the shoreline to avoid negative impacts on fragile coastal ecosystems and reduce the risk of future tsunami and storm surge disasters are routinely ignored by most operators (cf. Wong 2012). Those who have sufficient capital and connections to local or national power structures are in the best position to get around environmental regulations and secure approval of controversial developments that are in conflict with ecological objectives and planning regulations (Calgaro et al. 2009; Larsen et al. 2011).

31.3.2 Economic Transformations in the Aftermath of the Tsunami Disaster: The Case of Baan Nam Khem

Baan Nam Khem is a fishing and port village that had been inhabited by former labourers who had migrated from various parts of the country to work in the area's tin mines. When the mining concessions came to an end, they established their own community, but much of the land on which they settled had been claimed by outside entrepreneurs and investors (Rajah 2006). More than 1,400 people—at least a quarter of its population—were reported dead or missing after the disaster, but the number was likely to be much higher as many residents had been Burmese migrants that were not registered.[2] The majority of surviving villagers in Baan Nam Khem suffered not only from the loss of their household members and damage to property, but also from mental trauma, triggering a high rate of out-migration (Paphavasit et al. 2006). Unemployment rates after the tsunami were as high as 25–33 %, household debts increased significantly, and most people had to change their occupations, particularly former fishermen (Siriboon et al. 2006). One year after the tsunami only about half of the households were located in the area where they were living before the disaster (Paphavasit et al. 2006). The Department of Public Works and Town Planning had issued a reconstruction plan that proposed to divide the village into public, fishery, living, and monument & sightseeing area, with the intention to move the residential area further inland (Takada et al. 2010). Yet most residents of Baan Nam Khem rejected the plan and its underlying 'setback policy', as they wanted to keep their traditional lifestyle and live close to the sea.

> *The suggestion was good for the reasons of protecting life and property but I think it was not appropriate for the local community lifestyle. You know, fishermen need to live near the shore. They cannot live far from the shore because they have to look after their boats. (Baan Nam Khem resident, quoted in Paphavasit* et al. 2006, *p. 218)*

> *In fact, community members did not agree with this policy (the setback policy) because most of them are fishermen. They need to take care of their boats and instruments. I don't think the distance from the shore matters. Did the policy-makers face the tsunami? Did they know how big the wave length was? The length was not less than 500 meters. Even if we moved our houses rearward by about 200 meters as they advised, we would not escape. We would all die. In this case, we just stay. (Baan Nam Khem resident, quoted in Paphavasit* et al. 2006, *p. 218)*

Searching for alternative economic opportunities, the wealthy and politically well-connected among the survivors started to invest in the emerging bird's nest business. Originally, the collection of bird's nests[3]—one of the most highly priced food commodities in the world—was a tedious and dangerous activity undertaken mostly by indigenous coastal people in the limestone caves along the coast of Southern Thailand. In Baan Nam Khem, villagers had observed a few years before the tsunami that swiftlets had started to build nests in an abandoned workshop (Fig. 31.7a, b). While a few villagers had constructed birdhouses prior to the disaster, the bird's nest business became a lucrative business for the local economic and

[2] Burmese migrant workers have been referred to as the "Tsunami's invisible victims", as many deaths were never recorded, and most of the survivors felt left out of the relief efforts (e.g. Maw 2006; Rigg et al. 2012).

[3] The edible-bird swiftlet (*Aerodramus fuciphagus*) is a member of the swift family found in coastal areas in Southeast Asia. The nest is made of hardened saliva from the male swiftlet and is soaked and steamed to make bird's nest soup, considered a delicacy with medicinal and aphrodisiac properties in East and Southeast Asia.

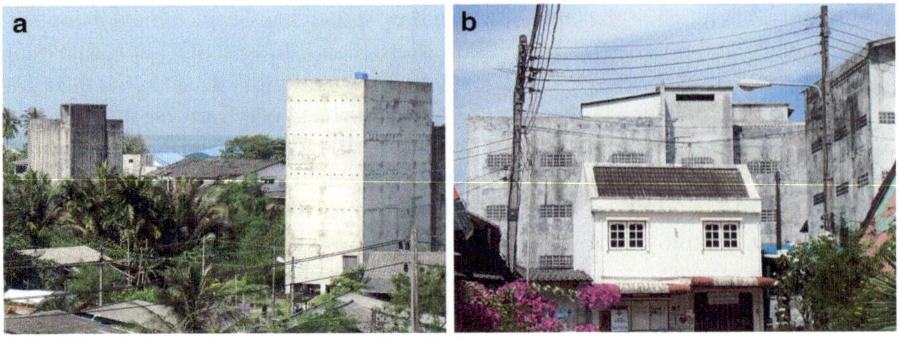

Fig. 31.7 (**a**) and (**b**) Swiftlet birdhouses dominating the village scenery in Baan Nam Khem. Photos by K. Neef

political elite in the post-tsunami recovery phase, as described by the village leader in the following quote:

> *After the tsunami came, all villagers who had enough money started exploring what they could do, so they could help themselves. […] When those who have money and power want to build the swift houses in this village, they start to find land and the land prices start to increase. (Village leader, Baan Nam Khem, interview, cited in Koh 2012)*

With the growing interest in this new investment opportunity, land prices have risen dramatically. Villagers reported that some plots have quadrupled in value within only a few years (Koh 2012).

> *If your place is not facing the bird's nest area, you can't really have a high price for it. But if your house has swift houses around and your house hasn't been sold away, I ask you, if they offer 10 million Baht, would you sell? So we discussed in the family that it's better to sell [the house], because I feel sorry for the children. What if someday [another] tsunami comes? Then we have to flee [again]. (Female resident of Baan Nam Khem, interview, cited in Koh 2012).*

Investment cost for birdhouses have also risen substantially, excluding those villagers with insufficient financial capital and lack of access to formal credit from the bird's nest business. Birdhouse owners compete against each other by installing sound technology that is aimed at attracting the swiftlets to the property. Villagers who do not own birdhouses complain about persistent noise pollution from the installed speakers (female resident, Baan Nam Khem, interview, January 2013).

In expectations of high returns on investment, the number of birdhouses in Baan Nam Khem more than doubled between 2009 and 2012 (Fig. 31.8). Yet some local investors reported that they are disappointed that their birdhouse did not attract many swiftlets:

> *My father used to work in a tin mine and is a very diligent man. He built a birdhouse in his own style, but it has not been successful. There are only two nests in it. I don't think it will become a profitable investment. (female resident, Baan Nam Khem, interview, January 2013)*

Nearly 10 years after the tsunami, the bird's nest business is largely controlled by wealthy local elites—such as the *kamnan*, the head of the sub-district or *tambon*—and

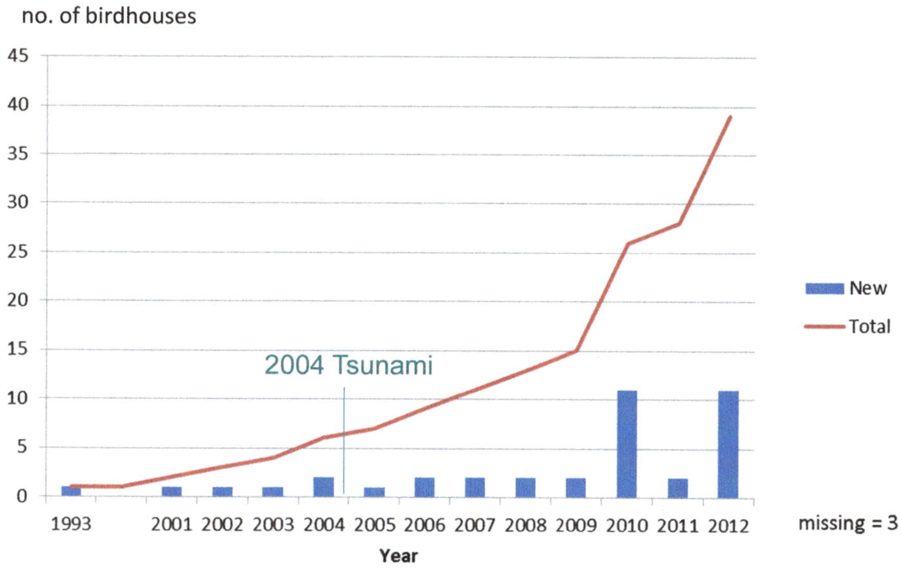

Fig. 31.8 Evolution of number of birdhouses in Baan Nam Khem (1993–2012). *Source*: Authors' survey 2013

investors from other provinces. Our survey of birdhouse ownership found that the majority of the houses were owned by business people who do not reside in Baan Nam Khem (Fig. 31.9). Most outside investors have their place of residence in Phuket (15 out of 24), five in the district capital Takua Pa and the other four in various neighbouring or nearby provinces (Table 31.1).

In several interviews, residents of Baan Nam Khem and representatives of local NGOs expressed their anger and frustration about what they regarded as an unequal economic recovery process and exploitative practices by absentee business owners, as the following quotes show:

> *I am very disappointed in local authorities in Baan Nam Khem. They got wealthier after the tsunami because they have been corrupted by the influx of money. They have enriched themselves through the bird's nest and fishing business. (Khun Chatchai, Thai NGO, Baan Nam Khem, interview, January 2013)*

> *Before the tsunami, the village economy was very good. Now our village has become a place for people from other provinces to exploit and take advantage of. (Female resident of Baan Nam Khem, interview, January 2013)*

In sum, the narratives of local residents in Baan Nam Khem attest to a disaster recovery process that was controlled by those with good connections to various levels of political decision-making and with sufficient financial resources to invest in lucrative business opportunity. As the bird's nest enterprise is a high-capital, but low labour-input business, it has also failed to create jobs among the local population, further aggravating socio-economic divisions between wealthy entrepreneurs and the poorer members of the community.

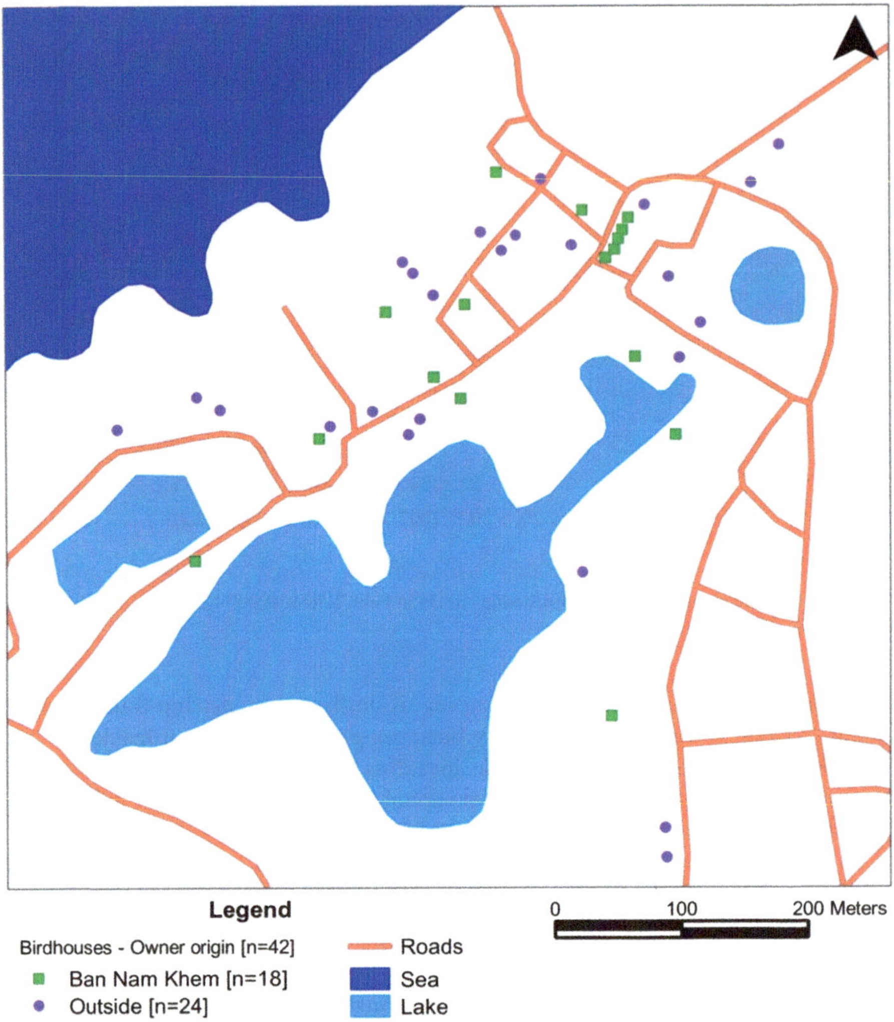

Fig. 31.9 Location of birdhouses and origin of birdhouse owners in Baan Nam Khem. *Source*: Authors' survey 2013

Table 31.1 Ownership structure of birdhouses in Baan Nam Khem

Origin/place of residence of the owner	No. of birdhouses	In % (of total)
Ban Nam Khem	18	42.9
Outside	24	57.1
Phuket	15	35.7
Takua Pa	5	11.9
Krabi	1	2.3
Surat Thani	1	2.3
Nakhon Si Thammarat	1	2.3
Ye Nok	1	2.3

Source: Authors' survey 2013

31.4 Discussion and Conclusion

Our findings challenge the common, overly simplistic view that post-disaster recovery processes follow general patterns that can be described as distinct phases and broken down into specific time periods (e.g. Birkmann 2010; Paul 2011). There is a risk that the focus on general recovery patterns diverts both researchers' and aid practitioners' attention from micro-level disruptions, inequalities and disparities that are important features of disaster recovery processes. Our study further confirms the suggestion of Brown et al. (2008) that the different scales at which recovery can occur (individual, household, business, community) should be analysed collectively and that studies at the household level should be well contextualized. It is equally important that people's own perspectives should be taken into consideration rather than imposing an outsider's framing of the recovery process. Our results further align with findings that "the fate of individual households or businesses may not be determined until several years after the disaster" (Brown et al. 2008, p. 2).

Another conclusion from our study is that economic disparities and power differentials are likely to be exacerbated by a tsunami disaster of such scale. Politically well-connected people, wealthy local elites and external investors are encouraged to exploit distorted recovery governance mechanisms and take advantage of the legal and institutional uncertainties in the disaster's aftermath to pursue their own economic interests. The negative externalities associated with exclusive patronage networks that are historically embedded in local community and leadership structures (Larsen et al. 2011) and in the wider cultural and social fabric of Thai society are a vivid example of the type of social capital that Aldrich (2012, p. 24) calls the "janus-faced resource for recovery".

The major policy implication of our study is that aid organizations and civil society need to pay more attention to the local politics of post-disaster recovery and rehabilitation processes in order to avoid widening social and economic disparities, enhance accountability of all actors involved in the governance of post-disaster recovery processes and mitigate the disempowerment of small entrepreneurs.

References

Aldrich DP (2012) Building resilience: social capital in post-disaster recovery. The University of Chicago Press, Chicago/London

Alovisi J, Hassam N, Logon D, Milligan A (2005) Tsunami: Indian ocean event and investigation into potential global risks. Guy Carpenter, London

Bell R, Cowan H, Dalziell E, Evans N, O'Leary M, Rush B, Yule L (2005) Survey of impacts on the Andaman coast, southern Thailand following the great Sumatra-Andaman earthquake and tsunami of December 26, 2004. Bull New Zeal Soc Earthq Eng 38:123–148

Birkmann J (2010) Global disaster response and reconstruction: stabilization versus destabilization – challenges of the global disaster response to reduce vulnerability and risk following disasters. In: Dölemeyer A, Zimmer J, Tetzlaff G (eds) Risk and planet earth: vulnerability, natural hazards, integrated adaptation strategies. Schweizerbart Science Publishers, Stuttgart, pp 43–55

Brown D, Saito K, Spence R, Chenvidyakarn T (2008) Indicators for measuring, monitoring and evaluating post-disaster recovery. In: Sixth international workshop on remote sensing for disaster response, University of Pavia, Italy, 11–12 September 2008. http://mceer.buffalo.edu/publications/workshop/RS-6/Papers/16Brown.pdf. Accessed 7 Mar 2014

Calgaro E, Lloyd K (2008) Sun, sea, sand and tsunami: examining disaster vulnerability in the tourism community of Khao Lak, Thailand. Singa J Trop Geogr 29:288–306

Calgaro E, Pongponrat K, Naruchaikusol S (2009) Destination vulnerability assessment for Khao Lak. Stockholm Environment Institute, Stockholm

Chamlee-Wright E, Storr VH (2011) Social capital as collective narratives and postdisaster community recovery. Sociol Rev 56(2):266–282

Hüger M, Kraus A, Wilhelm M (2006) Livelihoods and local economy: a descriptive report on the post-tsunami reconstruction processes in Khao Lak, Thailand and Banda Aceh, indonesia. Passau Study Group on Disaster Research, Passau

Koh M (2012) Swift business. Documentary video. http://vimeo.com/45710041. Accessed 7 Mar 2014

Larsen RK, Calgaro E, Thomalla F (2011) Governing resilience building in Thailand's tourism-dependent coastal communities: conceptualising stakeholder agency in social–ecological systems. Glob Environ Chang 21:481–491

Maw TZ (2006) The tsunami's invisible victims. In: SEAPA (ed) Disasters in Southeast Asia: stories of struggle, resilience, greed. Southeast Asian Press Alliance (SEAPA), Bangkok, pp 32–47

Neef A, Shaw R (2013) Local responses to natural disasters: issues and challenges. In: Neef A, Shaw R (eds) Risks and conflicts: local responses to natural disasters. Emerald Publishers, Bingley, pp 1–8

Nidhiprabha B (2007) Adjustment and recovery in Thailand two years after the tsunami. ADBI Discussion Paper No 72. Asian Development Bank Institute, Tokyo. http://www.econstor.eu/bitstream/10419/53440/1/547237782.pdf. Accessed 21 Mar 2014

Paphavasit N, Chotiyaputta C, Siriboon S (2006) Pre- and post-tsunami coastal planning and land-use policies and issues in Thailand. In: Proceedings of the workshop on coastal area planning and management in asian tsunami-affected countries, 27–29 September 2006, Bangkok, Thailand. FAO Regional Office for Asia and the Pacific, Bangkok, pp 199–227

Paul BK (2011) Environmental hazards and disasters: contexts, perspectives and management. Wiley, Chichester

Rajah D (2006) Land not for the taking. In: SEAPA (ed) Disasters in Southeast Asia: stories of struggle, resilience, greed. Southeast Asian Press Alliance (SEAPA), Bangkok, pp 21–31

Rice (2005) Post-tsunami reconstruction and tourism: a second disaster? Tourism Concern, Stapleton House, London

Rigg J, Law L, Tan-Mullins M, Grundy-Warr C, Horton B (2012) In the wake of the tsunami: researching across disciplines and development spaces in southern Thailand. In: Kearnes M, Klauser F, Lane S (eds) Critical risk research: practices, politics and ethics, 1st edn. Wiley, Chichester, pp 173–196

Römer H, Kaiser G, Sterr H, Ludwig R (2012) Using remote sensing to assess tsunami-induced impacts on coastal forest ecosystems at the Andaman Sea coast of Thailand. Nat Hazards Earth Syst Sci 10:729–745

Scheper E, Parakrama A, Patel S (2006) Impact of the tsunami response on local and national capacities: Thailand country report. Tsunami Evaluation Coalition. http://pdf.usaid.gov/pdf_docs/pdacn514.pdf. Accessed 21 Mar 2014

Siriboon S, Bangkaeo C, Milintangkul C, Paphavasit N (2006) Post-tsunami: monitoring on impact assessment on demographic and socioeconomic status. In: Post-tsunami monitor on impact assessment on mangrove ecosystem on Andaman coastline. Thailand Environment Institute and UNOCAL, Bangkok, pp 144–171

Takada S, Kuwata Y, Pinta A (2010) Damage and reconstruction of lifelines in Phang Nga Province, Thailand after the 2004 Indian ocean earthquake and tsunami. J Earthq Tsunami 4(2):83–93

UNCTT (2005) Tsunami Thailand one year later: national response and contribution of international partners. United Nations Country Team Thailand (UNCTT), Bangkok

Willroth P, Revilla Diez J, Arunothai N (2011) Modelling the economic vulnerability of households in the Phang-Nga Province (Thailand) to natural disasters. Nat Hazards 58:753–769

Willroth P, Massmann F, Wehrmann R, Revilla Diez J (2012) Socio-economic vulnerability of coastal communities in southern Thailand: the development of adaptation strategies. Nat Hazards Earth Syst Sci 12:2647–2658

Wong PP (2012) Impacts, recovery and resilience of Thai tourist coasts to the 2004 Indian Ocean Tsunami. Geological Society, London, Special Publications. doi:10.1144/SP361.11

CPSIA information can be obtained at www.ICGtesting.com
Printed in the USA
LVOW01*1916031014

407171LV00001B/42/P

9 784431 551164